T0362100

The Mathematics of Finite Elements and Applications X

and Applications X

MAFELAP 1999

Elsevier Science Internet Homepage

http://www.elsevier.nl (Europe)
http://www.elsevier.com (America)
http://www.elsevier.co.jp (Asia)

Consult the Elsevier homepage for full catalogue information on all books, journals and electronic products and services.

Elsevier Titles of Related Interest

APCOM '99 – 4th Asia Pacific Conference on Computational Mechanics
Ed. K H Lee
ISBN: 008-0432093

Deformation and Progressive Failure in Geomechanics
Ed. A Asaoka, T Adachi, F. Oka
ISBN: 008-042838X

CREAM – Cognitive Reliability and Error Analysis Method
Erik Hollnagel
ISBN: 008-0428487

Engineering Rock Mechanics: An Introduction to Principles
John A Hudson & John P Harrison
ISBN: 008-0419127

Related Journals
Free specimen copy gladly sent on request: Elsevier Science Ltd, The Boulevard, Langford Lane, Kidlington, Oxford, OX5 1GB, UK

Computer Methods in Applied Mechanics and Engineering
Computers and Fluids
Finite Elements in Analysis and Design
Engineering Analysis with Boundary Elements
Advances in Engineering Software
Soil Dynamics and Earthquake Engineering
Computers and Geotechnics
Engineering Failure Analysis
Probabilistic Engineering Mechanics
Reliability Engineering and Systems Safety
Structural Safety

To Contact the Publisher
Elsevier Science welcomes enquiries concerning publishing proposals: books, journal special issues, conference proceedings, etc. All formats and media can be considered. Should you have a publishing proposal you wish to discuss, please contact, without obligation, the publisher responsible for Elsevier's numerical methods in engineering programme

Dr James Milne
Publisher, Engineering and Technology
Elsevier Science Ltd
The Boulevard, Langford Lane   Phone: +44 1865 843891
Kidlington, Oxford   Fax: +44 1865 843920
OX5 1GB, UK   E.mail: j.milne@elsevier.co.uk

General enquiries, including placing orders, should be directed to Elsevier's Regional Sales Offices - please access the Elsevier homepage for full contact details (homepage details at top of this page).

The Mathematics of Finite Elements and Applications X

MAFELAP 1999

Edited by
J.R. Whiteman

BICOM, Institute of Computational Mathematics,
Brunel University, Uxbridge, Middlesex UB8 3PH, UK

2000
ELSEVIER

Amsterdam · Lausanne · New York · Oxford · Shannon · Singapore · Tokyo

ELSEVIER SCIENCE Ltd
The Boulevard, Langford Lane
Kidlington, Oxford OX5 1GB, UK

First edition 2000

Library of Congress Cataloging in Publication Data
A catalog record from the Library of Congress has been applied for.

British Library Cataloguing in Publication Data
A catalogue record from the British Library has been applied for.

ISBN: 008-043568-8

Transferred to digital printing 2006
Printed and bound by Antony Rowe Ltd, Eastbourne

Preface

The tenth conference on The Mathematics of Finite Elements and Applications, MAFELAP 1999, was held at Brunel University during the period 22-25 June, 1999. This book seeks to highlight certain aspects of the state-of-the-art of the theory and applications of finite element methods as of that time. All the papers result from presentations which were made at MAFELAP 1999.

In rather a serendipity manner the MAFELAP conferences have developed into a series. This latest conference followed the now well established MAFELAP pattern of bringing together mathematicians, engineers and others interested in the field to discuss finite element techniques. In the MAFELAP context finite elements have always been interpreted in a broad and inclusive manner, and have been taken to include techniques such as finite difference, finite volume and boundary element methods as well as actual finite element methods. The papers in this book reflect this. The increasing importance of modelling, in addition to numerical discretization, error estimation and adaptivity, was again evident in MAFELAP 1999.

The 1999 Zienkiewicz Lecture was presented by Roland Glowinski, who was also a Distinguished Lecturer of the Institute of Mathematics and its Applications during his visit to the UK. The other invited speakers were U. Langer, J. T. Oden, D. R. J. Owen, R. Rannacher, C. Schwab, E. Süli, M. F. Wheeler, J. R. Whiteman, P. Wriggers and L. C. Wrobel. Papers by these authors are included in this book, together with a number of additional papers which have been chosen to highlight areas of current research in the field which featured at MAFELAP 1999. Over 180 presentations were made at the conference and it has been possible to include the papers of only a fraction of these in this book.

The innovation of including a large number of mini-symposia in the conference programme was continued for MAFELAP 1999. These were organised by colleagues who were either invited or who volunteered to undertake the task, and made a large contribution to the conference. In addition to these organisers the success of the conference depended on the programme committee, the persons who chaired the sessions, the speakers, the poster session authors and, as always, on the extensive efforts of the BICOM Fellows and Research Students. My sincere thanks go to all these people and especially to Mary Hayiannis, who undertook much of the administration of the registration and accommodation for the conference. The organisation of MAFELAP 1999 was greatly simplified by the conference having an excellent web site, which was produced by Simon Shaw. Many thanks go to Simon for master minding and implementing this. I am pleased also to acknowledge financial support for the conference from the United Stated Army European Research Office and the Institute of Mathematics and its Applications. Once again Michael Warby performed his wizardry in producing on the computer all the documentation for the conference, and by turning all manner of forms of LATEX chapter manuscripts into the camera ready form of the manuscript of this book. It is a pleasure to express my thanks to him. Finally my heartfelt thanks go for the tenth time to my wife, Caroline, who not only produced the index for this book, but as always provided immense support and encouragement during the lead up period to the conference.

John Whiteman
BICOM, Brunel University
October 1999

Contents

Contents

1 FICTITIOUS DOMAIN METHODS FOR PARTICULATE FLOW IN TWO AND THREE DIMENSIONS

Roland Glowinski[a]*, Tsorng-Whay Pan[a] and Daniel D. Joseph[b]

[a]Department of Mathematics
University of Houston, Houston, Texas 77204, U.S.A.

[b]Department of Aerospace Engineering & Mechanics
University of Minnesota, Minneapolis, Minnesota 55455, U.S.A.

ABSTRACT

In this article we discuss a methodology for undertaking the direct numerical simulation of the flow of mixtures of rigid solid particles and incompressible viscous fluids, possibly non-Newtonian. The simulation methods are essentially combinations of:

(a) Lagrange multiplier based fictitious domain methods which allow the fluid flow computations to be done in a fixed flow region.

(b) Finite element approximations of the Navier-Stokes equations occurring in the global model.

(c) Time discretizations by operator splitting schemes in order to treat optimally the various operators present in the model.

We conclude this article by presenting of the results of various numerical experiments, including the simulation of sedimentation and fluidization phenomena in two- and three-dimensions.

Key words. particulate flow, liquid-solid mixtures, fictitious domain methods, Lagrange multipliers, Navier-Stokes equations, sedimentation, fluidization, Rayleigh-Taylor instabilities.

---

*Fourth Zienkiewicz Lecture, presented by Professor Glowinski.

1.1   INTRODUCTION

During MAFELAP 1996 the first author of this article presented a computational method
well suited to the simulation of the unsteady flow of an *incompressible viscous fluid,* around
a *moving rigid body,* when the law of motion of the moving object is known in advance.
This method (discussed in [1]) is based on a *Lagrange multiplier* based *fictitious domain
method,* the multiplier being defined on the boundary of the moving body. Since then,
motivated by applications from *Chemical* and *Petroleum Engineering,* the authors of this
article and their collaborators have investigated the solution of much more difficult problems such as the *direct numerical simulation* of *sedimentation* and *fluidization* phenomena,
including those situations where the fluid is non-Newtonian; for such problems the particle motion is not known in advance and results from the fluid-solid interaction and also
from particle-particle or particle-wall collisions or near-collisions. The methodology that
we employ for this class of problems still relies on Lagrange multipliers, but, unlike that
in [1], these multipliers are defined on the volume occupied by the particles. The goals of
this article are two-fold, namely:

(a) To review the distributed Lagrange multiplier based fictitious domain methodology
and to take this opportunity to introduce new ideas concerning for example the
treatment of advection and collisions.

(b) To present numerical results concerning, in particular, the direct numerical simulation
of sedimentation and fluidization phenomena for small and large ($> 10^3$) populations
of particles in two- and three-dimensions and for Newtonian and non-Newtonian
(Oldroyd-B) incompressible viscous fluids.

This article completes [2, 3, 4, 5, 6]

1.2   MODELLING OF THE FLUID-RIGID PARTICLE INTER-ACTION

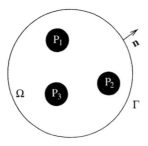

Figure 1.1: An example of a two-dimensional flow region with three rigid bodies

Let $\Omega \subset \mathbb{R}^d (d = 2, 3)$ be a space region; we suppose that Ω is filled with an *incompressible viscous fluid* of density ρ_f and contains J moving rigid particles $P_1, P_2, \ldots, P_J$

(see Figure 1.1 for a particular case where $d = 2$ and $J = 3$). We denote by $\mathbf{n}$ the unit normal vector on the boundary of $\Omega \backslash \overset{J}{\underset{j=1}{\cup}} \overline{P_j}$, in the outward direction from the flow region. Assuming that the only external force acting on the mixture is *gravity*, then, between *collisions* (assuming that collisions take place), the *fluid flow* is modelled by the following *Navier-Stokes equations*

$$
\begin{cases}
\rho_f[\dfrac{\partial \mathbf{u}}{\partial t} + (\mathbf{u} \cdot \boldsymbol{\nabla})\mathbf{u}] = \rho_f \mathbf{g} + \boldsymbol{\nabla} \cdot \boldsymbol{\sigma} \ \text{ in } \ \Omega \backslash \overset{J}{\underset{j=1}{\cup}} \overline{P_j(t)}, \\[2mm]
\boldsymbol{\nabla} \cdot \mathbf{u} = 0 \ \text{ in } \ \Omega \backslash \overset{J}{\underset{j=1}{\cup}} \overline{P_j(t)}, \\[2mm]
\mathbf{u}(\mathbf{x}, 0) = \mathbf{u}_0(\mathbf{x}), \ \forall \mathbf{x} \in \Omega \backslash \overset{J}{\underset{j=1}{\cup}} \overline{P_j(0)}, \ \ \boldsymbol{\nabla} \cdot \mathbf{u}_0 = 0,
\end{cases}
\tag{1.1}
$$

to be completed by

$$
\mathbf{u} = \mathbf{g}_0 \ \text{ on } \ \Gamma \ \text{ with } \ \int_\Gamma \mathbf{g}_0 \cdot \mathbf{n} d\Gamma = 0
\tag{1.2}
$$

and by the following *no-slip boundary condition* on the boundary ∂P_j of P_j,

$$
\mathbf{u}(\mathbf{x}, t) = \mathbf{V}_j(t) + \boldsymbol{\omega}_j(t) \times \overrightarrow{\mathbf{G}_j(t)\mathbf{x}}, \ \forall \mathbf{x} \in \partial P_j(t),
\tag{1.3}
$$

where, in (1.3), $\mathbf{V}_j$ (resp., $\boldsymbol{\omega}_j$) is the *velocity of the center of mass* $\mathbf{G}_j$ (resp., the *angular velocity*) of the jth particle, $\forall j = 1, \ldots, J$. In (1.1), the *stress-tensor* $\boldsymbol{\sigma}$ satisfies

$$
\boldsymbol{\sigma} = \boldsymbol{\tau} - p\mathbf{I},
\tag{1.4}
$$

typical situations for $\boldsymbol{\tau}$ being

$$
\boldsymbol{\tau} = 2\nu \mathbf{D}(\mathbf{u}) = \nu(\boldsymbol{\nabla}\mathbf{u} + \boldsymbol{\nabla}\mathbf{u}^t) \ \ (\textit{Newtonian case}),
\tag{1.5}
$$
$$
\boldsymbol{\tau} \ \textit{is a nonlinear function of } \boldsymbol{\nabla}\mathbf{u} \ \ (\textit{non-Newtonian case}).
\tag{1.6}
$$

The motion of the particles is modelled by the following *Newton-Euler equations*

$$
\begin{cases}
M_j \dfrac{d\mathbf{V}_j}{dt} = M_j \mathbf{g} + \mathbf{F}_j, \\[2mm]
\mathbf{I}_j \dfrac{d\boldsymbol{\omega}_j}{dt} + \boldsymbol{\omega}_j \times \mathbf{I}_j \boldsymbol{\omega}_j = \mathbf{T}_j,
\end{cases}
\tag{1.7}
$$

for $j = 1, \ldots, J$, where in (1.7):

- M_j is the *mass* of the jth rigid particle.

- $\mathbf{I}_j$ is the *inertia tensor* at $\mathbf{G}_j$ of the jth rigid particle.

- $\mathbf{F}_j$ is the resultant of the *hydrodynamical forces* acting on the jth particle, i.e.

$$
\mathbf{F}_j = \int_{\partial P_j} \boldsymbol{\sigma} \mathbf{n} \, d(\partial P_j).
\tag{1.8}
$$

- $\mathbf{T}_j$ is the torque at $\mathbf{G}_j$ of the *hydrodynamical forces* acting on the jth particle, i.e.

$$\mathbf{T}_j = \int_{\partial P_j} \overrightarrow{\mathbf{G}_j \mathbf{x}} \times (\boldsymbol{\sigma}\mathbf{n}) \, d(\partial P_j). \tag{1.9}$$

- We have

$$\frac{d\mathbf{G}_j}{dt} = \mathbf{V}_j. \tag{1.10}$$

Equations (1.7)-(1.10) have to be completed by the following *initial conditions*

$$P_j(0) = P_{0j}, \ \mathbf{G}_j(0) = \mathbf{G}_{0j}, \ \mathbf{V}_j(0) = \mathbf{V}_{0j}, \ \boldsymbol{\omega}_j(0) = \boldsymbol{\omega}_{0j}, \ \forall j = 1, \dots, J. \tag{1.11}$$

Remark 1.2.1. If P_j consists of an *homogeneous* material of *density* ρ_j, we have

$$M_j = \rho_j \int_{P_j} d\mathbf{x}, \quad \mathbf{I}_j = \begin{pmatrix} I_{11,j} & -I_{12,j} & -I_{13,j} \\ -I_{12,j} & I_{22,j} & -I_{23,j} \\ -I_{13,j} & -I_{23,j} & I_{33,j} \end{pmatrix} \tag{1.12}$$

where, in (1.12), $d\mathbf{x} = dx_1 dx_2 dx_3$ and

$$I_{11,j} = \rho_j \int_{P_j} (x_2^2 + x_3^2) \, d\mathbf{x}, \ I_{22,j} = \rho_j \int_{P_j} (x_3^2 + x_1^2) \, d\mathbf{x}, \ I_{33,j} = \rho_j \int_{P_j} (x_1^2 + x_2^2) \, d\mathbf{x},$$

$$I_{12,j} = \rho_j \int_{P_j} x_1 x_2 \, d\mathbf{x}, \ I_{23,j} = \rho_j \int_{P_j} x_2 x_3 \, d\mathbf{x}, \ I_{13,j} = \rho_j \int_{P_j} x_3 x_1 \, d\mathbf{x},$$

with the usual simplification for two-dimensional phenomena.

Remark 1.2.2. If the flow-rigid body motion is *two-dimensional*, or if P_j is a *spherical ball* made of an *homogeneous* material, then the quadratic term $\boldsymbol{\omega}_j \times \mathbf{I}_j \boldsymbol{\omega}_j$ in (1.7) vanishes.

Remark 1.2.3. Suppose that the particles do not touch at $t = 0$; then it has been shown by Desjardins and Esteban (ref. [7]) that the system of equations describing the flow of the above fluid-rigid particle mixture has a (weak) solution on a time interval $[0, t_*)$, $t_* (> 0)$ depending on the initial conditions; uniqueness is an open problem.

1.3 A GLOBAL VARIATIONAL FORMULATION OF THE FLUID-SOLID INTERACTION VIA THE VIRTUAL POWER PRINCIPLE

Let us denote by $P(t)$ the space region occupied at time t by the particles; we thus have $P(t) = \bigcup_{j=1}^{J} P_j(t)$. To obtain a *variational formulation* for the system of equations described in Section 1.2, we introduce the following *functional space* of *compatible* test functions:

$$\begin{aligned} W_0(t) = \{ \ \{\mathbf{v}, \mathbf{Y}, \boldsymbol{\theta}\} \mid \mathbf{v} \in H^1(\Omega \backslash \overline{P(t)})^d, \ \mathbf{v} = 0 \ \text{on} \ \Gamma, \ \mathbf{Y} = \{\mathbf{Y}_j\}_{j=1}^{J}, \\ \boldsymbol{\theta} = \{\boldsymbol{\theta}_j\}_{j=1}^{J}, \ \text{with} \ \mathbf{Y}_j \in \mathbb{R}^d, \ \boldsymbol{\theta}_j \in \mathbb{R}^3, \\ \mathbf{v}(\mathbf{x}, t) = \mathbf{Y}_j + \boldsymbol{\theta}_j \times \overrightarrow{\mathbf{G}_j(t)\mathbf{x}} \ \text{on} \ \partial P_j(t), \ \forall j = 1, \dots, J\}; \end{aligned} \tag{1.13}$$

in (1.13) we have $\boldsymbol{\theta}_j = \{0, 0, \theta_j\}$ if $d = 2$.

Applying the *virtual power* principle to the *whole* mixture (i.e., to the fluid *and* the particles) yields the following *global* variational formulation

$$
\left\{
\begin{array}{l}
\rho_f \displaystyle\int_{\Omega\backslash\overline{P(t)}} [\frac{\partial\mathbf{u}}{\partial t} + (\mathbf{u} \cdot \boldsymbol{\nabla})\mathbf{u}] \cdot \mathbf{v}\, d\mathbf{x} + 2\nu \int_{\Omega\backslash\overline{P(t)}} \mathbf{D}(\mathbf{u}) : \mathbf{D}(\mathbf{v})\, d\mathbf{x} \\[4mm]
- \displaystyle\int_{\Omega\backslash\overline{P(t)}} p\boldsymbol{\nabla} \cdot \mathbf{v}\, d\mathbf{x} + \sum_{j=1}^{J} M_j \dot{\mathbf{V}}_j \cdot \mathbf{Y}_j + \sum_{j=1}^{J} (\mathbf{I}_j\dot{\boldsymbol{\omega}}_j + \boldsymbol{\omega}_j \times \mathbf{I}_j\boldsymbol{\omega}_j) \cdot \boldsymbol{\theta}_j \\[4mm]
= \rho_f \displaystyle\int_{\Omega\backslash\overline{P(t)}} \mathbf{g} \cdot \mathbf{v}\, d\mathbf{x} + \sum_{j=1}^{J} M_j \mathbf{g} \cdot \mathbf{Y}_j, \;\; \forall\{\mathbf{v}, \mathbf{Y}, \boldsymbol{\theta}\} \in W_0(t),
\end{array}
\right.
\tag{1.14}
$$

$$
\int_{\Omega\backslash\overline{P(t)}} q\boldsymbol{\nabla} \cdot \mathbf{u}(t)\, d\mathbf{x} = 0, \;\; \forall q \in L^2(\Omega\backslash\overline{P(t)}),
\tag{1.15}
$$

$$
\mathbf{u} = \mathbf{g}_0 \;\; \text{on} \;\; \Gamma,
\tag{1.16}
$$

$$
\mathbf{u}(\mathbf{x}, t) = \mathbf{V}_j + \boldsymbol{\omega}_j \times \overrightarrow{\mathbf{G}_j(t)\mathbf{x}}, \;\; \forall\mathbf{x} \in \partial P_j(t), \;\; \forall j = 1, \ldots, J,
\tag{1.17}
$$

$$
\dot{\mathbf{G}}_j = \mathbf{V}_j, \;\; \forall j = 1, \ldots, J,
\tag{1.18}
$$

to be completed by the following initial conditions

$$
\mathbf{u}(\mathbf{x}, 0) = \mathbf{u}_0(\mathbf{x}), \;\; \mathbf{x} \in \Omega\backslash\overline{P(0)},
\tag{1.19}
$$

$$
P_j(0) = P_{0j}, \; \mathbf{G}_j(0) = \mathbf{G}_{0j}, \; \mathbf{V}_j(0) = \mathbf{V}_{0j}, \; \boldsymbol{\omega}_j(0) = \boldsymbol{\omega}_{0j}, \; \forall j = 1, \ldots, J.
\tag{1.20}
$$

In relations (1.14)-(1.20), it is reasonable to assume that $\mathbf{u}(t) \in (H^1(\Omega\backslash\overline{P(t)}))^d$ and $p(t) \in L^2(\Omega\backslash\overline{P(t)})$. Also, $\boldsymbol{\omega}_j(t) = \{0, 0, \omega_j(t)\}$ if $d = 2$ and the following notation has been used

$$
\mathbf{A} : \mathbf{B} = \sum_{i=1}^{d} \sum_{j=1}^{d} a_{ij} b_{ij}, \;\; \forall\mathbf{A} = (a_{i,j})_{1\le i,j\le d} \;\; \text{and} \;\; \mathbf{B} = (b_{ij})_{1\le i,j\le d}.
$$

Formulations such as (1.14)-(1.20) (or closely related ones) have been used by several authors (see, e.g., [8, 9, 10]) to simulate particulate flow via *arbitrary Lagrange-Euler (ALE)* methods using moving meshes. Our goal in this article is to discuss an alternative based on *fictitious domain methods* (also called *domain embedding* methods). The main advantage of this new approach is the possibility of achieving the flow related computations on a *fixed* space region, thus allowing the use of a fixed (finite difference or finite element) mesh, which is a significant simplification.

1.4   A DISTRIBUTED LAGRANGE MULTIPLIER BASED FICTITIOUS DOMAIN FORMULATION

In general terms our goal is to find a methodology such that:

(a) A fixed mesh can be used for flow computations.

(b) The particle position is obtained via the solution of the Newton-Euler equations of motion.

(c) The time discretization will be done by operator splitting methods in order to treat individually the various operators occurring in the mathematical model.

To achieve such a goal we proceed as follows:

(i) *We fill the particles with the surrounding fluid.*

(ii) *We assume that the fluid inside each particle has a rigid body motion.*

(iii) *We use (i) and (ii) to modify the variational formulation (1.14)-(1.20).*

(iv) *We force the rigid body motion inside each particle via a Lagrange multiplier defined (distributed) over the particle.*

(v) *We combine (iii) and (iv) to derive a variational formulation involving Lagrange multipliers to force the rigid body motion inside the particles.*

We suppose (for simplicity) that each particle P_j is made of an *homogeneous material* of density ρ_j; then, taking into account the fact that any rigid body motion velocity field $\mathbf{v}$ satisfies $\nabla \cdot \mathbf{v} = 0$ and $\mathbf{D}(\mathbf{v}) = \mathbf{0}$, steps (i) to (iii) yield the following variant of formulation (1.14)-(1.20):

For a.e. $t > 0$, find $\mathbf{u}(t), p(t), \{\mathbf{V}_j(t), \mathbf{G}_j(t), \boldsymbol{\omega}_j(t)\}_{j=1}^J$, such that

$$
\left\{
\begin{aligned}
&\rho_f \int_\Omega [\frac{\partial \mathbf{u}}{\partial t} + (\mathbf{u} \cdot \nabla)\mathbf{u}] \cdot \mathbf{v} \, dx - \int_\Omega p\nabla \cdot \mathbf{v} \, dx + 2\nu \int_\Omega \mathbf{D}(\mathbf{u}) : \mathbf{D}(\mathbf{v}) \, dx + \\
&\sum_{j=1}^J (1 - \rho_f/\rho_j) M_j \frac{d\mathbf{V}_j}{dt} \cdot \mathbf{Y}_j + \sum_{j=1}^J (1 - \rho_f/\rho_j)(\mathbf{I}_j \frac{d\boldsymbol{\omega}_j}{dt} + \boldsymbol{\omega}_j \times \mathbf{I}_j\boldsymbol{\omega}_j) \cdot \boldsymbol{\theta}_j \quad (1.21) \\
&= \rho_f \int_\Omega \mathbf{g} \cdot \mathbf{v} \, dx + \sum_{j=1}^J (1 - \rho_f/\rho_j) M_j \mathbf{g} \cdot \mathbf{Y}_j, \; \forall \{\mathbf{v}, \mathbf{Y}, \boldsymbol{\theta}\} \in \widetilde{W}_0(t),
\end{aligned}
\right.
$$

$$
\int_\Omega q\nabla \cdot \mathbf{u} \, dx = 0, \; \forall q \in L^2(\Omega), \tag{1.22}
$$

$$
\mathbf{u}(\mathbf{x}, t) = \mathbf{V}_j(t) + \boldsymbol{\omega}_j(t) \times \overrightarrow{\mathbf{G}_j(t)\mathbf{x}}, \; \forall \mathbf{x} \in P_j(t), \; \forall j = 1, \dots, J, \tag{1.23}
$$

$$
\mathbf{u} = \mathbf{g}_0 \text{ on } \Gamma, \tag{1.24}
$$

$$
\frac{d\mathbf{G}_j}{dt} = \mathbf{V}_j, \; \forall j = 1, \dots, J, \tag{1.25}
$$

$$
\mathbf{V}_j(0) = \mathbf{V}_{0j}, \; \mathbf{G}_j(0) = \mathbf{G}_{0j}, \; \boldsymbol{\omega}_j(0) = \boldsymbol{\omega}_{0j}, \; P_j(0) = P_{0j}, \; \forall j = 1, \dots, J, \tag{1.26}
$$

$$
\mathbf{u}(\mathbf{x}, 0) = \mathbf{u}_0(\mathbf{x}), \; \forall \mathbf{x} \in \Omega \backslash \overline{\bigcup_{j=1}^J P_j(0)} \text{ and } \mathbf{u}(\mathbf{x}, 0) = \mathbf{V}_{0j} + \boldsymbol{\omega}_{0j} \times \overrightarrow{\mathbf{G}_{0j}\mathbf{x}}, \; \forall \mathbf{x} \in \overline{P_{0j}}, (1.27)
$$

with, in formulation (1.21), the space $\widetilde{W}_0(t)$ defined by

$$\widetilde{W}_0(t) = \{ \, \{\mathbf{v}, \mathbf{Y}, \boldsymbol{\theta}\} \mid \mathbf{v} \in H_0^1(\Omega)^d, \ \mathbf{Y} = \{\mathbf{Y}_j\}_{j=1}^J, \boldsymbol{\theta} = \{\boldsymbol{\theta}_j\}_{j=1}^J, \ \text{with} \ \mathbf{Y}_j \in \mathbb{R}^d,$$
$$\boldsymbol{\theta}_j \in \mathbb{R}^3, \ \mathbf{v}(\mathbf{x}, t) = \mathbf{Y}_j + \boldsymbol{\theta}_j \times \overrightarrow{\mathbf{G}_j(t)\mathbf{x}} \ \text{in} \ P_j(t), \ \forall \, j = 1, \dots, J\}.$$

Concerning $\mathbf{u}$ and p it makes sense to assume that $\mathbf{u} \in H^1(\Omega)^d$ and $p \in L^2(\Omega)$.

In order to relax the *rigid body motion constraints* (1.23) we employ a family $\{\boldsymbol{\lambda}_j\}_{j=1}^J$ of *Lagrange multipliers* so that $\boldsymbol{\lambda}_j \in \Lambda_j(t)$ with

$$\Lambda_j(t) = H^1(P_j(t))^d, \ \forall j = 1, \dots, J. \tag{1.28}$$

We obtain, thus, the following *fictitious domain formulation with Lagrange multipliers*:

For a.e. $t > 0$, find $\mathbf{u}(t), p(t), \{\mathbf{V}_j(t), \mathbf{G}_j(t), \boldsymbol{\omega}_j(t), \boldsymbol{\lambda}_j(t)\}_{j=1}^J$, such that

$$\begin{cases} \mathbf{u}(t) \in H^1(t)^d, \ \mathbf{u}(t) = \mathbf{g}_0(t) \ \text{on} \ \Gamma, \ p(t) \in L^2(\Omega), \\ \mathbf{V}_j(t) \in \mathbb{R}^d, \ \mathbf{G}_j(t) \in \mathbb{R}^d, \ \boldsymbol{\omega}_j(t) \in \mathbb{R}^3, \ \boldsymbol{\lambda}_j(t) \in \Lambda_j(t), \forall j = 1, \dots, J, \end{cases} \tag{1.29}$$

and

$$\begin{cases} \rho_f \int_\Omega [\dfrac{\partial \mathbf{u}}{\partial t} + (\mathbf{u} \cdot \boldsymbol{\nabla})\mathbf{u}] \cdot \mathbf{v} \, d\mathbf{x} - \int_\Omega p \boldsymbol{\nabla} \cdot \mathbf{v} \, d\mathbf{x} + 2\nu \int_\Omega \mathbf{D}(\mathbf{u}) : \mathbf{D}(\mathbf{v}) \, d\mathbf{x} \\ \quad - \displaystyle\sum_{j=1}^J < \boldsymbol{\lambda}_j, \mathbf{v} - \mathbf{Y}_j - \boldsymbol{\theta}_j \times \overrightarrow{\mathbf{G}_j\mathbf{x}} >_j + \sum_{j=1}^J (1 - \rho_f/\rho_j) M_j \dfrac{d\mathbf{V}_j}{dt} \cdot \mathbf{Y}_j \\ \quad + \displaystyle\sum_{j=1}^J (1 - \rho_f/\rho_j)(\mathbf{I}_j \dfrac{d\boldsymbol{\omega}_j}{dt} + \boldsymbol{\omega}_j \times \mathbf{I}_j\boldsymbol{\omega}_j) \cdot \boldsymbol{\theta}_j = \rho_f \int_\Omega \mathbf{g} \cdot \mathbf{v} \, d\mathbf{x} \\ \quad + \displaystyle\sum_{j=1}^J (1 - \rho_f/\rho_j) M_j \mathbf{g} \cdot \mathbf{Y}_j, \ \forall \mathbf{v} \in H_0^1(\Omega)^d, \ \forall \mathbf{Y}_j \in \mathbb{R}^d, \ \forall \boldsymbol{\theta}_j \in \mathbb{R}^3, \end{cases} \tag{1.30}$$

$$\int_\Omega q \boldsymbol{\nabla} \cdot \mathbf{u} \, d\mathbf{x} = 0, \ \forall q \in L^2(\Omega), \tag{1.31}$$

$$< \boldsymbol{\mu}_j, \mathbf{u} - \mathbf{V}_j(t) - \boldsymbol{\omega}_j(t) \times \overrightarrow{\mathbf{G}_j(t)\mathbf{x}} >_j = 0, \ \forall \boldsymbol{\mu}_j \in \Lambda_j(t), \ \forall j = 1, \dots, J, \tag{1.32}$$

$$\dfrac{d\mathbf{G}_j}{dt} = \mathbf{V}_j, \ \forall j = 1, \dots, J, \tag{1.33}$$

$$\mathbf{V}_j(0) = \mathbf{V}_{0j}, \ \mathbf{G}_j(0) = \mathbf{G}_{0j}, \ \boldsymbol{\omega}_j(0) = \boldsymbol{\omega}_{0j}, \ P_j(0) = P_{0j}, \ \forall j = 1, \dots, J, \tag{1.34}$$

$$\mathbf{u}(\mathbf{x}, 0) = \mathbf{u}_0(\mathbf{x}), \ \forall \mathbf{x} \in \Omega \backslash \overset{J}{\underset{j=1}{\cup}} \overline{P_j(0)} \ \text{and} \ \mathbf{u}(\mathbf{x}, 0) = \mathbf{V}_{0j} + \boldsymbol{\omega}_{0j} \times \overrightarrow{\mathbf{G}_{0j}\mathbf{x}}, \ \forall \mathbf{x} \in \overline{P_{0j}}. \tag{1.35}$$

The two most natural choices for $< \cdot, \cdot >_j$ are

$$< \boldsymbol{\mu}, \mathbf{v} >_j = \int_{P_j(t)} (\boldsymbol{\mu} \cdot \mathbf{v} + \delta_j^2 \boldsymbol{\nabla}\boldsymbol{\mu} : \boldsymbol{\nabla}\mathbf{v}) \, d\mathbf{x}, \ \forall \, \boldsymbol{\mu} \ \text{and} \ \mathbf{v} \in \Lambda_j(t), \tag{1.36}$$

$$< \boldsymbol{\mu}, \mathbf{v} >_j = \int_{P_j(t)} (\boldsymbol{\mu} \cdot \mathbf{v} + \delta_j^2 \mathbf{D}(\boldsymbol{\mu}) : \mathbf{D}(\mathbf{v})) \, dx, \ \forall \, \boldsymbol{\mu} \ \text{and} \ \mathbf{v} \in \Lambda_j(t), \qquad (1.37)$$

with δ_j a *characteristic length* (the diameter of P_j, for example). Other possible choices are

$$< \boldsymbol{\mu}, \mathbf{v} >_j = \int_{\partial P_j(t)} \boldsymbol{\mu} \cdot \mathbf{v} \, d(\partial P_j) + \delta_j \int_{P_j(t)} \boldsymbol{\nabla} \boldsymbol{\mu} : \boldsymbol{\nabla} \mathbf{v} \, dx, \ \forall \, \boldsymbol{\mu} \ \text{and} \ \mathbf{v} \in \Lambda_j(t),$$

$$< \boldsymbol{\mu}, \mathbf{v} >_j = \int_{\partial P_j(t)} \boldsymbol{\mu} \cdot \mathbf{v} \, d(\partial P_j) + \delta_j \int_{P_j(t)} \mathbf{D}(\boldsymbol{\mu}) : \mathbf{D}(\mathbf{v}) \, dx, \ \forall \, \boldsymbol{\mu} \ \text{and} \ \mathbf{v} \in \Lambda_j(t).$$

Remark 1.4.1. The fictitious domain approach, described above, has clearly many similarities with the *immersed boundary* approach of Peskin (see refs. [11, 12, 13]). However, the systematic use of Lagrange multipliers seems to be new in this context.

Remark 1.4.2. An approach with many similarities to the present one has been developed by Schwarzer *et al.* (see ref. [14]) in a finite difference framework; in the above reference the rigid body motion inside the particles is forced via a penalty method, instead of the multiplier technique used in the present article.

Remark 1.4.3. In order to force the rigid body motion inside the particles we can use the fact that $\mathbf{v}$ defined over Ω is a rigid body motion velocity field inside each particle if and only if $\mathbf{D}(\mathbf{v}) = \mathbf{0}$ in $P_j(t)$, $\forall \, j = 1, \ldots, J$; i.e.,

$$\int_{P_j(t)} \mathbf{D}(\mathbf{v}) : \mathbf{D}(\boldsymbol{\mu}_j) \, d\mathbf{x} = 0, \ \forall \, \boldsymbol{\mu}_j \in \Lambda_j(t), \ \forall \, j = 1, \ldots, J. \qquad (1.38)$$

A computational method based on this approach is discussed in [15].

Remark 1.4.4. Since, in (1.30), $\mathbf{u}$ is *divergence free* and satisfies Dirichlet boundary conditions on Γ, we have

$$2 \int_\Omega \mathbf{D}(\mathbf{u}) : \mathbf{D}(\mathbf{v}) dx = \int_\Omega \boldsymbol{\nabla} \mathbf{u} : \boldsymbol{\nabla} \mathbf{v} dx, \ \forall \mathbf{v} \in H_0^1(\Omega)^d,$$

a substantial simplification, indeed, from a *computational point of view*, which is another plus for the fictitious domain approach used here.

Remark 1.4.5. Using High Energy Physics terminology, the multiplier $\boldsymbol{\lambda}_j$ can be viewed as a *gluon* whose role is to force the rigidity inside P_j. More prosaically, the multipliers $\boldsymbol{\lambda}_j$ are mathematical objects of the *mortar* type, very similar to those used in *domain decomposition methods* to match local solutions at interfaces or on overlapping regions. Indeed the $\boldsymbol{\lambda}_j$'s in this article have genuine mortar properties since their role is to force a fluid to behave like a rigid solid inside the particles.

1.5   ON THE TREATMENT OF COLLISIONS

In the above sections, we have considered the motion of fluid/particle mixtures and given various mathematical models of this phenomenon, assuming that there were no particle/particle or particle/boundary collisions. Actually, with the mathematical model we have considered, it is not known if collisions can take place in finite time (in fact several scientists strongly believe that lubrication forces prevent these collisions in the case of viscous fluids). However collisions take place in Nature and also in actual numerical simulations if special precautions are not taken. In the particular case of particles flowing in a viscous fluid we shall assume that the collisions taking place are *smooth* ones in the sense that if two particles collide (resp., if a particle hits the boundary) the particle velocities (resp., the particle and wall velocities) coincide at the points of contact. From the nature of these collisions the only precaution to be taken will be to avoid particle-particle or particle-boundary interpenetration. To achieve this goal we include in the right-hand sides of the *Newton-Euler equations* modelling particle motions a *short range repulsing force*. If we consider the particular case of *circular* particles (in 2-D) or *spherical* particles (in 3-D), and if P_i and P_j are such two particles, with radii R_i and R_j and centers of mass $\mathbf{G}_i$ and $\mathbf{G}_j$, we shall require the repulsion force $\overrightarrow{F_{ij}}$ between P_i and P_j to satisfy the following properties:

(i) To be parallel to $\overrightarrow{\mathbf{G}_i\mathbf{G}_j}$.

(ii) To satisfy

$$\begin{cases} |\,\overrightarrow{F_{ij}}\,| = 0 & \text{if } d_{ij} \geq R_i + R_j + \rho, \\ |\,\overrightarrow{F_{ij}}\,| = c/\varepsilon & \text{if } d_{ij} = R_i + R_j, \end{cases} \tag{1.39}$$

with $d_{ij} = |\overrightarrow{\mathbf{G}_i\mathbf{G}_j}|$, c a *scaling factor* and ε a "*small*" *positive number*.

(iii) $|\,\overrightarrow{F_{ij}}\,|$ has to behave as in Figure 1.2, below, for

$$R_i + R_j \leq d_{ij} \leq R_i + R_j + \rho.$$

The parameter ρ is the *range* of the repulsion force; for the simulations discussed in the following sections, we have taken $\rho \simeq h_\Omega$ (h_Ω is the *space discretization step* used for approximating the *velocity*). Boundary-particle collisions can be treated in a similar way.

Remark 1.5.1. For those readers wondering how to adjust h_Ω and c/ϵ, we make the following comments: clearly, the space discretization parameter h_Ω is adjusted so that the finite element approximation can resolve the boundary and shear layers occurring in the flow. Next, it is clear that ρ can be taken of the order of h_Ω. The choice of c/ϵ is more subtle; let us say that simple model problems for harmonic oscillators with rigid obstacles (see ref. [16] for details) show that we can expect interpenetrations of the order of $\sqrt{\epsilon/c}$; this suggests therefore that $\rho \gg \sqrt{\epsilon/c}$, which is what we took in our calculations.

Remark 1.5.2. In order to treat the collisions we can use repulsion forces derived by truncation of the *Lennard-Jones* potentials from *Molecular Dynamics*; we intend to investigate the applicability of these repulsion forces for the treatment of collisions in particulate flow.

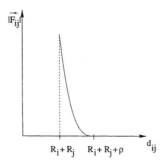

Figure 1.2: Repulsion force behavior

1.6 FINITE ELEMENT APPROXIMATION

For simplicity, we assume that $\Omega \subset \mathbb{R}^2$ (i.e., $d = 2$) and is polygonal; we have then $\boldsymbol{\omega}(t) = \{0, 0, \omega(t)\}$ and $\boldsymbol{\theta} = \{0, 0, \theta\}$ with $\omega(t)$ and $\theta \in \mathbb{R}$. For the *space approximation* of problem (1.29)-(1.35) by a finite element method, we shall proceed as follows:

With h a *space discretization step* we introduce a finite element triangulation $\mathcal{T}_h$ of $\overline{\Omega}$ and then $\mathcal{T}_{2h}$ a triangulation twice coarser (in practice we should construct $\mathcal{T}_{2h}$ first and then $\mathcal{T}_h$ by joining the midpoints of the edges of $\mathcal{T}_{2h}$, thus dividing each triangle of $\mathcal{T}_{2h}$ into 4 similar subtriangles, as shown in Figure 1.3, below).

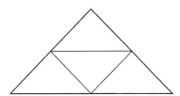

Figure 1.3: Subdivision of a triangle of $\mathcal{T}_{2h}$

We define the following finite dimensional spaces which approximate $H^1(\Omega)^2$, $H_0^1(\Omega)^2$, $L^2(\Omega)$, respectively, by:

$$V_h = \{\mathbf{v}_h \mid \mathbf{v}_h \in (C^0(\overline{\Omega}))^2, \ \mathbf{v}_h|_T \in P_1 \times P_1, \ \forall T \in \mathcal{T}_h\}, \tag{1.40}$$

$$V_{0h} = \{\mathbf{v}_h \mid \mathbf{v}_h \in V_h, \ \mathbf{v}_h = \mathbf{0} \ \text{on} \ \Gamma\}, \tag{1.41}$$

$$L_h^2 = \{q_h \mid q_h \in C^0(\overline{\Omega}), \ q_h|_T \in P_1, \ \forall T \in \mathcal{T}_{2h}\}; \tag{1.42}$$

in (1.40)-(1.42), P_1 is the space of the polynomials in two variables of degree ≤ 1.

Let $\overline{P_{jh}(t)}$ be a polygonal domain inscribed in $\overline{P_j(t)}$ and $\mathcal{T}_h^j(t)$ be a finite element triangulation of $\overline{P_{jh}(t)}$, like the one shown in Figure 1.4, below, where P_j is a disk.

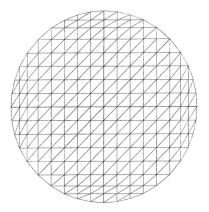

Figure 1.4: Triangulation of a disk.

Then, a finite dimensional space approximating $\Lambda_j(t)$ is

$$\Lambda_{jh}(t) = \{\boldsymbol{\mu}_h \mid \boldsymbol{\mu}_h \in C^0(\overline{P_{jh}(t)})^2, \ \boldsymbol{\mu}_h|_T \in P_1 \times P_1, \ \forall T \in \mathcal{T}_h^j(t)\}. \tag{1.43}$$

An alternative to $\Lambda_{jh}(t)$ defined by (1.43) is as follows: let $\{\mathbf{x}_i\}_{i=1}^{N_j}$ be a set of points from $\overline{P_j(t)}$ which cover $\overline{P_j(t)}$ (uniformly, for example); we define then

$$\Lambda_{jh}(t) = \{\boldsymbol{\mu}_h \mid \boldsymbol{\mu}_h = \sum_{i=1}^{N_j} \boldsymbol{\mu}_i \delta(\mathbf{x} - \mathbf{x}_i), \ \boldsymbol{\mu}_i \in \mathbb{R}^2, \ \forall i = 1, ... N_j\}, \tag{1.44}$$

where $\delta(\cdot)$ is the *Dirac measure* at $\mathbf{x} = \mathbf{0}$. Then instead of the scalar product of $H^1(P_{jh}(t))^2$ we shall use $< \cdot, \cdot >_{jh}$ defined by

$$< \boldsymbol{\mu}_h, \mathbf{v}_h >_{jh} = \sum_{i=1}^{N_j} \boldsymbol{\mu}_i \cdot \mathbf{v}_h(\mathbf{x}_i), \ \forall \boldsymbol{\mu}_h \in \Lambda_{jh}(t), \mathbf{v}_h \in V_h. \tag{1.45}$$

The approach, based on (1.44), (1.45), makes little sense for the continuous problem, but is meaningful for the discrete problem; it amounts to forcing the rigid body motion of $P_j(t)$ via a *collocation method*. A similar technique has been used to enforce Dirichlet boundary conditions by Bertrand et al. (ref. [17]).

Remark 1.6.1. The bilinear form in (1.45) has definitely the flavor of a *discrete $L^2(P_j(t))$*-scalar product. Let us insist on the fact by taking $\Lambda_j(t) = L^2(\Omega)^d$, and then

$$< \boldsymbol{\mu}, \mathbf{v} >_j = \int_{P_j(t)} \boldsymbol{\mu} \cdot \mathbf{v} d\mathbf{x}, \ \forall \boldsymbol{\mu} \ \text{and} \ \mathbf{v} \in \Lambda_j(t),$$

makes no sense for the continuous problem. On the other hand, it makes sense for the finite element variants of (1.29)-(1.35), but one should not expect $\boldsymbol{\lambda}_{jh}(t)$ to converge to an L^2-function as $h \to 0$ (it will converge to some element of the dual space $(H^1(P_j(t))^2)'$ of $H^1(P_j(t))^2$).

Using the above finite dimensional spaces leads to the following approximation of problem (1.29)-(1.35):

For $t > 0$ find $\mathbf{u}_h(t), p_h(t), \{\mathbf{V}_j(t), \mathbf{G}_{jh}(t), \omega_j(t), \boldsymbol{\lambda}_{jh}(t)\}_{j=1}^J$ such that

$$
\begin{cases}
\mathbf{u}_h(t) \in V_h, p_h(t) \in L_h^2, \\
\mathbf{V}_j(t) \in \mathbb{R}^2, \mathbf{G}_{jh}(t) \in \mathbb{R}^2, \omega_j(t) \in \mathbb{R}, \boldsymbol{\lambda}_{jh}(t) \in \Lambda_{jh}(t), \forall j = 1, \dots, J,
\end{cases}
\tag{1.46}
$$

and

$$
\begin{cases}
\rho_f \int_\Omega \left[\dfrac{\partial \mathbf{u}_h}{\partial t} + (\mathbf{u}_h \cdot \boldsymbol{\nabla})\mathbf{u}_h \right] \cdot \mathbf{v} \, dx - \int_\Omega p_h \boldsymbol{\nabla} \cdot \mathbf{v} \, dx + 2\nu \int_\Omega \mathbf{D}(\mathbf{u}_h) : \mathbf{D}(\mathbf{v}) dx \\
\quad + \displaystyle\sum_{j=1}^J (1 - \rho_f/\rho_j) M_j \dfrac{d\mathbf{V}_j}{dt} \cdot \mathbf{Y}_j + \sum_{j=1}^J (1 - \rho_f/\rho_j) I_j \dfrac{d\omega_j}{dt} \theta_j \\
\quad - \displaystyle\sum_{j=1}^J < \boldsymbol{\lambda}_{jh}, \mathbf{v} - \mathbf{Y}_j - \boldsymbol{\theta}_j \times \overrightarrow{\mathbf{G}_{jh}\mathbf{x}} >_{jh} = \rho_f \int_\Omega \mathbf{g} \cdot \mathbf{v} \, dx \\
\quad + \displaystyle\sum_{j=1}^J (1 - \rho_f/\rho_j) M_j \mathbf{g} \cdot \mathbf{Y}_j, \forall \mathbf{v} \in V_{0h}, \forall \mathbf{Y}_j \in \mathbb{R}^2, \forall \theta_j \in \mathbb{R},
\end{cases}
\tag{1.47}
$$

$$
\int_\Omega q \boldsymbol{\nabla} \cdot \mathbf{u}_h(t) \, dx = 0, \ \forall q \in L_h^2,
\tag{1.48}
$$

$$
\mathbf{u}_h = \mathbf{g}_{0h} \text{ on } \Gamma,
\tag{1.49}
$$

$$
< \boldsymbol{\mu}_{jh}, \mathbf{u}_h(t) - \mathbf{V}_j(t) - \boldsymbol{\omega}_j(t) \times \overrightarrow{\mathbf{G}_{jh}(t)\mathbf{x}} >_{jh} = 0, \ \forall \boldsymbol{\mu}_{jh} \in \Lambda_{jh}(t), \ \forall j = 1, \dots, J,
\tag{1.50}
$$

$$
\dfrac{d\mathbf{G}_{jh}}{dt} = \mathbf{V}_j, \ \forall j = 1, \dots, J,
\tag{1.51}
$$

$$
\mathbf{V}_j(0) = \mathbf{V}_{0j}, \ \mathbf{G}_{jh}(0) = \mathbf{G}_{0jh}, \ \omega_j(0) = \omega_{0j}, \ P_{jh}(0) = P_{0jh}, \ \forall j = 1, \dots, J,
\tag{1.52}
$$

$$
\mathbf{u}_h(\mathbf{x}, 0) = \mathbf{u}_{0h}(\mathbf{x}), \ \forall \mathbf{x} \in \Omega \backslash \overset{J}{\underset{j=1}{\cup}} \overline{P_{jh}(0)}, \mathbf{u}_h(\mathbf{x}, 0) = \mathbf{V}_{0j} + \boldsymbol{\omega}_{0j} \times \overrightarrow{\mathbf{G}_{0jh}\mathbf{x}}, \ \forall \mathbf{x} \in \overline{P_{0jh}}.
\tag{1.53}
$$

In (1.49), $\mathbf{g}_{0h}$ is an approximation of $\mathbf{g}_0$ belonging to

$$
\gamma V_h = \{\mathbf{z}_h \mid \mathbf{z}_h \in C^0(\Gamma)^2, \mathbf{z}_h = \tilde{\mathbf{z}}_h|_\Gamma, \text{ with } \tilde{\mathbf{z}}_h \in V_h\}
$$

and satisfying $\int_\Gamma \mathbf{g}_{0h} \cdot \mathbf{n} \, d\Gamma = 0$.

Remark 1.6.2. The *discrete pressure* in (1.46)-(1.53) is defined to within an additive constant. In order to "fix" the pressure we shall require it to satisfy

$$
\int_\Omega p_h(t) \, dx = 0, \ \forall t > 0,
$$

i.e., $p_h(t) \in L_{0h}^2$, with L_{0h}^2 defined by

$$
L_{0h}^2 = \{q_h | q_h \in L_h^2, \int_\Omega q_h dx = 0\}.
$$

Remark 1.6.3. From a practical point of view, the semi-discrete model (1.46)-(1.53) is incomplete since we still have to include the *virtual power* associated with the collision forces. Assuming that the particles are circular ($d = 2$) or spherical ($d = 3$) we shall add to the right-hand side of equation (1.47) the following term

$$\sum_{j=1}^{J} \mathbf{F}_j^r \cdot \mathbf{Y}_j, \tag{1.54}$$

where the repulsion force $\mathbf{F}_j^r$ is defined as in Section 1.5. If the particles were non-circular or non-spherical we would have to take into account the virtual power associated with the torque of the collision forces.

Remark 1.6.4. For the definition of the *multiplier space* $\Lambda_{jh}(t)$ several options are possible:

(i) If P_j is *rotationally invariant* (this will be the case for a circular or a spherical particle) we define $\Lambda_{jh}(t)$ from a triangulation $\mathcal{T}_h^j(t)$ obtained from $\mathcal{T}_h^j(0)$ by translation.

(ii) If P_j is *not rotationally invariant* we can define $\Lambda_{jh}(t)$ from a triangulation $\mathcal{T}_h^j(t)$ *rigidly attached* to P_j.

(iii) We can also define $\Lambda_{jh}(t)$ from the following set of points

$$\Sigma_{jh}(t) = \Sigma_{jh}^{\mathbf{v}}(t) \cup \Sigma_{jh}^{\partial}(t), \tag{1.55}$$

where, in (1.55), $\Sigma_{jh}^{\mathbf{v}}(t)$ is the set of vertices of the velocity grid $\mathcal{T}_h$ which are contained in $P_j(t)$ and where $\Sigma_{jh}^{\partial}(t)$ is a set of control points located on $\partial P_j(t)$. This hybrid approach is (relatively) easy to implement and is particularly well suited to those simulations where the boundary ∂P_j has corners or edges.

Remark 1.6.5. In relation (1.47), we can replace $2\int_{\Omega} \mathbf{D}(\mathbf{u}_h){:}\mathbf{D}(\mathbf{v})dx$ by $\int_{\Omega} \boldsymbol{\nabla}\mathbf{u}_h{:}\boldsymbol{\nabla}\mathbf{v}d\mathbf{x}$, by taking Remark 1.4.4 into account.

Remark 1.6.6. Let h_{Ω} (resp., h_j) be the mesh size associated with the velocity mesh $\mathcal{T}_h$ (resp., with the particle mesh $\mathcal{T}_h^j$). Then a relation such as

$$h_{\Omega} < \kappa h_j < h_j < 2h_{\Omega}, \tag{1.56}$$

with $0 < \kappa < 1$, seems to be needed – from a theoretical point of view – in order to satisfy some kind of *stability condition* (for generalities on the approximation of mixed variational problems, such as (1.29)-(1.35), involving Lagrange multipliers, see, for example, the publications by Brezzi and Fortin (ref. [18]) and Roberts and Thomas (ref. [19]). Actually, taking $h_{\Omega} = h_j$ seems to work fine in practice.

Remark 1.6.7. In order to avoid at each time step the solution of complicated *triangulation intersection problems* we advocate the use of

$$< \boldsymbol{\lambda}_{jh}, \pi_j\mathbf{v} - \mathbf{Y}_j - \boldsymbol{\theta}_j \times \overrightarrow{\mathbf{G}_{jh}(t)\mathbf{x}} >_{jh} \tag{1.57}$$

(resp.,

$$< \boldsymbol{\mu}_{jh}, \pi_j \mathbf{u}_h(t) - \mathbf{V}_j(t) - \boldsymbol{\omega}_j(t) \times \overrightarrow{\mathbf{G}_{jh}(t)\mathbf{x}} >_{jh})$$ (1.58)

in (1.47) (resp., (1.50)), instead of

$$< \boldsymbol{\lambda}_{jh}, \mathbf{v} - \mathbf{Y}_j - \boldsymbol{\theta}_j \times \overrightarrow{\mathbf{G}_{jh}(t)\mathbf{x}} >_{jh}$$

(resp.,

$$< \boldsymbol{\mu}_{jh}, \mathbf{u}_h(t) - \mathbf{V}_j(t) - \boldsymbol{\omega}_j(t) \times \overrightarrow{\mathbf{G}_{jh}(t)\mathbf{x}} >_{jh}),$$

where, in (1.57) and (1.58), $\pi_j : C^0(\overline{\Omega}))^2 \to \Lambda_{jh}(t)$ is the *piecewise linear interpolation operator* which to each function $\mathbf{w}$ belonging to $C^0(\overline{\Omega}))^2$ associates the unique element of $\Lambda_{jh}(t)$ defined from the values taken by $\mathbf{w}$ at the vertices of $\mathcal{T}_h^j(t)$

Remark 1.6.8. In general, the function $\mathbf{u}(t)$ has no more than the $(H^{3/2}(\Omega))^2$-regularity. This low regularity implies that we cannot expect more than $O(h^{3/2})$ convergence for the approximation error $\|\mathbf{u}_h(t) - \mathbf{u}(t)\|_{L^2(\Omega)}$.

1.7 TIME DISCRETIZATION BY OPERATOR–SPLITTING

1.7.1 Generalities

Following Chorin (refs. [20]-[22]), most "modern" Navier-Stokes solvers are based on operator splitting algorithms (see, e.g., refs. [23], [24]) in order to force the incompressibility condition via a Stokes solver or an L^2-projection method. This approach still applies to the initial value problem (1.46)-(1.53) which contains four numerical difficulties to each of which can be associated a specific operator, namely:

(a) The incompressibility condition and the related unknown pressure.

(b) An advection-diffusion term.

(c) The rigid body motion of $P_j(t)$ and the related multiplier $\boldsymbol{\lambda}_j(t)$.

(d) The collision terms $\mathbf{F}_j^r$.

The operators in (a) and (c) are essentially *projection operators*. From an abstract point of view, problem (1.46)-(1.53) is a particular case of the following class of initial value problems

$$\frac{d\varphi}{dt} + A_1(\varphi, t) + A_2(\varphi, t) + A_3(\varphi, t) + A_4(\varphi, t) = f, \quad \varphi(0) = \varphi_0,$$ (1.59)

where the operators A_i can be *multivalued*. From the many operator–splitting methods which can be employed to solve (1.59), we advocate (following, e.g., [25]) the very simple one below; it is only first order accurate but its low order accuracy is compensated by good stability and robustness properties. Actually, this scheme can be made *second order accurate* by *symmetrization* (see, e.g., [26] and [27] for the application of *symmetrized* splitting schemes to the solution of the Navier-Stokes equations).

A fractional step scheme à la Marchuk-Yanenko:
With $\triangle t(> 0)$ a *time discretization step*, applying the *Marchuk-Yanenko scheme* to the initial value problem (1.59) leads to

$$\varphi^0 = \varphi_0; \tag{1.60}$$

and for $n \geq 0$, compute φ^{n+1} from φ^n via

$$\frac{\varphi^{n+j/4} - \varphi^{n+(j-1)/4}}{\triangle t} + A_j(\varphi^{n+j/4}, (n+1)\triangle t) = f_j^{n+1}, \tag{1.61}$$

for $j = 1, 2, 3, 4$, with $\sum_{j=1}^{4} f_j^{n+1} = f^{n+1}$.

Remark 1.7.1. Recently, we have introduced a five operator decomposition obtained by treating *diffusion and advection* separately. Some of the numerical results presented in this article have been obtained with this new approach.

1.7.2 Application of the Marchuk-Yanenko scheme to particulate flow

Applying scheme (1.60), (1.61) to problem (1.46)-(1.53), we obtain (after dropping some of the subscripts h and denoting $\{\mathbf{G}_j^n\}_{j=1}^J$ by $\mathbf{G}^n$):

$$\mathbf{u}^0 = \mathbf{u}_{0h}, \ \{\mathbf{V}_j^0\}_{j=1}^J, \ \{\omega_j^0\}_{j=1}^J, \ \{P_j(0)\}_{j=1}^J \ \text{and} \ \mathbf{G}^0 \ \text{are given}; \tag{1.62}$$

for $n \geq 0$, knowing $\{\mathbf{V}_j^n\}_{j=1}^J$, $\{\omega_j^n\}_{j=1}^J$, $\{P_j^n\}_{j=1}^J$ and $\mathbf{G}^n$, we compute $\mathbf{u}^{n+1/4}$, $p^{n+1/4}$ via the solution of

$$\begin{cases} \rho_f \int_\Omega \frac{\mathbf{u}^{n+1/4} - \mathbf{u}^n}{\triangle t} \cdot \mathbf{v} \, d\mathbf{x} - \int_\Omega p^{n+1/4} \boldsymbol{\nabla} \cdot \mathbf{v} \, d\mathbf{x} = 0, \ \forall \mathbf{v} \in V_{0h}, \\ \int_\Omega q \boldsymbol{\nabla} \cdot \mathbf{u}^{n+1/4} \, d\mathbf{x} = 0, \ \forall q \in L_h^2; \\ \mathbf{u}^{n+1/4} \in V_h, \mathbf{u}^{n+1/4} = \mathbf{g}_{0h}^{n+1} \ \text{on} \ \Gamma, p^{n+1/4} \in L_{0h}^2. \end{cases} \tag{1.63}$$

Next we compute $\mathbf{u}^{n+2/4}$ via the solution of

$$\begin{cases} \rho_f \int_\Omega \frac{\mathbf{u}^{n+2/4} - \mathbf{u}^{n+1/4}}{\triangle t} \cdot \mathbf{v} \, d\mathbf{x} + \nu \int_\Omega \boldsymbol{\nabla}\mathbf{u}^{n+2/4} : \boldsymbol{\nabla}\mathbf{v} \, d\mathbf{x} \\ + \rho_f \int_\Omega (\mathbf{u}^{n+1/4} \cdot \boldsymbol{\nabla})\mathbf{u}^{n+2/4} \cdot \mathbf{v} \, d\mathbf{x} = \rho_f \int_\Omega \mathbf{g} \cdot \mathbf{v} \, d\mathbf{x}, \ \forall \mathbf{v} \in V_{0h}; \\ \mathbf{u}^{n+2/4} \in V_h, \mathbf{u}^{n+2/4} = \mathbf{g}_{0h}^{n+1} \ \text{on} \ \Gamma, \end{cases} \tag{1.64}$$

and then, predict the position and the translation velocity of the center of mass as follows, for $j = 1, \ldots, J$:

Take $\mathbf{V}_j^{n+2/4,0} = \mathbf{V}_j^n$ and $\mathbf{G}_j^{n+2/4,0} = \mathbf{G}_j^n$; then predict the new position and translation velocity of P_j via the following subcycling and predictor-corrector technique

For $k = 1, \ldots, N$, compute

$$\widehat{\mathbf{V}}_j^{n+2/4,k} = \mathbf{V}_j^{n+2/4,k-1} + (\triangle t/N)(\mathbf{g} + 0.5(1 - \rho_f/\rho_j)^{-1}M_j^{-1}\mathbf{F}_j^r(\mathbf{G}^{n+2/4,k-1})), \quad (1.65)$$

$$\widehat{\mathbf{G}}_j^{n+2/4,k} = \mathbf{G}_j^{n+2/4,k-1} + (\triangle t/4N)(\widehat{\mathbf{V}}_j^{n+2/4,k} + \mathbf{V}_j^{n+2/4,k-1}), \quad (1.66)$$

$$\mathbf{V}_j^{n+2/4,k} = \mathbf{V}_j^{n+2/4,k-1} + (\triangle t/N)\mathbf{g}$$

$$+(\triangle t/4N)(1 - \rho_f/\rho_j)^{-1}M_j^{-1}(\mathbf{F}_j^r(\widehat{\mathbf{G}}^{n+2/4,k}) + \mathbf{F}_j^r(\mathbf{G}^{n+2/4,k-1})), \quad (1.67)$$

$$\mathbf{G}_j^{n+2/4,k} = \mathbf{G}_j^{n+2/4,k-1} + (\mathbf{V}_j^{n+2/4,k} + \mathbf{V}_j^{n+2/4,k-1})(\triangle t/4N), \quad (1.68)$$

enddo;

and let $\mathbf{V}_j^{n+2/4} = \mathbf{V}_j^{n+2/4,N}, \ \mathbf{G}_j^{n+2/4} = \mathbf{G}_j^{n+2/4,N}.$ \hfill (1.69)

Now, compute $\mathbf{u}^{n+3/4}, \ \{\boldsymbol{\lambda}_j^{n+3/4}, \ \mathbf{V}_j^{n+3/4}, \ \omega_j^{n+3/4}\}_{j=1}^J$ via the solution of

$$\begin{cases} \rho_f \int_\Omega \dfrac{\mathbf{u}^{n+3/4} - \mathbf{u}^{n+2/4}}{\triangle t} \cdot \mathbf{v} \, d\mathbf{x} + \sum_{j=1}^J (1 - \rho_f/\rho_j)M_j \dfrac{\mathbf{V}_j^{n+3/4} - \mathbf{V}_j^{n+2/4}}{\triangle t} \cdot \mathbf{Y}_j \\[2ex] + \sum_{j=1}^J (1 - \rho_f/\rho_j)I_j \dfrac{\omega_j^{n+3/4} - \omega_j^n}{\triangle t}\theta_j \\[2ex] = \sum_{j=1}^J < \boldsymbol{\lambda}_j^{n+3/4}, \mathbf{v} - \mathbf{Y}_j - \theta_j \times \overrightarrow{\mathbf{G}_j^{n+2/4}\mathbf{x}} >_j, \forall \mathbf{v} \in V_{0h}, \mathbf{Y}_j \in \mathbb{R}^2, \theta_j \in \mathbb{R}, \end{cases} \quad (1.70)$$

$$< \boldsymbol{\mu}_j, \mathbf{u}^{n+3/4} - \mathbf{V}_j^{n+3/4} - \omega_j^{n+3/4} \times \overrightarrow{\mathbf{G}_j^{n+2/4}\mathbf{x}} >_j = 0, \ \forall \boldsymbol{\mu}_j \in \Lambda_{jh}^{n+2/4}. \quad (1.71)$$

Finally, take $\mathbf{V}_j^{n+1,0} = \mathbf{V}_j^{n+3/4}$ and $\mathbf{G}_j^{n+1,0} = \mathbf{G}_j^{n+2/4}$; then predict the final position and translation velocity of P_j as follows, for $j = 1, \ldots, J$:

For $k = 1, \ldots, N$, compute

$$\widehat{\mathbf{V}}_j^{n+1,k} = \mathbf{V}_j^{n+1,k-1} + (\triangle t/2N)(1 - \rho_f/\rho_j)^{-1}M_j^{-1}\mathbf{F}_j^r(\mathbf{G}^{n+1,k-1}), \quad (1.72)$$

$$\widehat{\mathbf{G}}_j^{n+1,k} = \mathbf{G}_j^{n+1,k-1} + (\triangle t/4N)(\widehat{\mathbf{V}}_j^{n+1,k} + \mathbf{V}_j^{n+1,k-1}), \quad (1.73)$$

$$\mathbf{V}_j^{n+1,k} = \mathbf{V}_j^{n+1,k-1} + (\triangle t/4N)(1 - \rho_f/\rho_j)^{-1}M_j^{-1}(\mathbf{F}_j^r(\widehat{\mathbf{G}}^{n+1,k}) + \mathbf{F}_j^r(\mathbf{G}^{n+1,k-1})), (1.74)$$

$$\mathbf{G}_j^{n+1,k} = \mathbf{G}_j^{n+1,k-1} + (\mathbf{V}_j^{n+1,k} + \mathbf{V}_j^{n+1,k-1})(\triangle t/4N), \quad (1.75)$$

enddo;

and let $\mathbf{V}_j^{n+1} = \mathbf{V}_j^{n+1,N}, \ \mathbf{G}_j^{n+1} = \mathbf{G}_j^{n+1,N}.$ \hfill (1.76)

Complete the final step by setting

$$\mathbf{u}^{n+1} = \mathbf{u}^{n+3/4}, \ \{\omega_j^{n+1}\}_{j=1}^J = \{\omega_j^{n+3/4}\}_{j=1}^J. \quad (1.77)$$

As shown above, one of the main advantages of operator splitting is that it allows the use of time steps much smaller than $\triangle t$ to predict and correct the position of the centers of mass. For our calculation we have taken $N = 10$ in relations (1.65)-(1.69) and (1.72)-(1.76)

1.7.3 On the solution of subproblems (1.63), (1.64), and (1.70)-(1.71). Further remarks

The iterative solution of the (linear) subproblems (1.63), (1.64), and (1.70)-(1.71) has been discussed with many details in refs. [5] and [6] and we refer interested readers to these two publications. Actually we would like to take advantage of this section to make some additional comments such as:

Remark 1.7.2. The *neutral buoyant* case $\rho_j = \rho_f$ is particularly easy to treat; we shall return to this issue in the review article on the direct numerical simulation of particulate flow that we have been asked to write for the *Journal of Computational Physics*.

Remark 1.7.3. We complete Remark 1.7.1 by observing that, via *further splitting*, we can replace the *advection-diffusion* step (1.64) by

$$
\begin{cases}
\int_\Omega \dfrac{\partial \mathbf{u}}{\partial t} \cdot \mathbf{v}\, dx + \int_\Omega (\mathbf{u}^{n+1/5} \cdot \boldsymbol{\nabla})\mathbf{u} \cdot \mathbf{v}\, dx = 0, \\
\qquad \forall \mathbf{v} \in V_{0h}^{n+1,-}, \ a.e. \ \text{on} \ (n\triangle t, (n+1)\triangle t), \\
\mathbf{u}(n\triangle t) = \mathbf{u}^{n+1/5}, \\
\mathbf{u}(t) \in V_h, \mathbf{u}(t) = \mathbf{g}_{0h}^{n+1} \ \text{on} \ \Gamma_-^{n+1} \times (n\triangle t, (n+1)\triangle t),
\end{cases}
\tag{1.78}
$$

$$
\mathbf{u}^{n+2/5} = \mathbf{u}((n+1)\triangle t),
\tag{1.79}
$$

$$
\begin{cases}
\rho_f \int_\Omega \dfrac{\mathbf{u}^{n+3/5} - \mathbf{u}^{n+2/5}}{\triangle t} \cdot \mathbf{v}\, dx + \nu \int_\Omega \boldsymbol{\nabla}\mathbf{u}^{n+3/5} {:} \boldsymbol{\nabla}\mathbf{v}\, dx = \rho_f \int_\Omega \mathbf{g} \cdot \mathbf{v}\, dx, \\
\forall \mathbf{v} \in V_{0h}; \ \mathbf{u}^{n+3/5} \in V_h, \mathbf{u}^{n+3/5} = \mathbf{g}_{0h}^{n+1} \ \text{on} \ \Gamma,
\end{cases}
\tag{1.80}
$$

with:

(a) $\mathbf{u}^{n+1/5}$ obtained from $\mathbf{u}^n$ via the "incompressibility" step (1.63).

(b) $\Gamma_-^{n+1} = \{\mathbf{x} \mid \mathbf{x} \in \Gamma, \mathbf{g}_{0h}^{n+1}(\mathbf{x}) \cdot \mathbf{n}(\mathbf{x}) < 0\}$.

(c) $V_{0h}^{n+1,-} = \{\mathbf{v} \mid \mathbf{v} \in V_h, \mathbf{v} = \mathbf{0} \ \text{on} \ \Gamma_-^{n+1}\}$.

Problem (1.80) is a discrete symmetric elliptic system for which iterative or direct solution is a quite classical problem. On the other hand, solving the *pure advection problem* (1.78) is a more delicate issue. Clearly, problem (1.78) can be solved by a *method of characteristics* (see, e.g., ref. [28] and the references therein). An easy way to implement an alternative to the method of characteristics is provided by the *wave-like equation* method briefly discussed below (see [26], [27] for more details):

Returning to (1.78) we observe that this problem is the semi-discrete analogue of

$$
\begin{cases}
\dfrac{\partial \mathbf{u}}{\partial t} + (\mathbf{u}^{n+1/5} \cdot \boldsymbol{\nabla})\mathbf{u} = 0 \ \text{in} \ \Omega \times (n\triangle t, (n+1)\triangle t), \\
\mathbf{u}(n\triangle t) = \mathbf{u}^{n+1/5}, \\
\mathbf{u} = \mathbf{g}_0^{n+1}(= \mathbf{u}^{n+1/5}) \ \text{on} \ \Gamma_-^{n+1} \times (n\triangle t, (n+1)\triangle t),
\end{cases}
\tag{1.81}
$$

with $\Gamma_-^{n+1} = \{\mathbf{x} \mid \mathbf{x} \in \Gamma, \mathbf{g}_0^{n+1}(\mathbf{x}) \cdot \mathbf{n}(\mathbf{x}) < 0\}$.

It follows from (1.81) that - after translation and dilation on the time axis - each component of $\mathbf{u}$ is the solution of a transport problem of the following type:

$$\begin{cases} \dfrac{\partial \phi}{\partial t} + \mathbf{V} \cdot \boldsymbol{\nabla} \phi = 0 & \text{in } \Omega \times (0,1), \\ \phi(0) = \phi_0, \\ \phi = g & \text{on } \Gamma_- \times (0,1), \end{cases} \tag{1.82}$$

with $\Gamma_- = \{\mathbf{x} \mid \mathbf{x} \in \Gamma, \mathbf{V}(\mathbf{x}) \cdot \mathbf{n}(\mathbf{x}) < 0\}$ and $\boldsymbol{\nabla} \cdot \mathbf{V} = 0$ and $\dfrac{\partial \mathbf{V}}{\partial t} = \mathbf{0}$. We can easily see that (1.82) is "equivalent" to the (formally) well-posed problem:

$$\begin{cases} \dfrac{\partial^2 \phi}{\partial t^2} - \boldsymbol{\nabla} \cdot ((\mathbf{V} \cdot \boldsymbol{\nabla} \phi)\mathbf{V}) = 0 & \text{in } \Omega \times (0,1), \\ \phi(0) = \phi_0, \ \dfrac{\partial \phi}{\partial t}(0) = -\mathbf{V} \cdot \boldsymbol{\nabla} \phi_0, \\ \phi = g \ \text{ on } \Gamma_- \times (0,1), \ \mathbf{V} \cdot \mathbf{n}(\dfrac{\partial \phi}{\partial t} + \mathbf{V} \cdot \boldsymbol{\nabla} \phi) = 0 \ \text{ on } (\Gamma \setminus \overline{\Gamma}_-) \times (0,1). \end{cases} \tag{1.83}$$

Solving the *wave-like equation* (1.83) by a classical finite element/time stepping method is quite easy, since a *variational formulation* of (1.83) is given by

$$\begin{cases} \displaystyle\int_\Omega \dfrac{\partial^2 \phi}{\partial t^2} v \, d\mathbf{x} + \int_\Omega (\mathbf{V} \cdot \boldsymbol{\nabla} \phi)(\mathbf{V} \cdot \boldsymbol{\nabla} v) \, d\mathbf{x} \\ \qquad + \displaystyle\int_{\Gamma \setminus \overline{\Gamma}_-} \mathbf{V} \cdot \mathbf{n} \dfrac{\partial \phi}{\partial t} v \, d\mathbf{x} = 0, \ \forall v \in W_0, \\ \phi(0) = \phi_0, \ \dfrac{\partial \phi}{\partial t}(0) = -\mathbf{V} \cdot \boldsymbol{\nabla} \phi_0, \\ \phi = g \ \text{ on } \Gamma_-, \end{cases} \tag{1.84}$$

with

$$W_0 = \{v \mid v \in H^1(\Omega), v = 0 \ \text{ on } \Gamma_-\}.$$

Solution methods for the Navier-Stokes equations, taking advantage of the "equivalence" between (1.82) and (1.83), (1.84) are discussed in [26], [27]; see also [29] (and Section 1.8) for further applications including the simulation of *viscoelastic fluid flow* à la Oldroyd-B.

1.8   NUMERICAL EXPERIMENTS

We now present the results of numerical experiments for two-dimensional and three-dimensional flow.

1.8.1   A Sedimentation phenomenon with a Rayleigh-Taylor instability

We consider the sedimentation of 504 circular particles in the closed channel $\Omega = (0,2) \times (0,2)$. We suppose all the particles to be of the same size with diameter $d = .0625$ and

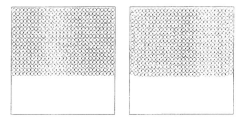

Figure 1.5: The initial position (left) and the position at $t = 1$ (right) of 504 particles.

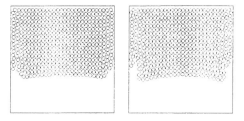

Figure 1.6: 504 particles sedimenting in a closed channel at time t =1.7 (left) and 2 (right).

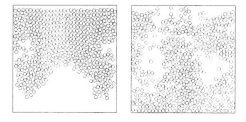

Figure 1.7: 504 particles sedimenting in a closed channel at time t =3 (left) and 5 (right).

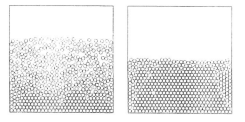

Figure 1.8: 504 particles sedimenting in a closed channel at time t =12 (left) and 24 (right).

a density $\rho_s = 1.01$, while the fluid density and viscosity are $\rho_f = 1$ and $\nu_f = 0.01$ respectively. The initial positions of the particles are shown on Figure 1.5; we suppose that at $t = 0$ fluid and particles are at rest. The solid fraction in this test case is 38.66%. The mesh size used for the velocity field is $h_v = 1/256$, while the one used for pressure is $h_p = 2h_v$. For the parameters discussed in (1.39) we have taken $\rho = h_v$, $c = 1$ and ϵ is in the order of 10^{-5}.

In Figures 1.5–1.8 we have illustrated the location of the particles at ($t = 1$, 1.7, 2, 3, 5, 12, 24). The slightly wavy shape of the interface observed at $t = 1$ is typical of the onset of a Rayleigh-Taylor instability which actually takes place from - approximately - $t = 1$ to $t = 7$ after which slow sedimentation becomes the dominating phenomenon.

1.8.2   A three dimensional case with two identical spherical particles

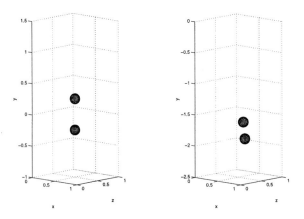

Figure 1.9: Particle position at $t = 0, 1$ (left, right).

The second test problem that we consider here concerns the simulation of the motion of two sedimenting balls in a rectangular cylinder. A 2-D analogue of this test case problem has been (successfully) investigated in [4] using similar techniques. The initial computational domain is $\Omega = (0,1) \times (-1, 1.5) \times (0, 1)$, after which it moves with the center of the lower ball. The diameter d of the two balls is 1/6 and the position of the balls at time $t = 0$ is shown in Figure 1.9. The initial and angular velocities of the balls are zero. The density of the fluid is $\rho_f = 1.0$ and the density of the balls is $\rho_s = 1.04$. The viscosity of the fluid is $\nu_f = 0.01$. The initial condition for the fluid flow is $\mathbf{u} = \mathbf{0}$. The mesh size for the velocity field is $h_v = 1/60$ and the mesh size for the pressure is $h_p = 1/30$. The time step is $\Delta t = 0.001$. For the parameters discussed in (1.39) we have taken $\rho = 1.5h_v$, $c = 1$, and ϵ is in the order of 10^{-3}. The maximal particle Reynolds number in the entire evolution is 47.57. Figures 1.9–1.11 follow the positions of these two balls and demonstrate the fundamental features of two sedimenting balls, i.e., drafting, kissing and tumbling [30]. We observe that a symmetry breaking occurs before the kissing; with a smaller Re, this symmetry breaking would occur after the kissing.

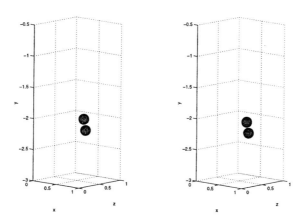

Figure 1.10: Particle position at $t = 1.149, 1.169$ (left, right).

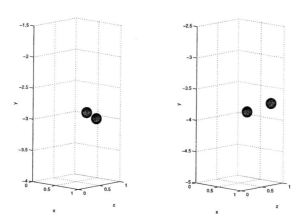

Figure 1.11: Particle position at $t = 1.5, 2$ (left, right).

1.8.3 Fluidization of a bed of 1204 spherical particles

We consider here the simulation of the fluidization in a bed of 1204 spherical particles. The computational domain is $\Omega = (0, 0.6858) \times (0, 20.29968) \times (0, 44.577)$ (or $0.27'' \times 7.992'' \times 17.55''$). The depth of this bed is slightly larger than the diameter of the 1208 balls in the simulation which is $0.25''$, so there is only one layer of balls in this bed. Many experimental results related to this type of "two-dimensional" bed are presented in [30]. The density of the fluid is $\rho_f = 1.0$ and the viscosity is $\nu_f = 0.01$. The initial condition for the fluid velocity is $\mathbf{u} = \mathbf{0}$. The boundary condition for the velocity field is

$$\mathbf{u} = \begin{cases} 0, & \text{on the four vertical walls,} \\ \begin{pmatrix} 0 \\ 0 \\ 5(1.0 - e^{-50t}) \end{pmatrix}, & \text{on the two horizontal walls.} \end{cases}$$

The initial translation velocities and angular velocities of the balls are zero and the density of the balls is $\rho_p = 1.14$. The mesh size for the velocity field is $h_v = 0.027'' = 0.06858$ (2,126,817 nodes). The mesh size for the pressure is $h_p = 2h_v$ (291,444 nodes). The time step is $\Delta t = 0.001$. The parameters ϵ used for the repulsion forces is $\epsilon_p = 5 \times 10^{-7}$ and we take $\rho = h_v$. The initial position of the balls is shown in Figure 1.12. After starting pushing the balls up, we can observe the propagation of cavities among the balls in the bed. Since the in-flow velocity is much greater than the critical fluidization velocity, many balls are pushed directly to the top of the bed. Those balls at the top of bed are stable and close packed, while the others circle around at the bottom of the bed. These numerical results are very close to experimental results and are illustrated in Figures 1.12–1.15. In the simulation, the maximal particle Reynolds number is 1512 and the maximal averaged particle Reynolds number is 285. This work was done on a SGI Origin2000 using a partially parallelized code.

1.8.4 Sedimentation of two disks in an Oldroyd-B viscoelastic fluid

For the fourth case we consider the simulation of two disks falling in a two-dimensional channel filled with an Oldroyd-B viscoelastic fluid. The computational domain is $\Omega = (0, 2) \times (0, 6)$. The initial condition for the fluid velocity field is $\mathbf{u} = \mathbf{0}$. The boundary condition for the velocity field is $\mathbf{u} = \mathbf{0}$ on $\partial\Omega$. The density of the fluid is $\rho_f = 1$ and the viscosity is $\nu_f = 0.25$. The relaxation time is $\lambda_1 = 1.4$ and the retardation time is $\lambda_2 = 0.7$. We place the two disks at the center of the channel at $(1, 5.25)$ and $(1, 4.75)$. The diameter of the disks is 0.25. The initial velocities and angular velocities of the disks are zero. The density of the disks is $\rho_p = 1.01$. In the simulation, the mesh size for the velocity field is $h_v = 1/128$ and the mesh size for the extra stress tensor is $h_\tau = 1/128$. The mesh size for the pressure is $h_p = 2h_v$. The time step is $\Delta t = 0.001$. We let the two disks fall in the closed channel. Before touching the bottom we can see in Figure 1.16 the fundamental features of two sedimenting disks in an Oldroyd-B viscoelastic fluid [31], i.e., drafting, kissing and chaining. The averaged terminal velocity is 0.29 in this simulation, so the Deborah number is $De=1.624$, the Reynolds number is $Re=0.29$, the viscoelastic Mach number value is $M=0.686$, and the elasticity number is $E=5.6$. This simulation

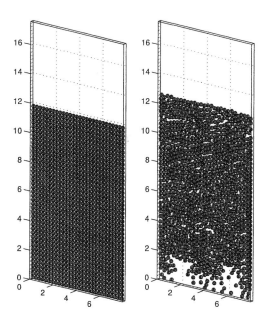

Figure 1.12: Particle position at $t = 0, 1.5$ (left, right).

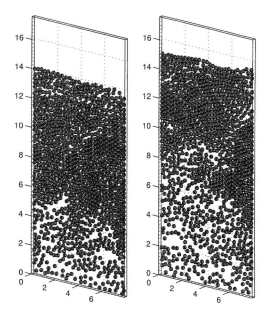

Figure 1.13: Particle position at $t = 3, 4.5$ (left, right).

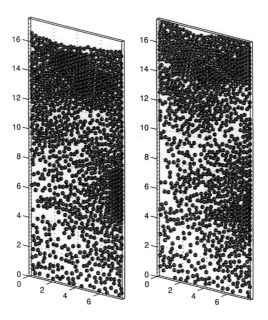

Figure 1.14: Particle position at $t = 6, 7$ (left, right).

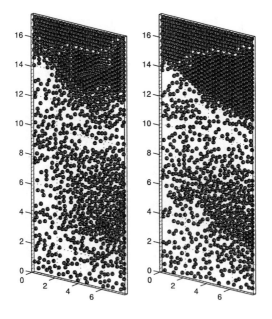

Figure 1.15: Particle position at $t = 8, 10$ (left, right).

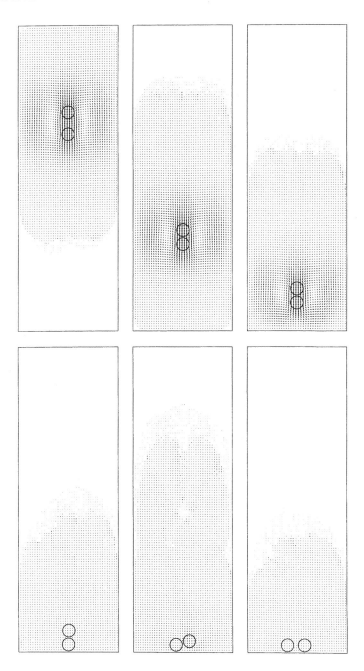

Figure 1.16: Sedimentation and chaining of two particles in an Oldroyd-B viscoelastic fluid.

has been done using the wave-like equation approach described in Section 1.7.

1.9 CONCLUSION

We have presented in this article a distributed Lagrange multiplier based fictitious domain method for the simulation of flow with moving boundaries. Compared to the one discussed earlier in [1], it allows the simulation of fairly complicated phenomena, such as particulate flow, including sedimentation. Some preliminary experiments have shown the potential of this method for the direct simulation of fluidization which is in some sense the inverse phenomenon of sedimentation; the results already obtained look promising. Other goals include: 3D particulate flow with a large number of particles of different sizes and shapes and particulate flow for viscoelastic liquids such as Oldroyd-B, etc..

1.10 ACKNOWLEDGMENTS

We acknowledge the helpful comments and suggestions of E. J. Dean, V. Girault, J. He, T.I. Hesla, Y. Kuznetsov, J. Périaux, G. Rodin, A. Sameh, V. Sarin, and P. Singh, and also the support of the Supercomputing Institute at the University of Minnesota concerning the use of a SGI Origin2000. We acknowledge also the support of the NSF (Grants ECS-9527123 and CTS-9873236) and Dassault Aviation.

REFERENCES

[1] R. Glowinski, T.-W. Pan, J. Périaux (1997). Fictitious domain methods for incompressible viscous flow around moving rigid bodies. In J.R. Whiteman (ed.), *The Mathematics of Finite Elements and Applications, Highlights 1996*, Wiley, Chichester, 155-174.

[2] R. Glowinski, T.I. Hesla, D.D. Joseph, T.W. Pan and J. Périaux (1997). Distributed Lagrange multiplier methods for particulate flows. In *Computational Science for the 21st Century*, M.O. Bristeau, G. Etgen, W. Fitzgibbon, J.L. Lions, J. Périaux and M.F. Wheeler, eds., Wiley, Chichester, 270-279.

[3] R. Glowinski, T.-W. Pan, T. Hesla, D.D. Joseph and J. Périaux (1998). A fictitious domain method with distributed Lagrange multipliers for the numerical simulation of particulate flows. In J. Mandel, C. Farhat, and X.-C. Cai (eds.), *Domain Decomposition Methods 10*, AMS, Providence, RI, 121-137.

[4] R. Glowinski, T.-W. Pan, T. Hesla and D.D. Joseph (1999). A distributed Lagrange multiplier/fictitious domain method for particulate flows. *Int. J. Multiphase Flow*, 25:755-794.

[5] R. Glowinski, T.-W. Pan, T. I. Hesla, D. D. Joseph and J. Périaux. A distributed Lagrange multiplier/fictitious domain method for flows around moving rigid bodies: Application to particulate flow. *Int. J. Numer. Meth. Fluids* (to appear).

[6] R. Glowinski, T.-W. Pan, T. I. Hesla, D. D. Joseph and J. Périaux. A distributed Lagrange multiplier/fictitious domain method for the simulation of flows around moving rigid bodies: Application to particulate flow. *Comput. Meth. Appl. Mech. Engrg.* (to appear).

[7] B. Desjardins and M.J. Esteban (1999). On weak solution for fluid-rigid structure interaction: compressible and incompressible models, *Cahiers du Ceremade*, No. 9908, University Paris-Dauphine.

[8] A. Johnson and T. Tezduyar (1997). 3D Simulation of fluid-particle interactions with the number of particles reaching 100. *Comput. Meth. Appl. Mech. Engrg.*, 145:301-321.

[9] H.H. Hu (1996). Direct simulation of flows of solid-liquid mixtures. *Int. J. Multiphase Flow*, 22:335-352.

[10] B.A. Maury and R. Glowinski (1997). Fluid-particle flow: a symmetric formulation. *C. R. Acad. Sci. Paris, Serie I*, t. 324:1079-1084.

[11] C.S. Peskin (1977). Numerical analysis of blood flow in the heart. *J. Comput. Phys.* 25:220-252.

[12] C.S. Peskin and D.M. McQueen (1980). Modeling prosthetic heart valves for numerical analysis of blood flow in the heart. *J. Comput. Phys.* 37:113-132.

[13] C.S. Peskin (1981). Lectures on mathematical aspects of Physiology. *Lectures in Appl. Math.*, 19:69-107.

[14] K. Höfler, M. Müller, S. Schwarzer and B. Wachmann (1999). Interacting particle-liquid systems. In E. Krause, W. Jäger, (eds.), *High Performance Computing in Science and Engineering*, Springer, Berlin, 54–64.

[15] N.A. Patankar, P. Singh, D.D. Joseph, R. Glowinski and T.-W. Pan. A new formulation of the distributed Lagrange multiplier/fictitious domain method for particulate flows. Submitted to *Int. J. Multiphase Flow*.

[16] E.J. Dean, R. Glowinski, Y. M. Kuo and M. G. Nasser (1990). On the discretization of some second order in time differential equations. Applications to nonlinear wave problems. In A. V. Balakrishnan ed., *Computational Techniques in Identification and Control of Flexible Flight Structures*, Optimization Software Inc., Los Angeles, 199-246.

[17] F. Bertrand, P.A. Tanguy and F. Thibault (1997). A three-dimensional fictitious domain method for incompressible fluid flow problem. *Int. J. Numer. Meth. Fluids* 25:719-736.

[18] F. Brezzi and M. Fortin (1991). *Mixed and Hybrid Finite Element Methods*, Springer-Verlag, New York.

[19] J.E. Roberts and J.M. Thomas (1991). Mixed and Hybrid Methods. In P.G. Ciarlet and J.L. Lions (eds.), *Handbook of Numerical Analysis*, Vol. II, North-Holland, Amsterdam, 523-639.

[20] A.J. Chorin (1967). A numerical method for solving incompressible viscous flow problems. *J. Comput. Phys.* 2:12-26.

[21] A.J. Chorin (1968). On the convergence and approximation of discrete approximation to the Navier-Stokes equations. *Math. Comput.*, 23:341-353.

[22] A.J. Chorin (1973). Numerical study of slightly viscous flow. *J. Fluid Mech.*, 57:785-796.

[23] R. Glowinski and O. Pironneau (1992). Finite element methods for Navier-Stokes equations. *Annu. Rev. Fluid Mech.*, 24:167-204.

[24] S. Turek (1996). A comparative study of time-stepping techniques for the incompressible Navier-Stokes equations: from fully implicit non-linear schemes to semi-implicit projection methods. *Int. J. Numer. Meth. Fluids*, 22:987-1011.

[25] G.I. Marchuk (1990). Splitting and alternate direction methods. In P.G. Ciarlet and J.L. Lions (eds.), *Handbook of Numerical Analysis*, Vol. I, North-Holland, Amsterdam, 197-462.

[26] E. Dean and R. Glowinski (1997). A wave equation approach to the numerical solution of the Navier-Stokes equations for incompressible viscous flow. *C. R. Acad. Sci. Paris, Série I*, t. 325:783-791.

[27] E. Dean, R. Glowinski and T.-W. Pan (1998). A wave equation approach to the numerical simulation of incompressible viscous fluid flow modeled by the Navier-Stokes equations. In J. De Santo (ed.), *Mathematical and Numerical Aspects of Wave Propagation*, SIAM, Philadelphia, 65-74.

[28] O. Pironneau (1989). *Finite element methods for fluids*, Wiley, Chichester.

[29] P. Parthasarathy (1999). Applications of a wave-like equation method for Newtonian and non-Newtonian flows. Master thesis, University of Houston.

[30] A.F. Fortes, D.D. Joseph and T.S. Lundgren (1987). Nonlinear mechanics of fluidization of beds of spherical particles. *J. Fluid Mech.* 177:467-483.

[31] D.D. Joseph and Y.J. Liu (1993). Orientation of long bodies falling in a viscoelastic liquid. *J. Rheol.* 37:961-983.

2 LOCALLY CONSERVATIVE ALGORITHMS FOR FLOW

Béatrice Rivière and Mary F. Wheeler

The Center for Subsurface Modeling,
Texas Institute for Computational and Applied Mathematics,
The University of Texas, Austin TX 78712, U.S.A.

ABSTRACT

Locally mass conservative methods for subsurface flow are presented and applied to the unstable miscible displacement problem arising in porous media. The pressure equation is solved either by the discontinuous Galerkin method or the mixed finite element method. The concentration equation is solved using a higher order Godunov method. Theoretical estimates are given for the approximation of the pressure obtained by the discontinuous Galerkin method. A velocity projection method based on the discontinuous Galerkin method is introduced. Numerical simulations comparing the two methods are presented.

Key words. Discontinuous Galerkin Method, Mixed Method, Projection Method, Unstable Miscible Displacement.

2.1 INTRODUCTION

In either surface water or subsurface water environmental quality modelling, the flow and multi-species transport may be solved separately using different numerical methods and grids due to differences in time and length scales involved. For example, in surface water, the flow grid usually needs to incorporate high resolution near land boundaries and should extend into the ocean to avoid spurious boundary effects. The transport code may only simulate transport over a small portion of the flow domain, and may use much coarser resolution. Therefore, for efficient coupling of flow and transport codes, it is critical to be able to take velocities from an arbitrary flow grid, and project them onto an arbitrary transport grid. For accurate multi-species transport, it is desirable for the velocities to be locally conservative on the transport grid.

Groundwater contaminant transport typically involves flow of one or more phases through a highly heterogeneous porous media, mass transfer between phases including the solid phases, advection with dispersion, and reaction of chemical and biological species.

29

These phenomena, augmented by others such as heat transfer, and local injection and/or extraction of fluid, are closely coupled. At the discrete level, accuracy requires locally conservative schemes to maintain mass balances.

The computational complexity of both these applications demands a highly efficient approach. In general, it is not possible to solve the entire system of governing equations on the scale of the fastest reactions. Thus, out of necessity one must look at employing dynamic adaptive grids and using higher order methods.

For simplicity in this paper we restrict our attention to subsurface flows, although much of what is described can be applied to hydrodynamic flow problems where Chorin splitting or pressure stabilization techniques are used. We shall consider two schemes, mixed finite element methods and discontinuous Galerkin methods, which can be defined with arbitrary orders and are locally conservative.

This paper is divided into five additional sections. In Section 2.2, we describe the physical problem of miscible displacement of one fluid by another in porous media. In Section 2.3, we formulate the discontinuous Galerkin method (DG) for solving the pressure equation. Theoretical convergence results are summarized. In addition, a projection algorithm for obtaining a mass conservative velocity field is introduced. The latter provides a viable approach for the bay and estuary problem discussed above.

In Section 2.4, we briefly describe, the mixed finite element method (MFE) for solving the pressure equation. Computational results for modelling miscible displacement, both stable and unstable, as well as results for flow in fractures are presented for DG and MFE in Section 2.5. In Section 2.6 we conclude with some remarks and future research possibilities.

2.2   SUBSURFACE PROBLEM

The subsurface flow problem we are focusing on is the unstable miscible displacement of one incompressible fluid by another in porous media. Numerical simulations of miscible displacement play an important role in engineering decisions about the exploitation of petroleum reservoirs. Groundwater pollution is another application of miscible displacement. Simulations of groundwater flow are important in the process of developing reliable predictions of the transport of dissolved contaminants within the flow systems. The classical governing equations in a domain $\Omega \subset I\!\!R^d, d = 1, 2, 3$ over the time interval $J = (0, T]$, are:

$$-\nabla \cdot (\frac{k}{\mu(c)}\nabla p) \;\equiv\; \nabla \cdot \boldsymbol{u} = q \qquad \text{in} \quad (\boldsymbol{x}, t) \in \Omega \times J \qquad (2.1)$$

$$\phi\frac{\partial c}{\partial t} + \nabla \cdot (\boldsymbol{u}c - D(\boldsymbol{x}, \boldsymbol{u})\nabla c) \;=\; \tilde{c}q + R(c) \qquad \text{in} \quad (\boldsymbol{x}, t) \in \Omega \times J \qquad (2.2)$$

where the dependent variables are p, the pressure in the fluid mixture, and c, the concentration of a solvent injected into the resident fluid. The permeability k of the medium measures the resistance of the medium to fluid flow; the viscosity μ measures the resistance to flow of the fluid mixture; $\boldsymbol{u}$ represents the Darcy velocity of the mixture (volume flowing across a unit cross-section per unit time); the porosity ϕ is the fraction of the volume of the medium occupied by pores and $\boldsymbol{D}$ is the coefficient of molecular diffusion and mechanical dispersion of one fluid into the other. The imposed external total flow rate q is a sum of point sources and sinks (combination of Dirac measures) and in the case of oil recovery, q represents the flow rates at injection and production wells; $\tilde{c}$ is the injected concentration at injection wells and the resident concentration at production wells. R is the rate of appearance and disappearance of the solvent due to chemical reaction, decay, and/or mass transfer. Here D is a tensor and can be velocity dependent. In two dimensional cases, for $u = (u_x, u_y)$, D can be written as

$$D(x, u) = D_m I + \frac{\alpha_l}{|u|} \begin{bmatrix} u_x^2 & u_x u_y \\ u_x u_y & u_y^2 \end{bmatrix} + \frac{\alpha_t}{|u|} \begin{bmatrix} u_y^2 & -u_x u_y \\ -u_x u_y & u_x^2 \end{bmatrix},$$

where D_m is the molecular diffusivity and α_l and α_t are the longitudinal and transverse dispersivities, respectively. The viscosity of the fluid mixture will be assumed to follow the quarter-power mixing law, commonly applicable to hydrocarbon mixtures:

$$\mu = [c\mu_s^{-0.25} + (1 - c)\mu_o^{-0.25}]^{-4},$$

where μ_s and μ_o are the viscosity of the solvent and the resident fluid respectively. The first elliptic equation (2.1), referred to as the pressure equation, is a formulation of Darcy's law. The second equation (2.2), the concentration equation, describes the convection-diffusion process, written in the divergence form: it is parabolic but normally convection-dominated.

We assume the following boundary conditions for the pressure and the transport problem:

$$\begin{aligned} \frac{k}{\mu(c)} \nabla p \cdot \nu &= 0, & (x, t) \in \partial\Omega \times J, \\ -D\nabla c \cdot \nu &= 0, & (x, t) \in \Gamma_N \times J, \\ c(x, t) &= c_D(x, t), & (x, t) \in \Gamma_D \times J, \end{aligned}$$

where the boundary $\partial\Omega$ is the union of the Dirichlet, no-flow and outflow parts denoted by Γ_D, Γ_N and Γ_O respectively.

The initial concentration must be specified, but the initial pressure distribution is obtained by (2.1). We note that the pressure is determined up to an additive constant. The initial condition for the concentration is the following:

$$c(x, 0) = c_0(x), \quad x \in \Omega.$$

2.3   DISCONTINUOUS FINITE ELEMENT METHODS

2.3.1   Functional Settings

Let $\mathcal{E}_h = \{E_1, E_2, \ldots, E_{N_h}\}$ be a nondegenerate subdivision of Ω, where E_j is a triangle or a quadrilateral. The nondegeneracy requirement is that there exists $\rho > 0$ such that if $h_j = \text{diam}(E_j)$, then E_j contains a ball of radius ρh_j in its interior. Let $h =$

$\max\{h_j, \quad j = 1 \dots N_h\}$. The edges of the polygon are denoted by $\{e_1, \dots, e_{P_h}, \dots, e_{M_h}\}$ where $e_k \subset \Omega, 1 \leq k \leq P_h$,and $e_k \subset \partial\Omega$, $P_h + 1 \leq k \leq M_h$. With each edge e_k , we associate a unit normal vector $\boldsymbol{\nu}_k$. For $k > P_h$, $\boldsymbol{\nu}_k$ is taken to be the unit outward vector normal to $\partial\Omega$.

For $s \geq 0$, let

$$H^s(\mathcal{E}_h) = \{v \in L^2(\Omega) : v|_{E_j} \in H^s(E_j), j = 1 \dots N_h\}.$$

We now define the average and the jump for $\phi \in H^s(\mathcal{E}_h)$, $s > \frac{1}{2}$. Let $1 \leq k \leq P_h$. For $e_k = \partial E_i \cap \partial E_j$ with $\boldsymbol{\nu}_k$ exterior to E_i for $i > j$ for some pair of elements (E_i, E_j), set

$$\{\phi\} = \frac{1}{2}(\phi|_{E_i})|_{e_k} + \frac{1}{2}(\phi|_{E_j})|_{e_k}, \quad [\phi] = (\phi|_{E_i})|_{e_k} - (\phi|_{E_j})|_{e_k}.$$

We denote the usual Sobolev norm by $\|p\|_{s,E}$ for $p \in H^s(E)$ and $E \subset \mathbb{R}^d$. We also define the following seminorm

$$|\!|\!|\phi|\!|\!|^2 = \sum_{j=1}^{N_h} \left\| \left(\frac{k}{\mu}\right)^{\frac{1}{2}} \nabla\phi \right\|_{0,E_j}^2$$

Let r be a positive integer. The finite element subspace is taken to be

$$\mathcal{D}_r(\mathcal{E}_h) = \prod_{j=1}^{N_h} P_r(E_j)$$

where $P_r(E_j)$ denotes the set of polynomials of (total) degree less than or equal to r on E_j, where E_j is a triangle or a quadrilateral.

2.3.2 Scheme

We present a DG formulation for the pressure equation, that is based on the work done by Baumann & al. [3], [6] for elliptic equations. Let us consider the non-symmetric bilinear form:

$$a_{NS}(\chi; \varphi, \psi) = \sum_{j=1}^{N_h} \int_{E_j} \frac{k}{\mu(\chi)} \nabla\varphi \cdot \nabla\psi - \sum_{k=1}^{P_h} \int_{e_k} \left\{ \frac{k}{\mu(\chi)} \nabla\varphi \cdot \nu_k \right\} [\psi]$$
$$+ \sum_{k=1}^{P_h} \int_{e_k} \left\{ \frac{k}{\mu(\chi)} \nabla\psi \cdot \nu_k \right\} [\varphi]$$

The continuous-time approximation of the pressure is given by the map $P : [0, T] \to \mathcal{D}_r(\mathcal{E}_h)$ determined by the relations

$$a_{NS}(C(t); P(t), w) = (q(t), w), \quad w \in \mathcal{D}_r(\mathcal{E}_h), \quad 0 \leq t \leq T,$$
$$(P(t), 1) = 0$$

where C is the continuous-time approximation of the concentration. The concentration equation is solved by using a higher order Godunov method that is described in [4]. For the purpose of this paper, we only focus on the approximation of the pressure equation.

2.3.3 A Priori Error Estimates

The convergence of the method is guaranteed by the following *hp* error estimates which are proved in [2].

Theorem 2.1 *There is a constant C independent of h, r, p such that for $s \geq 2$ and for $t \in [0, T]$*

$$\|p(t) - P(t)\| \leq C \frac{h^{\mu-1}}{r^{s-2}} \left(\sum_{j=1}^{N_h} \|p(t)\|^2_{s,E_j} \right)^{\frac{1}{2}},$$

where $\mu = \min(r+1, s)$.

2.3.4 A Posteriori Error Estimates

The estimation of the H^1 seminorm has been analyzed. We fix $t \in (0, T]$. Let $V = H^{\frac{3}{2}+\epsilon}(\mathcal{E}_h)$ for $\epsilon > 0$. Clearly $(V, \|\cdot\|)$ is a normed linear space. We consider the residual to be the linear form R defined as follows:

$$R(v) = L(v) - a_{NS}(C(t); P(t), v), \quad \forall v \in V.$$

We define the following norm of the residual:

$$\|R\| = \sup_{v \in V \setminus \{0\}} \frac{R(v)}{\|v\|}$$

and have the error estimate for t in $(0, T]$:

Theorem 2.2

$$\|p(t) - P(t)\| \leq \|R\|$$

It suffices now to estimate the norm of the residual. In [1] we have shown the following results. For a fixed t, consider the problem: find $\Phi \in V$ such that

$$\sum_{j=1}^{N_h} \int_{E_j} \frac{k}{\mu} \nabla \Phi \cdot \nabla v = R(v), \quad \forall v \in V. \tag{2.3}$$

We discretize (2.3) in the following fashion: for q a positive integer, find $\Phi_h \in \mathcal{D}_{r+q}(\mathcal{E}_h)$ such that

$$\sum_{j=1}^{N_h} \int_{E_j} \frac{k}{\mu} \nabla \Phi_h \cdot \nabla v = R(v), \quad \forall v \in \mathcal{D}_{r+q}(\mathcal{E}_h). \tag{2.4}$$

The relationships between the residual and the solutions to the problems (2.3) and (2.4) are stated in the following theorem.

Theorem 2.3

$$\begin{aligned}
\|R\| &= \|\Phi\|, \\
\|p(t) - P(t)\| &\leq \|\Phi\|, \\
\sqrt{1 - \sigma^2} \|R\| &\leq \|\Phi_h\| \leq \|R\|,
\end{aligned}$$

where σ is a positive constant in $[0, 1)$.

We note that the problem (2.4) is equivalent to solving local problems on each finite element E_j, for $j = 1, \cdots, N_h$. Thus, the computation of Φ_h is quite cheap. We impose a Neumann boundary condition for the local problem, and thus we also have the constraint $\int_{E_j} \Phi_h = 0$ for all finite elements E_j. The constant σ depends on the enrichment of the discrete space $\mathcal{D}_{r+q}(\mathcal{E}_h)$. The higher q is, the smaller σ is, and the closer $\|\Phi_h\|$ is to $\|R\|$. Of course, increasing q also means increasing the computational cost for the error estimator. The choice of a good value for q is problem dependent.

2.3.5   DG Application: Locally Conservative Projection Algorithm for Changing Meshes

Let W_{h^o} and V_{h^n} be finite dimensional subspaces corresponding to the old and new meshes. Given $U^o \in W_{h^o}$, the problem is to find $U^n \in V_{h^n}$ such that the new velocity field is locally mass conservative. The new velocity can be expressed as

$$U^n = P_{h^n} U^o + \Gamma,$$

where P_{h^n} is a projection operator onto V_{h^n}. The local mass conservation implies

$$0 = \nabla \cdot U^n = \nabla \cdot P_{h^n} U^o + \nabla \cdot \Gamma.$$

By writing $\Gamma = \nabla \Phi$, the following elliptic problem is obtained

$$-\Delta \Phi = \nabla \cdot P_{h^n} U^o.$$

The elliptic problem is solved by using the discontinous Galerkin method: find Φ_h in $V_{h^n} \equiv \mathcal{D}_r(\mathcal{E}_h)$ such that

$$\sum_{j=1}^{N_h} \int_{E_j} \nabla \Phi_h \cdot \nabla \psi - \sum_{k=1}^{P_h} \int_{e_k} \{\nabla \Phi_h \cdot \nu_k\} [\psi] + \sum_{k=1}^{P_h} \int_{e_k} \{\nabla \psi \cdot \nu_k\} [\Phi_h]$$
$$= \int_{\Omega} \nabla \cdot U^o \psi, \qquad \forall \psi \in \mathcal{D}_r(\mathcal{E}_h).$$

2.4   MIXED FINITE ELEMENT METHODS

Let $W = L^2(\Omega)$ denote the set of square integrable functions and $H(\Omega; \mathrm{div}) = \{v \in (L^2(\Omega))^d \mid \nabla \cdot v \in L^2(\Omega)\}$. Let $V = H^0(\Omega; \mathrm{div}) = \{v \in H(\Omega; \mathrm{div}) \mid v \cdot \nu = 0 \text{ on } \partial\Omega\}$.

For spatial discretization, we employ the lowest order Raviart-Thomas spaces $(W_h \times V_h)$ [7] defined over a rectangular grid of Ω with maximal grid spacing $h > 0$. $W_h \subset W$ consists of the space of piecewise constants and $\tilde{V}_h \subset H(\Omega; \mathrm{div})$ is the space of functions $v = (v_1, v_2, v_3)$ (if $d = 3$) such that v_i is continuous, piecewise linear over the grid in the ith direction and discontinuous, piecewise constant over the grid in the other two directions. We also need the subspace $V_h = \tilde{V}_h \cap V$.

We briefly describe the mixed finite element method for approximating (2.1). With $(\cdot, \cdot)$ denoting the $L^2(\Omega)$-inner product, we write (2.1) in variational form as

$$\left(\left(\frac{k}{\mu}\right)^{-1} u, v\right) - (p, \nabla \cdot v) = 0, \quad v \in V,$$
$$(\nabla \cdot u, w) = (q, w), \quad w \in W.$$

In the mixed finite element formulation, we seek the pair $(\boldsymbol{U}_h, P_h) \in V_h \times W_h$ satisfying

$$\left(\left(\frac{k}{\mu}\right)^{-1} \boldsymbol{U}_h, \boldsymbol{v}_h\right) - (P_h, \nabla \cdot \boldsymbol{v}_h) = 0, \quad \boldsymbol{v}_h \in V_h,$$

$$(\nabla \cdot \boldsymbol{U}_h, w_h) = (q, w_h), \quad w_h \in W_h.$$

Chippada, Dawson, Martinez, and one of the authors formulated and analyzed a conservative projection method [8] based on a mixed hybrid finite element method for constructing mass conservation velocity fields. These results were for the lowest order mixed spaces; however, the analysis applies also to higher order approximating spaces.

2.5 NUMERICAL RESULTS

2.5.1 Single phase flow in a fractured rock

We compared the flow patterns in a fractured rock using both the mixed finite element method and the discontinuous Galerkin method. In this test case, two inclined fracture zones intersect one another at depth. The medium is isotropic but the value of the hydraulic conductivity (K) is higher in the fracture zones $(K = 10^{-6} ms^{-1})$ than in the surrounding rock $(K = 10^{-8} ms^{-1})$. The boundary of the region is assumed to be impermeable to flow except for the top boundary (Γ_D) (see Fig. 2.1 (a)), where we impose the following condition:

$$p(x, y) = y, \quad \forall (x, y) \in \Gamma_D.$$

A more detailed description can be found in [5]. Fig. 2.1 shows the pressure field. The discontinuous Galerkin method is applied with a cubic order of approximation. The lowest order Raviart-Thomas space is used for the mixed method. Fig. 2.2 shows the velocity field obtained with both methods. The results are very similar and show that the flow is concentrated in the fracture zones as expected.

2.5.2 Random permeability fields

For the concentration we consider a rectangular domain with size $(0, 320) \times (0, 160)$. The permeability field has been randomly generated on the coarse grid made with 256 rectangular elements. The variable MR will denote the mobility ratio, which is the ratio of the viscosity of the resident fluid to the viscosity of the solvent. The simulations are done with the following parameters:

$$T = 6e^6 s, \quad \alpha_l = 0.1m, 16m, \quad \alpha_t = 0.1\alpha_l,$$
$$D_m = 1.16e^{-9} m^2 s^{-1}, \quad MR = 1, 10, 41, 100.$$

Concentration profiles are shown for the values $c = 0.25$ and $c = 0.50$ in the Fig. 2.3-2.10. In the case where the pressure equation is approximated by the DG method, the figures show that as the degree of approximation increases, the front converges. In the case of the mixed finite element method approximation, uniform refinements have also been considered. For a mobility ratio equal to one (Fig. 2.3, 2.7), an improvement of

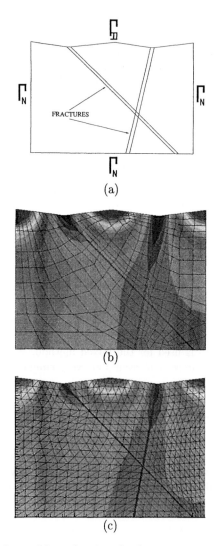

Figure 2.1: Diagram of the problem showing the fracture zones and the boundary conditions: (a). Pressure Field: (b) DG $r = 3$ and (c) MFE

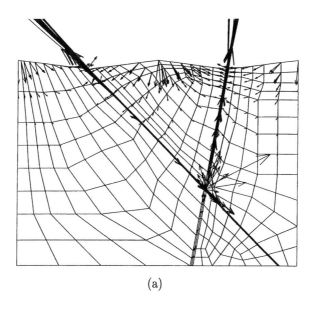

(a)

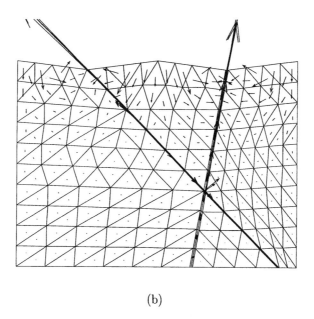

(b)

Figure 2.2: Velocity Field: (a) DG $r = 3$ and (b) MFE

the concentration profile can be seen if the mesh is refined once or twice: here $h0, h1$ and $h2$ denote the coarse and refined meshes respectively. The DG velocities on a coarser grid are as accurate as the MFE velocities on a refined grid. This numerical result holds true for the other values of the mobility ratio if the mesh has been refined only once. This is shown in Fig. 2.4- 2.6, 2.8- 2.10.

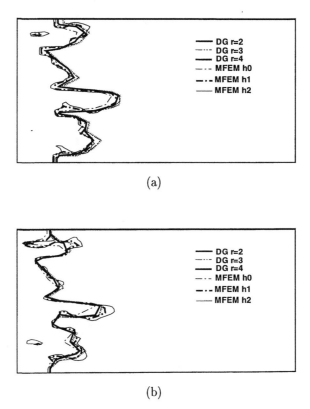

(a)

(b)

Figure 2.3: Front contours. Mobility ratio $= 1.$ $\alpha_l = 0.1.$ (a) c=0.25 and (b) c=0.50. For this case, the viscosity of the fluid mixture μ is independent of the concentration. Here two levels of refinement of the MFE appear to converge to the DG solution.

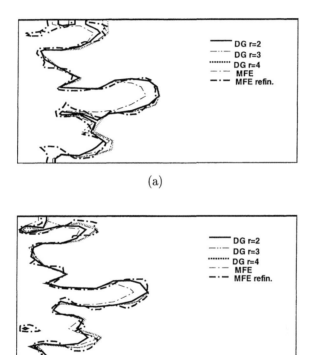

(a)

(b)

Figure 2.4: Front contours. Mobility ratio = 10. $\alpha_l = 0.1$. (a) c=0.25 and (b) c=0.50. For this case, the viscosity μ depends on the concentration; hence the concentration approximations affect the coarse grid DG flow approximations.

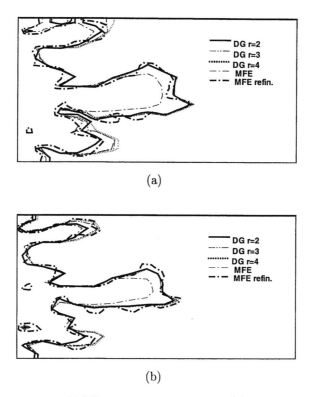

(a)

(b)

Figure 2.5: Front contours. Mobility ratio = 41. $\alpha_l = 0.1$. (a) c=0.25 and (b) c=0.50. In this case, the flow depends more strongly on the effect of concentration on viscosity and more fingering and instability occurs. Thus both MFW and DG require finer grids.

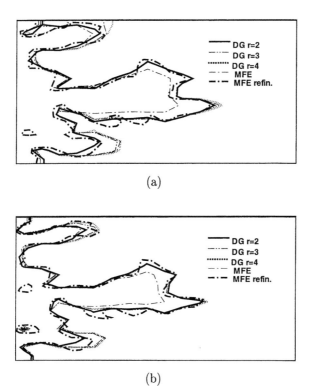

(a)

(b)

Figure 2.6: Front contours. Mobility ratio = 100. $\alpha_l = 0.1$. (a) c=0.25 and (b) c=0.50. The same phenomenon is more pronounced than in Fig. 2.5.

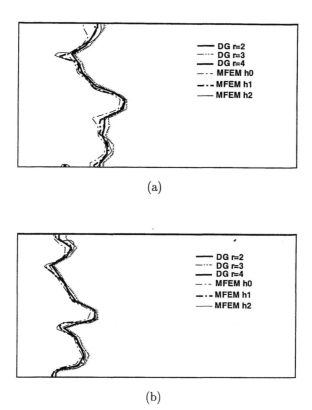

Figure 2.7: Front contours. Mobility ratio = 1. $\alpha_l = 16$. (a) c=0.25 and (b) c=0.50. The effect of longitudinal dispersivity is shown here, namely the fronts are less pronounced. Again as in Fig. 2.3, the viscosity is independent of concentration and two levels of refinement of the MFE appear to converge to the DG solution.

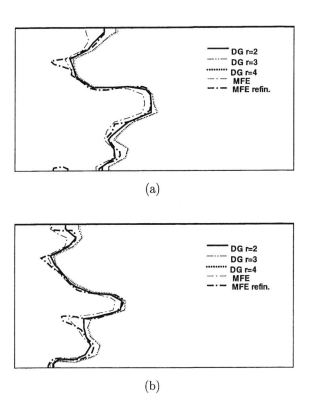

Figure 2.8: Front contours. Mobility ratio = 10. $\alpha_l = 16$. (a) c=0.25 and (b) c=0.50. This increase in longitudinal dispersivity as compared with Fig. 2.4 shows a smearing of the front. The nonlinearity of viscosity again affects the DG approximations.

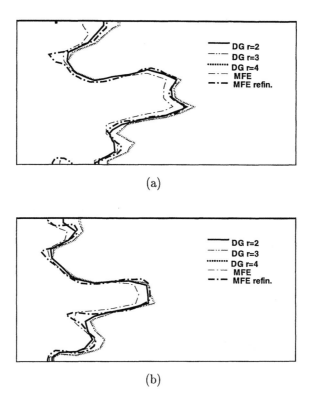

(a)

(b)

Figure 2.9: Front contours. Mobility ratio $= 41$. $\alpha_l = 16$. (a) c=0.25 and (b) c=0.50. Results are similar to Fig. 2.5.

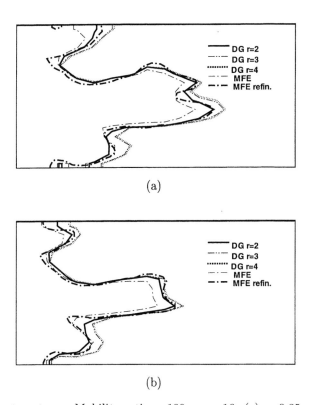

Figure 2.10: Front contours. Mobility ratio $= 100$. $\alpha_l = 16$. (a) c=0.25 and (b) c=0.50. Results are similar to Fig. 2.6.

2.6 CONCLUSION

A new higher order locally conservative scheme for computing Darcy flow has been presented. Computational results indicate that concentration grids can be coarser for higher order velocity approximations. Our current research is to investigate the use of different grids for concentration and for flow.

REFERENCES

[1] B.Rivière and M.F.Wheeler. A-posteriori error estimates for a discontinuous Galerkin method. In preparation.

[2] B.Rivière, M.F.Wheeler and V. Girault (1999). Part I. Improved energy estimates for interior penalty, constrained and discontinuous Galerkin methods for elliptic problems. Technical Report 09, TICAM.

[3] C.E.Baumann (1997). *An h-p adaptive discontinuous finite element method for computational fluid dynamics.* Ph.D. thesis, The University of Texas at Austin.

[4] C.N.Dawson (1993). Godunov-mixed methods for advection-diffusion equations in multidimensions. *SIAM J.Numer.Anal.*, 30(5):11315–1332.

[5] E.F.Kaasschieter (1995). Mixed finite elements for accurate particle tracking in saturated groundwater flow. *Advances in Water Resources*, 18(5):277–294.

[6] J.T.Oden, I.Babuška and C.E. Baumann (1998). A discontinuous hp finite element method for diffusion problems. *Journal of Computational Physics*, 146:491–519.

[7] P.A.Raviart and J.M.Thomas (1977). A mixed finite element method for second order elliptic problems. *Mathematical Aspects of the Finite Element Method, Lecture Notes in Mathematics*, 606:292–315.

[8] S.Chippada, C.N.Dawson, M.L.Martinez and M.F.Wheeler (1998). A projection method for constructing a mass conservative velocity field. In *Computer Methods in Applied Mechanics and Engineering: Proceedings of the 1997 Symposium on Advances in Computational Mechanics*, volume 157, pages 1–10.

3 RECENT ADVANCES IN ADAPTIVE MODELLING OF HETEROGENEOUS MEDIA

J. Tinsley Oden and Kumar Vemaganti

Texas Institute for Computational and Applied Mathematics,
The University of Texas, Austin TX 78712, U.S.A.

ABSTRACT

The concept of *hierarchical modelling* has been put forward [30, 25] as an adaptive methodology for the efficient and accurate solution of problems of heterogeneous media. This approach consists of using *a posteriori* estimates of the modelling error to adaptively select a model that best meets the requirements of the simulation. The ability to accurately estimate the modelling error in a measure consistent with the goal of the simulation, therefore, is of paramount importance.

In earlier works [30, 25], linearly elastic heterogeneous materials were modelled adaptively, with energy estimates of the modelling error guiding the adaptive process. We extend the theory of *a posteriori* estimation of modelling error to "quantities of interest" – quantities that are typically highly local in nature and are characterized by continuous linear functionals. We establish computable upper and lower bounds and estimates of such quantities. The use of this technique is demonstrated in conjunction with a new adaptive algorithm that we refer to as the Hierarchical Adaptive Method for Model Enhancement and Refinement (HAMMER). Results of numerical experiments are given.

Key words. Heterogeneous Materials, Hierarchical Modelling, Modelling Error

3.1 INTRODUCTION

The use of *adaptivity* – the process of dynamically changing some or all of the parameters that govern a simulation – is now widely accepted as necessary for the accurate and efficient solution of engineering problems. This is especially true in the context of finite element analysis, where the mesh size and approximation order, denoted h and p, respectively, have been manipulated independently to produce exponential rates of convergence for certain classes of problems [27, 21, 1].

47

In the simulation of heterogeneous or composite media, the material model used is an important feature. However, virtually all methods of analysis of heterogeneous media choose the model *a priori* and this model remains unchanged during the simulation. To cite some examples :

- *Macro* models : Here, one assumes the existence of a Representative Volume Element (RVE), a piece of material whose behavior is representative of the entire system. The overall response of the RVE to standard boundary conditions (uniform displacements, for example) determines the "overall properties" of the system. The drawback of this classical approach is that all information about the field variables at the fine-scale is lost because the material is "smeared" or "homogenized". Moreover, the existence of an RVE for all composite media is not guaranteed. A considerable volume of work exists on this approach; see [13, 10, 11, 12, 9].

- *Micro* models : In these models, fine-scale features such as inclusions and voids are taken into account in the material model. However, the use of a micro model in a simulation that employs, say, finite elements, leads to algebraic systems with millions of unknowns. Thus, the use of these models is restricted to problems with a small number of fine-scale features. Some examples of methods that incorporate fine-scale features, either directly or through the use of special basis functions, are the VCFEM method [7, 8], the Fast Multipole Method (FMM) [6], the composite finite element method [17] and the Multiscale Finite Element Method (MFEM) [14, 15].

- *Asymptotic/Periodic* models : These models assume, as the name suggests, that the microstructure is periodic. In classical asymptotic methods, the solution is described as the sum of long wavelength and short wavelength components. The long wavelength component of the solution can be viewed as the response of a "homogenized" medium to the same set of loads and boundary conditions as the original system. The homogenized properties are obtained by solving a set of periodic boundary value problems on a unit cell. The behavior of these models as the size of the assumed period approaches zero has been studied in detail; see [4, 16, 28, 2, 19, 20]. The main drawback of this approach is the restrictive assumption of periodicity placed on the microstructure.

In an attempt to enable the adaptive selection of models, Oden and Zohdi introduced the concept of *hierarchical modelling* [30, 25, 29]. In this adaptive framework, the most appropriate model for a given problem is picked from a hierarchical set of models, based on *a posteriori* estimates of the modelling error – the difference between the solution produced by a given model and the solution produced by the finest-scale model in the hierarchy.

Based on this concept, an adaptive domain-decomposition method known as the Homogenized Dirichlet Projection Method (HDPM) [30, 25] was developed for linearly elastic heterogeneous materials. In this method, the proximity of the homogenized solution to the fine-scale solution is estimated in the energy norm. In regions where the homogenized solution is considered inadequate, the fine-scale model is used to represent the material with the homogenized solution providing the local boundary conditions. For a complete description of the HDPM, see [30]. For related work, see [18, 24].

In this work, we present a new method of model adaptation for linearly elastic composites that we refer to as the Hierarchical Adaptive Method for Model Enhancement and Refinement (HAMMER). In this method, a sequence of elasticity tensors is used to approximate the fine-scale microstructure so that the resulting sequence of solutions converges to the fine-scale solution. We describe this method in detail later on.

Obviously, the final model resulting from adaptive strategies like the HDPM and the HAMMER depends on (a) the measure used to quantify the modelling error and (b) the ability to accurately estimate the modelling error in the chosen measure. The ability to accurately estimate the modelling error in different measures gives the analyst the ability to arrive adaptively at a model that meets the specific goals of the simulation.

Towards this end of developing goal-oriented model adaptation techniques, we extend the theory of *a posteriori* estimation of modelling error to a new class of measures referred to as "quantities of interest". This represents a significant departure from traditional methods of error estimation since these quantities of interest can be designed to measure the modelling error in such local quantities as interfacial stress and displacement components, among others. The only requirement is that the quantity be expressed as a continuous linear functional on the space of admissible displacements. An important feature of our theory is that the modelling error in a quantity of interest can be bounded above and below. We demonstrate that it is possible to develop goal-oriented model adaptation strategies based on this new theory of error estimation.

Our paper is organized as follows : We first define some notation and set up the model class of problems in Section 3.2. In Section 3.3, we review, following [25] and [23], energy estimates of the modelling error. This is followed by the theory of modelling error estimation for "quantities of interest". In Section 4, we introduce the HAMMER, a new model adaptation strategy. Finally, we present results from several numerical experiments demonstrating the performance of the proposed error bounds and estimates and our adaptive strategy.

3.2   MODEL CLASS OF PROBLEMS

Let $\Omega \subset \mathbb{R}^N$, $N = 1, 2, 3$, denote an open bounded domain with a piecewise smooth boundary which is the region in space occupied by a material body. We denote by $H^1(\Omega)$ the space of functions with distributional derivatives of order ≤ 1 in $L^2(\Omega)$. Let $\mathbf{H}^1(\Omega) \stackrel{\text{def}}{=} (H^1(\Omega))^N$ and $\mathbf{L}^2(\Omega) \stackrel{\text{def}}{=} (L^2(\Omega))^N$ with the usual norms.

The body is assumed to be in static equilibrium under the action of body forces $\mathbf{f} \in \mathbf{L}^2(\Omega)$ and tractions $\mathbf{t} \in \mathbf{L}^2(\Gamma_t)$, where $\Gamma_t \subset \partial\Omega$. Zero displacements are prescribed on $\Gamma_u = \partial\Omega \setminus \Gamma_t$. Denote by $\mathbf{V}(\Omega)$ the space of admissible displacements :

$$\mathbf{V}(\Omega) \stackrel{\text{def}}{=} \left\{ \mathbf{v} : \mathbf{v} \in \mathbf{H}^1(\Omega), \mathbf{v}|_{\Gamma_u} = \mathbf{0} \right\}. \tag{3.1}$$

3.2.1   The fine-scale problem

The heterogeneous body is assumed to be characterized by an elasticity tensor $\mathbf{E} \in (L^\infty(\Omega))^{N^2 \times N^2}$ which satisfies the standard uniform ellipticity and symmetry conditions. It is understood that $\mathbf{E}$ is a highly oscillatory function of position $\mathbf{x}$ over Ω, not necessarily periodic. According to the principle of virtual work, the elastostatics problem is characterized as follows :

$$
\begin{aligned}
&\text{Find } \mathbf{u} \in \mathbf{V}(\Omega) \text{ such that} \\
&\mathcal{B}(\mathbf{u}, \mathbf{v}) = \mathcal{F}(\mathbf{v}) \quad \forall\, \mathbf{v} \in \mathbf{V}(\Omega),
\end{aligned}
\tag{3.2}
$$

where the bilinear and linear forms in (3.2) are defined as

$$
\mathcal{B}(\mathbf{u}, \mathbf{v}) \stackrel{\text{def}}{=} \int_\Omega \boldsymbol{\nabla}\mathbf{v} : \mathbf{E}\,\boldsymbol{\nabla}\mathbf{u}\,d\mathbf{x} = \int_\Omega \mathbf{tr}\left[(\boldsymbol{\nabla}\mathbf{v})^T \mathbf{E}\boldsymbol{\nabla}\mathbf{u}\right] d\mathbf{x}
\tag{3.3}
$$

and

$$
\mathcal{F}(\mathbf{v}) \stackrel{\text{def}}{=} \int_\Omega \mathbf{f} \cdot \mathbf{v}\,d\mathbf{x} + \int_{\Gamma_t} \mathbf{t} \cdot \mathbf{v}\,ds.
\tag{3.4}
$$

Under the stated assumptions, it is well known that (3.2) possesses a unique solution $\mathbf{u} \in \mathbf{V}(\Omega)$. We shall refer to (3.2) as the *fine-scale problem* and to its solution $\mathbf{u}$ as the *fine-scale solution*.

3.2.2   The "homogenized" problem

Owing to the complexity of (3.2), $\mathbf{E}$ is usually replaced by a homogenized elasticity tensor, most often a constant, that is designed to characterize the macroscopic behavior of the body. However, without restricting ourselves to a constant function, we assume that the elasticity tensor $\mathbf{E}$ is replaced by a suitable approximation that satisfies the uniform ellipticity and symmetry conditions. We denote this approximation by $\mathbf{E}^{(i)}$[1]. The "homogenized" problem then reads

$$
\begin{aligned}
&\text{Find } \mathbf{u}^{(i)} \in \mathbf{V}(\Omega) \text{ such that} \\
&\mathcal{B}^{(i)}(\mathbf{u}^{(i)}, \mathbf{v}) = \mathcal{F}(\mathbf{v}) \quad \forall\, \mathbf{v} \in \mathbf{V}(\Omega).
\end{aligned}
\tag{3.5}
$$

with

$$
\mathcal{B}^{(i)}(\mathbf{u}^{(i)}, \mathbf{v}) \stackrel{\text{def}}{=} \int_\Omega \boldsymbol{\nabla}\mathbf{v} : \mathbf{E}^{(i)}\,\boldsymbol{\nabla}\mathbf{u}^{(i)}\,d\mathbf{x}.
\tag{3.6}
$$

The right hand side is as defined in (3.4). Under the stated conditions, there exists a unique *homogenized solution* $\mathbf{u}^{(i)} \in \mathbf{V}(\Omega)$ to the homogenized problem (3.5).

---

[1]Later in this paper, we consider a sequence of approximations to $\mathbf{E}$. Hence the notation.

3.3 SUMMARY OF A-POSTERIORI MODELLING ERROR ESTIMATES

The process of approximating the fine-scale elasticity tensor $\mathbf{E}$ by $\mathbf{E}^{(i)}$ obviously produces a solution $\mathbf{u}^{(i)}$ that is in error. The difference between the fine-scale solution and the homogenized solution is referred to as the modelling error function

$$\mathbf{e}^{(i)} \stackrel{\text{def}}{=} \mathbf{u} - \mathbf{u}^{(i)}. \tag{3.7}$$

In this section, we summarize various results on the estimation of this error. After introducing some notation required for the definition of these estimates, we review theorems on the estimation of modelling error in (a) the energy norm, and (b) quantities of interest.

3.3.1 Further Notation

The energy norm of a function $\mathbf{v} \in \mathbf{V}(\Omega)$ is defined as

$$\|\mathbf{v}\|_{E(\Omega)} = \mathcal{B}(\mathbf{v}, \mathbf{v})^{1/2} \tag{3.8}$$

where $\mathcal{B} : \mathbf{V}(\Omega) \times \mathbf{V}(\Omega) \to \mathbb{R}$ is defined in (3.3). Next, we define the inner product space

$$\mathbf{L}^2_{\text{sym}}(\Omega) \stackrel{\text{def}}{=} \{\mathbf{A} \in [L^2(\Omega)]^{N \times N} : \mathbf{A}^T = \mathbf{A}\}, \tag{3.9}$$

with the inner product

$$((\mathbf{A}, \mathbf{B})) = \int_\Omega \mathbf{A} : \mathbf{B} \, d\mathbf{x}. \tag{3.10}$$

The elasticity operator $\mathbf{E} \in [L^\infty(\Omega)]^{N^2 \times N^2}$ can be viewed as an $\mathbf{L}^2_{\text{sym}}(\Omega)$-map and, with the assumptions of symmetry and uniform ellipticity on $\mathbf{E}$, we can define the following weighted inner product on $\mathbf{L}^2_{\text{sym}}(\Omega)$:

$$((\mathbf{A}, \mathbf{B}))_E = ((\mathbf{A}, \mathbf{EB})) = ((\mathbf{EA}, \mathbf{B})). \tag{3.11}$$

Finally, let

$$\mathcal{I}_{(i)} = (\mathbf{I} - \mathbf{E}^{-1} \mathbf{E}^{(i)}). \tag{3.12}$$

Then, for $\mathbf{g} \in \mathbf{V}(\Omega)$, we define the associated linear *residual functional* $\mathcal{R}_\mathbf{g} : \mathbf{V}(\Omega) \to \mathbb{R}$,

$$\mathcal{R}_\mathbf{g}(\mathbf{v}) = -\int_\Omega \boldsymbol{\nabla}\mathbf{v} : \mathbf{E}\mathcal{I}_{(i)}\boldsymbol{\nabla}\mathbf{g} \, d\mathbf{x}, \quad \mathbf{v} \in \mathbf{V}(\Omega). \tag{3.13}$$

3.3.2 Energy Bounds on the Modelling Error

Theorem 3.1 *Let $\mathbf{u}$ and $\mathbf{u}^{(i)}$ be the solutions to problems (3.2) and (3.5) respectively. Then the following holds :*

$$\zeta^{(i)}_{\text{low}} \leq \|\mathbf{e}^{(i)}\|_{E(\Omega)} = \|\mathbf{u} - \mathbf{u}^{(i)}\|_{E(\Omega)} \leq \zeta^{(i)}_{\text{upp}}, \tag{3.14}$$

where

$$\zeta_{\text{low}}^{(i)} \overset{\text{def}}{=} \frac{|\mathcal{R}_{\mathbf{u}^{(i)}}(\mathbf{u}^{(i)})|}{\|\mathbf{u}^{(i)}\|_{E(\Omega)}}, \quad \zeta_{\text{upp}}^{(i)} \overset{\text{def}}{=} \left(\left(\mathcal{I}_{(i)}\nabla\mathbf{u}^{(i)}, \mathcal{I}_{(i)}\nabla\mathbf{u}^{(i)}\right)\right)_E^{1/2}. \tag{3.15}$$

□

The proof for the assertions $\|\mathbf{e}^{(i)}\|_{E(\Omega)} \leq \zeta_{\text{upp}}^{(i)}$ and $\|\mathbf{e}^{(i)}\|_{E(\Omega)} \geq \zeta_{\text{low}}^{(i)}$ can be found in [30] and [23], respectively. These assertions follow from the fact that the modelling error $\mathbf{e}^{(i)}$ is governed by

$$\mathcal{B}(\mathbf{e}^{(i)}, \mathbf{v}) = \mathcal{R}_{\mathbf{u}^{(i)}}(\mathbf{v}) = -\int_\Omega \nabla\mathbf{v} : \mathbf{E}\mathcal{I}_{(i)}\nabla\mathbf{u}^{(i)}\,\mathrm{d}\mathbf{x}, \ \forall\,\mathbf{v} \in \mathbf{V}(\Omega). \tag{3.16}$$

Thus, the modelling error is bounded above and below by quantities that depend only on the homogenized solution $\mathbf{u}^{(i)}$, the fine-scale microstructure $\mathbf{E}$, the homogenized elasticity tensor $\mathbf{E}^{(i)}$ and the domain Ω.

3.3.3 Modelling Error in Quantities of Interest

Recent work in error estimation in the context of finite element analysis has focused on obtaining bounds on the numerical error in quantities of interest other than the energy norm [3, 26]. Here, we use a generalization of the approach of Prudhomme and Oden [26] to analyze the modelling error in other quantities. We are interested in estimating $L(\mathbf{e}^{(i)}) = L(\mathbf{u}) - L(\mathbf{u}^{(i)})$, where L is a continuous linear functional on $\mathbf{V}(\Omega)$, $L \in \mathbf{V}'(\Omega)$. For instance, L may represent something more localized than the global estimate (3.14), such as the average error in $\mathbf{u}$ or $\nabla\mathbf{u}$ over a small region in Ω. Some examples of linear functionals are

- $L(\mathbf{v}) = \int_\omega \mathbf{v} \cdot \mathbf{n}\,\mathrm{d}s$ where $\omega \subset \partial\Omega$ and $\mathbf{n}$ is the unit outward normal.

- $L(v) = v(x_0)$, $x_0 \in \Omega$ for one-dimensional problems.

- $L(\mathbf{v}) = \int_{B(\mathbf{x}_0,\epsilon)} k_\epsilon(\mathbf{x};\mathbf{x}_0)\sigma_{ij}(\mathbf{v})\,\mathrm{d}\mathbf{x}$, $1 \leq i,j \leq N, \epsilon > 0, \mathbf{x}_0 \in \Omega$, where k_ϵ is a mollifier kernel (see [22], Chapter 5) and $B(\mathbf{x}_0, \epsilon)$ is an open ball of radius ϵ centered at $\mathbf{x}_0$.

In order to relate $L(\mathbf{e}^{(i)})$ to the source of the modelling error, i.e., the functional $\mathcal{R}_{\mathbf{u}^{(i)}}$, we assume the existence of an "influence function" $\mathbf{w}$ such that $L(\mathbf{e}^{(i)}) = \mathcal{R}_{\mathbf{u}^{(i)}}(\mathbf{w})$ and using (3.16), we obtain $L(\mathbf{e}^{(i)}) = \mathcal{B}(\mathbf{e}^{(i)}, \mathbf{w})$. Noting that $\mathbf{e}^{(i)} \in \mathbf{V}(\Omega)$, we pose the following global *adjoint fine-scale problem* in order to obtain $\mathbf{w}$:

$$\begin{aligned} \text{Find } \mathbf{w} \in \mathbf{V}(\Omega) \text{ such that} \\ \mathcal{B}(\mathbf{v}, \mathbf{w}) = L(\mathbf{v}) \qquad \forall\,\mathbf{v} \in \mathbf{V}(\Omega) \end{aligned} \tag{3.17}$$

The adjoint problem (3.17) is of the same computational complexity as the original fine-scale problem (3.2) since both problems are governed by the same bilinear form. A natural way to simplify the adjoint problem is to pose the *adjoint homogenized problem*

$$\begin{aligned} \text{Find } \mathbf{w}^{(i)} \in \mathbf{V}(\Omega) \text{ such that} \\ \mathcal{B}^{(i)}(\mathbf{v}, \mathbf{w}^{(i)}) = L(\mathbf{v}) \qquad \forall\,\mathbf{v} \in \mathbf{V}(\Omega). \end{aligned} \tag{3.18}$$

The solution to this homogenized adjoint problem will be referred to as the *homogenized influence function*. In what follows, we sometimes refer to the problems (3.2) and (3.5) as the *primal fine-scale problem* and *primal homogenized problem*, respectively. It is obvious that, under the stated assumptions on $\mathbf{E}$ and $\mathbf{E}^{(i)}$, the functions $\mathbf{w}$ and $\mathbf{w}^{(i)}$ exist and are uniquely defined.

It immediately follows that the modelling error in the influence function

$$\bar{\mathbf{e}}^{(i)} \stackrel{\text{def}}{=} \mathbf{w} - \mathbf{w}^{(i)} \tag{3.19}$$

satisfies (recall (3.16))

$$\mathcal{B}(\mathbf{v}, \bar{\mathbf{e}}^{(i)}) = \mathcal{R}_{\mathbf{w}^{(i)}}(\mathbf{v}) \qquad \forall \mathbf{v} \in \mathbf{V}(\Omega). \tag{3.20}$$

We also note that $\bar{\mathbf{e}}^{(i)}$ satisfies the following relationship (analogous to (3.14)) :

$$\bar{\zeta}^{(i)}_{\text{low}} \leq \|\bar{\mathbf{e}}^{(i)}\|_{E(\Omega)} = \|\mathbf{w} - \mathbf{w}^{(i)}\|_{E(\Omega)} \leq \bar{\zeta}^{(i)}_{\text{upp}} \tag{3.21}$$

where

$$\bar{\zeta}^{(i)}_{\text{low}} \stackrel{\text{def}}{=} \frac{|\mathcal{R}_{\mathbf{w}^{(i)}}(\mathbf{w}^{(i)})|}{\|\mathbf{w}^{(i)}\|_{E(\Omega)}}; \quad \bar{\zeta}^{(i)}_{\text{upp}} \stackrel{\text{def}}{=} \left((\mathcal{I}_{(i)}\boldsymbol{\nabla}\mathbf{w}^{(i)}, \mathcal{I}_{(i)}\boldsymbol{\nabla}\mathbf{w}^{(i)})\right)^{1/2}_E. \tag{3.22}$$

These preliminaries bring us to the following result :

Theorem 3.2 *Let $\mathbf{u}^{(i)}$ and $\mathbf{w}^{(i)}$ be the solutions to problems (3.5) and (3.18), respectively. Then,*

$$\eta^{(i)}_{\text{low}} \leq L(\mathbf{e}^{(i)}) \leq \eta^{(i)}_{\text{upp}} \tag{3.23}$$

where

$$\eta^{(i)}_{\text{low}} \stackrel{\text{def}}{=} \frac{1}{4}(\eta^{+(i)}_{\text{low}})^2 - \frac{1}{4}(\eta^{-(i)}_{\text{upp}})^2 + \mathcal{R}_{\mathbf{u}^{(i)}}(\mathbf{w}^{(i)}), \tag{3.24}$$

$$\eta^{(i)}_{\text{upp}} \stackrel{\text{def}}{=} \frac{1}{4}(\eta^{+(i)}_{\text{upp}})^2 - \frac{1}{4}(\eta^{-(i)}_{\text{low}})^2 + \mathcal{R}_{\mathbf{u}^{(i)}}(\mathbf{w}^{(i)}), \tag{3.25}$$

with arbitrary $s \in \mathbb{R}^+$,

$$\eta^{\pm(i)}_{\text{upp}} \stackrel{\text{def}}{=} \sqrt{s^2(\bar{\zeta}^{(i)}_{\text{upp}})^2 \pm 2\left((\mathcal{I}_{(i)}\boldsymbol{\nabla}\mathbf{u}^{(i)}, \mathcal{I}_{(i)}\boldsymbol{\nabla}\mathbf{w}^{(i)})\right)_E + s^{-2}(\bar{\zeta}^{(i)}_{\text{upp}})^2}, \tag{3.26}$$

and

$$\eta^{\pm(i)}_{\text{low}} \stackrel{\text{def}}{=} \frac{|\mathcal{R}_{s\mathbf{u}^{(i)}\pm s^{-1}\mathbf{w}^{(i)}}(\mathbf{u}^{(i)} + \theta^\pm\mathbf{w}^{(i)})|}{\|\mathbf{u}^{(i)} + \theta^\pm\mathbf{w}^{(i)}\|_{E(\Omega)}}, \tag{3.27}$$

where $\zeta^{(i)}_{\text{upp}}$ and $\bar{\zeta}^{(i)}_{\text{upp}}$ are defined by (3.15) and (3.22), respectively, and $\theta^\pm$ is given by

$$\theta^\pm = \frac{\mathcal{B}(\mathbf{u}^{(i)}, \mathbf{w}^{(i)})\mathcal{R}_{\mathbf{u}^{(i)}}(s\mathbf{u}^{(i)} \pm s^{-1}\mathbf{w}^{(i)}) - \mathcal{B}(\mathbf{u}^{(i)}, \mathbf{u}^{(i)})\mathcal{R}_{\mathbf{w}^{(i)}}(s\mathbf{u}^{(i)} \pm s^{-1}\mathbf{w}^{(i)})}{\mathcal{B}(\mathbf{u}^{(i)}, \mathbf{w}^{(i)})\mathcal{R}_{\mathbf{w}^{(i)}}(s\mathbf{u}^{(i)} \pm s^{-1}\mathbf{w}^{(i)}) - \mathcal{B}(\mathbf{w}^{(i)}, \mathbf{w}^{(i)})\mathcal{R}_{\mathbf{u}^{(i)}}(s\mathbf{u}^{(i)} \pm s^{-1}\mathbf{w}^{(i)})} \tag{3.28}$$

Proof. Here, we only present an outline of the proof which can be found in [23] in its entirety. The main idea is to decompose the modelling error in the quantity of interest as

$$L(\mathbf{e}^{(i)}) = \mathcal{B}(\mathbf{e}^{(i)}, \mathbf{w}) = \mathcal{B}(\mathbf{e}^{(i)}, \bar{\mathbf{e}}^{(i)}) + \mathcal{B}(\mathbf{e}^{(i)}, \mathbf{w}^{(i)})$$

$$= \mathcal{B}(s\mathbf{e}^{(i)}, s^{-1}\bar{\mathbf{e}}^{(i)}) + \mathcal{R}_{\mathbf{u}^{(i)}}(\mathbf{w}^{(i)}), \tag{3.29}$$

where $s \in \mathbb{R}^+$ is an arbitrary positive scaling factor. Now, using a simple property of an inner product, we rewrite the expression (3.29) as

$$L(\mathbf{e}^{(i)}) = \frac{1}{4}\|s\mathbf{e}^{(i)} + s^{-1}\bar{\mathbf{e}}^{(i)}\|^2_{E(\Omega)} - \frac{1}{4}\|s\mathbf{e}^{(i)} - s^{-1}\bar{\mathbf{e}}^{(i)}\|^2_{E(\Omega)} + \mathcal{R}_{\mathbf{u}^{(i)}}(\mathbf{w}^{(i)}). \tag{3.30}$$

The first two terms on the right hand side of (3.30) can be bounded above and below by noting that the quantity $s\mathbf{e}^{(i)} \pm s^{-1}\bar{\mathbf{e}}^{(i)}$ satisfies

$$\mathcal{B}(s\mathbf{e}^{(i)} \pm s^{-1}\bar{\mathbf{e}}^{(i)}, \mathbf{v}) = \mathcal{R}_{s\mathbf{u}^{(i)}\pm s^{-1}\mathbf{w}^{(i)}}(\mathbf{v}) \qquad \forall \mathbf{v} \in \mathbf{V}(\Omega), \tag{3.31}$$

and then using Theorem 1.3.1. The third term $\mathcal{R}_{\mathbf{u}^{(i)}}(\mathbf{w}^{(i)})$ can be computed exactly. The term $\theta^{\pm}$ is introduced to maximize the lower bounds on $s\mathbf{e}^{(i)} \pm s^{-1}\bar{\mathbf{e}}^{(i)}$. □

Remark 3.1 *The scaling factor s is designed to balance the contributions of the primal (original) and adjoint problems to the modelling error in L. It can be easily shown that the value $s^* = \sqrt{\|\bar{\mathbf{e}}^{(i)}\|_{E(\Omega)}/\|\mathbf{e}^{(i)}\|_{E(\Omega)}}$ is optimal in that it minimizes the quantities $\|s\mathbf{e}^{(i)} + s^{-1}\bar{\mathbf{e}}^{(i)}\|_{E(\Omega)}$ and $\|s\mathbf{e}^{(i)} - s^{-1}\bar{\mathbf{e}}^{(i)}\|_{E(\Omega)}$. But, this optimal scaling factor s^* cannot be computed exactly since the modelling errors $\mathbf{e}^{(i)}$ and $\bar{\mathbf{e}}^{(i)}$ are not known exactly. Hence, in our numerical experiments, we use $s^* = \sqrt{\bar{\zeta}^{(i)}_{\mathrm{upp}}/\zeta^{(i)}_{\mathrm{upp}}}$ as the upper bounds (3.14) and (3.21) have been observed to provide exceptionally accurate estimates of the modelling errors in the primal and adjoint solutions, respectively.* □

Remark 3.2 *Based on (3.30), we propose the following estimates of the modelling error in the quantity of interest :*

$$L(\mathbf{e}^{(i)}) \approx \eta^{(i)}_{\mathrm{est,upp}} \stackrel{\mathrm{def}}{=} \frac{1}{4}(\eta^{+(i)}_{\mathrm{upp}})^2 - \frac{1}{4}(\eta^{-(i)}_{\mathrm{upp}})^2 + \mathcal{R}_{\mathbf{u}^{(i)}}(\mathbf{w}^{(i)}), \tag{3.32}$$

and

$$L(\mathbf{e}^{(i)}) \approx \eta^{(i)}_{\mathrm{est,low}} \stackrel{\mathrm{def}}{=} \frac{1}{4}(\eta^{+(i)}_{\mathrm{low}})^2 - \frac{1}{4}(\eta^{-(i)}_{\mathrm{low}})^2 + \mathcal{R}_{\mathbf{u}^{(i)}}(\mathbf{w}^{(i)}). \tag{3.33}$$

□

Evidently, the accuracy of the bounds (3.23) and the estimates (3.32) and (3.33) of the modelling error in a given quantity of interest depend on the accuracy of the estimates $\eta^{\pm(i)}_{\mathrm{low}}$ and $\eta^{\pm(i)}_{\mathrm{upp}}$. This issue is addressed in detail in [23].

3.4 HAMMER : HIERARCHICAL ADAPTIVE METHOD FOR MODEL ENHANCEMENT AND REFINEMENT

We now present a new method for adaptive model selection and refinement, and reduction of the modelling error. Our approach here is to generate a sequence of elasticity tensors $\mathbf{E}^{(i)}$ that are uniformly elliptic and symmetric so that the solutions $\mathbf{u}^{(i)}$ they generate converge to the fine-scale solution $\mathbf{u}$. The sequence $\mathbf{E}^{(i)}$ of elasticity tensors is generated as follows : We begin with a sufficiently fine partition of the domain into non-overlapping subdomains and choose a homogenized elasticity tensor $\mathbf{E}^0$ as the first element of the sequence $\mathbf{E}^{(i)}$, i.e., $\mathbf{E}^{(0)} = \mathbf{E}^0$. If the homogenized elasticity tensor does not produce an acceptable solution, we change $\mathbf{E}^0$ to $\mathbf{E}$ *locally* on the subdomains that contribute the most to the modelling error (according to some error indicator). This results in a non-uniform material model, with the exact microstructural representation in certain regions and the homogenized model in the rest of the domain. This non-uniform characterization of the material is denoted $\mathbf{E}^{(1)}$ and used as the next element of the sequence of elasticity tensors. We then solve (3.5) (this entails a global solve) to obtain $\mathbf{u}^{(1)}$, the next element of the sequence $\mathbf{u}^{(i)}$. This process is repeated until the global error tolerance is satisfied.

The advantage that the HAMMER has over the HDPM [30, 25] is that the homogenized solution $\mathbf{u}^0$ is used only to select the subdomains that need model refinement and not to enforce boundary conditions. Thus, the final model generated by this method is expected to show less dependence on the homogenized solution (the initial guess). Also, this method lends itself to easy iterative solution since $\mathbf{u}^{(i)}$ can be used as the initial guess to obtain $\mathbf{u}^{(i+1)}$.

3.4.1 The HAMMER Algorithm

Step 1. Initialization. Given the initial data Ω, Γ_u, Γ_t, $\mathbf{E}$, $\mathbf{f}$ and $\mathbf{t}$, construct a non-overlapping partition of the domain $\mathcal{P} = \{\Theta_k\}$, $k = 1, 2 \ldots N(\mathcal{P})$. Choose a homogenized material tensor $\mathbf{E}^0$. Specify error tolerance parameters α_{tol} and β_{tol}.

Step 2. Begin Adaptive Loop. Set $\mathbf{E}^{(0)} = \mathbf{E}^0$ (the homogenized material properties), and for $i = 0, 1, \ldots,$

Step 3. Solution and Error Estimation. Compute $\mathbf{u}^{(i)}$ by solving

$$\mathcal{B}^{(i)}(\mathbf{u}^{(i)}, \mathbf{v}) = \mathcal{F}(\mathbf{v}) \quad \forall \mathbf{v} \in \mathbf{V}. \tag{3.34}$$

and compute

$$(\zeta_{\text{upp}}^{(i)})^2 = \sum_k (\zeta_{k,\text{upp}}^{(i)})^2 = \sum_k \int_{\Theta_k} \mathcal{I}_{(i)} \boldsymbol{\nabla} \mathbf{u}^{(i)} : \mathbf{E} \, \mathcal{I}_{(i)} \boldsymbol{\nabla} \mathbf{u}^{(i)} \, d\mathbf{x} \tag{3.35}$$

Step 4. Tolerance Test. If $\zeta_{\text{upp}}^{(i)} \le \alpha_{\text{tol}} \|\mathbf{u}^{(i)}\|_{E(\Omega)}$, STOP.

Step 5. Model Refinement. For $k = 1, 2 \ldots N(\mathcal{P})$, if

$$\zeta_{k,\text{upp}}^{(i)} > \beta_{\text{tol}} \left(\max_k (\zeta_{k,\text{upp}}^{(i)}) - \min_k (\zeta_{k,\text{upp}}^{(i)}) \right), \tag{3.36}$$

set $\mathbf{E}^{(i+1)}(\mathbf{x}) = \mathbf{E}(\mathbf{x}), \mathbf{x} \in \Theta_k$. Else, set $\mathbf{E}^{(i+1)}(\mathbf{x}) = \mathbf{E}^{(i)}(\mathbf{x}), \mathbf{x} \in \Theta_k$.

Step 6. End Adaptive Loop. Set $i \leftarrow i + 1$ and GOTO Step 3.

Remark 3.3 *In the algorithm presented above, we use an energy criterion (3.36) to select the subdomains that need model refinement. The criterion is designed such that whenever the global error tolerance is not met, at least one subdomain undergoes model refinement.*

If the goal of the analysis is to reduce the error in a given quantity of interest $L \in \mathbf{V}'(\Omega)$, we note that

$$\begin{aligned} |L(\mathbf{e}^{(i)})| &\leq |\mathcal{B}(\mathbf{e}^{(i)}, \bar{\mathbf{e}}^{(i)})| + |\mathcal{B}(\mathbf{e}^{(i)}, \mathbf{w}^{(i)})| \\ &\leq \|\mathbf{e}^{(i)}\|_{E(\Omega)} \left(\|\bar{\mathbf{e}}^{(i)}\|_{E(\Omega)} + \|\mathbf{w}^{(i)}\|_{E(\Omega)} \right) \\ &\leq \zeta_{\mathrm{upp}}^{(i)} \left(\bar{\zeta}_{\mathrm{upp}}^{(i)} + \|\mathbf{w}^{(i)}\|_{E(\Omega)} \right), \end{aligned} \tag{3.37}$$

and hence the quantity $\gamma_k = \zeta_{k,\mathrm{upp}}^{(i)}(\bar{\zeta}_{k,\mathrm{upp}}^{(i)} + \|\mathbf{w}^{(i)}\|_{E(\Theta_k)})$ can be used as an error indicator in Step 5 of the HAMMER algorithm in place of $\zeta_{k,\mathrm{upp}}^{(i)}$.

3.5   NUMERICAL EXAMPLE

We present an example to illustrate the performance of the various error estimates, as well as the HAMMER algorithm, presented earlier. The problem considered, depicted in Figure 3.1, consists of a partially loaded fiber-reinforced composite body that is essentially infinite in the direction of the fibers. Assuming that the fibers are aligned, we can approximate the state of stress in the body by plane-strain conditions. This idealization is shown in Figure 3.2, along with the positions of the fibers. The figure also shows the partition of the domain into 10 equal subdomains.

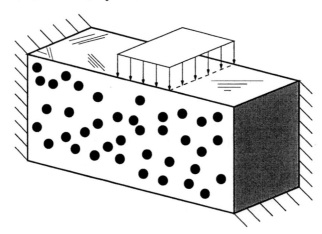

Figure 3.1: A three-dimensional fiber-reinforced composite material.

The engineering material properties of the fiber and matrix materials are taken to be : $E_{\mathrm{matirx}} = 100.0$ MPa, $E_{\mathrm{fiber}} = 10.0 \times E_{\mathrm{matirx}}$, and $\nu_{\mathrm{matirx}} = \nu_{\mathrm{fiber}} = 0.2$. We consider three different volume fractions of the fiber material : 0.1, 0.2 and 0.3.

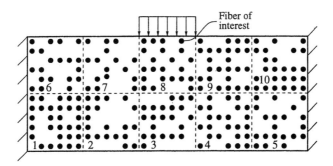

Figure 3.2: A plane-strain idealization of the fiber-reinforced composite material with the positions of the fibers shown. The dotted lines indicate the partitioning of the domain into subdomains.

In order to compute the effectivity of the various error estimates, we need a reference solution against which the estimates can be compared. Since the problem considered does not have a closed form solution, we generate an "exact" solution for each volume fraction by means of a very high fidelity h-p mesh using the adaptive finite element code ProPHLEX^{TM} [5]. These "overkill" meshes are highly refined and employ polynomials of degree upto 5. The overkill mesh used to generate the reference primal solution $\mathbf{u}$ for a fiber volume fraction of 0.3, shown in Figure 3.3, has 110,000 degrees of freedom.

3.5.1 Adaptation using the Energy Error Criterion

Here, we use the HAMMER algorithm presented in Section 3.4 to adaptively reduce the error $\|\mathbf{u} - \mathbf{u}^{(i)}\|_{E(\Omega)}$. For $i = 0$, we generate the initial homogenized elasticity tensor $\mathbf{E}^{(0)}$ using the average of the Hashin-Shtrikman bounds. We then generate a sequence of solutions $\mathbf{u}^{(i)}$ as outlined in the algorithm, taking care to ensure that these "homogenized" solutions have negligible numerical error. The results for the three fiber volume fractions considered are shown in Table 3.1. It is seen that the bound $\zeta_{\text{upp}}^{(i)}$ performs very well irrespective of the relative modelling error $\|\mathbf{e}^{(i)}\|_{E(\Omega)}/\|\mathbf{u}\|_{E(\Omega)}$. After 3 iterations of the HAMMER algorithm, the modelling error is reduced to 10 % or less.

3.5.2 Adaptation using the Quantity of Interest Criterion

We suppose that the quantity that is of interest in this computation is the normal stress component σ_{yy} in a fiber under the load. This fiber is indicated in Figure 3.2. Since, point-wise values of stresses are mathematically not meaningful, we express the quantity of interest in terms of a mollifier kernel as follows :

$$L(\mathbf{v}) = \int_{B(\mathbf{x}_0, \epsilon)} k_\epsilon(\mathbf{x}; \mathbf{x}_0) \, \sigma_{yy}(\mathbf{v}) \, d\mathbf{x}, \qquad (3.38)$$

where the mollifier $k_\epsilon(\mathbf{x}; \mathbf{x}_0)$, centered at $\mathbf{x}_0 \in \Omega$ and characterized by the parameter ϵ, has the following properties :

- $k_\epsilon(\mathbf{x}; \mathbf{x}_0)$ has continuous derivatives of all orders on $\mathbb{R}^N$.

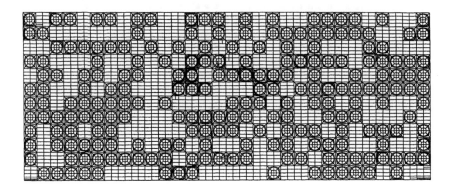

Figure 3.3: Overkill mesh used to generate the reference primal solution $\mathbf{u}$ for a fiber volume fraction of 0.3.

- $k_\epsilon(\mathbf{x}; \mathbf{x}_0) = 0$ for $|\mathbf{x} - \mathbf{x}_0| \geq \epsilon$ and $k_\epsilon(\mathbf{x}; \mathbf{x}_0) > 0$ for $|\mathbf{x} - \mathbf{x}_0| < \epsilon$. Thus, the support of $k_\epsilon(\mathbf{x}; \mathbf{x}_0)$ is $B(\mathbf{x}_0, \epsilon)$, a ball of radius ϵ centered at $\mathbf{x}_0$. Also, all partial derivatives of $k_\epsilon(\mathbf{x}; \mathbf{x}_0)$ are zero outside $B(\mathbf{x}_0, \epsilon)$.

- The mollifier kernel satisfies the condition

$$\int_{B(\mathbf{x}_0, \epsilon)} k_\epsilon(\mathbf{x}, \mathbf{x}_0) = 1. \tag{3.39}$$

Mollification can be viewed as a weighted averaging process over the support of the mollifier. A standard example of a mollifier is

$$k_\epsilon(\mathbf{x}; \mathbf{x}_0) = \begin{cases} c_0 \exp^{\dfrac{\epsilon^2}{|\mathbf{x} - \mathbf{x}_0|^2 - \epsilon^2}} & |\mathbf{x} - \mathbf{x}_0| < \epsilon, \\ 0 & |\mathbf{x} - \mathbf{x}_0| \geq \epsilon, \end{cases} \tag{3.40}$$

where the constant c_0 is chosen such that the condition (3.39) is satisfied. For the fiber chosen, we have $\mathbf{x}_0 = (2.75, 1.91\bar{6})$ and $\epsilon = 0.0657$. This results in $c_0 = 4.9653 \times 10^2$.

The HAMMER algorithm is modified so that the model is refined according to the modelling error in the quantity of interest; see Remark 3.3. The results for this case are presented in Table 3.2. It can be seen that while the bounds on the modelling error $L(\mathbf{e}^{(i)})$ are not tight, the estimate $\eta_{\text{est,upp}}^{(i)}$ has an accuracy between 82% and 88%.

Table 3.1: Results from the HAMMER algorithm applied to the plane-strain problem using the energy error criterion for model refinement. NREF is the number of subdomains that have fine-scale representation of the microstructure.

Step (i)	$\dfrac{\|e^{(i)}\|_{E(\Omega)}}{\|u\|_{E(\Omega)}}$	$\dfrac{\zeta^{(i)}_{upp}}{\|e^{(i)}\|_{E(\Omega)}}$	NREF
Volume Fraction = 0.1			
0	0.54	1.05	0
1	0.20	1.06	3
2	0.08	1.06	5
Volume Fraction = 0.2			
0	0.66	1.07	0
1	0.23	1.08	4
2	0.09	1.10	6
Volume Fraction = 0.3			
0	0.71	1.08	0
1	0.24	1.08	4
2	0.10	1.11	6

Table 3.2: Results from the HAMMER algorithm applied to the plane-strain problem using the accuracy in the quantity of interest L as the criterion for model refinement. NREF is the number of subdomains that have fine-scale representation of the microstructure.

Step (i)	$\dfrac{L(e^{(i)})}{L(u)}$	$\dfrac{\eta^{(i)}_{low}}{L(e^{(i)})}$	$\dfrac{\eta^{(i)}_{upp}}{L(e^{(i)})}$	$\dfrac{\eta^{(i)}_{est,upp}}{L(e^{(i)})}$	NREF
Volume Fraction = 0.1					
0	-1.26	-4.70	6.12	0.88	0
1	0.09	-7.01	8.60	0.83	1
Volume Fraction = 0.2					
0	-1.31	-5.80	7.44	0.87	0
1	0.08	-7.71	9.20	0.86	1
Volume Fraction = 0.3					
0	-1.35	-6.44	8.12	0.82	0
1	0.08	-9.08	10.68	0.85	1

3.6   CONCLUSIONS

The use of the concept of adaptive modelling in the solution of multiscale problems provides a natural way of resolving the most important scales that affect the outcome of a simulation since neither periodicity of microscale constituents nor the existence of RVEs is assumed. In this work, we have introduced the HAMMER, a new adaptive strategy for model refinement.

The success of any adaptive modelling strategy lies in the ability to estimate the inherent modelling error in a given model (compared to the finest-scale model) in a variety of measures. We have introduced a new theory for the estimation of the modelling error in quantities of interests described by continuous linear functionals and presented preliminary results. These results show that it is possible to estimate the modelling error in such quantities with good accuracy even when the bounds on these errors are not very tight.

Acknowledgments We gratefully acknowledge the support of this work by the Office of Naval Research under grant N00014-95-1-0401 and by the NSF through NPACI, the National Partnership for Advanced Computational Infrastructure.

REFERENCES

[1] I. Babuška and B. Q. Guo (1996). Approximation properties of the h–p version of the finite element method. *Comput. Meth. Appl. Mech. Engrg.*, 133(3-4):319–346.

[2] I. Babuška and R. C. Morgan (1985). Composites with a periodic structure : Mathematical analysis and numerical treatment. *Comp. and Maths. with Appls.*, 11(10):995–1005.

[3] I. Babuška, T. Strouboulis, C. S. Upadhyay and S. K. Gangaraj (1995). A-posteriori estimation and adaptative control of the pollution error in the h-version of the finite element method. *Int. J. Numer. Meth. Engng.*, 38:4207–4235.

[4] A. Bensoussan, J. L. Lions and G. Papanicolaou (1978). *Asymptotic analysis for periodic structures*, volume 5 of *Studies in Mathematics and its Applications*. North-Holland, Amsterdam.

[5] Computational Mechanics Company, Inc., Austin, TX (1996). *ProPHLEX User Manual for Version 2.0.*

[6] Y. Fu, K. J. Klimkowski, G. J. Rodin, E. Berger, J. C. Browne, J. K. Singer, R. A. Van De Geijn and K. Vemaganti (1998). A fast solution method for three-dimensional many-particle problems of linear elasticity. *Int. J. Numer. Meth. Engng.*, 42:1215–1229.

[7] S. Ghosh and Y. Liu (1995). Voronoi cell finite element model based on micropolar theory of thermoelasticity for heterogeneous materials. *Int. J. Numer. Meth. Engng.*, 38:1361–1398.

[8] S. Ghosh and S. Moorthy (1995). Elastic-plastic analysis of arbitrary heterogeneous materials with the voronoi cell finite element method. *Comput. Meth. Appl. Mech. Engrg.*, 121:373–409.

[9] Z. Hashin (1983). Analysis of composite materials, A survey. *J. Appl. Mech.*, 50:481–505.

[10] Z. Hashin and S. Shtrikman (1962). On some variational principles in anisotropic and non-homogeneous elasticity. *J. Mech. Phys. Solids*, 10:335–342.

[11] Z. Hashin and S. Shtrikman (1962). A variational approach to the theory of the elastic behaviour of polycrystals. *J. Mech. Phys. Solids*, 10:343–352.

[12] Z. Hashin and S. Shtrikman (1963). A variational approach to the theory of the elastic behaviour of multiphase materials. *J. Mech. Phys. Solids*, 11:127–140.

[13] R. Hill (1952). The elastic behavior of a crystalline aggregate. *Proc. Phys. Soc. London*, 65:349–354.

[14] T. Y. Hou and X. Wu (1997). A multiscale finite element method for elliptic problems in composite materials and porous media. *J. Comp. Phys.*, 134:169–189.

[15] T. Y. Hou, X. Wu and Z. Cai (1999). Convergence of a multiscale finite element method for elliptic problems with rapidly oscillating coefficients. *Math. Comput.* posted on March 3, 1999, PII : S0025–5718(99)01077–7 (to appear in print).

[16] V. V. Jikov, S. M. Kozlov and O. A. Oleinik (1994). *Homogenization of differential operators and integral functionals*. Springer-Verlag, Heidelberg.

[17] S. A. Meguid and Z. H. Zhu (1995). A novel finite element for treating inhomogeneous solids. *Int. J. Numer. Meth. Engng.*, 38:1579–1592.

[18] N. Moës, J. T. Oden and T. I. Zohdi (1998). Investigation of the interactions between the numerical and the modeling errors in the homogenized Dirichlet projection method. *Comput. Meth. Applied Mech. Engrg.*, 159:79–101.

[19] R. C. Morgan and I. Babuška (1991). An approach for constructing families of homogenized equations for periodic media. I: An integral representation and its consequences. *SIAM J. Math. Anal.*, 22:1–15.

[20] R. C. Morgan and I. Babuška (1991). An approach for constructing families of homogenized equations for periodic media. II: Properties of the kernel. *SIAM J. Math. Anal.*, 22:16–33.

[21] J. T. Oden (1994). Error Estimation and Control in Computational Fluid Dynamics. In *The Mathematics of Finite Elements and Applications Highlights, 1993*, Wiley, Chichester, 1–23.

[22] J. T. Oden and J. N. Reddy (1976). *An Introduction to the Mathematical Theory of Finite Elements*. Wiley, New York.

[23] J. T. Oden and K. Vemaganti (1999). Adaptive modeling of composite structures :
     Modeling error estimation. Technical Report TR 1999-11, Texas Institute for Com-
     putational and Applied Mathematics, Austin, Texas. Also to appear in *Int. J. Comp.
     Civil Str. Engrg.*

[24] J. T. Oden, K. Vemaganti and N. Moës (1999). Hierarchical modeling of heteroge-
     neous solids. *Comput. Meth. Appl. Mech. Engrg.*, 172(1–4):3–25.

[25] J. T. Oden and T. I. Zohdi (1997). Analysis and adaptive modeling of highly hetero-
     geneous elastic structures. *Comput. Meth. Applied Mech. Engrg.*, 148(3–4):367–391.

[26] S. Prudhomme and J. T. Oden (1999). On Goal-oriented error estimation for elliptic
     problems : Application to the control of pointwise errors. *Comput. Meth. Appl. Mech.
     Engrg.* to appear.

[27] W. Rachowicz, J. T. Oden and L. F. Demkowicz (1989). Towards a universal h–p
     finite element strategy. Part 3. Design of h–p meshes. *Comput. Meth. Appl. Mech.
     Engrg.*, 77:181–212.

[28] E. Sanchez-Palencia (1980). *Non-homogeneous media and vibration theory.* Number
     127 in Lecture Notes in Physics. Springer-Verlag, Berlin.

[29] T. I. Zohdi (1997). *Analysis and adaptive modeling of highly heterogeneous elastic
     structures.* Ph.D. thesis, The University of Texas at Austin.

[30] T. I. Zohdi, J. T. Oden and G. J. Rodin (1996). Hierarchical modeling of heteroge-
     neous bodies. *Comput. Meth. Appl. Mech. Engrg.*, 138:273–298.

4 MODELLING AND FINITE ELEMENT ANALYSIS OF APPLIED POLYMER VISCOELASTICITY PROBLEMS

Simon Shaw[a], Arthur R. Johnson[b] and J. R. Whiteman[a]

[a]BICOM, Brunel University, Uxbridge, UB8 3PH, UK

[b]Army Research Laboratory, Computational Structures Branch,
NASA Langley Research Center, Hampton, VA 23681-0001, USA

ABSTRACT

Constitutive models for viscoelastic deformation are presented. Numerical schemes, based on these, are described for quasistatic problems, and *a posteriori* error estimates are presented. Numerical results on test problems illustrate spatial adaptivity.

Key words. finite element method, viscoelasticity, adaptivity, Volterra equation

4.1 INTRODUCTION

Polymers are extensively employed in the design of high performance lightweight composite structures to provide both structural damping and sound mitigation. Passive polymer damping systems have been in use for more than fifty years. Typical applications include lag dampers in helicopter rotors, engine mounts, equipment shock mounts, suspension system components, and structural panels that mitigate noise.

The importance of the role of computational modelling in the development, understanding and design of damping systems is being increasingly recognized, and conferences and workshops[1] are being held to bring together engineers, materials scientists and mathematicians to discuss current issues in damping technology.

However, appropriate mathematical models and efficient numerical schemes for analyzing the material level, time dependent, stresses that occur in these applications are still under active development. Issues of current concern include: constitutive models that

---

[1]**First US Army Research Office Workshop on Novel Structural Damping Concepts and Materials.** Center for Intelligent Material Systems and Structures, Virginia Tech, USA. October 19–22, 1998.

relate to the chemistry of the material, and that (without modification) apply to relaxation, creep, and dynamics; and the development of numerical methods that are efficient in terms of computer memory and CPU time. In addition, the numerical schemes should provide *a posteriori* error control via the application of adaptive methods, and we note that achieving such control for problems of viscoelastic damping involving inertia terms is a major challenge. In this article our aim, as a step toward treating dynamic problems, is to present models, numerical schemes and error analysis in the context of quasistatic problems. These improved analytical methods will allow designers to more accurately evaluate the performance of, and reduce the weight of, future polymer damping systems.

This paper addresses the basic mathematical issues that arise when error control is considered within the context of viscoelastic history integral, and internal variable, finite element formulations.

To model the dynamic response of (linear) viscoelastic structures we use the *hyperbolic Volterra equation*,

$$\varrho u''(t) + Au(t) = f(t) + \int_0^t B(t-s)u(s)\,ds, \qquad (4.1)$$

more details on which are given below. Here the Volterra integral arises from the memory dependence in the constitutive relationship between stress and strain, and A and B are partial differential operators closely related to the linear elasticity operator. This article contains a short summary of the various constitutive relationships that have been used to model viscoelastic behaviour, and also of recent efforts to construct adaptive finite element solvers for quasistatic viscoelasticity problems. These quasistatic problems are modelled by the *Elliptic Volterra problem*,

$$Au(t) = f(t) + \int_0^t B(t-s)u(s)\,ds. \qquad (4.2)$$

The plan of the article is as follows. In section 4.2 we outline the basic theory of linear viscoelasticity in terms of constitutive relationships linking stress to strain. These can be viewed either as hereditary integrals or as internal variable models. We also give a brief review of various internal variable theories that have been used to model viscoelastic effects. The brief Section 4.3 sets up the *elliptic Volterra* equation that is used to model quasistatic viscoelasticity and then, using a pure-time prototype problem, we discuss a finite element discretization and *a posteriori* error estimates in Section 4.4. A space-time finite element discretization of the space-time problem is outlined in Section 4.5, and we give some numerical results to illustrate the convergence of the scheme, and then in Section 4.6 we describe our progress so far on *a posteriori* error estimation and adaptivity. Section 4.7 concludes the article.

4.2 CONSTITUTIVE THEORIES

Let $u = (u_i)_{i=1}^n$ denote the displacement at time $t \in \mathcal{J} := [0, T]$, of a point x in a loaded viscoelastic body, $\mathcal{G}$, occupying a region $\Omega \subset \mathbb{R}^n$. In classical linear viscoelasticity theory the stress tensor, $\sigma := (\sigma_{ij})_{i,j=1}^n$, is given in terms of the history of the strain tensor, $\varepsilon(u) := (\varepsilon_{ij})_{i,j=1}^n$, by the constitutive equation,

$$\sigma(x,t) = D(t)\varepsilon(u(x,0)) + \int_0^t D(t-s)\varepsilon(u_s(x,s))\,ds. \qquad (4.3)$$

Here $\boldsymbol{\varepsilon} := (\varepsilon_{ij})_{i,j=1}^n$ denotes the (linear) strain tensor, and the subscript indicates partial differentiation. Alternatively, assuming that $\boldsymbol{D}(t)$ is smooth we can arrive at an alternate form by partial integration,

$$\boldsymbol{\sigma}(\boldsymbol{x}, t) = \boldsymbol{D}(0)\boldsymbol{\varepsilon}(\boldsymbol{u}(\boldsymbol{x}, t)) - \int_0^t \boldsymbol{D}_s(t - s)\boldsymbol{\varepsilon}(\boldsymbol{u}(\boldsymbol{x}, s))\,ds. \tag{4.4}$$

Either of these equations may be used as the constitutive relationship, and each demonstrates clearly the role of memory in viscoelastic modelling. In these $\boldsymbol{D} := (D_{ijkl})_{i,j,k,l=1}^n$ is a fourth order tensor of *stress relaxation functions* which is positive definite at $t = 0$ over all symmetric second order tensors. Also, $t = 0$ is a reference time prior to which it is assumed that $\boldsymbol{\varepsilon} = \boldsymbol{0}$.

The Maxwell solid

The linear relationships (4.3), (4.4), may be derived by a Boltzmann superposition technique, but a perhaps more illuminating method for deriving them is to imagine $\mathcal{G}$ as composed of a network of linear-elastic springs and Newtonian (viscous) dashpots. A simple but effective building block in this style is the *Maxwell solid* used by Johnson and Tessler in [8], and shown in Figure 4.1. For the moment we restrict ourselves to one-dimensional stress-strain states, indicated by σ and ε.

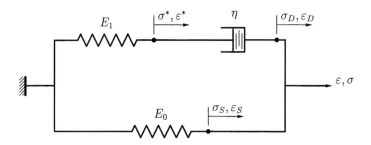

Figure 4.1: The Maxwell solid

In this model E_0 and E_1 are spring stiffnesses and σ^*, ε^* are internal stress and strain variables. Evidently, $\sigma^* = E_1\varepsilon^*$, $\varepsilon_D = \varepsilon - \varepsilon^*$ and $\sigma_S = E_0\varepsilon_S$. Also $\sigma^* = \sigma_D$ and this gives,

$$E_1\varepsilon^* = \eta\frac{d}{dt}(\varepsilon - \varepsilon^*) \quad \Longrightarrow \quad \frac{d\varepsilon^*}{dt} + \frac{\varepsilon^*}{\tau} = \frac{d\varepsilon}{dt}, \tag{4.5}$$

where $\tau := \eta/E_1$. Solving this for ε^* (with $\varepsilon^*(0) = \varepsilon(0)$), and using

$$\sigma = \sigma_S + \sigma^* = E_0\varepsilon + E_1\varepsilon^* \quad (\text{since } \varepsilon_S = \varepsilon),$$

gives

$$\sigma(t) = D(0)\varepsilon(t) - \int_0^t D_s(t - s)\varepsilon(s)\,ds, \tag{4.6}$$

where we define the *stress relaxation function*,

$$D(t) := E_0 + E_1 e^{-t/\tau},$$

as the scalar analogue to the tensor $\boldsymbol{D}(t)$ in (4.3) and (4.4). This expression for σ is the scalar analogue of equation (4.4) and suggests that in the general case we model $\boldsymbol{D}$ with the *Dirichlet-Prony series*,

$$\boldsymbol{D}(t) = \varphi(t)\boldsymbol{D}(0), \tag{4.7}$$

where $\varphi(t)$ is a generic stress relaxation function given by,

$$\varphi(t) = \varphi_0 + \sum_{i=1}^{N} \varphi_i e^{-\alpha_i t}. \tag{4.8}$$

Here the (possibly $\boldsymbol{x}$ dependent) coefficients $\{\varphi_i\}_{i=0}^{N}$ are non-negative and normalized so that $\varphi(0) = 1$, and the (possibly $\boldsymbol{x}$ dependent) $\{\alpha_i\}_{i=1}^{N}$ are non-negative. More generally one could of course write (with summation not implied),

$$D_{ijkl}(t) := (D_{ijkl})_0 + \sum_{m=1}^{N_{ijkl}} (D_{ijkl})_m \exp(-(\alpha_{ijkl})_m t).$$

We now look at a slight variation on this theme.

The Reversed Maxwell Solid

One can also arrive at a Maxwell solid by switching the order in which the spring and dashpot appear in the viscous arm of the network, as shown in Figure 4.2. We call this the *Reversed Maxwell Solid*.

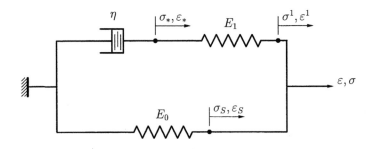

Figure 4.2: The Reversed Maxwell Solid

Balancing stresses such that $\sigma_* = \sigma^1$ and using $\varepsilon^1 = \varepsilon - \varepsilon_*$ now gives the differential equation for the new internal variable ε_* as,

$$\frac{d\varepsilon_*}{dt} + \frac{\varepsilon_*}{\tau} = \frac{\varepsilon}{\tau}, \qquad \text{where } \tau := \eta/E_1. \tag{4.9}$$

Note that ε_* is a viscous strain and so is quite different from ε^* as introduced above for the Maxwell model, although ε^* and ε_* are very simply related by,

$$\varepsilon(t) = \varepsilon^*(t) + \varepsilon_*(t).$$

This is obvious since the total strain in the viscous arm of the networks is the sum of the strains in each component.

Note also that (4.9) has the current strain on the right, and not the strain derivative as in (4.5). This reversed model is thus more useful for temporally discontinuous numerical approximations, as introduced below. Also, with

$$\sigma(t) = E_0\varepsilon + E_1(\varepsilon - \varepsilon_*),$$

we again get (4.6) and so from a physical point of view these models are interchangeable.

Alternative models

The Dirichlet-Prony series is an extremely convenient form to take for large scale computational approximations to problems (4.1) and (4.2) since one can exploit the simple recurrence,

$$\psi(t) := e^{-\alpha t} \quad \Longrightarrow \quad \psi(t+k) = e^{-\alpha k}\psi(t).$$

to update the history term arising from a discretization of the Volterra integral. Alternatively, one may use the evolution equations for the internal strain variables, (4.5) or (4.9), to capture the history locally in time, and then employ a time stepping approximation.

On the other hand, for general Volterra problems one must usually store the entire solution history as the computation advances through the time levels and, moreover, at each time level this history needs to be summed to approximate the integral. For such methods the number of operations required at time level N is of the order $O(N^2)$. The Dirichlet-Prony series provides a very useful short cut around this "N^2 problem". A short summary, drawn from [7], of related internal variable theories now follows.

Strains and stresses are commonly used as internal variables. Models that result from employing these variables are closely related to each other and are also closely related to plasticity models, see Rice [18]. The overviews presented below describe several recently proposed models. They are described here in a simple one-dimensional linear form to allow the relationships between the models to be easily understood. Readers are encouraged to consult the cited literature to obtain additional details on the special aspects of each model. For the models in this section: E, E_1 are elastic moduli, ε, σ are the physical strain and stress, ε^*, σ^* are internal strain and stress variables, η is a viscosity coefficient, and τ is a relaxation time constant.

Keren, Partom, and Rosenberg in [12] employ strains as internal variables. They allow the dashpots in a generalized Maxwell model to be nonlinear. In the case of the standard linear solid with a nonlinear dashpot in the Maxwell element, their constitutive model and evolution equation can be written as follows,

$$\sigma\left(\varepsilon, \varepsilon^*\right) = E\varepsilon + E_1\left(\varepsilon - \varepsilon^*\right), \tag{4.10}$$

and,

$$\frac{d\varepsilon^*}{dt} + g\left(\varepsilon - \varepsilon^*\right) = \frac{d\varepsilon}{dt},\tag{4.11}$$

where g is a nonlinear function. Lubliner [16] extended Green and Tobolsky's [6] nonlinear model for rubber viscoelasticity. The linear isothermal small strain version of Lubliner's model is similar to Keren, Partom, and Rosenberg's model. Lubliner's constitutive model for the standard linear solid in the isothermal small strain case has the form,

$$\sigma\left(\varepsilon, \varepsilon^*\right) = E\varepsilon + E_1\left(\varepsilon - \varepsilon^*\right),\tag{4.12}$$

and,

$$\frac{d\varepsilon^*}{dt} + \frac{\varepsilon^*}{\tau} = \frac{\varepsilon}{\tau},\tag{4.13}$$

which, we note, is identical to (4.9).

Less often, stresses are employed as internal variables. Simo [27] introduced a model for large strain viscoelastic deformations of rubber which included damage mechanics. Additional literature on Simo's model can be found in Govindjee and Simo [5], and Govindjee and Reese [3, 4]. The model is based on an energy function of the form,

$$\Psi\left(\varepsilon, \sigma^*\right) = \frac{1}{2}E\varepsilon^2 - \sigma^*\varepsilon + \frac{c_1}{2}(\sigma^*)^2,\tag{4.14}$$

where c_1 is a material parameter. The total stress, $\dfrac{\partial\Psi\left(\varepsilon, \sigma^*\right)}{\partial\varepsilon}$, is given by,

$$\sigma\left(\varepsilon, \sigma^*\right) = E\varepsilon - \sigma^*.\tag{4.15}$$

To complete the constitutive model the internal stress variable, σ^* is connected to the deformation of the solid by an evolution equation as follows,

$$\frac{d\sigma^*}{dt} + \frac{\sigma^*}{\tau} = \left(\frac{1-\gamma}{\tau}\right)\sigma_d,\tag{4.16}$$

where σ_d is the rate of change of the deviatoric part of the elastic energy with respect to strain (a stress quantity), and γ is a material parameter. A similar model was employed by Lesieutre and coworkers in [15, 13, 14]. Their energy function has the form,

$$\Psi\left(\varepsilon, \varepsilon^*\right) = \frac{1}{2}E\varepsilon^2 - \left(c_2\varepsilon^*\right)\varepsilon + \frac{c_1}{2}(\varepsilon^*)^2,\tag{4.17}$$

The constitutive equations can be written as,

$$\sigma\left(\varepsilon, \varepsilon^*\right) = E\varepsilon - c_2\varepsilon^*,\tag{4.18}$$

and,

$$\frac{d\varepsilon^*}{dt} + \frac{\varepsilon^*}{\tau} = \left(\frac{c_2}{c_1\tau}\right)\varepsilon,\tag{4.19}$$

where c_1, c_2 are material parameters.

Motivated by work on internal variable models for large strain rubber viscoelasticity Johnson, Tessler and Dambach in [9] employed strains as internal variables to model the small strain deformations of thick viscous composite beams. Their constitutive model is of the form,

$$\sigma\left(\varepsilon, \varepsilon^*\right) = E\varepsilon + E_1\varepsilon^*, \tag{4.20}$$

and

$$\frac{d\varepsilon^*}{dt} + \frac{\varepsilon^*}{\tau} = \frac{d\varepsilon}{dt}, \tag{4.21}$$

which corresponds to (4.5).

In their simple one-dimensional linear forms, the above models are all similar. Additive elastic and viscous stresses, and first-order differential equations for the evolution of the internal variables are common to each model. The nonlinear versions of these models are different. However, the differences are mainly due to the form of the nonlinear differential equations that specify the evolution of the internal variables.

4.3 THE MATHEMATICAL MODELS

As is usual in continuum mechanics the governing equations for viscoelastic deformation are based on Newton's second law. Let $\mathcal{I} := (0, T)$ denote a time interval, and let $\boldsymbol{u}(\boldsymbol{x}, t) := (u_i)_{i=1}^n$, for $\boldsymbol{x} \in \overline{\Omega}$ and $t \in \mathcal{I}$, denote the displacement of $\mathcal{G}$. Then: for $i = 1, \ldots, n$,

$$\varrho u_i''(t) - \sigma_{ij,j}(\boldsymbol{u}) = f_i(t) \qquad \text{in } \mathcal{I} \times \Omega. \tag{4.22}$$

Here: ϱ is the mass density of $\mathcal{G}$; the primes denote time differentiation; the summation convention is in force; we have suppressed the $\boldsymbol{x}$ dependence; and, $\boldsymbol{f} := (f_i)_{i=1}^n$ are body forces acting throughout $\mathcal{G}$ due, for example, to rotational inertia or gravity.

To prescribe boundary conditions we suppose now that $\{\Gamma_D, \Gamma_N\}$ forms a partition of $\partial\Omega$ such that $\Gamma_N := \partial\Omega \setminus \Gamma_D$ and $\mathrm{meas}(\Gamma_D) > 0$. Then, we can have,

$$\boldsymbol{u} = \boldsymbol{0} \quad \text{on} \quad \mathcal{J} \times \Gamma_D, \tag{4.23}$$

$$\sigma_{ij}\widehat{n}_j = g_i \quad \text{on} \quad \mathcal{J} \times \Gamma_N. \tag{4.24}$$

Here $\widehat{\boldsymbol{n}}$ is the unit outward directed normal to Γ_N, and $\boldsymbol{g} := (g_i)_{i=1}^n$ is a prescribed system of surface tractions. Also required are the initial conditions,

$$\boldsymbol{u}(0) = \boldsymbol{u}_0 \quad \text{in} \quad \Omega, \tag{4.25}$$

$$\boldsymbol{u}'(0) = \boldsymbol{u}_1 \quad \text{in} \quad \Omega, \tag{4.26}$$

and, as is common, we also assume that the initial and boundary data are compatible on $\partial\Omega$ at $t = 0$.

In the linear theory the strain tensor is given in terms of the displacements as,

$$\varepsilon_{ij}(\boldsymbol{u}) := \frac{1}{2}\left(\frac{\partial u_i}{\partial x_j} + \frac{\partial u_j}{\partial x_i}\right), \tag{4.27}$$

and using this with equation (4.4) in (4.22) we arrive at the problem (4.1).

An assumption often made by practitioners is that in (4.22) the inertia forces are small in comparison to the elastic and viscous forces. In such cases we set $\varrho u''(t) := \mathbf{0}$ and this produces the *quasistatic* problem (4.2). For this problem we replace $\mathcal{I}$ with $\mathcal{J} := [0, T]$ and discard the initial conditions (4.25) and (4.26).

In the next two sections we describe a space-time finite element discretization of this quasistatic problem and outline some *a posteriori* error estimates.

4.4   FEM FOR A PROTOTYPE QUASISTATIC PROBLEM

If we semidiscretize the quasistatic problem, represented by (4.2), in the space variables using a temporally constant finite element space of (say) piecewise linear functions, we obtain a system of second-kind Volterra equations. A simple scalar prototype for this semidiscrete system is the problem: find $u \in L_p(\mathcal{J})$ such that,

$$u(t) = f(t) + \int_0^t \phi(t - s)u(s) \, ds,$$

for $f \in L_p(\mathcal{J})$ and $\phi \in L_1(\mathcal{J})$ given, and some $p \in [1, \infty]$. Using $(\cdot, \cdot)$ to denote the $L_2(\mathcal{J})$ inner product we can write this problem in "variational" form as follows,

$$(u, v) = (f, v) + (\Lambda u, v) \qquad \forall v \in L_q(\mathcal{J}), \tag{4.28}$$

where q is the conjugate Hölder index to p, given by $p^{-1} + q^{-1} = 1$, and, for brevity, we define,

$$\Lambda u(t) := \int_0^t \phi(t - s)u(s) \, ds.$$

Note that $\Lambda u \in L_p(\mathcal{J})$ since $\phi \in L_1(\mathcal{J})$.

To effect a finite element discretization of this problem we define $N > 0$ subintervals $\mathcal{J}_i := [t_{i-1}, t_i]$, with $t_0 = 0$ and $t_N = T$, and the associated time steps, $k_i = t_i - t_{i-1}$. Letting $V^k \subset L_\infty(\mathcal{J})$ denote the space of piecewise constant functions with respect to this mesh we pose the finite element approximation as: find $U \in V^k$ such that,

$$(U, v) = (f, v) + (\Lambda U, v) \qquad \forall v \in V^k. \tag{4.29}$$

Subtracting (4.29) from (4.28) results in the familiar Galerkin orthogonality relationship,

$$(R(U), v) = 0 \qquad \forall v \in V^k,$$

where $R(U) := f - U + \Lambda U \in L_p(\mathcal{J})$ is the residual, and is computable.

To develop an *a posteriori* error estimate for this approximation we follow the techniques described by Johnson *et al.* in, for example, [1, 10]. Firstly we introduce the dual backward problem: find $\chi \in L_q(\mathcal{J})$ such that,

$$(v, \chi) = (v, g) + (v, \Lambda^* \chi) \qquad \forall v \in L_p(\mathcal{J}),$$

where $g \in L_q(\mathcal{J})$ is arbitrary, and Λ^* is dual to Λ in the sense that $(v, \Lambda^* w) = (\Lambda v, w)$. Introducing an interpolant $\pi\chi \in V^k$ to χ, and taking $v = e := u - U$ in the dual problem we have by Galerkin orthogonality that,

$$(e, g) = (R(U), \chi - \pi\chi), \tag{4.30}$$

since $R(U) \equiv e - \Lambda e$. Assuming now the existence of a constant C_π and a stability factor $S(T)$ such that,

$$\|\chi - \pi\chi\|_{L_q(\mathcal{J})} \le C_\pi \|\chi\|_{L_q(\mathcal{J})}, \qquad (4.31)$$

and,

$$\|\chi\|_{L_q(\mathcal{J})} \le S(T) \|g\|_{L_q(\mathcal{J})}, \qquad (4.32)$$

we have,

$$\|\chi - \pi\chi\|_{L_q(\mathcal{J})} \le C_\pi S(T) \|g\|_{L_q(\mathcal{J})}.$$

Now, regarding $L_p(\mathcal{J})$ as the topological dual to $L_q(\mathcal{J})$, for $p > 1$, we use this estimate along with (4.30) in,

$$\|e\|_{L_p(\mathcal{J})} := \sup \left\{ \frac{|(e,g)|}{\|g\|_{L_q(\mathcal{J})}} : g \in L_q(\mathcal{J}) \setminus \{0\} \right\},$$

and so obtain the *a posteriori* error estimate,

$$\|u - U\|_{L_p(\mathcal{J})} \le C_\pi S(T) \|R(U)\|_{L_p(\mathcal{J})} \qquad \text{for } p > 1. \qquad (4.33)$$

This result is given in [20] along with an *a priori* error estimate, an upper bound on the residual, and numerical experiments. For viscoelasticity problems as described earlier, optimal estimates for the stability factor are derived in [24]. In particular, Gronwall's lemma is not used and for viscoelastic solids we have $S(T) = O(1)$ independently of T.

Note that in (4.31) we cannot in general have $C_\pi < 1$ and so by setting $\pi\chi := 0$ we can take $C_\pi = 1$ in (4.33).

The obvious drawback with (4.33) is that the time steps do not appear and so it cannot be used to derive an adaptive time step control. To include the time steps in an *a posteriori* error estimate we use a sharper version of (4.31),

$$\|\kappa^{-1}(\chi - \pi\chi)\|_{L_q(\mathcal{J})} \le \|\chi'\|_{L_q(\mathcal{J})}, \qquad (4.34)$$

where $\kappa|_{\mathcal{J}_i} := k_i$ is the piecewise constant time step function. Also (and for convolution equations), in place of (4.32) we have,

$$\|\chi'\|_{L_q(\mathcal{J})} \le S(T) \|g'\|_{L_q(\mathcal{J})} \qquad \forall g \in \mathring{W}_q^1(\mathcal{J}), \qquad (4.35)$$

from which we now obtain,

$$\|\kappa^{-1}(\chi - \pi\chi)\|_{L_q(\mathcal{J})} \le S(T) \|g'\|_{L_q(\mathcal{J})}.$$

We can use this to derive an alternative *a posteriori* error estimate, where the residual is weighted with the time steps, provided we measure the error in a weaker norm. The appropriate space is $W_p^{-1}(\mathcal{J})$ with norm,

$$\|w\|_{W_p^{-1}(\mathcal{J})} := \sup \left\{ \frac{|(w,g)|}{\|g'\|_{L_q(\mathcal{J})}} : g \in \mathring{W}_q^1(\mathcal{J}) \setminus \{0\} \right\} \qquad \text{for } p > 1.$$

With these, and replacing (4.30) with,

$$(e,g) = (\kappa R(U), (\chi - \pi\chi)/\kappa),$$

we obtain the alternative estimate,

$$\|u - U\|_{W_p^{-1}(\mathcal{J})} \leq S(T) \|\kappa R(U)\|_{L_p(\mathcal{J})} \qquad \text{for } p > 1. \tag{4.36}$$

Full details are given in [21].

The presence of the time steps on the right now allows for adaptive time step control. For example, we can guarantee that $\|u - U\|_{W_\infty^{-1}(\mathcal{J})} \leq \mathsf{TOL}$, where $\mathsf{TOL} > 0$ is a user-specified tolerance, by ensuring that,

$$k_i = \frac{\mathsf{TOL}}{S(T) \|R(U)\|_{L_\infty(\mathcal{J}_i)}}$$

on each time level. Some numerical experiments using an apparently more efficient time step control are also given in [21].

4.5   THE SPACE-TIME PROBLEM

The material in this and the next section is drawn from the technical reports [22, 23, 26] and the summary paper [25].

The space-time quasistatic problem represented by (4.2) is given by combining the constitutive and linear strain relations, (4.4) and (4.27), and the boundary data (4.23) and (4.24), with the equilibrium (non-inertial) equations taken from (4.22):

$$-\sigma_{ij,j}(\boldsymbol{u}) = f_i(t) \qquad \text{in } \mathcal{J} \times \Omega. \tag{4.37}$$

To effect a space-time finite element approximation of this problem we first cast it into weak form in the standard way by partially integrating against an appropriate test function. The resulting problem is then: find $\boldsymbol{u} \in L_\infty(\mathcal{J}; H)$ such that,

$$A(\boldsymbol{u}(t), \boldsymbol{v}) = L(t; \boldsymbol{v}) + \int_0^t B(t, s; \boldsymbol{u}(s), \boldsymbol{v}) \, ds \qquad \forall \boldsymbol{v} \in H, \text{ a.e. in } \mathcal{J}. \tag{4.38}$$

Here: L is a time dependent linear form containing the loads and $H \subset \boldsymbol{H}^1(\Omega)$ is a test space incorporating the essential boundary condition (4.23). Also, A and B are continuous bilinear forms generated by the operators A and B (with A symmetric and H-coercive):

$$A(\boldsymbol{w}, \boldsymbol{v}) \quad := \quad \int_\Omega D_{ijkl}(0) \varepsilon_{kl}(\boldsymbol{w}) \varepsilon_{ij}(\boldsymbol{v}) \, d\Omega, \tag{4.39}$$

$$B(t, s; \boldsymbol{w}, \boldsymbol{v}) \quad := \quad \int_\Omega \frac{\partial D_{ijkl}(t - s)}{\partial s} \varepsilon_{kl}(\boldsymbol{w}) \varepsilon_{ij}(\boldsymbol{v}) \, d\Omega, \tag{4.40}$$

Detailed background assumptions for this formulation are given in each of [22, 23, 25].

For each time stage $\mathcal{J}_i$ we let $H_i \subset H$ be a finite element space of continuous piecewise linear functions with respect to some triangulation of Ω. Note that at this stage we assume no relationship between these H_i. The semidiscretization of (4.38) then proceeds in the usual way. To discretize in the time variable we integrate the semidiscrete version of (4.38) over $\mathcal{J}$, and then introduce the space-time finite element space,

$$V_r := \left\{ \boldsymbol{v} \in L_\infty(\mathcal{J}; H) : v|_{\mathcal{J}_i} \in \mathbb{P}_r(\mathcal{J}_i; H_i) \; \forall i \in \mathbb{N}(1, N) \right\}.$$

Here $\mathbb{P}_r(\mathcal{J}_i; H_i)$ is the vector space of polynomials of degree at most r defined on $\mathcal{J}_i$ with coefficients in H_i. Note that our approximating functions in V_r are continuous in space but in general discontinuous at the knots $\{t_i\}_{i=1}^{N-1}$. These discontinuities allow the space-meshes to change with time.

With this approximating space we generate a finite element approximation U to u of functions that are continuous piecewise linear in space and discontinuous piecewise constant $(r = 0)$ or linear $(r = 1)$ in time. In the numerical experiments below we restrict ourselves to the simpler $r = 0$ case. The *a priori* error estimate for this problem is derived in [22]; it takes the following form.

Theorem 4.1 (*A priori* energy-error estimate) *With natural assumptions, and for approximation in V_r, for $r = 0, 1$, the Galerkin error $e := u - U$ satisfies the a priori error estimate,*

$$\|u - U\|_{L_\infty(\mathcal{J};H)} \leq C(T) \left(\Pi_h \left\| h D^2 u \right\|_{L_\infty(\mathcal{J};L_2)} + \Pi_k \left\| k^{r+1} \frac{\partial^{r+1} u}{\partial t^{r+1}} \right\|_{L_\infty(\mathcal{J};H)} \right),$$

where Π_h and Π_k are constants, and $C(T)$ is a stability factor. This estimate holds for $r = 1$ only if each k_q is small enough, and depends also on the ratios k_q/k_{q-1}.

Below, and in the next section, we give some numerical results (for two-dimensional problems) illustrating this approximation. In all cases we consider only isotropic materials where, if D is the matrix representation of the tensor $\underline{D}$, we have,

$$D(t) = \lambda(t) I_\lambda + \mu(t) I_\mu,$$

where,

$$I_\lambda := \begin{pmatrix} 1 & 1 & 0 \\ 1 & 1 & 0 \\ 0 & 0 & 0 \end{pmatrix} \quad \text{and} \quad I_\mu := \begin{pmatrix} 1 & 0 & 0 \\ 0 & 1 & 0 \\ 0 & 0 & 1/2 \end{pmatrix},$$

and $\lambda(t)$, $\mu(t)$ are given by Prony series-type relaxation functions,

$$\lambda(t) := \lambda_0 + \sum_{i=1}^{N_\lambda} \lambda_i e^{-l_i t} \quad \text{and} \quad \mu(t) := \mu_0 + \sum_{i=1}^{N_\mu} \mu_i e^{-m_i t},$$

where the λ_i, l_i, μ_i, m_i and N_λ, N_μ are constants. For the stress we then have,

$$\sigma(t) = \varsigma(t) - \Sigma(u; t),$$

where,

$$\varsigma(t) := D(0)\varepsilon(u(t)) = \text{instantaneous elastic stress,}$$

$$\Sigma(u; t) := \int_0^t D_s(t - s)\varepsilon(u(s))\, ds = \text{inherited viscous stress.}$$

Setting $\lambda_i := \lambda_i I_\lambda$ and $\mu_i := \mu_i I_\mu$ and defining the internal stress variables,

$$\Sigma_{\lambda_i}(u; t) := \lambda_i l_i \int_0^t e^{-l_i(t-s)} \varepsilon(u(s))\, ds$$

$$\Sigma_{\mu_i}(u; t) := \mu_i m_i \int_0^t e^{-m_i(t-s)} \varepsilon(u(s))\, ds,$$

we have from (4.4) that,

$$\Sigma(\boldsymbol{u};t) = \int_0^t \boldsymbol{D}_s(t-s)\varepsilon(\boldsymbol{u}(s))\,ds = \sum_{i=1}^{N_\lambda}\Sigma_{\lambda_i}(\boldsymbol{u};t) + \sum_{i=1}^{N_\mu}\Sigma_{\mu_i}(\boldsymbol{u};t).$$

Note that these internal variables correspond to the reversed Maxwell model and so satisfy evolution equations of the form,

$$\Sigma'_{\lambda_i}(\boldsymbol{u};t) + l_i\Sigma_{\lambda_i}(\boldsymbol{u};t) = \lambda_i l_i\varepsilon(\boldsymbol{u}(t)).$$

Note also that the following "update recurrences" apply,

$$\Sigma_{\lambda_i}(\boldsymbol{u};t) = e^{-l_i(t-\tau)}\Sigma_{\lambda_i}(\boldsymbol{u};\tau) + \lambda_i l_i \int_\tau^t e^{-l_i(t-s)}\varepsilon(\boldsymbol{u}(s))\,ds,$$

$$\Sigma_{\mu_i}(\boldsymbol{u};t) = e^{-m_i(t-\tau)}\Sigma_{\mu_i}(\boldsymbol{u};\tau) + \mu_i m_i \int_\tau^t e^{-m_i(t-s)}\varepsilon(\boldsymbol{u}(s))\,ds.$$

Using these new definitions with (4.39) and (4.40) we get an alternative representation of the history as,

$$\int_0^t B(t,s;\boldsymbol{u}(s),\boldsymbol{v})\,ds = \int_\Omega \left(\int_0^t \boldsymbol{D}_s(t-s)\varepsilon(\boldsymbol{u}(s))\,ds\right) \cdot \varepsilon(\boldsymbol{v})\,d\Omega,$$

$$= \int_\Omega \Sigma(\boldsymbol{u};t) \cdot \varepsilon(\boldsymbol{v})\,d\Omega.$$

The point to note here is that the history integral has now vanished and has been replaced by a local (in time) term. We use a similar approach in the numerical scheme. Full details of the practical implementation and numerical algorithm are given in [26, § 7.4]).

4.5.1 Convergence tests

To demonstrate the convergence of the scheme we compare the computed solution against a known (but artificial) exact solution. We take the exact solution for the displacements to be of the form,

$$u_1(x,y,t) = T(t)X(x,y) \qquad \text{and} \qquad u_2(x,y,t) = T(t)Y(x,y),$$

and then substitute these into the governing equations to determine the required loads and tractions. In the following examples we use the data,

$$E = 2.776\,090\,556\,\text{GPa} \qquad \text{and} \qquad \nu = 0.4, \tag{4.41}$$

which imply for plane stress that,

$$\lambda(0) = 1.321\,947\,884\,\text{GPa} \qquad \text{and} \qquad \mu(0) = 1.982\,921\,826\,\text{GPa}. \tag{4.42}$$

These data are appropriate to Maranyl Nylon 6.6 (see later), but for the moment we use the arbitrary relaxation functions,

$$\lambda(t) := \lambda(0)\left(0.3 + 0.2e^{-t} + 0.5e^{-0.1t}\right), \tag{4.43}$$

$$\mu(t) := \mu(0)\left(0.2 + 0.1e^{-3t} + 0.2e^{-0.7t} + 0.3e^{-2t} + 0.2e^{-0.2t}\right). \tag{4.44}$$

We now look at results for two problems: in the first there is no temporal error, and in the second there is no spatial error. The domain (arbitrarily chosen) is shown in Figure 4.3.

Table 4.1: Spatial convergence results ($k = 0.5$; $h = 0.4, 0.2, 0.1, \ldots$).

Number of elements	Energy norms		Energy errors	
	$\|u\|_{L_\infty(0,1;H)}$	$\|U\|_{L_\infty(0,1;H)}$	$\|u - U\|_{L_\infty(0,1;H)}$	$S(1)\mathcal{E}_\Omega(1;U)$
10	8.551(+03)	5.856(+03)	6.244285(+03)	7.072830(+03)
45	8.561(+03)	8.065(+03)	2.898224(+03)	3.233216(+03)
177	8.561(+03)	8.440(+03)	1.448311(+03)	1.616137(+03)
762	8.561(+03)	8.531(+03)	7.154630(+02)	7.818353(+02)
3109	8.561(+03)	8.554(+03)	3.422946(+02)	3.783819(+02)
12628	8.561(+03)	8.559(+03)	1.663714(+02)	1.856497(+02)

Spatial convergence

We set $T(t) := 1$ and put,

$$X(x, y) := -0.03x^7 + 0.025 \sin(2\pi x) \sin(\pi y + \pi/2),$$
$$Y(x, y) := -0.04y^9 + 0.035 \sin(\pi x + \pi/2) \sin(2\pi y).$$

The time step is $k = 0.5$ and we loop through the sequence of "regular" meshes corresponding to $h = 0.4, 0.2, 0.1, \ldots$ (the first two of which are shown in Figure 4.3). The results are shown in Table 4.1 and it is evident that the energy error is $O(h)$ as expected. The quantity $S(1)\mathcal{E}_\Omega(1; U)$ is an *a posteriori* spatial-error estimate, and is explained in the next section.

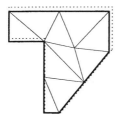
11 nodes, 10 elements

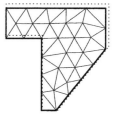
33 nodes, 45 elements

Figure 4.3: Regular meshes for $h = 0.4$ and $h = 0.2$

Temporal convergence

To examine the time discretization error in isolation we now set,

$$T(t) := 1 + t/2 + \sin(2\pi t), \quad X(x, y) := -0.03x \quad \text{and} \quad Y(x, y) := -0.04y.$$

We expect to see the errors converge to zero as $k \to 0$ independently of h, and so we choose $h = 0.4$—corresponding to the mesh shown on the left of Figure 4.3. (Note that

Table 4.2: Temporal convergence results ($k = 1.0, 0.5, 0.25, \ldots$).

k	$\|\mathbf{u}\|_{L_\infty(0,1;H)}$	$\|\mathbf{U}\|_{L_\infty(0,1;H)}$	$\|\mathbf{u} - \mathbf{U}\|_{L_\infty(0,1;H)}$	$S(1)\mathcal{E}_\Omega(1;\mathbf{U})$
1.0000(+00)	3.945(+03)	3.178(+03)	7.677814(+02)	3.119966(+02)
5.0000(−01)	3.945(+03)	4.641(+03)	1.986738(+03)	7.177276(+02)
2.5000(−01)	5.589(+03)	4.764(+03)	1.798382(+03)	6.519429(+02)
1.2500(−01)	5.589(+03)	5.404(+03)	1.050321(+03)	3.812408(+02)
6.2500(−02)	5.589(+03)	5.562(+03)	5.476365(+02)	1.988857(+02)
3.1250(−02)	5.589(+03)	5.593(+03)	2.771164(+02)	1.006664(+02)
1.5625(−02)	5.597(+03)	5.595(+03)	1.390760(+02)	5.052768(+01)
7.8125(−03)	5.597(+03)	5.597(+03)	6.962842(+01)	2.529828(+01)
3.9062(−03)	5.598(+03)	5.597(+03)	3.483190(+01)	1.265596(+01)
1.9531(−03)	5.598(+03)	5.598(+03)	1.741975(+01)	6.329462(+00)

the mesh diagrams have these computed displacements superposed—the true boundary is shown as a dashed line.) The results are shown in Table 4.2, and we see that the energy error is $O(k)$, again as expected.

Note that in this table the quantity $S(1)\mathcal{E}_\Omega(1;\mathbf{U})$ does not estimate the error since there is no space discretization error in this example. The fact that $S(1)\mathcal{E}_\Omega(1;\mathbf{U})$ is of order $O(k)$ is due to the time error in the traction jump, $\mathbf{g} - \tilde{\mathbf{g}}$, on Γ_N.

In both of these tables we showed an *a posteriori* estimate of the spatial error. In the next section we summarize the background to this.

4.6   ADAPTIVITY FOR THE QUASISTATIC PROBLEM

This section is only a summary of our results, once again we refer to [22, 23, 26] and the summary paper [25] for full details.

In [23] we give the following basic *a posteriori* Galerkin energy-error estimate: for each discrete time $t_1, t_2, \ldots, t_p, \ldots$,

$$\|\mathbf{u} - \mathbf{U}\|_{L_\infty(0,t_p;H)} \leq S(t_p)\big(\mathcal{E}_\Omega(t_p;\mathbf{U}) + \mathcal{E}_J(t_p;\mathbf{U}) + \mathcal{E}_V(t_p;\mathbf{U})\big), \qquad (4.45)$$

where $\mathcal{E}_\Omega$, $\mathcal{E}_J$ and $\mathcal{E}_V$ are residuals which are computable in terms of the data and the finite element solution $\mathbf{U}$, and $S(t)$ is the stability factor, introduced before in (4.32), and discussed further below.

In this section we will be concerned mostly with $\mathcal{E}_\Omega$. This term contains the spatial discretization error (in the case where we allow only *nested* mesh refinements) and can be used to guide adaptive space mesh refinement. It is essentially identical to the residual derived for linear elasticity by Johnson and Hansbo in [11].

The residual $\mathcal{E}_J$ is either unstable (useless) as $h \to 0$ or—when written in a different form—prohibitively expensive to implement (see [23]). We eventually hope to provide an alternative error estimate in which $\mathcal{E}_\Omega$ and $\mathcal{E}_V$ are essentially the same, while $\mathcal{E}_J$ is stabilized at the expense of estimating the error in a weaker norm, as for the scalar case described earlier, (4.36).

It is the term $\mathcal{E}_V$ that causes the greatest difficulty in this estimate. The spatial residuals in $\mathcal{E}_\Omega$ are constructed by integrating the discrete solution over each element to arrive at a *distributional* divergence of the discrete stress (compare (4.37)). This divergence comprises two parts: the smooth function inside the element (which is zero in our case of piecewise linear approximation), and the stress jumps across inter-element boundaries. The difficulty arises because the stress is history dependent. This means that we have to integrate by parts over not just the elements in the current mesh, but also over all elements in all previous meshes. The internal edges that appeared in previous meshes but are no longer present in the current mesh (e.g. due to de-refinement) are therefore "left behind" when forming the standard residual $f + \nabla \cdot \sigma^h$ (which constitutes $\mathcal{E}_\Omega$), and so we consign the stress jumps across these edges to the term $\mathcal{E}_V$. In the particular case where only nested refinements are permitted, no edges are left behind in this way and we have $\mathcal{E}_V \equiv 0$. This is the case in our examples below.

To deal with mesh de-refinement would appear to require fairly complex data structures in the computer code in order to track all the resulting previous edges. Also, it is not likely that $\mathcal{E}_V$ will act in any way other than to degrade the quality of the estimate since it contains historical contributions to the current stress. These can then act only to reinforce one another in the estimate when in fact the residual could be much smaller due to cancellation. This "loss of cancellation" problem has been noted by others in the context of stress-jump residual-type estimators, and the memory in the Volterra integral acts here only to exacerbate the problem. Our feeling at the moment is that a representation of the algorithm in terms of internal variables could go some way toward removing the $\mathcal{E}_V$ residual, since then all hereditary information is automatically represented on the current mesh.

We now look in more detail at the residual term $\mathcal{E}_\Omega$. This is defined at each discrete time level $t_1, t_2, \ldots, t_p, \ldots$ as:

$$\mathcal{E}_\Omega(t_p; U) := \max_{1 \leq q \leq p} \left\{ \Pi_{\Omega_q} \|h_q f\|_{L_\infty(\mathcal{J}_q; L_2(\Omega))} + \Pi_\ell \|h_q \mathcal{G}\|_{L_\infty(\mathcal{J}_q; L_2(\Omega))} \right\}, \qquad (4.46)$$

where Π_{Ω_q} and Π_ℓ are constants appearing in certain interpolation-error bounds, and $h_q = h_q(x)$ is the piecewise constant mesh function for the mesh during $\mathcal{J}_q$.

The first term in the estimate is straightforward to interpret since it involves only the $L_2(\Omega)$ norm of the body forces weighted with the mesh function. (Note that this term is known *a priori* and could be used to determine the initial mesh.) To define the second term we first define $r|_{\mathcal{J}_q} = (r_k)_{k=1}^n$ on each time level $\mathcal{J}_q$ by,

$$r_k(t; U(t)) := \begin{cases} \frac{1}{2} \left| [\![\tilde{g}_k(t; U(t))]\!]_\ell \right|, & \text{on each edge } \ell, \\[2mm] \left| g_k(t) - \tilde{g}_k(t; U(t)) \right|, & \text{on } \Gamma_N, \\[2mm] 0, & \text{on } \Gamma_D, \end{cases}$$

where, on each internal element edge ℓ, $\tilde{g}_k$ is a discrete traction component resolved normal to the edge, and where $[\![\tilde{g}_k]\!]$ denotes the jump in the component across the edge.

With these definitions we define $\mathcal{G} \in L_2(\Omega)$ by,

$$\mathcal{G}|_{\Omega_{qj}} := \frac{\|r(t; U(t))\|_{L_2(\partial\Omega_{qj})}}{\sqrt{h_{qj} \text{meas}(\Omega_{qj})}},$$

where Ω_{qj} is the jth element in the mesh during $\mathcal{J}_q$ and $h_{qj} := \text{diam}(\Omega_{qj})$. Note that,

$$\|h_q\mathcal{G}\|_{L_2(\Omega)} \equiv \left(\sum_{\Omega_{qj} \subset \overline{\Omega}_q} \left\| h_{qj}^{\frac{1}{2}} \boldsymbol{r}(t; \boldsymbol{U}(t)) \right\|_{\boldsymbol{L}_2(\partial\Omega_{qj})}^2 \right)^{\frac{1}{2}}. \tag{4.47}$$

We now use $\mathcal{E}_\Omega$ to drive adaptive mesh refinement.

4.6.1 Adaptive mesh control

Following the usual practice we develop an adaptive mesh size control based on equidistribution. The goal is to design a mesh during each $\mathcal{J}_q$ for which $S(t_q)\mathcal{E}_\Omega(t_q; \boldsymbol{U}) \leq \mathsf{TOL}$, where $\mathsf{TOL} > 0$ is a user-defined tolerance. Further, this mesh should be optimal in the sense that the error control is achieved with as few degrees of freedom as possible. Although impossibly difficult to obtain in an exact sense (see for example the discussion in [2]), such a mesh can be approximated by using an *a posteriori* error estimate. We aim for a mesh size modification strategy for each element Ω_j in the mesh.

To achieve this error control we require,

$$\Pi_{\Omega_q}\|h_q\boldsymbol{f}\|_{L_\infty(\mathcal{J}_q; \boldsymbol{L}_2(\Omega))} + \Pi_\ell\|h_q\mathcal{G}\|_{L_\infty(\mathcal{J}_q; \boldsymbol{L}_2(\Omega))} \leq \frac{\mathsf{TOL}}{S(T)},$$

at each time level t_q. Assuming a local tolerance tol, equidistribution yields the adaptive mesh size selector,

$$h_{qj}^{\text{new}} := \sqrt{\frac{\text{tol}}{\Pi_\Omega^2\|\boldsymbol{f}\|_{L_\infty(\mathcal{J}_q; \boldsymbol{L}_2(\Omega_j))}^2 + \Pi_\ell^2\|\mathcal{G}\|_{L_\infty(\mathcal{J}_q; \boldsymbol{L}_2(\Omega_j))}^2}}, \tag{4.48}$$

and then $S(t_p)\mathcal{E}_\Omega(t_p; \boldsymbol{U}) \leq \mathsf{TOL}$ is guaranteed by choosing,

$$\text{tol} = \frac{\mathsf{TOL}^2}{2MS^2(T)}.$$

Below, in the implementation, we replace the indicated norms with practical quadrature approximations. Note that in the above we have set $\Pi_{\Omega_q} = \Pi_\Omega$ for each q: i.e. we take the interpolation-error constants as time independent (even though the mesh is not), and then use the values for Π_Ω and Π_ℓ given below.

The interpolation-error constants Π_Ω and Π_ℓ

To complete the description of the *a posteriori* error estimate we need to specify the interpolation-error constants Π_Ω and Π_ℓ. For this we adapt the approximate values calculated by Ludwig in [17, Tables 6.2 and 6.4]. Here the interpolation constants are given for a variety of Poisson ratios, for plane strain with $E = 1$, and assuming a mesh of right-angled triangles. The values are reproduced here in Table 4.3.

The value for Π_ℓ for $\nu \to 0.5$ is not given by Ludwig. To calculate it here we assume a scaling of $\Pi_\Omega/\Pi_\ell = 0.42$ and then work from the tabulated value of Π_Ω. These values

Table 4.3: Interpolation-error constants

ν	Π_Ω	Π_ℓ	Π_Ω/Π_ℓ
0.1	1.143177	2.683068	0.426071
0.2	1.172415	2.715093	0.431814
0.3	1.197708	2.734093	0.438064
0.4	1.219404	2.741196	0.444844
$0.5 - \epsilon$	1.237806	(2.947157)	0.42*

can be incorporated into a computer code as a look-up table and then values for any ν obtained by linear interpolation. The constants scale with E in the following way:

$$\frac{\Pi_\Omega}{\widehat{\Pi}_\Omega} = \frac{\Pi_\ell}{\widehat{\Pi}_\ell} = \sqrt{\frac{\widehat{E}}{E}}.$$

(Note also that Ludwig uses 2μ in Hooke's law where we use μ, but since this does not affect E and ν this difference is immaterial).

We need now to obtain the corresponding interpolation constants for plane stress, note first that:

$$\left.\begin{array}{l} \lambda = \dfrac{\nu E}{(1-2\nu)(1+\nu)} \\[2mm] \mu = \dfrac{E}{1+\nu} \end{array}\right\} \quad\Longrightarrow\quad \left\{\begin{array}{l} E = \dfrac{(\mu+3\lambda)\mu}{\mu+2\lambda} \\[2mm] \nu = \dfrac{\lambda}{\mu+2\lambda} \end{array}\right. \qquad \text{for plane strain,}$$

$$\left.\begin{array}{l} \lambda = \dfrac{\nu E}{(1-\nu)^2} \\[2mm] \mu = \dfrac{E}{1+\nu} \end{array}\right\} \quad\Longrightarrow\quad \left\{\begin{array}{l} E = \dfrac{(\mu+2\lambda)\mu}{\mu+\lambda} \\[2mm] \nu = \dfrac{\lambda}{\mu+\lambda} \end{array}\right. \qquad \text{for plane stress.}$$

So, given E and ν for plane stress we can first calculate λ and μ, and then work backwards with these values to find the corresponding E and ν for plane strain—$\widehat{E}$ and $\widehat{\nu}$ say. From these we can then determine the interpolation constants from the table and the scaling given above.

In our numerical results we take $\nu = 0.4$ and $E = 2.776\,090\,556$ GPa which give the plane stress values,

$$\lambda = 0.476\,190\,476E \qquad \text{and} \qquad \mu = 0.714\,285\,714E.$$

These give the corresponding $\widehat{E}$ and $\widehat{\nu}$ plane strain values,

$$\widehat{E} = 0.918\,367\,346E \qquad \text{and} \qquad \widehat{\nu} = 0.2857\ldots.$$

Assuming $\widehat{E} = 1$ the table gives for this Poisson ratio,

$$\widehat{\Pi}_\Omega \approx 1.2 \qquad \text{and} \qquad \widehat{\Pi}_\ell \approx 2.8,$$

and then using the scaling for $\widehat{E}$ we finally get the values,

$$\Pi_\Omega = \frac{1.252198}{\sqrt{E}} \qquad \text{and} \qquad \Pi_\ell = \frac{2.9218}{\sqrt{E}}.$$

However, there is some doubt as to whether this exercise is worthwhile. These constants represent the worse possible case in interpolation error and often end up making the *a posteriori* error estimate significantly over-estimate the finite element error. To address this difficulty we have *calibrated* these constants against exact solutions in order to render the *a posteriori* estimates more realistic. The end result is that we divide the values given above by a factor of ten and twenty respectively. These values are suggested by calculations based on an exact solution (given in [26]). Clearly, it is desirable to determine a more systematic calibration technique.

The stability factor $S(T)$

For an isotropic viscoelastic material in two dimensions the stability factor $S(t)$ has been derived by Shaw and Whiteman in [24]. We give here only the main result and refer to the reference for details.

For a viscoelastic solid we have,

$$S(t) := \left(1 - \int_0^t \phi(s)\,ds\right)^{-1} \qquad \text{where} \qquad \phi(t) := \max\{\omega_1(t), \omega_2(t)\},$$

and,

$$\omega_1(t) := -\frac{\lambda'(t) + \mu'(t)}{\lambda(0) + \mu(0)}, \qquad \omega_2(t) := -\frac{\mu'(t)}{\mu(0)}.$$

For example, using our test data from (4.43) and (4.44) we have,

$$\phi(t) := \begin{cases} \omega_2(t), & \text{for } 0 \leq t \leq t^* := 2.28476\ldots, \\ \omega_1(t), & \text{for } t \geq t^*. \end{cases}$$

these give,

$$S(t) := \begin{cases} \dfrac{\mu(0)}{\mu(t)}, & \text{for } 0 \leq t \leq t^*, \\[2ex] \left(\dfrac{\mu(t^*)}{\mu(0)} + \dfrac{\lambda(t) - \lambda(t^*) + \mu(t) - \mu(t^*)}{\lambda(0) + \mu(0)}\right)^{-1}, & \text{for } t \geq t^* \end{cases}$$

Note also (thinking ahead to the "Maranyl" Nylon 6.6 data given below) that for a synchronous (solid) viscoelastic material where there exists a generic relaxation function $\varphi(t)$, normalized to $\varphi(0) = 1$, satisfying,

$$\frac{\lambda(t)}{\lambda(0)} = \frac{\mu(t)}{\mu(0)} = \varphi(t),$$

we have the simpler result: $S(t) = 1/\varphi(t)$. In the next subsection we give a physically realistic example.

4.6.2 An L-shaped lever arm

In this section we demonstrate mesh adaptivity for an L-shaped lever arm (Figure 4.4). We assume a synchronous material wherein λ and μ exhibit the same time dependence (which implies a constant Poisson ratio) and take the data as given by (4.41) and (4.42). For the single stress relaxation function we take $E\varphi(t)$ where,

$$\varphi(t) = \sum_{i=0}^{2} \varphi_i e^{-\alpha_i t},$$

with:

$$
\begin{array}{llll}
\varphi_0 & = & 0.183\,429\,971 & \quad \alpha_0 & = & 0.0 \\
\varphi_1 & = & 0.385\,804\,129 & \quad \alpha_1 & = & 53.821\,223\,820 \\
\varphi_2 & = & 0.430\,765\,899 & \quad \alpha_2 & = & 1.592\,948\,754
\end{array}
$$

Here the φ_i are dimensionless while the α_i have units $(\text{years})^{-1}$. These data are derived from experimental creep response curves for the Nylon 6.6 compound "Maranyl", and are taken from [19, Equation (5.39)]. (Note that we have "modernized" the units by using the conversion factor $1\,\text{psi} = 6894.8\,\text{Pa}$.)

We now consider an example of an L-shaped lever arm as shown in Figure 4.4. The arm is fixed rigidly in both displacements along its leftmost vertical edge. In addition a vertical traction of $-5\,\text{MPa}$ is applied along the horizontal top edge, and a horizontal traction of $-500y\,\text{kPa}$ is applied along the rightmost vertical edge. All other data is as before (e.g. $k = 1$ and $T = 1$).

The results in Table 4.4 are for uniform refinements with $h = 0.4, 0.2, 0.1, \ldots$, while those in Table 4.5 are for adapted solutions with $\mathsf{TOL} = 32, 24, 16, 8$ (and where we start with an initial mesh of $h = 0.1$). Some of the meshes are shown in Figure 4.4, along with a plot of the shear stress surface.

Table 4.4: L-shaped lever arm: uniform refinements ($k = 1.0$; $h = 0.4, 0.2, 0.1, \ldots$).

#elements	$\|U\|_{L_\infty(0,1;H)}$	$S(1)\mathcal{E}_\Omega(1;U)$
11	2.566(+02)	3.472313(+01)
50	3.080(+02)	4.257241(+01)
204	3.543(+02)	3.285800(+01)
890	3.693(+02)	2.215892(+01)
3744	3.764(+02)	1.080548(+01)
15208	3.781(+02)	7.078156(+00)

Here it requires over 15,000 elements to achieve a similar estimated error as the adaptive solution produces with only 6460 elements.

Other examples given in the same format are contained in [26]. For more details on the error estimates we refer to [22, 23, 26], and the summary paper [25]

4.7 CONCLUDING REMARKS

The procedure described above for adaptive mesh refinement appears to work well, even though $\mathcal{E}_\Omega$ is contaminated with time discretization error associated with the traction

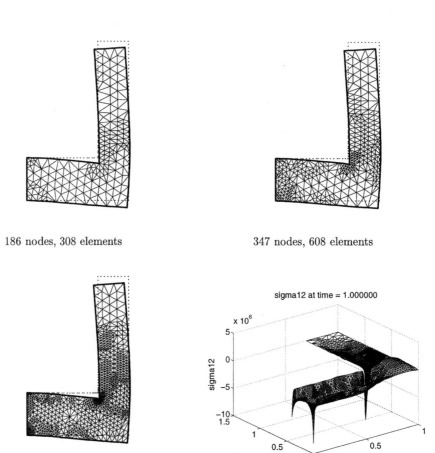

186 nodes, 308 elements   347 nodes, 608 elements

836 nodes, 1542 elements

Figure 4.4: Adapted meshes for the L-shaped lever arm with TOL = 32, 24 and 16 (initial mesh: $h = 0.1$). Also shown is a plot of the shear stress surface for TOL = 8.

Table 4.5: L-shaped lever arm: adaptive solutions ($k = 1.0$).

TOL	#elements	$\|U\|_{L_\infty(0,1;H)}$	$S(1)\mathcal{E}_\Omega(1;U)$
32	308	3.642(+02)	2.733388(+01)
24	608	3.715(+02)	2.074337(+01)
16	1542	3.757(+02)	1.441982(+01)
8	6460	3.782(+02)	7.942375(+00)

boundary condition jump $g - \tilde{g}$ (see Table 4.2). The next step in the development of this work is expected to be concerned with adaptive time step control, and the elimination of the troublesome $\mathcal{E}_V$—thus allowing for mesh de-refinement.

In addition to these mathematical issues, there is a need for a more complete data base against which new constitutive theories can be evaluated. The objective being to evaluate constitutive theories in multiple modes of deformation. The same model should be valid for relaxation, creep, and dynamics. Specifically, an example data base would fill a multidimensional space with the following coordinates: stress, strain (including tension, shear, and biaxial), strain rates for saw tooth cyclic tests, frequencies for harmonic cyclic tests, and temperature. It is expected that close coordination with chemists and polymer physicists will be needed to determine such a constitutive model for continuum mechanics.

ACKNOWLEDGEMENT

The research summarized herein has been sponsored in part by the United States Army through its European Research Office (contract number N68171-97-M-5763), and in part by the Brunel University Research Committee: 96–97 QR funding scheme.

REFERENCES

[1] K. Eriksson, D. Estep, P. Hansbo and C. Johnson (1995). Introduction to adaptive methods for differential equations. *Acta Numerica*, pages 105—158. Cambridge University Press.

[2] Kenneth Eriksson and Claes Johnson (1991). Adaptive finite element methods for parabolic problems. I: a linear model problem. *SIAM J. Numer. Anal.*, 28:43—77.

[3] S. Govindjee and S. Reece (1997). A presentation and comparison of two large deformation viscoelasticity models. *ASME Journal of Engineering Materials and Technology*, 119:251—255.

[4] S. Govindjee and S. Reece (1998). A theory of finite viscoelasticity and numerical aspects. *Int. J. Solids Structures*, 35:3455—3482.

[5] S. Govindjee and J. Simo (1991). A micro-mechanically based continuum damage model for carbon black-filled rubbers incorporating mullin's effect. *J. Mech. Phys. Solids*, 39:87—112.

[6] M. S. Green and A. V. Tobolsky (1946). A new approach to theory of relaxing ploymeric media. *J. Chem. Phys.*, 14:80—92.

[7] A. R. Johnson (1999). Modeling viscoelastic materials using internal variables. *The Shock and Vibration Digest*, 31:91—100.

[8] A. R. Johnson and A. Tessler (1997). A viscoelastic high order beam finite element. In J. R. Whiteman (ed.), *The Mathematics of Finite Elements and Applications.* MAFELAP *1996*, pages 333—345. Wiley, Chichester.

[9] A. R. Johnson, A. Tessler and M. Dambach (1997). Dynamics of thick viscoelastic beams. *Journal of Engineering Materials and Technology*, 119:273—278.

[10] C. Johnson (1994). A new paradigm for adaptive finite element methods. In J. R. Whiteman, editor, *The Mathematics of Finite Elements and Applications, highlights.* MAFELAP *1993*, pages 105—120. John Wiley and Sons Ltd., Chichester.

[11] C. Johnson and P. Hansbo (1992). Adaptive finite element methods in computational mechanics. *Comput. Methods Appl. Mech. Engrg.*, 101:143—181.

[12] B. Keren, Y. Partom and Z. Rosenberg (1984). Nonlinear viscoelastic response in two dimensions: numerical modeling and experimental verification. *Polymer Eng. Sci.*, 24:1409—1416.

[13] G. A. Lesieutre (1992). Finite elements for dynamic modeling of uniaxial rods with frequency-dependent material properties. *Int. J. Solids Structures*, 29:1567—1579.

[14] G. A. Lesieutre and E. Bianchini (1995). Time domain modeling of linear viscoelasticity using anelastic displacement fields. *J. Vibration and Acoustics*, 117:424—430.

[15] G. A. Lesieutre and D. L. Mingori (1990). Finite element modeling of frequency-dependent material damping using augmenting thermodynamic fields. *J. Guidance, Control, and Dynamics*, 13:1040—1050.

[16] J. Lubliner (1985). A model of rubber viscoelasticity. *Mech. Res. Communications*, 12:93—99.

[17] Marcus J. Ludwig (1998). *Finite element error estimation and adaptivity for problems of elasticity.* Ph.D. thesis, Brunel University, England. (See `www.brunel.ac.uk/~icsrbicm`).

[18] J. R. Rice (1971). Inelastic constitutive relations for solids: an internal variable theory and its applications to metal plasticity. *J. Mech. Phys. Solids*, 19:433—455.

[19] Simon Shaw, M. K. Warby, J. R. Whiteman, C. Dawson and M. F. Wheeler (1994). Numerical techniques for the treatment of quasistatic viscoelastic stress problems in linear isotropic solids. *Comput. Meth. Appl. Mech. Engrg.*, 118:211—237.

[20] Simon Shaw and J. R. Whiteman (1996). Discontinuous Galerkin method with *a posteriori* $L_p(0, t_i)$ error estimate for second-kind Volterra problems. *Numer. Math.*, 74:361—383.

[21] Simon Shaw and J. R. Whiteman (1998). Negative norm error control for second-kind convolution Volterra equations. To appear in *Numer. Math*; BICOM Tech. Rep. 98/6, see www.brunel.ac.uk/~icsrbicm.

[22] Simon Shaw and J. R. Whiteman (1998). Numerical solution of linear quasistatic hereditary viscoelasticity problems I: *a priori* estimates. BICOM Tech. Rep. 98/2, see www.brunel.ac.uk/~icsrbicm.

[23] Simon Shaw and J. R. Whiteman (1998). Numerical solution of linear quasistatic hereditary viscoelasticity problems II: *a posteriori* estimates. BICOM Tech. Rep. 98/3, see www.brunel.ac.uk/~icsrbicm.

[24] Simon Shaw and J. R. Whiteman (1998). Optimal long-time $L_p(0, T)$ data stability and semidiscrete error estimates for the Volterra formulation of the linear quasistatic viscoelasticity problem. Submitted to *Numer. Math*; BICOM Tech. Rep. 98/7 see: www.brunel.ac.uk/~icsrbicm.

[25] Simon Shaw and J. R. Whiteman (1999). Numerical solution of linear quasistatic hereditary viscoelasticity problems. Submitted to SIAM J. Numer. Anal.

[26] Simon Shaw and J. R. Whiteman (1999). Robust adaptive finite element schemes for viscoelastic solid deformation: an investigative study. Technical report, BICOM, Brunel University, Uxbridge, England. TR99/1 (US Army ERO Seed Project report).

[27] J. C. Simo (1987). On a fully three-dimensional finite-strain viscoelastic damage model: formulation and computational aspects. *Comput. Meth. Appl. Mech. Engrg.*, 60:153—173.

5   A VISCOELASTIC HYBRID SHELL FINITE ELEMENT

Arthur R. Johnson
Vehicle Technology Directorate, MS 240
Army Research Laboratory,
NASA Langley Research Center
Hampton, VA 23681-0001

ABSTRACT

An elastic large displacement thick-shell hybrid finite element is modified to allow for the calculation of viscoelastic stresses. Internal strain variables are introduced at the element's stress nodes and are employed to construct a viscous material model. First order ordinary differential equations relate the internal strain variables to the corresponding elastic strains at the stress nodes. The viscous stresses are computed from the internal strain variables using viscous moduli which are a fraction of the elastic moduli. The energy dissipated by the action of the viscous stresses is included in the mixed variational functional. Nonlinear quasistatic viscous equilibrium equations are then obtained. Previously developed Taylor expansions of the equilibrium equations are modified to include the viscous terms. A predictor-corrector time marching solution algorithm is employed to solve the algebraic-differential equations. The viscous shell element is employed to numerically simulate a stair-step loading and unloading of an aircraft tyre in contact with a frictionless surface.

Key words. Viscoelasticity, Hybrid Elements, Tyre Models

5.1   INTRODUCTION

Aircraft tyres are composite structures manufactured with viscoelastic materials such as carbon black filled rubber and nylon cords. When loaded, tyres experience large deflections and moderately large strains [2]. Finite element models of tyres typically employ either two-dimensional thick shell or three-dimensional solid elements [3, 15]. Elastic finite element shell models for tyres have been used to predict the shape of tyre footprints as a function of loading [10, 13, 14, 18, 19, 20]. Elastic models do not include the viscoelastic nature of the tyre which can have a significant effect on load-displacement curves. In this paper the quasistatic viscoelastic loading and unloading of an aircraft tyre is numerically simulated. An experimental effort is reported elsewhere [11].

In several previous papers viscoelastic constitutive models have been utilized to determine the dynamic deformations of tyres. The following references are provided as a starting point for readers interested in obtaining details about other viscoelastic finite element models for tyres. Padovan, et al. [9, 12, 16] performed an extensive study in which a finite element algorithm was developed for rolling tyres. Padovan's model included the effects of large deformations and contact. It also employed fractional derivatives to model the viscoelastic effects. The tyre models made by Oden et al. [3, 15] also included the effects of large deformations and contact. However, Oden's model employed the history integral formulation for the viscoelastic effects.

In this paper, internal strain variables are employed to convert an elastic hybrid shell element [14] into a viscous mixed shell element. The model is developed as follows. Internal strain variables are introduced at the stress nodes of the mixed element. First order differential equations relate the internal variables to the physical strain variables. The equations represent a Maxwell solid [4, 5, 6, 7, 21]. Viscous stresses are determined from the internal strains by using material parameters referred to as viscous moduli. An expression for the energy dissipated during deformation is computed from the viscous stresses. This is accomplished by employing the finite element interpolations that are used to compute the stresses from the strains in the elastic version of the shell element. The dissipation energy functional is added to the mixed variational statement for the elastic problem. Nonlinear algebraic equilibrium equations are determined and are numerically solved, simultaneously, with the internal variable differential equations. The numerical predictor-corrector solution procedure employs the Newton-Raphson method for the nonlinear algebraic equations and the trapezoidal method for the differential equations. The tangent matrix required in the Newton-Raphson scheme is a modified version of the previously determined [14, 18, 19, 20] tangent matrix for the nonlinear elastic problem.

At the end of the paper, the viscous shell element is employed in a computational simulation of a stair-step loading and unloading of an aircraft tyre.

5.2   VISCOELASTIC MIXED SHELL ELEMENT

An elastic shell element capable of modelling geometrically nonlinear deformations of thick laminated composites was developed by Noor, et al.,[10, 13, 14, 18, 19, 20]. Figure 5.1 shows the physical variables employed to describe the energy in the deformed shell. The elastic finite element has nine displacement nodes with five variables at each node, and four stress nodes with eight variables at each node, see Figure 5.2. The constitutive model [18, 19, 20] in the shell generalized coordinates is abbreviated as

$$\{\sigma\} = [C]\{\epsilon\} \tag{5.1}$$

where

$$\{\sigma\} = (N_s, N_\theta, N_{s\theta}, M_s, M_\theta, M_{s\theta}, Q_s, Q_\theta)^T$$

is a vector of stress variables,

$$\{\epsilon\} = (N_s, N_\theta, N_{s\theta}, M_s, M_\theta, M_{s\theta}, Q_s, Q_\theta)^T$$

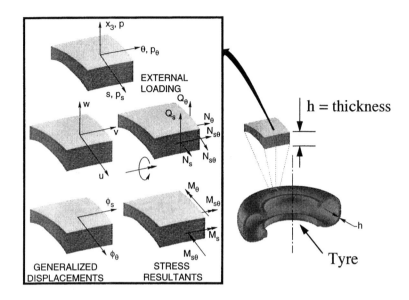

Figure 5.1: Shell displacement and force variables.

is a vector of the Sanders-Budiansky nonlinear strains[1, 17], and $[C]$ is a matrix of elastic stiffness constants. The elastic element employs the Hellinger-Reissner mixed variational principle which is constructed as follows. The complementary form of the energy is integrated over the volume of the shell and the total work done by external forces is subtracted. This results in Π_{HR} which is expressed as follows.

$$\Pi_{HR} = \int_{\Omega} \left(\{\sigma\}^T \{\varepsilon\} - \frac{1}{2} \{\sigma\}^T [F] \{\sigma\} \right) d\Omega - W \tag{5.2}$$

where Ω is the volume of the shell, $[F]$ is a flexibility matrix, and W is the work done by external forces.

Next, the energy functional, Π_{HR}, is discretized by the finite element method [14]. At the element level, the displacements and stress resultants are approximated by employing interpolation functions with the nodal values shown in Figure 5.2. The Sanders-Budiansky nonlinear strains are computed and substituted into equation (5.2) above. After the volume integration is performed for an element, the Hellinger-Reissner variational expression is given in short-hand notation as follows.

$$\Pi_{HR}^{elt} \left(\{x\}, \{h\} \right) = V - U^C - W \tag{5.3}$$

where

$$V = \int_{\Omega_{elt}} \{\sigma\}^T \{\varepsilon\} \, d\Omega \equiv \left\{ \widehat{h} \right\}^T \left(\left[\widehat{S}_{lx} \right] + \frac{1}{2} \left[\widehat{M}_{nlxx} \right] \right)$$

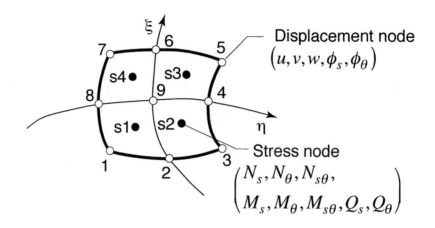

Figure 5.2: Shell element displacement and stress variables.

$$\equiv \{\widehat{h}\}^T \{\widehat{\varepsilon}\},$$

$$U^C = \int_{\Omega_{elt}} \frac{1}{2} \{\sigma\}^T [F] \{\sigma\} \, d\Omega \equiv \frac{1}{2} \{\widehat{h}\}^T \left[\widehat{F}\right] \{\widehat{h}\},$$

$W = \{\widehat{x}\}^T \{\widehat{p}\} = \{\widehat{p}\}^T \{\widehat{x}\}$, $\{\widehat{x}\}$ and $\{\widehat{\varepsilon}\}$ are vectors containing the element's nodal displacements and strains, $\{h\}$ is an element level vector, $\left[\widehat{S}_{lx}\right]$ and $\left[\widehat{M}_{nlxx}\right]$ are operators that produce the linear and nonlinear contributions to $\{\widehat{\varepsilon}\}$ from the nodal displacements, $\left[\widehat{F}\right]$ is an element level flexibility matrix, and $\{\widehat{p}\}$ represents the consistent applied load vector.

Internal strain variables are employed herein to modify the above formulation making it applicable to a Maxwell type viscoelastic material [6, 7]. In vector form, internal strain fields, $\{\varepsilon_{jv}\}$, and matrices of viscous stiffness constants, $[C_{jv}]$, are introduced within the element. The viscous stress vector at a point in the element is computed as follows.

$$\{\sigma_v\} = \sum_{j=1}^{n} [C_{jv}] \{\varepsilon_{jv}\} \tag{5.4}$$

The total value of the stress vector, elastic plus viscous, at a point is $\{\sigma_t\} = \{\sigma\} + \{\sigma_v\}$. Next, following the computation of the elastic potential, V, in equation (5.3), the energy dissipated by the viscous stresses, Q, throughout the element is computed as follows.

$$Q = \int_{\Omega_{elt}} \{\sigma_v\}^T \{\varepsilon\} \, d\Omega \equiv \{\widehat{h_v}\}^T \{\widehat{\varepsilon}\} = \{\widehat{h_v}\}^T \left([\widehat{S}_{lx}] + \frac{1}{2}[\widehat{M}_{nlxx}] \right) \tag{5.5}$$

where $\{\widehat{h_v}\}$ is an element level vector. The mixed variational functional for the viscous element, $\Pi_{HR}^{elt,v}$, is given by

$$\Pi_{HR}^{elt,v} = V + Q - U^C - W \tag{5.6}$$

The model is completed by relating the rate of change of the internal strain variables to the physical strain variables. Here we employ a simple form of the Maxwell solid theory by requiring the following differential equations to be satisfied [6, 7, 4].

$$\frac{d\{\widehat{\varepsilon}_{jv}\}}{dt} + \frac{\{\widehat{\varepsilon}_{jv}\}}{\tau_j} = \frac{d\{\widehat{\varepsilon}\}}{dt} \tag{5.7}$$

Equations (5.4) and (5.7) determine the viscous stresses for a time dependent deformation of the element, $\{\widehat{x}(t)\}$. Note, an advantage of this algorithm is that a variety of viscoelastic models can be employed by simply changing equation (5.7).

At each instant of time, the element equilibrium equations are given by the first variation of, equation (5.6). The equilibrium equations are

$$\{f_{\widehat{h}}(\{\widehat{x}\}, \{\widehat{h}\})\} = \left([\widehat{S}_{lx}] + \frac{1}{2}[\widehat{M}_{nlxx}] \right) - [\widehat{F}]\{\widehat{h}\} = \{0\} \tag{5.8}$$

and

$$\{f_{\widehat{x}}(\{\widehat{x}\}, \{\widehat{h}\})\} = \left([\widehat{S}_l] + \frac{1}{2}[\widehat{M}_{nlx}] \right)^T (\{\widehat{h}\} + \{\widehat{h_v}\}) - \{\widehat{p}\} = \{0\} \tag{5.9}$$

where $[\widehat{S}_l]$ and $[\widehat{M}_{nlx}]$ are the derivatives of the operators $[\widehat{S}_{lx}]$ and $[\widehat{M}_{nlxx}]$ with respect to the element displacement variables, $\{\widehat{x}\}$. Equations (5.8) and (5.9) are assembled by standard methods to obtain the global equilibrium equations. The global equations are then solved simultaneously with equation (5.7) (for all elements.)

The Taylor expansion of equations (5.8) and (5.9) produces the element's tangent matrix. The resulting element level Newton-Raphson equations for the increments of the variables $\Delta\{\widehat{x}\}$ and $\Delta\{\widehat{h}\}$ are given below.

$$\begin{bmatrix} -[\widehat{F}] & ([\widehat{S}_l] + \frac{1}{2}[\widehat{M}_{nlx}]) \\ ([\widehat{S}_l] + \frac{1}{2}[\widehat{M}_{nlx}])^T & [\widehat{M}_{nl}](\{\widehat{h}\} + \{\widehat{h_v}\}) \end{bmatrix} \left\{ \begin{matrix} \Delta\{\widehat{h}\} \\ \Delta\{\widehat{x}\} \end{matrix} \right\} = \left\{ \begin{matrix} \{f_{\widehat{h}}\} \\ \{f_{\widehat{x}}\} \end{matrix} \right\} \tag{5.10}$$

where $[\widehat{M}_{nl}]$ is the second derivative of the operator $[\widehat{M}_{nlxx}]$ with respect to the nodal displacements. Equations (5.10) are assembled for all elements and solved to provide estimates of the variables increments across a time step. The new elastic strains at the end of the time step are computed at all stress nodes. The internal strain variables are then estimated at the end of the time step by employing the trapezoidal method for equation (5.7). Next, global equilibrium is checked. If equilibrium is not satisfied the process is repeated. When equilibrium is satisfied the required output is computed and the time is advanced.

5.3 TYRE STAIR-STEP LOADING SIMULATION

The aircraft tyre modelled below is a 32 x 8.8, type VII, bias-ply Shuttle nose-gear tyre which has a 20-ply rated carcass and a maximum speed rating of 217 knots [11, 18, 19, 20]. The tread pattern consists of three circumferential grooves and the rated inflation pressure is 320 psi.

The details of the tyre's elastic material model are described by Tanner [18]. The tyre is a cord-rubber composite and was treated as a laminated material. It was divided into seven regions in the direction of the meridian (from the center of the tread region to the rim.) Tyre thickness, properties of the plies, etc. were measured and tabulated. Elastic constants were computed by the law of mixtures to obtain linear orthotropic stress-strain constitutive models for each layer. These properties were transformed to the shell coordinate system and integrated through the thickness of the shell elements.

The viscous material model employed was determined by least-square fitting tyre load-relaxation data [8] to a Prony series of the form

$$f_v\left(t\right) = \left(\sum_{j=1}^{3} \alpha_j \, e^{\frac{-t}{\tau_j}}\right) f_v\left(0\right) \tag{5.11}$$

where $f_v\left(t\right)$ is the relaxing component of the load, $f_v\left(0\right)$ is the initial load, τ_{jv} are the constants in equations (5.7), and α_j are factors of the products $[C_{jv}]\{\varepsilon_{jv}\}$ so that equation (5.4) becomes

$$\{\sigma_v\} = \sum_{j=1}^{n} \alpha_j \, [C] \, \{\varepsilon_{jv}\} \tag{5.12}$$

The values of the constants found by Johnson et al. [8] and used here are

$$\{(\alpha_j, \tau_j)\}_{j=1}^{3} = \{(0.01836, \ 10), (0.01630, \ 100), (0.03650, \ 1000)\}.$$

The tyre's finite element mesh and a sketch of the frictionless loading platform are shown in Figure 5.3. The mesh is similar to the "Model 1" mesh employed by Tanner [18]. The elastic model has 540 elements and 28,565 degrees of freedom (not including the Lagrange multipliers used for points that come into contact.) An additional 103,680 internal variables were added to program the solution algorithm for the material model described above. Computed elastic and viscoelastic load-displacement curves, obtained by enforcing a stair-step tyre rim displacement are shown in Figure 5.4. The finite element load-displacement hysteresis loop is shown in Figure 5.5.

5.4 CONCLUDING REMARKS

An algorithm for converting elastic structural elements based on the Hellinger-Reissner mixed variational principle to viscoelastic structural elements was presented. The thirteen node large displacement thick-shell element derived by Noor and Hartley [14] was employed to describe the algorithm. A finite element tyre model based on this shell element, and used by Tanner [18, 19, 20] to analyze tyre footprints, was modified so that the

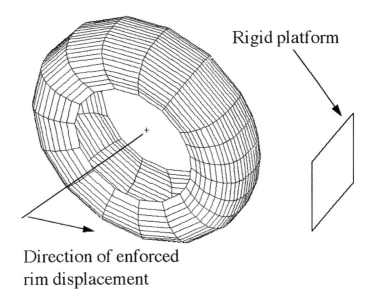

Rigid platform

Direction of enforced
rim displacement

Figure 5.3: Tyre finite element model.

tyre material would represent a Maxwell solid. A stair-step tyre loading and unloading was numerically simulated. The new computational algorithm functioned successfully.

ACKNOWLEDGMENT

The author would like to thank Ms. Jeanne M. Peters of the Center for Advanced Computational Technology, University of Virginia, NASA Langley Research Center for assisting with the modifications of the finite element code. Also, the author is grateful to the United States Army Research Standardization Group, Europe for supporting his participation in MAFELAP99 and to the Army Research Laboratory at the NASA Langley Research Center for supporting this research.

REFERENCES

[1] B. Budiansky (1968). Notes on Nonlinear Shell Theory. *J. Appl. Mech.*, 35, 2: 393–401.

[2] S. K. Clark (1981), Mechanics of Pneumatic Tires. *US Government Printing Office*, 475–540.

[3] L. O. Faria, J. M. Bass, and J. T. Oden (1989). A Three-Dimensional Rolling Contact

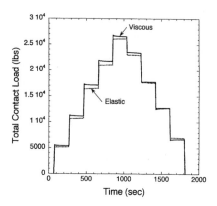

Figure 5.4: Computed viscous and elastic tyre loads.

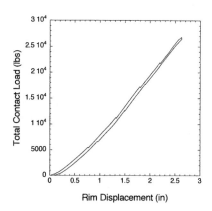

Figure 5.5: Finite element hysteresis curve.

Model for a Reinforced Rubber Tire. *Tire Science and Technology, TSTCA*, 17, 3: 217–233.

[4] A. R. Johnson (1999). Modeling Viscoelastic Materials Using Internal Variables. *The Shock and Vibration Digest*, 31, 2: 91–100.

[5] A. R. Johnson and R. G. Stacer (1993). Rubber viscoelasticity using the physically constrained system's stretches as internal variables. *ACS, Rubber Chemistry and Technology*, 66, 4: 567–577.

[6] A. R. Johnson and A. Tessler (1997). In J. R. Whiteman (ed.), A Viscoelastic Higher-Order Beam Finite Element. *The Mathematics of Finite Elements and Applications, Highlights 1996, Wiley, Chichester*, 333–345.

[7] A. R. Johnson, A. Tessler, and M. L. Dambach (1997). Dynamics of Thick Viscoelastic Beams. *ASME J. Engng and Matls Tech*, 119: 273–278.

[8] A. R. Johnson, J. A. Tanner, and A. J. Mason (1999). Quasi-Static Viscoelastic Finite Element Model of an Aircraft Tire. *NASA/TM-1999-209141*.

[9] R. Kennedy and J. Padovan (1987). Finite Element Analysis of Steady and Transiently Moving/Rolling Nonlinear Viscoelastic Structure - II, Shell and Three Dimensional Simulations. *Computers and Structures*, 27, 2: 259–273.

[10] K. O. Kim, J. A. Tanner, A. K. Noor, and M. P. Robertson (1991). Computational Methods for Frictionless Contact with Application to Space Shuttle Orbitor Nose-Gear Tires. *NASA/TP-1991-3073*.

[11] A. J. Mason, J. A. Tanner, and A. R. Johnson (1997). Quasi-Static Viscoelasticity Loading Measurements of an Aircraft Tire. *NASA/TM-1997-4779, ARL/TR-1402*.

[12] Y. Nakajima and J. Padovan (1987). Finite Element Analysis of Steady and Transiently Moving/Rolling Nonlinear Viscoelastic Structure - III, Impact/Contact Simulations. *Computers and Structures*, 27, 2: 275–286.

[13] A. K. Noor, C. M. Anderson, and J. A. Tanner (1987). Exploiting Symmetries in the Modeling and Analysis of Tires. *NASA/TP-1987-2649*.

[14] A. K. Noor and S. J. Hartley (1977). Nonlinear Shell Analysis via Mixed Isoparametric Elements. *Computers and Structures*, 7: 615–626.

[15] J. T. Oden, T. L. Lin, and J. M. Bass (1988). A Finite Element Analysis of the General Rolling Contact Problem for a Viscoelastic Rubber Cylinder. *Tire Science and Technology, TSTCA*, 16, 1: 18–43.

[16] J. Padovan (1987). Finite Element Analysis of Steady and Transiently Moving/Rolling Nonlinear Viscoelastic Structure - I Theory. *Computers and Structures*, 27, 2: 249–257.

[17] J. L. Sanders (1963). Nonlinear Theories for Thin Shells. *Q. Appl. Math.*, 21, 1: 21–36.

[18] J. A. Tanner (1996). Computational Methods for Frictional Contact with Application to the Space Shuttle Orbiter Nose-Gear Tire, Comparisons of Experimental Measurements and Analytical Predictions. *NASA/TP-1996-3573.*

[19] J. A. Tanner (1996). Computational Methods for Frictional Contact with Application to the Space Shuttle Orbiter Nose-Gear Tire, Development of Frictional Contact Algorithm. *NASA/TP-1996-3574.*

[20] J. A. Tanner, V. J. Martinson, and M. P. Robinson (1994). Static Frictional Contact of the Space Shuttle Nose-Gear Tire. *Tire Science and Technology, TSTCA*, 22, 4: 242–272.

[21] I. M. Ward (1983). *Mechanical Properties of Solid Polymers, Wiley and Sons.*, Chichester.

6   THE DUAL-WEIGHTED-RESIDUAL METHOD FOR ERROR CONTROL AND MESH ADAPTATION IN FINITE ELEMENT METHODS

Rolf Rannacher

Institute for Applied Mathematics,
University of Heidelberg,
INF 293, D-69120 heidelberg, Germany

ABSTRACT

We present a general concept of *a posteriori* error estimation and mesh size control in Galerkin finite element discretizations. The method is based on weighted *a posteriori* error estimates which are obtained by global duality arguments. The computed adjoint solutions contain quantitative information about the global dependence of the error quantity on the local cell residuals and can be used for guiding the mesh refinement process. This approach is particularly useful in cases where the local residuals do not control the local error, e.g., in the presence of global pollution effects, strongly varying coefficients and strongly coupled multi-physics processes. This is illustrated by some examples from fluid mechanics and astrophysics.

Key words. Error estimates, *a posteriori*, adaptivity, Navier-Stokes

6.1   THE MOTIVATION

A posteriori estimation of the discretization error in finite element models of complex physical systems has to deal with two major problems:

(i) Global error transport: The local error at some mesh cell K may be significantly affected by residuals at distant cells K' ("pollution effect").

(ii) Interaction of error components: The error in a single solution component may depend on all other error components with unknown coefficients of quite different size.

An effective method for error estimation should be able to capture all these functional dependencies. The effect of the cell residual components $\rho_{K,i}$ on the local error components $e_{K',i}$, at another cell K', is governed by a certain global Green tensor of the continuous problem. Capturing this dependence by numerical evaluation is the general philosophy underlying our approach to error control.

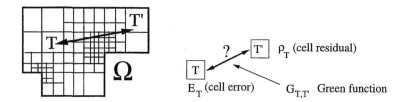

Figure 6.1: Scheme of error propagation

The mechanisms of error propagation can be rather different depending on the characteristics of the differential operator:

- Diffusion terms cause slow isotropic error decay, but global error pollution may occur from local irregularities.
- Advection terms do not allow for error decay in the transport direction, but errors decay exponentially in the crosswind direction.
- Reaction terms cause isotropic exponential error decay, but "stiff" behavior may occur in the coupling of error components.

For models in which all these mechanisms interact it is almost impossible to determine the complex error interaction by analytical means, but rather one has to be aided by computation. This automatically leads to a feed-back process in which error estimation and mesh adaptation go hand-in-hand leading to the most economical meshes for computing the quantity of interest. Below, we will present some examples from fluid flow and from radiation transfer for which such an approach seems indispensible.

6.2 THE CONCEPT OF THE "DUAL-WEIGHTED-RESIDUAL METHOD"

We begin by considering a simple linear algebra model, in order to exemplify our approach. Let $A \in \mathbb{R}^{mn \times mn}$ be a large square block-matrix and $b \in \mathbb{R}^{mn}$ be a corresponding block-vector, such that $Ax = b$, where n represents spatial distance and m local system dimension. The matrix A is assumed to be regular but may have a very complicated structure (indefinite, nonsymmetric, etc.). Further, let $\tilde{A} \in \mathbb{R}^{mn \times mn}$ and $\tilde{b} \in \mathbb{R}^{mn}$ be approximations satisfying $\tilde{A}\tilde{x} = \tilde{b}$. Suppose that we are interested in estimating only the error with respect to a certain (linear) functional of the solution

$$J(x) - J(\tilde{x}) = J(e) = \langle e, \psi \rangle,$$

where $e := x - \tilde{x}$, and $\psi \in \mathbb{R}^{mn}$ is fixed. In order to estimate this error, we use a "duality argument". Let $z \in \mathbb{R}^{mn}$ be the solution of the "dual problem" $A^* z = \psi$. Then,

$$J(e) = \langle e, A^* z \rangle = \langle Ae, z \rangle) = \langle b - A\tilde{x}, z \rangle \qquad (6.1)$$

holds which results in the *a posteriori* error bound

$$|J(e)| \leq \sum_{i,j=1}^{m,n} |\rho_{ij}| |z_{ij}|, \qquad (6.2)$$

with the local residuals $\rho_{ij} := (b - A\tilde{x})_{ij}$. Here, the terms $|z_{ij}|$ act as weights which describe the influence of the residuals ρ_{ij} on the error quantity $J(e)$. This argument can be extended to nonlinear equations. Let $F : \mathbb{R}^{mn} \to \mathbb{R}^{mn}$ be a (differentiable) vector function and solve $F(x) = b$. The perturbed equation is $\tilde{F}(\tilde{x}) = \tilde{b}$. Then, the residual $\rho := b - F(\tilde{x})$ satisfies

$$\rho = F(x) - F(\tilde{x}) = \int_0^1 F'(\tilde{x} + se)e\, ds =: L(x, \tilde{x})e,$$

with the Jacobian F'. The linear operator $L(x, \tilde{x}) : \mathbb{R}^{mn} \to \mathbb{R}^{mn}$ depends on the (unknown) solution x. Then, with the solution $z \in \mathbb{R}^{mn}$ of the corresponding "dual problem" $L(x, \tilde{x})^* z = \psi$, it follows that

$$J(e) = \langle e, L(x, \tilde{x})^* z \rangle = \langle L(x, \tilde{x})e, z \rangle = \langle \rho, z \rangle.$$

This again results in an error bound of the form (6.2), this time with the nonlinear residuals $\rho_{ij} := \tilde{b}_{ij} - F_{ij}(\tilde{x})$. Below, we will use this duality technique for generating *a posteriori* error estimates in Galerkin finite element methods for rather general differential systems. One of the main difficulties will be the computation of sufficiently accurate approximations to the dual solution z.

6.2.1 *A paradigm for* a posteriori *error estimation*

The approach to residual–based error estimation described above for algebraic systems can be extended to general nonlinear variational problems. We outline the underlying concept in an abstract setting following the general paradigm introduced in Johnson [17] and in Eriksson *et al.* [11]. For a detailed discussion of various aspects of this approach, we refer to [8] and [9]; a survey with applications to various problems in mechanics and physics has been given in [20].

Let V be a Hilbert space with inner product $(\cdot, \cdot)$ and corresponding norm $\| \cdot \|$, $A(\cdot; \cdot)$ a continuous semi-linear form and $F(\cdot)$ a continuous functional defined on V. We seek a solution $u \in V$ to the abstract variational problem

$$A(u; \phi) = F(\phi) \qquad \forall \phi \in V. \qquad (6.3)$$

This problem is approximated by a Galerkin method using a sequence of finite dimensional subspaces $V_h \subset V$ parameterized by a discretization parameter h. The discrete problems seek $u_h \in V_h$ satisfying

$$A(u_h; \phi_h) = F(\phi_h) \qquad \forall \phi_h \in V_h. \qquad (6.4)$$

The key feature of this approximation is the "Galerkin orthogonality" which in this non-linear case is expressed as

$$A(u; \phi_h) - A(u_h; \phi_h) = 0, \quad \phi_h \in V_h.$$

By elementary calculus, it follows that

$$A(u; \phi_h) - A(u_h; \phi_h) = \int_0^1 A'(su + (1-s)u_h; e, \phi_h) \, ds,$$

with $A'(v; \cdot, \cdot)$ denoting the tangent form of $A(\cdot; \cdot)$ at some $v \in V$. This leads us to introduce the bilinear form

$$L(u, u_h; \phi, \psi) := \int_0^1 A'(su + (1-s)u_h; \phi, \psi) \, ds,$$

which depends on the solutions u as well as u_h. Then, denoting the error by $e = u - u_h$, we have that

$$L(u, u_h; e, \phi_h) = \int_0^1 A'(su + (1-s)u_h; e, \phi_h) \, ds = A(u; \phi_h) - A(u_h; \phi_h) = 0, \quad (6.5)$$

for any $\phi_h \in V_h$. Suppose that the quantity $J(u)$ has to be computed, where $J(\cdot)$ is a linear functional defined on V. For representing the error $J(e)$, we use the solution $z \in V$ of the *dual problem*

$$L(u, u_h; \phi, z) = J(\phi) \qquad \forall \phi \in V. \quad (6.6)$$

Assuming that this problem is solvable and using the Galerkin orthogonality (6.5), we obtain the error representation

$$J(e) = L(u, u_h; e, z - z_h) = F(z - z_h) - A(u_h; z - z_h), \quad (6.7)$$

with any approximation $z_h \in V_h$. Since the bilinear form $L(u, u_h; \cdot, \cdot)$ contains the unknown solution u in its coefficient, the evaluation of (6.7) requires approximation. The simplest way is to replace u by u_h, yielding a perturbed dual problem

$$L(u_h, u_h; \phi, \tilde{z}) = J(\phi) \qquad \forall \phi \in V. \quad (6.8)$$

Controlling the effect of this perturbation on the accuracy of the resulting error estimator may be a delicate task and depends strongly on the particular problem considered. Our own experience with several different types of problems (including the Navier-Stokes equations) indicates that this problem is less critical as long as the continuous solution is stable. The crucial problem is the numerical computation of the perturbed dual solution $\tilde{z}$ by solving a discretized dual problem

$$L(u_h, u_h; \phi, \tilde{z}_h) = J(\phi) \qquad \forall \phi \in V_h. \quad (6.9)$$

This results in a practically useful error estimator $J(e) \approx \tilde{\eta}(u_h)$. We remark that the bilinear form $L(u_h, u_h; \cdot, \cdot)$ in (6.9) is identical to $A'(u_h; \cdot, \cdot)$.

6.2.2 Evaluation of the error estimates

The goal is to evaluate the right hand side of (6.7) numerically, in order to get a criterion for the local adjustment of the discretization. For the further discussion, we need to become more specific about the setting of the above problem. Let the variational problem (6.3) originate from a semi-linear second-order partial differential equation of the from

$$A(u) = -\sum_{i,j=1}^{2} \partial_i\{a_{ij}\partial_j u\} + \sum_{j=1}^{d} a_{0j}\partial_j u + a_{00}u, \qquad (6.10)$$

on a bounded domain $\Omega \subset \mathbb{R}^2$, with coefficients $a_{ij}(\cdot, u)$, and homogeneous Dirichlet boundary conditions, $u|_{\partial\Omega} = 0$. Hence, the natural solution space is $V = H_0^1(\Omega)$. The following discussion assumes (6.10) to be a scalar equation, but everything carries directly over to systems. Accordingly, the semi-linear form $A(\cdot; \cdot)$ and its linearization $A'(\cdot; \cdot, \cdot)$ have their natural meaning. We assume discretization by standard low-order conforming (linear or bilinear) finite elements on quasi-regular meshes $\mathcal{T}_h = \{K\}$, as described, e.g., in [16]. The local mesh width is denoted by $h_K = \text{diam}(K)$. In this setting, the error representation (6.7) has the following concrete form:

$$J(e) = \sum_{K\in\mathcal{T}_h} \left\{ (f + A(u_h), z - z_h)_K - \tfrac{1}{2}([\partial_n^A u_h], z - z_h)_{\partial K} \right\}, \qquad (6.11)$$

where $\partial_n^A = \sum_{i,j=1}^{d} n_i a_{ij}(\cdot, u_h)\,\partial_j$ denotes the derivative in the co–normal direction along the cell boundary ∂K, while $[\cdot]$ indicates the corresponding jump across ∂K. To evaluate this formula, one may replace the unknown solution u in the bilinear form $L(u, u_h; \cdot, \cdot)$ by the computed approximation u_h, and solve the corresponding perturbed dual problem by the same method as used in computing u_h, yielding an approximation $\tilde{z}_h \in V_h$ to the exact dual solution z,

$$\tilde{z}_h \in V_h: \quad L(u_h, u_h; \phi, \tilde{z}_h) = J(\phi) \qquad \forall\phi \in V_h. \qquad (6.12)$$

However, the use of the same meshes for computing the primal and dual solutions is by no means obligatory. In fact, for transport-oriented problems it may be advisable to compute the dual solution on a different mesh; see [19] and [13] for examples.

Then, one may try to evaluate (6.11) directly by replacing the local errors $z - z_h$ by $\tilde{z}_h - z_h$, where $\tilde{z}_h$ is some (possibly higher-order) approximation of z and z_h its interpolation in V_h. Alternatively, one may instead convert (6.11) into an estimate,

$$|J(e)| \leq \sum_{K\in\mathcal{T}_h} \alpha_K\{\rho_K^{(1)}\omega_K^{(1)} + \rho_K^{(2)}\omega_K^{(2)}\}, \qquad (6.13)$$

with parameters $\alpha_K = h_K^4$, and residuals and weights defined by

$$\rho_K^{(1)} := h_K^{-1}\|f + A(u_h)\|_K, \qquad \rho_K^{(2)} := \tfrac{1}{2}h_K^{-3/2}\|[\partial_n^A u_h]\|_{\partial K},$$
$$\omega_K^{(1)} := h_K^{-3}\|z - z_h\|_K, \qquad \omega_K^{(2)} := h_K^{-5/2}\|z - z_h\|_{\partial K}.$$

Having computed an approximation $\tilde{z}_h \in V_h$ to the exact dual solution z, the weights $\omega_K^{(k)}$ can be determined numerically in different ways:

- Approximation by second-order difference quotients of the discrete dual solution $\tilde{z}_h \in V_h$, e.g.,

$$\omega_K^{(1)} \leq C_I h_K^{-1} \|\nabla^2 z\|_K \approx C_I |\nabla_h^2 \tilde{z}_h(x_K)|, \tag{6.14}$$

x_K being the center point of K, with an interpolation constant $C_I \sim 0.1$ independent of h_K, where $\nabla^2 z$ is the tensor of second derivatives of z and $\nabla_h^2 \tilde{z}_h$ a suitable difference approximation.

- Computation of a discrete dual solution $\tilde{z}_{h'} \in V_{h'}$ in a richer space $V_{h'} \supset V_h$ (e.g., on a finer mesh or using higher-order finite elements) and setting, e.g.,

$$\omega_K^{(1)} \approx h_K^{-3} \|\tilde{z}_{h'} - I_h \tilde{z}_{h'}\|_K, \tag{6.15}$$

where $I_h \tilde{z}_{h'} \in V_h$ is the generic nodal interpolation.

- Interpolation of the discrete dual solution $\tilde{z}_h \in V_h$ by higher order polynomials on certain cell-patches, e.g., biquadratic interpolation $I_h^{(2)} \tilde{z}_h$:

$$\omega_K^{(1)} \approx h_K^{-3} \|I_h^{(2)} \tilde{z}_h - z_h\|_K. \tag{6.16}$$

The second option is quite expensive and rarely used. Since we normally do not want to spend more time in evaluating the error estimator than for solving the primal problem, we recommend the first or the third option. Notice that the third one does not involve an interpolation constant which needs to be specified. Our experience is that the use of biquadratic interpolation on patches of four quadrilaterals is more accurate than using the finite difference approximation (6.14).

One may try to further improve the quality of the error estimate by solving local (patchwise) defect equations, either Dirichlet problems (à la Babuska/Miller) or Neumann problems (à la Bank/Weiser); for details we refer to [9]. References for these approaches are Verfürth [23] and Ainsworth, Oden [1].

6.2.3   Strategies for mesh adaptation

We use the notation introduced above: Let u be the solution of the variational problem posed on a 2-dimensional domain Ω and u_h its piecewise linear (or bilinear) finite element approximation. Further, $e = u - u_h$ is the discretization error and $J(\cdot)$ a linear error functional for measuring e. We suppose that there is an a posteriori error estimate of the form

$$|J(e)| \leq \eta := \sum_{K \in \mathcal{T}_h} \eta_K, \tag{6.17}$$

with the parameters $\alpha_K := h_K^4$ and the cell-error indicators $\eta_K := \alpha_K \{\rho_K^{(1)} \omega_K^{(1)} + ...\}$. The mesh design strategies are oriented towards a prescribed tolerance TOL for the error quantity $J(e)$ and the number of mesh cells N which measures the complexity of the computational model. Usually the admissible complexity is constrained by some maximum value $N_{\max}$.

There are various strategies for organizing a mesh adaptation process on the basis of the *a posteriori* error estimate (6.17).

- *Error balancing strategy:* Cycle through the mesh and equilibrate the local error indicators,

$$\eta_K \approx \frac{TOL}{N} \quad \Rightarrow \quad \eta \approx TOL. \tag{6.18}$$

This process requires iteration with respect to the number of mesh cells N.

- *Fixed fraction strategy:* Order cells according to the size of η_K and refine a certain percentage (say 30%) of cells with largest η_K (or those which make up 30% of the estimator value η) and coarsen those cells with smallest η_K. By this strategy, we may achieve a prescribed rate of increase of N (or keep it constant as desirable in nonstationary computations).

- *Mesh optimized strategy:* Use the representation

$$\eta := \sum_{K \in \mathcal{T}_h} \alpha_K \{ \rho_K^{(1)} \omega_K^{(1)} + ... \} \approx \int_\Omega h(x)^2 A(x)\, dx \tag{6.19}$$

for generating a formula for an optimal mesh-size distribution $h_{opt}(x)$.

We want to discuss the "mesh optimization strategy" in more detail. As a side-product, we will also obtain the justification of the indicator equilibration strategy. Let $N_{\max}$ and TOL be prescribed. We assume that for $TOL \to 0$, the cell residuals and the weights approach certain limits, e.g.,

$$\rho_K^{(2)} \approx \tfrac{1}{2} h_K^{-1/2} \| h_K^{-1} [\partial_n^A u_h] \|_{\partial K} \to |D^2 u(x_K)|,$$
$$\omega_K^{(2)} \approx h_K^{-5/2} \| z - I_h z \|_{\partial K} \to |D^2 z(x_K)|.$$

These properties can be proven on uniformly refined meshes by exploiting superconvergence effects, but they still need theoretical justification on locally refined meshes as constructed by the strategies described above. This suggests the relations

$$\eta \approx \int_\Omega h(x)^2 A(x)\, dx, \qquad N = \sum_{K \in \mathcal{T}_h} h_K^2 h_K^{-2} \approx \int_\Omega h(x)^{-2}\, dx. \tag{6.20}$$

Consider the mesh optimization problem

$$\eta \to min!, \quad N \le N_{\max}. \tag{6.21}$$

Applying the usual Lagrangian approach results in

$$\frac{d}{dt} \left[\int_\Omega (h + t\phi)^2 A\, dx + (\lambda + t\mu) \left\{ \int_\Omega (h + t\mu)^{-2}\, dx - N_{\max} \right\} \right]_{t=0} = 0,$$

implying

$$2h(x)A(x) - 2\lambda h(x)^{-3} = 0, \qquad \int_\Omega h(x)^{-2}\, dx - N_{\max} = 0.$$

Consequently,

$$h(x) = \left(\lambda^{-1} A(x)\right)^{-1/4} \quad \Rightarrow \quad \eta \approx h^4 A = \lambda^{-1},$$

and

$$\lambda^{-1/2} \int_{\Omega} A(x)^{1/2} \, dx = N_{\max}, \quad W := \int_{\Omega} A(x)^{1/2} \, dx.$$

From this, we infer a formula for the "optimal" mesh-size distribution,

$$\lambda = \left(\frac{W}{N_{\max}}\right)^2 \quad \Rightarrow \quad h_{opt}(x) = \left(\frac{W}{N_{\max}}\right)^{1/2} A(x)^{-1/4}. \tag{6.22}$$

In an analogous way, we can also treat the optimization problem

$$N \to min!, \quad \eta \leq TOL. \tag{6.23}$$

We note that for "regular" functionals $J(\cdot)$ the quantity W is bounded, e.g.,

$$J(e) = \partial_i e(0) \quad \Rightarrow \quad A(x) \approx |x|^{-3} \quad \Rightarrow \quad W = \int_{\Omega} |x|^{-3/2} \, dx < \infty.$$

Hence, the described optimization approach is justified. However, the evaluation of *hyper-singular* error functionals (e.g., higher derivatives) may require regularization.

6.3   FIRST EXAMPLE: INCOMPRESSIBLE FLUID FLOW

The results in this section are collected from Becker [3], [4], and Becker and Rannacher [9]. We consider a viscous incompressible Newtonian fluid modelled by the stationary Navier-Stokes equations

$$-\nu \Delta v + v \cdot \nabla v + \nabla p = f, \quad \nabla \cdot v = 0, \tag{6.24}$$

in a bounded domain $\Omega \subset \mathbb{R}^2$. In the following, vector functions are also denoted by normal type and no distinction is made in the notation of the corresponding inner products and norms. The unknowns are the velocity v and the pressure p; ν is the normalized viscosity (density $\rho \equiv 1$) and the volume force is assumed as $f \equiv 0$. At the boundary $\partial\Omega$, the usual non-slip condition is imposed along rigid parts together with suitable inflow and free-stream outflow conditions,

$$v|_{\Gamma_{rigid}} = 0, \quad v|_{\Gamma_{in}} = v^{in}, \quad \nu \partial_n v - pn|_{\Gamma_{out}} = 0.$$

As an example, we consider the flow around the cross section of a cylinder in a channel shown in Figure 6.2. This is part of a set of benchmark problems discussed in [22]. Quantities of physical interest are, for example,

$$\text{pressure drop:} \quad J_{\Delta p} = p(a_{front}) - p(a_{back}),$$

$$\text{drag coefficient:} \quad J_{drag} = \frac{2}{\bar{U}^2 D} \int_S n \cdot \sigma(v, p) e_x \, ds,$$

$$\text{lift coefficient:} \quad J_{lift} = \frac{2}{\bar{U}^2 D} \int_S n \cdot \sigma(v, p) e_y \, ds,$$

where S is the surface of the cylinder, D its diameter, $\bar{U}$ the reference velocity, and $\sigma(v,p) = \frac{1}{2}\nu(\nabla v + \nabla v^T) + pI$ the stress force acting on S. In our example, the Reynolds number is $Re = \bar{U}^2 D/\nu = 20$, such that the flow is stationary.

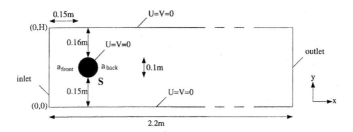

Figure 6.2: Geometry of the benchmark problem "Flow around a Cylinder" in 2D

For discretizing this problem, we use a finite element method based on the quadrilateral Q_1/Q_1-Stokes element with globally continuous (isoparametric) bilinear shape functions for both unknowns, pressure and velocity. To ease local mesh refinement and coarsening, "hanging" nodes are used with static condensation of variables for preserving conformity, as indicated in Figure 6.3.

$$\text{isopar}\{1, x_1, x_2, x_1 x_2\}$$

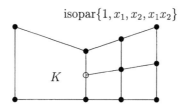

Figure 6.3: Quadrilateral mesh with "hanging nodes"

The discrete trial and test spaces for velocity and pressure are denoted by H_h and L_h, respectively. We set again $V := H \times L$ and $V_h = H_h \times L_h$, and the compact notation $u := \{v, p\} \in V$ is used for the continuous and $u_h := \{v_h, p_h\} \in V_h$ for the discrete solution. Accordingly, the Navier-Stokes system can be written in vector form as

$$Lu := \begin{bmatrix} -\nu\Delta v + v\cdot\nabla v + \nabla p \\ \nabla\cdot v \end{bmatrix} = \begin{bmatrix} f \\ 0 \end{bmatrix} =: F.$$

Further, for $u = \{v, p\}$ and $\varphi = \{\psi, \chi\}$, we define the semi-linear form

$$A(u; \varphi) := \nu(\nabla v, \nabla\psi) + (v\cdot\nabla v, \psi) - (p, \nabla\cdot\psi) + (\nabla\cdot v, \chi),$$

and for velocities v, w the stabilizing bilinear form

$$(v, w)_\delta := \sum_{T\in\mathcal{T}_h} \delta_K (\nabla v, \nabla w)_K.$$

Then, with a suitable finite element approximation $v_h^{in} \approx v^{in}$, the discrete problem seeks to determine $v_h \in v_h^{in} + H_h$ and $p_h \in L_h$, such that

$$A(u_h; \varphi_h) + (Lu_h, S\varphi_h)_\delta = (F, \varphi_h) + (F, S\varphi_h)_\delta \qquad \forall \varphi_h \in V_h, \qquad (6.25)$$

where $S\varphi := [\nu\Delta\psi + v\cdot\nabla\psi + \nabla\chi, 0]^T$, and $\delta_K = [\nu/h_K^2, |v_h|_K/h_K]^{-1}$. This formulation contains the least-squares terms for achieving velocity-pressure as well as transport stabilization as proposed, e.g., in [14] and [15]. The resulting (finite-dimensional) nonlinear problem is solved by a nested multilevel technique; for details we refer to [3].

We denote the discretization error for the velocity by $e_v := v - v_h$ and that for the pressure by $e_p := p - p_h$. The standard (global) energy-norm a posteriori error estimate reads

$$\|\nabla e_v\| + \|e_p\| \leq C_I C_S \Big(\sum_{K \in \mathcal{T}_h} \big\{ (h_K^2 + \delta_K)\|R(u_h)\|_K^2 + \|\nabla\cdot v_h\|_K^2$$

$$+ \nu h_K \|[\partial_n v_h]\|_{\partial K}^2 \big\} \Big)^{1/2}, \qquad (6.26)$$

with the residual $R(u_h) := f + \nu\Delta v_h - v_h\cdot\nabla v_h - \nabla p_h$. In this estimate the additional terms representing the errors in approximating the inflow data and the curved boundary S are neglected; they can be expected to be small compared to the other residual terms. The *interpolation constant* C_I can be determined and is of moderate size $C_I \sim 0.2$. The most critical point is the *stability constant* C_S which is completely unknown. It is related to the constant in the coercivity constant of the tangent form $A'(v; \cdot, \cdot)$ of $A(\cdot; \cdot)$ taken at the solution v,

$$\|\nabla w\| + \|q\| \leq \tilde{C}_S \sup_{\phi \in V} \left\{ \frac{A'(v; z, \phi)}{\|\nabla\psi\| + \|\chi\|} \right\},$$

where $z = \{w, q\}$ and $\phi = \{\psi, \chi\}$. In order to use the error bound (6.26) for mesh-size control, we have set $C_S = 1$.

The error estimate (6.26) is not appropriate for controlling the error in local quantities like drag and lift, because it measures the residual uniformly over the whole computational domain. One way of introducing more a priori information into the mesh refinement process based on (6.26) is to start from an initial mesh which is already refined towards the contour S. Alternatively, one may also use (on heuristic grounds) additional weighting factors which inforce stronger mesh refinement in the neighborhood of S. The resulting global error indicator reads as follows:

$$\|\nabla e_v\| + \|e_p\| \leq C_I C_S \Big(\sum_{K \in \mathcal{T}_h} \sigma_K \big\{ (h_K^2 + \delta_K)\|R(u_h)\|_K^2 + \|\nabla\cdot v_h\|_K^2$$

$$+ \nu h_K \|[\partial_n v_h]\|_{\partial K}^2 \big\} \Big)^{1/2}, \qquad (6.27)$$

where the weights σ_K are chosen large along S.

Corresponding *weighted a posteriori* error estimates can be obtained following the general line of argument described in the preceding section. The approximate dual problem seeks a couple $z := \{w, q\} \in V$ satisfying

$$A'(u_h; \varphi, z) + (L'(u_h)^* \varphi, Sz)_\delta = J(\varphi) \qquad \forall \varphi \in V. \tag{6.28}$$

The resulting weighted *a posteriori* estimate for the error $e := u - u_h$ becomes

$$|J(e)| \leq \sum_{K \in \mathcal{T}_h} \left\{ \sum_{i=1}^{3} \rho_K^{(i)} \omega_K^{(i)} + ... \right\}, \tag{6.29}$$

with the local residual terms $\rho_K^{(i)}$ and weights $\omega_K^{(i)}$ defined by

$$
\begin{aligned}
\rho_K^{(1)} &= \|R(u_h)\|_K, & \omega_K^{(1)} &= \|w - w_h\|_K + \delta_K \|v_h \cdot \nabla(w - w_h) + \nabla(q - q_h)\|_K, \\
\rho_K^{(2)} &= \tfrac{1}{2}\nu\|[\partial_n v_h]\|_{\partial K}, & \omega_K^{(2)} &= \|w - w_h\|_{\partial K}, \\
\rho_K^{(3)} &= \|\nabla \cdot v_h\|_K, & \omega_K^{(3)} &= \|q - q_h\|_K.
\end{aligned}
$$

The dots "..." stand for additional terms measuring the errors in approximating the inflow and the curved cylinder boundary. For more details on this aspect, we refer to [9] and [4]. The bounds for the dual solution $\{w, q\}$ are obtained computationally by replacing the unknown solution u in the convection term by its approximation u_h and solving the resulting linearized problem on the same mesh. From this approximate dual solution $\tilde{z}_h$, patchwise biquadratic interpolations are taken to approximate z in evaluating the weights $I_h^{(2)} \tilde{z}_h - z_h \approx z - z_h$. This frees us from choosing any interpolation constants. Table 6.1 shows the corresponding results for the pressure drop computed on four different types of meshes:

1. Hierarchically refined meshes starting from a coarse mesh of type Grid 1 with almost uniform mesh width; see Figure 6.4.

2. Hierarchically refined meshes starting from a coarse mesh of type Grid2 which is hand-refined towards the cylinder contour; see Figure 6.4.

3. Adapted meshes using the global energy-error estimator (6.26) with additional refinement along the contour S; see Figure 6.27.

4. Adapted meshes using the weighted error estimator (6.29) for the pressure drop; see Figure 6.26.

These results demonstrate clearly the superiority of the weighted error estimator in computing local quantities. It produces an error of less than 1% after only 6 refinement cycles on a mesh with about $20,000$ unknowns while the other algorithms need at least $75,000$ unknowns to achieve the same accuracy. Table 6.2 shows some results of the computation of drag and lift coefficients using the corresponding weighted error estimators. The effectivity index is defined by $I_{eff} := \eta(u_h)/|J(e)|$. Figure 6.5 and 6.6 show sequences of meshes generated by the (heuristically) weighted energy error estimator (6.26) and the fully weighted error estimator for the pressure drop (6.27).

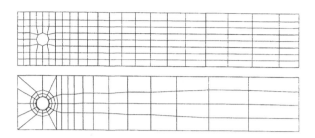

Figure 6.4: Coarse grids Grid1 and Grid2 for the flow benchmark

Table 6.1: Results for the pressure drop (ref. value $\Delta p = 0.11752016...$) on uniformly refined meshes of type Grid1 and Grid2 (upper row), and on adaptively refined meshes starting from Grid1 (bottom row)

Uniform Refinement, Grid1			Uniform Refinement, Grid2		
L	N	Δp	L	N	Δp
1	2268	0.109389	1	1296	0.106318
2	8664	0.110513	2	4896	0.112428
3	33840	0.113617	3	19008	0.115484
4	133728	0.115488	4	**74880**	**0.116651**
5	**531648**	**0.116486**	5	297216	0.117098

Adaptive Refinement, Grid1			Weighted Adaptive Refinement		
L	N	Δp	L	N	Δp
2	1362	0.105990	4	650	0.115967
4	5334	0.113978	6	**1358**	**0.116732**
6	**21546**	**0.116915**	9	2858	0.117441
8	86259	0.117379	11	5510	0.117514
10	330930	0.117530	12	8810	0.117527

Table 6.2: Results for drag and lift (ref. values $c_{drag} = 5.579535...$ and $c_{lift} = 0.0106189...$) on adaptively refined meshes starting from mesh Grid1

Computation of drag					Computation of lift				
L	N	c_{drag}	η_{drag}	I_{eff}	L	N	c_{lift}	η_{lift}	I_{eff}
4	984	5.66058	$1.1e-1$	0.76					
5	2244	5.59431	$3.1e-2$	0.47	4	2208	0.01318	$1.3e-2$	0.19
6	**4368**	**5.58980**	$1.8e-2$	0.58	5	5088	0.01100	$2.7e-3$	0.14
6	7680	5.58507	$8.0e-3$	0.69	7	**14016**	**0.01071**	$7.8e-4$	0.12
7	9444	5.58309	$6.3e-3$	0.55	8	53040	0.01065	$1.8e-4$	0.18
8	22548	5.58151	$2.5e-3$	0.77	9	142896	0.01064	$6.4e-5$	0.25
9	41952	5.58051	$1.2e-3$	0.76	11	489648	0.01063	$1.8e-5$	0.27

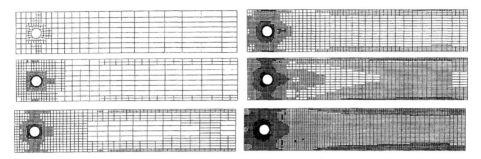

Figure 6.5: A sequence of refined meshes generated by the (heuristically) weighted global energy estimator which induces refinement also far behind the cylinder.

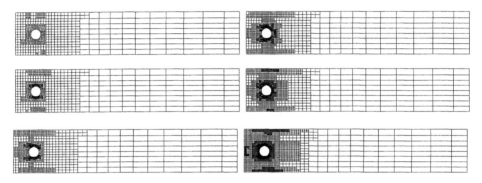

Figure 6.6: A sequence of refined meshes generated by the weighted error estimator for the pressure drop which induces refinement only in the vicinity of the cylinder surface.

6.4 SECOND EXAMPLE: COMPRESSIBLE REACTIVE GAS FLOW

The results in this section are taken from Waguet [24] and [5]. We consider a laminar flow reactor for determining the reaction velocity of the energy transfer process between Hydrogene molecules and Deuterium (slow chemistry) as sketched in Figure 6.7,

$$H_2^{(\nu=1)} + D_2^{(\nu=0)} \quad \rightarrow \quad H_2^{(\nu=0)} + D_2^{(\nu=1)}.$$

More complex combustion processes (fast chemistry) like the ozone recombination or an even more complex model of methane combustion (17 specie and 88 reactions) have been treated by the same methods in [10], see also [5]. The quantity to be computed is the CARS signal (Coherent Anti-Stokes Raman Spectroscopy)

$$J(c) = \kappa \int_{-R}^{R} \sigma(s)c(r-s)^2\,ds,$$

where $c(r)$ is, for example, the concentration of $D_2^{(\nu=1)}$ along the line of the laser measurement.

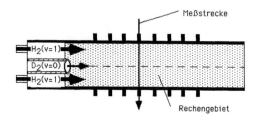

Figure 6.7: Configuration of flow reactor

Since the inflow velocity is small, a low-Mach approximation of the compressible Navier-Stokes equations is used, i.e., the pressure is split like $p(x,t) = P_*(t) + p(x,t)$ into a thermodynamical part $P_*(t)$ which is constant in space and used in the gas law, and a much smaller hydrodynamical part $p(x,t) \ll P_*(t)$ which occurs in the momentum equation. The governing system of equations consists of the (stationary) equation of mass, momentum and energy conservation supplemented by the equations of species mass conservation and the law of an ideal gas:

$$\nabla \cdot (\rho v) = 0,$$
$$(\rho v \cdot \nabla)v - \nabla \cdot (\mu \nabla v) + \nabla p = \rho f_e,$$
$$\rho v \cdot \nabla T - c_p^{-1} \nabla \cdot (\lambda \nabla T) = c_p^{-1} f_t(T, w), \qquad (6.30)$$
$$\rho v \cdot \nabla w_i - \nabla \cdot (\rho D_i \nabla w_i) = f_i(T, w), \quad i = 1, ..., n,$$
$$\rho = \frac{P_* \bar{M}}{RT}.$$

Due to exponential dependence on temperature (Arrhenius law) and polynomial dependence on w, the source terms $f_i(T, w)$ are highly nonlinear. In general, these zero-order terms lead to a coupling between all chemical species mass fractions. For robustness the resulting system of equations is to be solved by an implicit and fully coupled process which uses strongly adapted meshes.

The discretization of the flow system above uses continuous Q_1-finite elements for all unknowns and employs least-squares stabilization for the velocity-pressure coupling as well as for the transport terms. We do not state the corresponding discrete equations since they have the same abstract structure as already seen in the preceding section for the incompressible Navier-Stokes equations. The derivation of the related (linearized) dual problem and the resulting a posteriori error estimates follows the same line of argument. For details, we refer to [24] and [6].

The discretization of this problem uses continuous (isoparametric) Q_1-trial functions for all unknowns and employs least-squares stabilization for the velocity-pressure coupling as well as for the transport terms. We do not state the corresponding discrete equations since they have the same general structure as already seen in the preceding section for the incompressible Navier-Stokes equations. The derivation of the related (linearized) dual problem and the corresponding *a posteriori* error estimate follows the same line of argument. For reasons of economy, we do not use the full Jacobian of the coupled system in setting up the dual problem, but only include its dominant parts. The same simplification is used in the nonlinear iteration process.

For illustration, we state the strong form of the dual problem (suppressing terms related to the least-squares stabilization) used in our test computations. The dual solution $z = (z_v, z_p, z_T, z_w)$ is determined by the system of stationary equations:

$$\nabla \cdot z_v = J_p,$$
$$-\rho(v \cdot \nabla) z_v - \nabla \cdot \mu \nabla z_v - \rho \nabla z_p = J_v,$$
$$T^{-1} v \cdot \nabla z_p + T^{-2} v \cdot \nabla T \cdot z_p - \rho v \cdot \nabla z_T - c_p^{-1} \nabla \cdot (\lambda \nabla z_T) - (Df^T)_T(z_T, z_w)^T = J_T,$$
$$-\rho v \cdot \nabla z_{w_i} - \nabla \cdot (\rho D_i \nabla z_{w,i}) - (Df^T)_i(z_T, z_w)^T = J_{w_i},$$

for $i = 1, ..., n$, where $J = (J_p, J_v, J_T, J_w)$ is a suitable error functional defined on the solution space. This system is supplemented by appropriate boundary conditions induced by those of the primal problem. The resulting *a posteriori* error estimate

$$|J(e)| \approx \sum_{K \in \mathcal{T}_h} \sum_{\alpha \in \{p,u,T,w_i\}} h_K^4 \{\rho_{K,\alpha} + \sigma_{K,\alpha}\} \tilde{\omega}_{K,\alpha} \qquad (6.31)$$

involves the cell residuals

$$\rho_{K,p} := h_K^{-1} \|R_p(u_h)\|_K, \qquad \rho_{K,v} := h_K^{-1} \|R_v(u_h)\|_K + j_{\partial K}(v_h)$$
$$\rho_{K,T} := h_K^{-1} \|R_T(u_h)\|_K + j_{\partial K}(T_h), \qquad \rho_{K,w_i} := h_K^{-1} \|R_{w_i}(u_h)\|_K + j_{\partial K}(w_{h,i}),$$

in terms of the cell-wise equation residuals $R_\alpha(u_h)$ and certain normal-jumps across inter-element boundaries of the discrete solution $u_h = (p_h, v_h, T_h, w_h)^T$, the contributions from the least-squares stabilization

$$\sigma_{K,p} := \delta_{K;p} h_K^{-1} \|R_v(u_h)\|_K, \qquad \sigma_{K,v} := \delta_{K;v} \|\rho v\|_{\infty;K} h_K^{-1} \|R_v(u_h)\|_K,$$
$$\sigma_{K,T} := \delta_{K;T} \|\rho v\|_{\infty;K} h_K^{-1} \|R_T(u_h)\|_K, \qquad \sigma_{K,w_i} := \delta_{K;i} \|\rho v\|_{\infty;K} h_K^{-1} \|R_{w_i}(u_h)\|_K.$$

and the weights

$$\tilde{\omega}_{K,p} := h_K^{-3} \|z_p - z_{p,h}\|_K, \qquad \tilde{\omega}_{K,v} := h_K^{-5/2} \|z_v - z_{v,h}\|_{\partial K},$$
$$\tilde{\omega}_{K,T} := h_K^{-5/2} \|z_T - z_{T,h}\|_{\partial K}, \qquad \tilde{\omega}_{K,w_i} := h_K^{-5/2} \|z_{w_i} - z_{w_i,h}\|_{\partial K}.$$

In the case of bilinear trial functions as used in these computations, the jump terms $j_{\partial T}(\cdot)$ in the cell residuals $\rho_{T,\alpha}$ are dominant and determine the relevant size of the local error indicator. The weights in the *a posteriori* error bound (6.31) are again evaluated by solving the dual problem numerically on the current mesh and approximating the exact dual solution z by patchwise higher-order interpolation of the computed dual solution $\tilde{z}_h$, as described above. This technique shows sufficient robustness and does not require the determination of any interpolation constant.

The most important feature of the *a posteriori* error estimate (6.31) is that the local cell residuals related to the various physical effects governing flow and transfer of temperature and chemical species are systematically weighted according to their impact on the error quantity to be controlled. For illustration, let us consider control of the mean-velocity

$$J(u) = |\Omega|^{-1} \int_\Omega v \, dx.$$

Then, in the dual problem the right-hand-sides J_p, J_T and J_{w_i} vanish, but because of the coupling of the variables all components of the dual solution z will be non-zero. Consequently, the error term to be controlled is also affected by the cell-residuals of the energy equation and the balance equations of the chemical species. This sensitivity is quantitatively represented by the weights involving z_T and z_{w_i}. This avoids the need for any heuristic guessing in balancing the various residual terms in the error estimator.

Table 6.3 contains results obtained by our approach for the computation of the mass fraction of $D_2^{(\nu=1)}$ and $D_2^{(\nu=0)}$. The comparison is against computations on heuristically refined tensor-product meshes. We observe improved accuracy on the systematically adapted meshes; particularly monotone convergence of the quantities is achieved.

Table 6.3: Some results of simulation for the $H_2^{(\nu=1)}+D_2^{(\nu=0)} \rightarrow H_2^{(\nu=0)}+D_2^{(\nu=1)}$ experiment on hand-adapted (left) and on automatic-adapted (right) meshes

Heuristic refinement				Adaptive refinement			
L	N	$D_2^{(\nu=0)}$	$D_2^{(\nu=1)}$	L	N	$D_2^{(\nu=0)}$	$D_2^{(\nu=1)}$
1	137	0.776139	0.000000	1	137	0.776139	0.000000
2	481	0.742228	0.002541	2	244	0.738019	0.004020
3	1793	0.780133	0.002531	3	446	0.745037	0.002600
4	1923	0.782913	0.002729	4	860	0.756651	0.002010
5	2378	0.785116	0.001713	5	1723	0.780573	0.001390
6	3380	0.791734	0.001162	6	3427	0.785881	0.001130
7	5374	0.791627	0.001436	7	7053	0.799748	0.001090

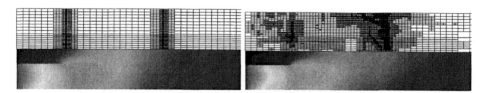

Figure 6.8: Mass fraction of $D_2^{(\nu=1)}$ in the flow reactor computed on a tensor-product mesh (left) and on a locally adapted mesh (right)

6.5   THIRD EXAMPLE: RADIATIVE TRANSFER

The results in this section are taken from Kanschat [18], [19], and Führer and Kanschat [12]. The emission of light of a certain wave length from a cosmic source is described by the *radiative transfer equation* (neglecting frequency coupling)

$$\theta \cdot \nabla_x u + (\kappa + \mu)u = \mu \int_{S_2} K(\theta, \theta')u \, d\theta' + B \quad \text{in } \Omega \times S_2, \qquad (6.32)$$

for the radiation intensity $u = u(x, \theta)$. Here, $x \in \Omega \subset \mathbb{R}^3$ is a bounded domain and $\theta \in S_2$ the unit-sphere in $\mathbb{R}^3$. The usual boundary condition is $u = 0$ on the "inflow" boundary $\Gamma_{\text{in},\theta} = \{x \in \partial\Omega, n{\cdot}\theta \leq 0\}$. The absorption and scattering coefficients κ, μ, the redistribution kernel $K(\cdot,\cdot)$, and the source term B (Planck function) are given. In interesting applications these functions exhibit strong variations in space requiring the use of locally refined meshes.

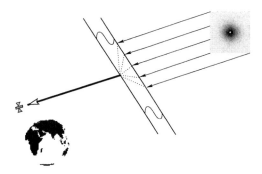

Figure 6.9: Observer configuration of radiation emission

We consider a proto-typical example from astrophysics. A satellite-based observer measures the light (at a fixed wave length) emitted from a cosmic source hidden in a dust cloud. A sketch of this situation is shown in Figure 6.9. The measurement is compared with results of a (two-dimensional) simulation which assumes certain properties of the coefficients in the underlying radiative transfer model (6.32). Because of the distance to the source, only the mean value of the intensity emitted in the observer direction θ_{obs} can be measured. Hence, the quantity to be computed is

$$J(u) = \int_{\{n \cdot \theta_{\text{obs}} \geq 0\}} u(x, \theta_{\text{obs}}) \, ds,$$

where $\{n \cdot \theta_{\text{obs}} \geq 0\}$ is the outflow boundary of the computational domain $\Omega \times S_1$ (here $\Omega \subset \mathbb{R}^2$ a square) containing the radiating object.

The Galerkin finite element formulation of (6.32) reads

$$((T + \Sigma)u_h, \phi_h)_{\Omega \times S_2} = (B, \phi_h)_{\Omega \times S_2} \quad \forall \phi_h \in V_h, \tag{6.33}$$

where

$$Tu_h := \theta{\cdot}\nabla_x u_h, \qquad \Sigma u_h := (\kappa + \mu)u_h - \mu \int_{S_2} K(\theta, \theta')u_h \, d\theta',$$

and $V_h \subset H^1(\Omega) \times L^2(\Omega)$ is a proper finite element subspace. The discretization uses standard (continuous) Q_1-finite elements in $x \in \Omega$, on meshes $\mathcal{T}_h = \{K\}$ with local width h_K, and (discontinuous) P_0-finite elements in $\theta \in S_2$, on meshes $\mathcal{D}_k = \{\Delta\}$ of uniform width k_Δ. The x-mesh is adaptively refined, while the θ-mesh is kept uniform (suggested by *a priori* error analysis). The refinement process is organized as described above. The associated dual problem reads

$$z \in V : \quad (z, (T + \Sigma)\phi)_{\Omega \times S_2} = J(\phi) \quad \forall \phi \in V. \tag{6.34}$$

Using this notation, we obtain the weighted *a posteriori* error estimate

$$|J(e)| \approx \eta_\omega := \sum_{\Delta \in \mathcal{D}_k} \sum_{K \in \mathcal{T}_h} \omega_K \, \|B - (T + \Sigma)u_h\|_{K \times \Delta}, \tag{6.35}$$

where $\omega_K := \|z - z_h\|_{K \times \Delta}$. This error bound has to be compared with a global (heuristic) L^2-error estimator

$$\|e\|_{\Omega \times S_2} \leq C_S \Big(\sum_{\Delta \in \mathcal{D}_k} \sum_{T \in \mathcal{T}_h} (h_T^2 + k_\Delta^2) \, \|B - (T + \Sigma)u_h\|_{T \times \Delta}^2 \Big)^{1/2}, \tag{6.36}$$

where the stability constant C_S is either computed by solving numerically the dual problem corresponding to the source term $\|e\|_{\Omega \times S_2}^{-1} e$, or simply set to $C_S = 1$. The results shown in Table 6.4 demonstrate the superiority of the weighted error estimator over the heuristic global L^2-error indicator. The effect of the presence of the weights on the mesh refinement is shown in Figure 6.10. In this example it appears appropriate to approximate the dual solution z on a mesh different from that used for the primal computation.

Table 6.4: Results obtained for the radiative transfer problem by the (heuristic) L^2-error indicator (left) and the weighted error estimator (middle and right), the total number of unknowns being $N_{\text{tot}} = N_x \cdot 32$

L	L^2-indicator		weighted estimator			
	N_x	$J(u_h)$	N_x	$J(u_h)$	η_ω	$\eta_\omega/J(e)$
1	564	0.181	576	0.417	3.1695	23.77
2	1105	0.210	1146	0.429	1.0804	8.62
3	2169	0.311	2264	0.461	0.7398	7.11
4	4329	0.405	4506	0.508	0.2861	3.94
5	8582	0.460	**9018**	**0.555**	0.1375	3.33
6	17202	0.488	18857	0.584	0.0526	2.39
7	34562	0.537	39571	0.599	0.0211	1.76
8	**68066**	**0.551**	82494	0.608	0.0084	1.41
			∞	0.618		

CONCLUSION

We have presented a residual-based approach to *a posteriori* error control and automatic mesh adaptation in Galerkin finite element methods which can be applied to any problem which is posed in a variational setting. By solving a global dual problem the relevant error dependencies of the quantity of physical interest are captured. The resulting *weighted a posteriori* error estimates are the basis of mesh-control algorithms which allow the most economical meshes to be constructed. The performance of this method has been demonstrated by two examples from Fluid Mechanics and one from Astrophysics. Further applications not discussed in this paper deal with problems in Structural Mechanics [21], in Acoustics [2], and in Optimization [7].

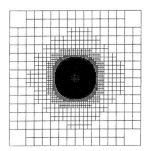 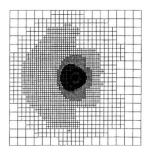

Figure 6.10: Optimized meshes for the radiative transfer problem generated by the (heuristic) L^2-error indicator (left) and the weighted error estimator (right)

REFERENCES

[1] M. Ainsworth and J. T. Oden (1997). *A posteriori* error estimation in finite element analysis. *Comput. Methods Appl. Mech. Engrg.*, 142:1–88.

[2] W. Bangerth and R. Rannacher (1999). Adaptive finite element techniques for the acoustic wave equation. Proc. Int. Conf. Theory and Computational Acoustics, Trieste 1999, *J. Comput. Acoustics*, to appear.

[3] R. Becker (1995). *An Adaptive Finite Element Method for the Incompressible Navier-Stokes Equations on Time-Dependent Domains*. Thesis, Preprint 95-44, SFB 359, U Heidelberg.

[4] R. Becker (1998). Weighted error estimators for finite element approximations of the incompressible Navier-Stokes equations Preprint 98-20, SFB 359, U Heidelberg. Also submitted to *Comput. Methods. Appl. Mech. Engrg.*.

[5] R. Becker, M. Braack, R. Rannacher, and C. Waguet (1999). Fast and reliable solution of the Navier-Stokes equations including chemistry. Proc. AMIF Conf. Applied Mathematics for Industrial Flow Problems, San Feliu de Guixols, Spain, 1-3 Oct., 1998, Preprint 99-03 (SFB 359), U Heidelberg, 1999. *Computing and Visualization in Science*, to appear.

[6] R. Becker, M. Braack, and R. Rannacher (1998). Numerical simulation of laminar flames at low Mach number with adaptive finite elements. Preprint, December 1998, SFB 359, U Heidelberg, *Combustion Theory and Modelling*, to appear.

[7] R. Becker, H. Kapp and R. Rannacher (1998). Adaptive finite element methods for optimal control of partial differential equations: basic concepts. *Preprint 98-55*, SFB 359, U Heidelberg,

[8] R. Becker and R. Rannacher (1998). Weighted *a posteriori* error control in FE methods. ENUMATH-95, Paris, 1995, *Proc. ENUMATH-97*, World Scient. Publ., Singapore.

[9] R. Becker and R. Rannacher (1996). A feed-back approach to error control in finite element methods: basic analysis and examples. *East-West J. Numer. Math.*, 4:237-264.

[10] M. Braack (1998). *An Adaptive Finite Element Method for Reactive Flow Problems*. Dissertation, U Heidelberg, 1998.

[11] K. Eriksson, D. Estep, P. Hanspo, and C. Johnson (1995). Introduction to adaptive methods for differential equations. *Acta Numerica* (A. Iserles, ed.), Cambridge University Press, pp. 105-158.

[12] C. Führer and G. Kanschat (1997). Error control in radiative transfer. *Computing*, 58:317-334.

[13] P. Houston, R. Rannacher, and E. Süli (1999). *A posteriori* error analysis for stabilised finite element approximation of transport problems. Report No. 99/04, Oxford University Computing Laboratory.

[14] T. J. R. Hughes and A. N. Brooks (1982). Streamline upwind/Petrov Galerkin formulations for convection dominated flows with particular emphasis on the incompressible Navier-Stokes equation. *Comp. Math. Appl. Mech. Eng.*, 32:199-259.

[15] T. J. R. Hughes, L. P. Franca, and M. Balestra (1986). A new finite element formulation for computational fluid dynamics: V. Circumvent the Babuska-Brezzi condition: A stable Petrov-Galerkin formulation for the Stokes problem accomodating equal order interpolation. *Comput. Methods. Appl. Mech. Engrg.*, 59:89-99.

[16] C. Johnson (1987). *Numerical Solution of Partial Differential Equations by the Finite Element Method*, Cambridge University Press, Cambridge-Lund.

[17] C. Johnson (1993). A new paradigm for adaptive finite element methods. In J. R. Whiteman(ed.), The Mathematics of Finite Elements and Applications, Highlights 1993. Wiley, Chichester, 105-120.

[18] G. Kanschat (1996). *Parallel and Adaptive Galerkin Methods for Radiative Transfer Problems*. Thesis, Preprint 96-29, SFB 359, U Heidelberg

[19] G. Kanschat (1997). Efficient and reliable solution of multi-dimensional radiative transfer problems. *Proc. Multiscale Phenomena and Their Simulation*, Bielefeld, Sept. 30 - Oct. 4, 1996, World Scient. Publ., Singapore.

[20] R. Rannacher (1998). Error control in finite element computations. Proc. Summer School on *Error Control and Adaptivity in Scientific Computing*, Antalya, Turkey, Aug. 9-12,1998, Kluwer, to appear.

[21] R. Rannacher and F.-T. Suttmeier (1999). *A posteriori* error estimation and mesh adaptation for finite element models in elasto-plasticity. *Comput. Methods. Appl. Mech. Engrg.*, to appear.

[22] M. Schäfer and S. Turek (1996). The benchmark problem "flow around a cylinder". *Flow Simulation with High-Performance Computers*, E.H. Hirschel, ed., Notes Comput. Fluid Mech., Vieweg, Stuttgart.

[23] R. Verfürth (1996). *A Review of A Posteriori Error Estimation and Adaptive Mesh-Refinement Techniques*, Wiley/Teubner, New York-Stuttgart.

[24] C. Waguet (1999). *Adaptive Finite Element Computation of Chemical Flow Reactors*, Thesis, U Heidelberg.

7 H-ADAPTIVE FINITE ELEMENT METHODS FOR CONTACT PROBLEMS

P. Wriggers, C.-S. Han and A. Rieger

Institute of Structural and Computational Mechanics IBNM,
University of Hannover, D-30167 Hannover, Germany

ABSTRACT

In this paper, several projection based *a posteriori* error indicators for frictionless contact problems are presented. Applications are to 3D solids and shells under the hypothesis of nonlinear elastic and elasto-plastic material behaviour associated with finite deformations. A penalization technique is applied to enforce unilateral boundary conditions due to contact. The approximate solution of this problem is obtained by using the Finite Element Method. Several numerical results are reported to show the applicability of the adaptive algorithm to the problems considered.

Key words. finite element method, frictionless contact, shells, plasticity, error estimators, adaptivity.

7.1 INTRODUCTION

In order to apply the finite element method (FEM) to simulate engineering problems, one has to choose a suitable model and discretize it numerically. Usually the most sophisticated model is computationally the most expensive one. Therefore reduced models, leading to so-called modelling errors are used. Discretization errors due to the discretization in space and time of a simulation also have to be considered. With regard to the reliability of the simulation, computationally efficient adaptive procedures have received special attention in recent years. In this paper we restrict our considerations to the discretization error in space. In any *h*- or *p*-adaptive procedure, *a posteriori* error estimators and indicators play an important role in delivering simulations with a preset numerical accuracy. In the linear case, several approaches are available to achieve mathematically sound error measures for the discretization using residual equations, see e.g. Babuška and Rheinboldt [1] or Johnson and Hansbo [15]. In Wriggers *et al* [28] a residual based error estimator has been developed for frictionless contact in linear elasticity, see also [32] and [5]. Following Rheinboldt [23], an approach for nonlinear contact problems by applying error estimators to the linearized problem is developed in Wriggers and Scherf [31].

Due to history dependence and the non-associative character of the tangential contact stresses, it has not been possible so far to derive mathematically rigorous error measures for frictional contact problems. However, introducing a heuristic extension to the residual based concept Wriggers and Scherf [34] achieved an error indicator suitable for contact with friction.

To deal also with non-elastic material behavior Johnson and Hansbo [16] proposed a residual based error estimator for J_2 elasto-plasticity and small strains. Wriggers and Scherf [29] developed a recovery procedure, which can be applied within a time increment of the solution algorithm using return mapping schemes. However, a rigorous theoretical analysis is still missing for frictional contact.

A modified version of the Z^2 error indicator, due to Zienkiewicz *et al* [35, 36, 37] can be applied to contact problems as shown in Wriggers and Scherf [33]. Some advantages and disadvantages of residual based error estimators and Z^2 indicators have been discussed in [33] for the frictionless case.

Here we summarize projection based error indicator methods for 3D solids undergoing large deformations as well as for J_2 elasto–plastic shells with contact. The presentation is organized as follows: In section 7.2, basic kinematic relations and the weak form of the frictionless contact problem undergoing large deformation are stated. We briefly review the five parameter shell formulation that has been used and point out some special aspects in the context of contact. In addition to a simple hyperelastic constitutive model we review the basic equations of elasto-plastic material behavior and their numerical treatment within shell problems. An *a posteriori* error indicator method, based on several projection techniques related to the problem type being considered is proposed in section 7.3. Some numerical examples are presented in section 7.4 in order to corroborate the applicability of the indicators.

7.2   FORMULATION OF FRICTIONLESS CONTACT PROBLEMS

Let us consider two bodies $\mathcal{B}^\alpha, \alpha = 1, 2$, each of them occupying the bounded domain $\Omega^\alpha \subset \mathbb{R}^3$. The mapping φ^α maps points in the initial configuration, described by the position vector $\mathbf{X}^\alpha$, to points in the deformed configuration $\mathbf{x}^\alpha = \varphi^\alpha(\mathbf{X}^\alpha)$.

In the following we will state the weak form for bodies undergoing large deformations which leads in case of contact to a variational inequalitiy.

7.2.1   Contact kinematics

Since the deformation of two bodies in space can be arbitrary, and may consist of finite rotations and large deflections, a global search procedure is needed to find the parts of the bodies, denoted by Γ_c, which come into contact. This search process can be based on methods such as bucket search or binary tree algorithms, but will not be discussed here in detail.

Within the *master-slave* concept we define Γ^1 as the master surface and Γ^2 as the slave surface. When assuming that the contact boundary describes, at least locally, a convex

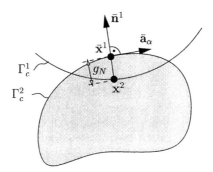

Figure 7.1: Minimal distance problem

region, we can relate to every point $\mathbf{x}^2$ on Γ^2 a point $\bar{\mathbf{x}}^1 = \mathbf{x}^1(\bar{\xi})$ on Γ^1 via the minimal distance problem

$$\|\mathbf{x}^2 - \bar{\mathbf{x}}^1\| = \min d^1(\xi) = \min_{\mathbf{x}^1 \subseteq \Gamma^1} \|\mathbf{x}^2 - \mathbf{x}^1(\xi)\|, \tag{7.1}$$

where ξ denotes the parametrization of the boundary Γ^1. The minimization process yields the condition $(\mathbf{x}^2 - \bar{\mathbf{x}}^1) \cdot \mathbf{a}^1 = 0$ which means that $(\mathbf{x}^2 - \bar{\mathbf{x}}^1)$ points in the direction of $\bar{\mathbf{n}}^1$, see Fig. 7.1 for the two-dimensional case.

Once the point $\bar{\mathbf{x}}^1$ is known, we can write the geometrical contact constraint inequality which permits penetration of one body into the other, as

$$g_N = (\mathbf{x}^2 - \bar{\mathbf{x}}^1) \cdot \bar{\mathbf{n}}^1 \geq 0. \tag{7.2}$$

In view of the penalty formulation which will be applied to solve the contact problem we introduce a penetration function since the method allows a small penetration in Γ_c:

$$g_{\bar{N}} = \begin{cases} (\mathbf{x}^2 - \bar{\mathbf{x}}^1) \cdot \bar{\mathbf{n}}^1 & \text{if } (\mathbf{x}^2 - \bar{\mathbf{x}}^1) \cdot \bar{\mathbf{n}}^1 < 0 \\ 0 & \text{otherwise}. \end{cases} \tag{7.3}$$

7.2.2 Contact formulation for Finite Deformations

For the formulation of the boundary value problem we need to discuss in detail only the additional terms due to contact. The equations describing the behaviour of the bodies coming into contact do not change. However, for completeness, the weak form is stated for finite deformations.

We can formulate the local momentum equation for a body $\mathcal{B}^\alpha$ as $\operatorname{Div} \mathbf{P}^\alpha + \bar{\boldsymbol{f}}^\alpha = \mathbf{0}$ in the case that inertia terms are neglected. $\mathbf{P}^\alpha$ denotes the first Piola–Kirchhoff stress tensor acting in the body $\mathcal{B}^\alpha$, whilst $\bar{\boldsymbol{f}}^\alpha$ are the body forces. Let $\bar{\mathbf{u}}^\alpha$ and $\bar{\mathbf{t}}^\alpha$ be prescribed quantities on Γ_φ^α and on Γ_σ^α, respectively. Due to the fact that the constraint condition (7.2) is represented by an inequality we obtain a variational inequality

$$\sum_{\alpha=1}^{2} \int_{\Omega^\alpha} \mathbf{P}^\alpha \cdot \nabla_X (\boldsymbol{\eta}^\alpha - \boldsymbol{\varphi}^\alpha) \, dV \geq \sum_{\alpha=1}^{2} \int_{\Omega^\alpha} \bar{\boldsymbol{f}}^\alpha \cdot (\boldsymbol{\eta}^\alpha - \boldsymbol{\varphi}^\alpha) \, dV + \int_{\Gamma^\alpha} \bar{\boldsymbol{t}}^\alpha \cdot (\boldsymbol{\eta}^\alpha - \boldsymbol{\varphi}^\alpha) \, dA, \tag{7.4}$$

where the integration is performed with respect to the domain Ω^α occupied by the body $\mathcal{B}^\alpha$ in the undeformed configuration. Here ∇_X is the gradient operator with respect to the reference coordinates. We now have to find the deformation $(\varphi^1, \varphi^2) \in \mathcal{K}$ such that (7.4) is satisfied for all $(\eta^1, \eta^2) \in \mathcal{K}$ with

$$\mathcal{K} = \{ (\eta^1, \eta^2) \in \mathcal{V} \,|\, [\eta^2 - \eta^1(\bar\xi)] \cdot \bar{\mathbf{n}}_1 \geq 0 \}. \tag{7.5}$$

The space $\mathbf{V}$ is defined as $\mathcal{V} = \{ \eta^\alpha \in [W^{1,p}(\Omega^\alpha)]^3 \,|\, \eta^\alpha = \mathbf{0} \text{ on } \Gamma_u \}$.

One of the most used approaches to solve (7.4) is an active set strategy combined with the penalty method, which will also be employed in this paper. Introducing a penalty constraint for the normal contact replaces, by assuming that the contact surface is known, the variational inequality (7.4) by the unconstrained problem

$$R(\varphi_\epsilon, \eta) = G(\varphi_\epsilon, \eta) - \lambda f(\eta) = 0, \tag{7.6}$$

where φ_ϵ and η denote the deformation fields of the two bodies and their variations of the penalized problem, respectively. The load parameter λ is introduced to allow scaling of the applied loads $F = \lambda f(\eta)$.

It can be shown that the solution of the unconstrained problem φ_ϵ will converge to the solution of the associated variational inequality φ, as ϵ_N tends to infinity. For simplicity we write φ instead of φ_ϵ in the following. The functional G can now be split into a part associated with the bodies G_i^α and a part which describes the contact interface G_c:

$$G(\varphi, \eta) = \sum_{\alpha=1}^{2} G_i^\alpha(\varphi^\alpha, \eta^\alpha) + G_c(\varphi^1, \varphi^2, \eta^1, \eta^2) \tag{7.7}$$

with the explicit expressions

$$G_i^\alpha(\varphi^\alpha, \eta^\alpha) = \int_{\Omega^\alpha} \mathbf{P}^\alpha \cdot \nabla_X \eta^\alpha \, dV, \tag{7.8}$$

$$G_c(\varphi^1, \varphi^2, \eta^1, \eta^2) = \int_{\Gamma_c^1} \epsilon_N g_N^-(\eta^2 - \bar\eta^1) \cdot \bar{\mathbf{n}}^1 \, dA, \quad \epsilon_N > 0 \tag{7.9}$$

where ϵ_N is the penalty parameter.

7.2.3 Shell Formulation

In this section the shell formulation is introduced briefly. This is based on a so called five parameter *quasi–Kirchhoff–theory* derived in detail in [6]. This finite element formulation is applicable to structural problems with finite rotations and large strains.

An arbitrary point in the deformed shell space of the current configuration is determined by

$$\mathbf{x}(\xi^1, \xi^2, \zeta) = \phi(\xi^1, \xi^2) + \zeta \mathbf{d}(\xi^1, \xi^2) \tag{7.10}$$

where ϕ denotes the shell mid-surface and $\mathbf{d}$ is an inextensible director of unit length. The curvilinear base vectors tangential to ξ^1, ξ^2, ζ are given by

$$\mathbf{g}_\alpha = \phi_{,\alpha} + \zeta \mathbf{d}_{,\alpha} = \mathbf{a}_\alpha + \zeta \mathbf{d}_{,\alpha} \quad , \quad \mathbf{g}_3 = \mathbf{d} \tag{7.11}$$

with $\mathbf{a}_\alpha = \phi_{,\alpha}$. A strain measure with respect to the reference configuration is provided by the right Cauchy-Green tensor $\mathbf{C}$ with $C_{ij} = \mathbf{g}_i \cdot \mathbf{g}_j$ being the covariant components of $\mathbf{C}$. The right Cauchy-Green tensor $\mathbf{C}$ can be split into membrane (m), shear (s) and bending (b) parts

$$\mathbf{C} = \mathbf{C}^m + \mathbf{C}^s + \zeta \mathbf{C}^b. \tag{7.12}$$

The quadratic terms in ζ are neglected due to the thin shell limit. The relevant components of the different parts of $\mathbf{C}$ can be written as

$$C^m_{\alpha\beta} = \mathbf{a}_\alpha \cdot \mathbf{a}_\beta \quad , \quad C^s_{\alpha 3} = 2\, \mathbf{a}_\alpha \cdot \mathbf{d} \quad \text{and} \quad C^b_{\alpha\beta} = \mathbf{a}_\alpha \cdot \mathbf{d}_{,\beta} + \mathbf{a}_\beta \cdot \mathbf{d}_{,\alpha} \ . \tag{7.13}$$

From these components we can easily compute the Green–Lagrange strain $\mathbf{E}$, needed in the weak form

$$E_{\alpha\beta} = \frac{1}{2}(C_{\alpha\beta} - \delta_{\alpha\beta}). \tag{7.14}$$

Many parameterizations of the director $\mathbf{d}$ can be found in the literature. The parameterization used here has been proposed in [21]. The director can be expressed in terms of two rotation angles β_1 and β_2, see [6].

For the *quasi–Kirchhoff*–approach the constraint $E_{\alpha 3} = C_{\alpha 3} = 0$ is imposed to suppress shear strains. This condition is incorporated in the finite element formulation by adding this constraint by a penalty term to the variational form

$$G^s_i = \int_{B_o} S_{\alpha\beta} \delta E_{\alpha\beta} dV + c_{pen} \int_{B_o} C_{\alpha 3} \delta C_{\alpha 3} dV \tag{7.15}$$

with the shear penalty parameter c_{pen}. Within this shell model we use the classical plane stress assumption $S_{33} = 0$ to describe the material behavior. Thus with the additional condition $C_{\alpha 3} = 0$ the constitutive equation of eq.(7.15) is fully described by the plane stress state where $S_{i3} = 0$, $i = 1, 2, 3$.

In the shell formulation which has been used the structure is considered thin and is therefore modelled by its shell mid-surface. The error introduced by the reduction of the continuum to a surface including contact is therefore also assumed to be small. Contact is considered with ϕ being the slave–surface

$$\mathbf{x}^s = \phi. \tag{7.16}$$

The description of the rigid master surface is done by a parameter formulation. The algorithm in combination with free form surfaces used here is illustrated in [9].

The regularity of shell problems in unilateral contact

If $g_n \geq 0$ is enforced exactly, different curvatures of deformed shell and the rigid master surface result in a reduction of the contact surface to a line or a point. Therefore in the penalty formulation for contact the penalty parameter should be viewed in a constitutive sense, since with arbitrary large parameters it is impossible catch the reduction of the contact surface with the numerical form of integral (7.9). If the deformed shell and the master surface have very different curvatures the contact area will still be thin and hence no good regularity can be expected.

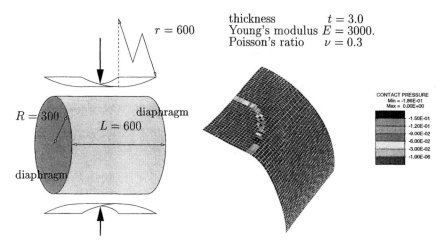

thickness $t = 3.0$
Young's modulus $E = 3000.$
Poisson's ratio $\nu = 0.3$

Figure 7.2: Cylindrical shell in contact with a rigid cylinder.

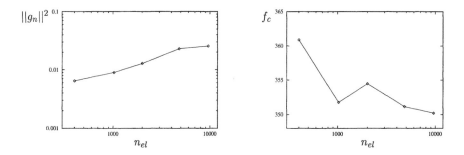

Figure 7.3: Contact norm and vertical contact force.

In Fig. 7.2 a cylindric structure is considered as an example which is deformed by a rigid cylinder perpendicular to it from both sides. The contact pressure of a solution with 40×50 elements is presented in Fig. 7.2. In Fig. 7.3 the norm of contact terms and the sum of the contact forces in the vertical direction is plotted against the number of elements used for the simulation. It can be seen that the convergence of the structural response with respect to the density of the mesh is quite slow.

7.2.4   Constitutive Models

Elasticity

The simplest constitutive model for isotropic hyperelasticity is a compressible neo–Hookian model. It relates the Kirchhoff stress $\boldsymbol{\tau}$ and the left Cauchy Green tensor $\mathbf{b} = \mathbf{F}\,\mathbf{F}^T$ by

$$\boldsymbol{\tau} = 2\frac{\partial W}{\partial \mathbf{b}}\,\mathbf{b} = \frac{\Lambda}{2}\,(\,J^2 - 1\,)\,\mathbf{1} + \mu\,(\,\mathbf{b} - \mathbf{1}\,) \tag{7.17}$$

where Λ and μ are the Lamé constants. Note that the Kirchhoff stress $\boldsymbol{\tau}$ is related to the first Piola–Kirchhoff stress via $\boldsymbol{\tau} = \mathbf{P}\,\mathbf{F}^T$, with $\mathbf{F} = \nabla_X\,\boldsymbol{\varphi}$ being the deformation gradient and $J = \det \mathbf{F}$ the Jacobian of the deformation.

Plasticity

The here relevant equations of the elasto–plastic continuum for the case of isotropic metal plasticity are recalled briefly in the following. For an explicit illustration the reader is referred to [12] or [8]. The elasto–plastic analysis is employed by the usual local multiplicative decomposition of $\mathbf{F}$

$$\mathbf{F} = \mathbf{F}^e\mathbf{F}^p. \tag{7.18}$$

A local free energy function Ψ is assumed in which the hardening mechanisms are decoupled from the elastic deformation, so that

$$\Psi(\boldsymbol{\varepsilon}^e, \alpha) = \mathcal{W}(\boldsymbol{\varepsilon}^e) + \mathcal{H}(\alpha), \tag{7.19}$$

with $\boldsymbol{\varepsilon}^e = \ln(\mathbf{b}^e)$ and $\mathbf{b}^e = \mathbf{F}^e\mathbf{F}^{e^T}$. For the terms in (7.19) moderately large elastic strains are considered common for metal plasticity with $\mathcal{W} = \frac{1}{2}\boldsymbol{\varepsilon}^e \cdot \mathbb{C}\boldsymbol{\varepsilon}^e$, where $\mathbb{C}$ is the elasticity matrix. In addition a potential function for the hardening variables is defined by $\mathcal{H} = \frac{1}{2}H\alpha^2$, with the hardening modulus H.

The principal Kirchhoff stresses τ_i are work conjugate to the elastic principal logarithmic strains ε_i^e and hence the stress response can be written as

$$\tau_i = \partial_{\varepsilon_i^e}w \ , \quad q = -H\alpha. \tag{7.20}$$

The von Mises yield criterion

$$\Phi(\mathbf{s}, q) = \frac{3}{2}tr(\mathbf{s}^2) - (\sigma_y - q)^2 \le 0 \tag{7.21}$$

is applied with the deviator of the Kirchhoff stresses $\mathbf{s}$ yielding the associated flow rule via the principle of maximum dissipation. After these steps (described i.e. in [18]) we arrive at the eigenvalue problem in the current configuration

$$\begin{aligned} \mathbf{C}^{p^{-1}}\mathbf{N}_i &= (\lambda_i^e)^2\mathbf{C}^{-1}\mathbf{N}_i \\ \mathbf{N}^i\mathbf{C} &= (\lambda_i^e)^2\mathbf{N}^i\mathbf{C}^p, \end{aligned} \tag{7.22}$$

where $\mathbf{N}_i$ and $\mathbf{N}^i$ are the dual co– and contravariant eigenvectors associated with $(\lambda_i^e)^2$. With the principal stretches $(\lambda_i^e)^2$ the return mapping procedure described below can be applied via $\varepsilon_i^e = \frac{1}{2}\ln(\lambda_i^e)^2$.

The strains and stresses can be represented with these eigenvectors in terms of the reference configuration. Hence we get the representations of the strains and of the 2nd Piola–Kirchhoff–tensor

$$\mathbf{C}^p = \sum_{i=1}^{3} \mathbf{N}_i \otimes \mathbf{N}_i \;, \quad \mathbf{C} = \sum_{i=1}^{3} (\lambda_i^e)^2 \mathbf{N}_i \otimes \mathbf{N}_i \;, \quad \mathbf{S} = \sum_{i=1}^{3} \frac{\tau_i}{(\lambda_i^e)^2} \mathbf{N}^i \otimes \mathbf{N}^i, \tag{7.23}$$

where $\mathbf{N}_i$ and $\mathbf{N}^i$ obey the relations $\mathbf{N}^i \mathbf{N}_k = \delta_k^i$ and $\mathbf{N}_i \cdot \mathbf{C}^{p^{-1}} \mathbf{N}_i = 1$.

The *exponential return mapping* is formulated in terms of principal axes for the plane stress state. We start from the right Cauchy–Green tensors $\mathbf{C}$ and $\mathbf{C}^p$ in the local element coordinate system, where $\mathbf{C}$ and $\mathbf{C}^p$ have now dimension 2×2. A time interval $[t_n, t_{n+1}]$ is considered where at t_n all variables and at t_{n+1} the total deformation are known. The trial state of t_{n+1} is set by $\mathbf{C}_{n+1}^{p^{tr}} = \mathbf{C}_n^p$ and $\alpha_{n+1}^{tr} = \alpha_n$ and with the right Cauchy–Green tensor $\mathbf{C}_{n+1}$ defining the current deformation we can solve

$$\mathbf{C}_{n+1}^{p^{tr^{-1}}} \mathbf{N}_{\beta_{n+1}}^{tr} = (\lambda_{\beta_{n+1}}^{e tr})^2 \mathbf{C}_{n+1}^{-1} \mathbf{N}_{\beta_{n+1}}^{tr} \tag{7.24}$$

to obtain $(\lambda_{\beta_{n+1}}^{e tr})^2$ and $\mathbf{N}_{\beta_{n+1}}^{tr}$. The logarithmic principal elastic strains are then calculated from $\varepsilon_{\alpha_{n+1}}^{e tr} = \frac{1}{2} \ln(\lambda_{\alpha_{n+1}}^{e tr})^2$.

Let us introduce the following vector notation for the Kirchhoff stresses and the elastic strains in principal form

$$\bar{\boldsymbol{\tau}} = [\tau_1, \tau_2]^T \quad , \quad \bar{\boldsymbol{\varepsilon}}^e = [\varepsilon_1^e, \varepsilon_2^e]^T. \tag{7.25}$$

The stress–strain relation and the hardening law then take the form

$$\bar{\boldsymbol{\tau}} = \bar{\mathbf{C}} \bar{\boldsymbol{\varepsilon}}^e, \quad q = -H\alpha. \tag{7.26}$$

Here $\bar{\mathbf{C}}$ contains the elastic moduli of the plane stress state with respect to the principal axes. With the trial state defined by $\bar{\boldsymbol{\tau}}_{n+1}^{tr} = \bar{\mathbf{C}} \bar{\boldsymbol{\varepsilon}}_{n+1}^{e tr}$, $\alpha_{n+1}^{tr} = \alpha_n$ the implicit integration $(t_{n+1} - t_n)\dot{\gamma} \longrightarrow \Delta\gamma$ leads after some steps to

$$\alpha(\Delta\gamma)_{n+1} = \frac{\alpha_n + 2\Delta\gamma\sigma_Y}{1 - 2\Delta\gamma H} \;, \quad \bar{\boldsymbol{\tau}}_{n+1} = (\bar{\mathbf{C}}^{-1} + \Delta\gamma\bar{\mathbf{P}})^{-1} \bar{\boldsymbol{\varepsilon}}_{n+1}^{e tr}, \tag{7.27}$$

with $\bar{\mathbf{P}} = \begin{bmatrix} 2 & -1 \\ -1 & 2 \end{bmatrix}$. The stresses $\bar{\boldsymbol{\tau}}_{n+1}$ and the hardening variable α_{n+1} are functions of $\Delta\gamma$. In contrast to the three dimensional case where $\Delta\gamma$ can be eliminated directly, $\Delta\gamma$ is computed here via Newton iteration from

$$\Phi(\bar{\boldsymbol{\tau}}(\Delta\gamma), \alpha(\Delta\gamma)) = 0. \tag{7.28}$$

7.3   ERROR INDICATION

In this section we wish to summarize projection based error indication methods for 3D solids undergoing large deformations, as well as for elastoplastic shells undergoing contact.

7.3.1   Superconvergent Patch Recovery

Let us recall the simple, but in many cases very efficient, stress–recovery–procedure SPR proposed by Zienkiewicz and Zhu [35]. The derivation of these error indicators is based on the fact that many finite element meshes have superconvergence properties which means that there exist points where the stresses are approximated with higher accuracy than that formed generally. By using a recovery procedure the stresses $\mathbf{S}^*$ can be computed from those at the superconvergent points, see e.g. Zienkiewicz and Taylor [36] or Zienkiewicz and Zhu [37]. It should be noted, that SPR also works well if the sampling points are not superconvergent points, see Babuška *et al.*

Let $\mathbf{u}$ denote the exact solution of (7.6) and $\mathbf{u}_h$ the associated discrete FEM–solution. With the error $\mathbf{e} = \mathbf{u} - \mathbf{u}_h$ in the displacements we can define the error energy in $\| \cdot \|_E^2$ as

$$\| \mathbf{u} - \mathbf{u}_h \|_E^2 = \int_\Omega (\mathbf{S} - \mathbf{S}^h) \cdot (\mathbf{E} - \mathbf{E}^h) \, d\Omega \tag{7.29}$$

where $\mathbf{S}$ denotes the 2nd Piola-Kirchhoff stresses and $\mathbf{E}$ is the Green-Lagrangian strain tensor. The error due to the penalty regularization will not be discussed here, for details see e.g. Kikuchi and Oden [17].

Denoting by $\mathbb{P}$ a projection operator we obtain $\mathbf{S}^*$ and $\mathbf{E}^*$ from

$$\int_\Omega \mathbb{P}\,[\,\mathbf{S}^* - \mathbf{S}^h\,]\, d\Omega = \mathbf{0} \quad , \quad \int_\Omega \mathbb{P}\,[\,\mathbf{E}^* - \mathbf{E}^h\,]\, d\Omega = \mathbf{0} \tag{7.30}$$

and can then compute an approximation of the error energy

$$\| \mathbf{u} - \mathbf{u}_h \|_E^2 \approx \int_\Omega (\mathbf{S}^* - \mathbf{S}^h) \cdot (\mathbf{E}^* - \mathbf{E}^h) \, d\Omega \tag{7.31}$$

Within the recovery scheme a simple polynomial expansion $S_i^* = \mathbf{p}\,\mathbf{a}$ of each component of the improved stresses $\mathbf{S}^*$ and strains $\mathbf{E}^*$, respectively, is applied. Here $\mathbf{p}$ contains appropriate polynomial terms, for instance $\mathbf{p} = [1, x, y, z, xy, yz, zx]^T$ for linear 3D-Tetrahedra and $\mathbf{a}$ is a set of unknown parameters. A least square fit minimization

$$\sum_{s=1}^n (S_{i_s}^h - S_s^*)^2 \longrightarrow MIN \tag{7.32}$$

leads to the linear equation system

$$\sum_{s=1}^n \mathbf{p}^T \mathbf{p}\,\mathbf{a} = \sum_{s=1}^n \mathbf{p}^T S_{i_s}^h \tag{7.33}$$

which has to be solved for every tensor component i. Here s are the sampling points within a patch assembly.

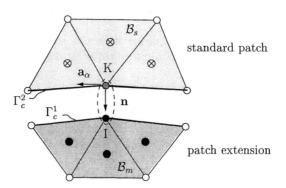

Figure 7.4: Element patches for stress recovery on Γ_c.
$\otimes$: sampling point standard patch, $\bullet$: sampling point extended patch

7.3.2 Superconvergent Patch Recovery for Frictionless Contact

The stress recovery for contact problems needs some special consideration, since the exact contact pressure is continous but not smooth over the contact interface. Local equilibrium requires the normal stresses $p_N = \mathbf{n} \cdot \boldsymbol{\sigma} \mathbf{n}$ to be continous over Γ_c: $[p_N] = 0$. The tangential stresses $t_\alpha = \mathbf{a}_\alpha \cdot \boldsymbol{\sigma} \mathbf{n}$ must vanish on Γ_c. This information will be used during the smoothing process of the stresses to consider implicitly the contact conditions. For this purpose, we search for the closest node I on Γ_c^1 related to node K of Γ_c^2, see Fig. 7.4. All elements connected to node K define the standard patch while all elements connected to node I define an extended patch.

Next we perform a least squares fit minimization in the standard patch as well as in the extended patch

$$\sum_{S_s}(\mathbf{S}_s^h - \mathbf{p}_s\,\mathbf{a}_s)^2 \longrightarrow MIN \quad , \quad \sum_{S_e}(\mathbf{S}_e^h - \mathbf{p}_e\,\mathbf{a}_e)^2 \longrightarrow MIN \qquad (7.34)$$

where the indices s and e denote values in the standard and the extended patches, respectively. Continuity of p_n^* across Γ_c can be formulated as

$$\mathbf{N} \cdot (\bar{\mathbf{p}}_s\,\mathbf{a}_s - \bar{\mathbf{p}}_e\,\mathbf{a}_e) = 0, \qquad (7.35)$$

with $\mathbf{N} = \mathbf{n} \otimes \mathbf{n}$. Analogously we formulate the vanishing tractions t_α^* for the standard and extended patch

$$\mathbf{T}_\alpha^T \cdot \bar{\mathbf{p}}_s\,\mathbf{a}_s = 0 \quad , \quad \mathbf{T}_\alpha^T \cdot \bar{\mathbf{p}}_e\,\mathbf{a}_e = 0 \quad \text{with} \ \mathbf{T}_\alpha = \mathbf{n} \otimes \mathbf{a}_\alpha \quad (\alpha = 1, 2) \ . \qquad (7.36)$$

Bars in (7.35) and (7.36) mark values which are evaluated at node K.
To enforce the boundary conditions (7.35) and (7.36) within the minimization (7.34) we

apply a penalty regularization and arrive at

$$\sum_{S_s}(S_s^h - \mathbf{p}_s\,\mathbf{a}_s)^2 + \epsilon_N[\mathbf{N}^T \cdot (\bar{\mathbf{p}}_s\,\mathbf{a}_s - \bar{\mathbf{p}}_e\,\mathbf{a}_e)]^2 + \epsilon_T\sum_{\alpha=1}^{2}(\mathbf{T}_\alpha^T \cdot \bar{\mathbf{p}}_s\,\mathbf{a}_s)^2 \longrightarrow MIN$$

$$\sum_{S_e}(S_e^h - \mathbf{p}_e\,\mathbf{a}_e)^2 + \epsilon_N[\mathbf{N}^T \cdot (\bar{\mathbf{p}}_s\,\mathbf{a}_s - \bar{\mathbf{p}}_e\,\mathbf{a}_e)]^2 + \epsilon_T\sum_{\alpha=1}^{2}(\mathbf{T}_\alpha^T \cdot \bar{\mathbf{p}}_e\,\mathbf{a}_e)^2 \longrightarrow MIN$$

(7.37)

Finally the coupled linear equation system

$$\left\{\sum_{S_s}\mathbf{p}_s^T\mathbf{p}_s + \epsilon_N\,\bar{\mathbf{p}}_s^T\mathbf{N}\mathbf{N}^T\bar{\mathbf{p}}_s + \epsilon_T\,\bar{\mathbf{p}}_s^T\mathbf{T}\mathbf{T}^T\bar{\mathbf{p}}_s\right\}\,\mathbf{a}_s = \sum_{S_s}\mathbf{p}_s^T\,S_s^h + \epsilon_N\,\bar{\mathbf{p}}_s^T\mathbf{N}\mathbf{N}^T\bar{\mathbf{p}}_e\,\mathbf{a}_e$$

$$\left\{\sum_{S_e}\mathbf{p}_e^T\mathbf{p}_e + \epsilon_N\,\bar{\mathbf{p}}_e^T\mathbf{N}\mathbf{N}^T\bar{\mathbf{p}}_e + \epsilon_T\,\bar{\mathbf{p}}_e^T\mathbf{T}\mathbf{T}^T\bar{\mathbf{p}}_e\right\}\,\mathbf{a}_e = \sum_{S_e}\mathbf{p}_e^T\,S_e^h + \epsilon_N\,\bar{\mathbf{p}}_e^T\mathbf{N}\mathbf{N}^T\bar{\mathbf{p}}_s\,\mathbf{a}_s$$

(7.38)

has to be solved to obtain the unknown parameter $\mathbf{a}_s$ and $\mathbf{a}_e$.

7.3.3 Local Error Indication Based on Dual Principles

The recovery procedures described in the previous sections are based on error measures in the global energy norm (7.31). However for practical purposes the variables of interest may be displacements or stresses at some particular points. Recently procedures for estimation of various kinds of local error functionals have been introduced in a general framework by Becker and Rannacher in [4]. Under certain assumptions this approach can also be applied to contact problems involving large elastic strains. In the following we wish to outline only the basic results.

By $a(\mathbf{e}, \mathbf{e})$ we refer to the error energy norm as defined in (7.29). In order to estimate the error of a specific displacement in the component i at point $\mathbf{x} = \bar{\mathbf{x}}$ we additionally consider the dual problem

$$a(\mathbf{G}, \boldsymbol{\eta}) = (\boldsymbol{\delta}_i, \boldsymbol{\eta})\,, \tag{7.39}$$

where $\boldsymbol{\delta}_i$ is the Dirac delta (unit pointload vector) in the direction i and $\mathbf{G}$ denotes the Green's function. The test functions $\boldsymbol{\eta}$ are from the space $\mathcal{V}$ as defined above. Applying the principle of Betti-Maxwell the local error $e_i(\bar{\mathbf{x}}) = (\mathbf{e}, \boldsymbol{\delta}_i)$ can be expressed by the bilinear form

$$e_i(\bar{\mathbf{x}}) = a(\mathbf{e}, \mathbf{G}) \tag{7.40}$$

or after using the Galerkin orthogonality

$$e_i(\bar{\mathbf{x}}) = a(\mathbf{e}, \mathbf{G} - \mathbf{G}_h) \tag{7.41}$$

with $\mathbf{G}_h$ as finite element approximation of $\mathbf{G}$. The Cauchy-Schwarz inequality finally leads to the local estimate

$$e_i^2(\bar{\mathbf{x}}) \leq a(\mathbf{e}, \mathbf{e})\,a(\mathbf{G} - \mathbf{G}_h, \mathbf{G} - \mathbf{G}_h)\,. \tag{7.42}$$

The second term in (7.42) serves as a weighting function and filters out the influence of the global error over the local displacement error of interest. It can be obtained by computing a numerical approximation to **G** based on the same discretization as for the primal problem. The recovery procedures described above simply have to be applied to a second load case. This concept of local error control can easily be extended to arbitrary integral variables, such as pressure quantities at the contact interface, see section 7.4.

Remark: Errors for stress variables at a specific point $\bar{\mathbf{x}}$ can be estimated in a straight forward manner using the same concept as above. In contrast to the local displacement error control, a discontinuity on the related displacement variable must be applied in the dual problem: $a(\mathbf{G}, \boldsymbol{\eta}) = (\frac{\partial}{\partial x_j} \boldsymbol{\delta}_i(\bar{\mathbf{x}}), \boldsymbol{\eta})$.

7.3.4 Error Indication for Elasto-Plastic Shells

The elasto–plastic error indicator is applied to large strain problems and derived with respect to the current configuration, in which the return mapping procedure is referred to by ε_i^e and τ_i. All quantities are then pulled back again into the reference configuration where the indicator, evaluated via SPR, is applied.

First we consider the free elastic energy $\mathcal{W}$ as norm - equivalent to the complementary energy norm, which in our case can be written using the plane stress assumption as

$$\|\boldsymbol{\tau}\|_{\mathcal{W}}^2 = \int_{\Omega} \bar{\boldsymbol{\tau}} \cdot \bar{\mathbf{C}}^{-1} \bar{\boldsymbol{\tau}} \, d\Omega \tag{7.43}$$

where $\bar{\boldsymbol{\tau}}$ has been defined by (7.25).

To determine the error of the finite element solution $\boldsymbol{\tau}_h$ we compare with the exact solution $\boldsymbol{\tau}$ and define $\mathbf{e}_{\tau} = \boldsymbol{\tau}_h - \boldsymbol{\tau}$. Here $\bar{\boldsymbol{\tau}}$ and $\bar{\boldsymbol{\varepsilon}}^e$ have to be thought of in 3D–tensorial form, since the local coordinate systems where the plane stress state is defined are in general different from each other. This can be done by completing the principal values with $\tau_3 = 0$ and $\varepsilon_3^e = -\frac{\nu}{1-\nu}(\varepsilon_1^e + \varepsilon_2^e)$, ν is the Poisson's ratio. The operators $\bar{\mathbf{C}}$ and $\bar{\mathbf{P}}$ have to be converted correspondingly.

An upper bound can be found following [16] by reformulating (7.27) as

$$\boldsymbol{\varepsilon}^{e^{tr}} = (\mathbf{C}^{-1} + \Delta\gamma\mathbf{P})\boldsymbol{\tau} \tag{7.44}$$

and we obtain

$$\boldsymbol{\varepsilon}^{e^{tr}} - \boldsymbol{\varepsilon}_h^{e^{tr}} = \mathbf{C}^{-1}(\boldsymbol{\tau} - \boldsymbol{\tau}_h) + \Pi(\boldsymbol{\tau}) - \Pi(\boldsymbol{\tau}_h)$$

with

$$\Pi(\boldsymbol{\tau}) = \begin{cases} \mathbf{0} & , f \leq 0 \wedge \gamma = 0 \\ \Delta\gamma\mathbf{P}\boldsymbol{\tau} & , f = 0 \wedge \gamma > 0 \end{cases} .$$

By multiplying this equation by $(\boldsymbol{\tau} - \boldsymbol{\tau}_h)$, integrating over Ω and using the monotonicity $[\Pi(\mathbf{q}) - \Pi(\mathbf{p})] \cdot (\mathbf{q} - \mathbf{p}) \geq 0$ we arrive at

$$\|\mathbf{e}_{\tau}\|_{\mathcal{W}}^2 \leq \int_{\Omega} (\boldsymbol{\varepsilon}^{e^{tr}} - \boldsymbol{\varepsilon}_h^{e^{tr}}) \cdot (\boldsymbol{\tau} - \boldsymbol{\tau}_h) d\Omega. \tag{7.45}$$

To be able to evaluate all quantities in the reference configuration we have to retrieve $\boldsymbol{\tau}$ and $\boldsymbol{\varepsilon}^{e^{tr}}$ yielding the second Piola-Kirchhoff stress tensor $\mathbf{S}$ and a strain measure $\tilde{\mathbf{E}}^{e^{tr}}$

$$\mathbf{S} = \mathbf{F}^{-1}\boldsymbol{\tau}\mathbf{F}^{-T} \quad , \quad \tilde{\mathbf{E}}^{e^{tr}} = \mathbf{F}^T\boldsymbol{\varepsilon}^{e^{tr}}\mathbf{F} \tag{7.46}$$

with the discrete forms $\mathbf{S}_h$ and $\tilde{\mathbf{E}}_h^{etr}$ respectively. With the tensors (7.46) the estimator (7.45) can be written as

$$\|\mathbf{e}_\tau\|_{\mathcal{W}}^2 \leq \int_\Omega (\tilde{\mathbf{E}}^{etr} - \tilde{\mathbf{E}}_h^{etr}) \cdot (\mathbf{S} - \mathbf{S}_h)d\Omega. \tag{7.47}$$

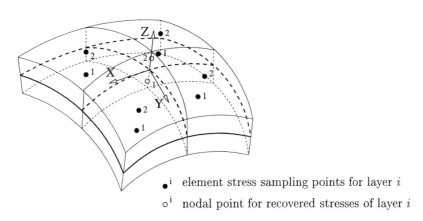

$\bullet^i$ element stress sampling points for layer i

$\circ^i$ nodal point for recovered stresses of layer i

Figure 7.5: Superconvergent Patch Recovery for the shell

To evaluate (7.47) the exact values $\mathbf{S}$ and $\tilde{\mathbf{E}}^{etr}$ are replaced by values determined by SPR. For this the relevant tensor representations in the plane stress state are completed to become 3D representations which are then supplied to the SPR procedure.

Thus additional to (7.47) we use SPR to approximate $\mathbf{S}$ and $\tilde{\mathbf{E}}^{etr}$ and arrive at

$$\|\mathbf{e}_\tau\|_{\mathcal{W}}^2 \approx \int_\Omega (\tilde{\mathbf{E}}_\star^{etr} - \tilde{\mathbf{E}}_h^{etr}) \cdot (\mathbf{S}_\star - \mathbf{S}_h)d\Omega. \tag{7.48}$$

with the recovered values $\tilde{\mathbf{E}}_\star^{etr}$ and $\mathbf{S}_\star$. The recovery procedure is done in the tangential coordinate system for the $X-$ and $Y-$ coordinate for every layer i_{layer}, see Fig 7.5. The recovery is applied for each component of $\mathbf{S}$ and $\tilde{\mathbf{E}}^{etr}$ evaluated at the midpoints of the elements. Since in the finite element solutions $\mathbf{S}_h$ and $\tilde{\mathbf{E}}_h^{etr}$ are obtained at the Gauss-points (2×2 integration) the history variables $(\mathbf{C}^p, \alpha)$ in this procedure are interpolated at $(0,0)$. With (7.48) the error is indicated more strictly than by simply pulling back ε^e instead of ε^{etr}. This can be motivated by the errors caused by the return mapping itself (see [25]). An expansion of (7.48) to the free energy Ψ (7.19) as error norm is easily done with

$$\|\mathbf{e}_{(\tau,\alpha)}\|_{\Psi}^2 \approx \int_\Omega (\tilde{\mathbf{E}}_\star^{etr} - \tilde{\mathbf{E}}_h^{etr}) \cdot (\mathbf{S}_\star - \mathbf{S}_h) + H(\alpha_\star - \alpha_h)^2 d\Omega. \tag{7.49}$$

7.3.5 Adaptive Strategies

A criterion for remeshing has to be defined to control the discretization error. This can be done by evaluating the relative error

$$\eta = (\|\mathbf{e}\|^2/\|\mathbf{e}\|^2 + \|\mathbf{u}\|^2)^{\frac{1}{2}} \tag{7.50}$$

thickness $t = 3.0$
Young's modulus $E = 3000.$
Poisson's ratio $\nu = 0.3$
yield stress $\sigma_Y = 24.3$
lin. hardening $h = 300.$

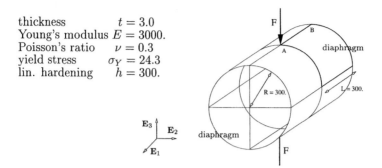

Figure 7.6: Pinched cylinder.

which has to be under a given tolerance η_{tol}. To obtain a mesh satisfying the required tolerance $\eta \leq \eta_{tol}$ a local predicted element size yielding an even distribution of the error is computed from

$$h^e_{new} = h^e_{old} \frac{e_m}{\|e_\tau\|_e} \quad \text{with} \quad \|e\|_e \leq e_m = \eta_{tol}[(\|e_\tau\|^2 + \|\tau_h\|^2)/n_{el}]^{\frac{1}{2}}. \qquad (7.51)$$

Since the optimal mesh changes with the deformation, the data of the current stage of the structure has to be transferred from one mesh to the other in order to obtain an optimal exploitation of the computational resources. If only primal variables have to be transferred this can be done without severe problems by projecting the deformation to the new mesh and an iteration to the equilibrium.

If history dependent problems are considered the internal variables have to be transferred, which cannot be done in general without loss of accuracy. To avoid a new source of error a second strategy is applied in which no transfer of variables is needed. Here for every remeshing the computation starts from the first load step. The density of the elements can then be defined by the smallest local element size during the computation of all load steps $k = 1, ..., k_{max}$

$$h^e = \min_k\{h^e_k\} . \qquad (7.52)$$

The procedure ends if all load steps are computed with a mesh satisfying the prescribed tolerance

$$\eta_k \leq \eta_{tol} , \quad \forall k. \qquad (7.53)$$

7.4   NUMERICAL EXAMPLES

Pinched cylinder

In this problem the elasto–plastic deformation of the pinched cylinder (Fig. 7.6) widely known from the literature is considered. Before the adaptive simulations are described some modifications of the discretization of this problem are pointed out.

To the discretization with linear shape functions. The meshes generated by the used method [22] are constructed from elements without any peculiar orientation. In order to use robust and efficient elements linear quadrilaterals are preferred here. With a discretization of the structure by linear elements only plane and hyperbolic curvature is possible inside the element. Thus elliptic surfaces as depicted in Fig. 7.7 are approximated in an unstructured mesh by locally hyperbolic curvatures.

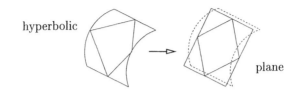

hyperbolic

plane

Figure 7.7: correction of the curvature.

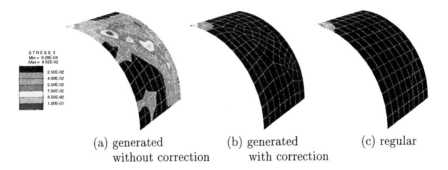

(a) generated without correction   (b) generated with correction   (c) regular

Figure 7.8: Stress plots of different meshes.

In [13] examinations for linear problems considering thin shell theory ($S_{3\alpha} = S_{33} = 0$) have been made for local hyperbolic curvatures in the element yielding the conclusion that the error made by these warped elements can be of the same order as the strain energy itself. Fig.7.8 illustrates this behavior (only one-eighth of the structure is discretized). In the numerical analysis, where the generated mesh has been used, excessive stresses, Fig.7.8a, compared to the stresses of the regular mesh, Fig.7.8c, can be observed. To overcome this problem, in [13] and [27] special modes have been introduced which can only be generated successfully for small and moderate strain states. A more simple correction of the element curvature is applied (Fig.7.7) here to compute the element stiffness matrix and residual vector. As can be seen in Fig.7.8b the excessive stresses are avoided with this correction and hence a reasonable discretization can be obtained for the error indication.

Adaptive computations. The adaptive solutions are carried out for tolerances of 15 and 10% with respect to $\|.\|_W$. Since the minimal element sizes of the adaptive meshes are restricted to 1/8 of the thickness, the loading can be viewed as a sort of regularization in the close area of the point A. As a reference solution the computation with a regular mesh of 5000 elements is considered. The applied load and the relative error of the computations are presented in Fig. 7.9, additionally in the right diagram of Fig. 7.9 the number of elements used in each adaptive mesh is shown. Four adaptive steps were made for each 15 and 10% relative tolerance.

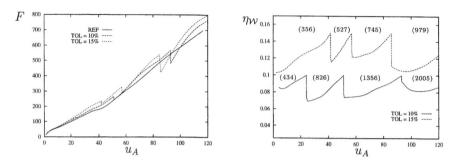

Figure 7.9: Applied loads and relative error over the load steps with transfer.

For displacement loading up to $u_M = 80.0$ we can see that the adaptive computations are in good agreement with the reference solution, especially after remeshing when the structure gets softer and the applied force decreases. Between $u_M = 80$ and 100 all adaptive runs show a decrease which is over–proportional in relation to the applied forces. For reasons to explain this strong decrease we have of course to think about the error made by transferring the plastic metric $\mathbf{C}_p$ and the hardening variable α described in [8]. This error is especially severe in this problem due to the single loading and the plastification in the kink which is characteristic for this pinched cylinder problem. The results of the computations are strongly dependent on the size of the elements used in the region of this kink, see also [10]. The structure exhibits large bending strains at this kink yielding a plastic 'hinge' for large u_M. Hence the results of the computations will be very sensitive to the errors made by transferring $\mathbf{C}_p$ and α within this region. If the mesh is finer the errors made by transferring the history data will be smaller and hence the descent of the applied force is also smaller, see Fig.7.9 left diagram.

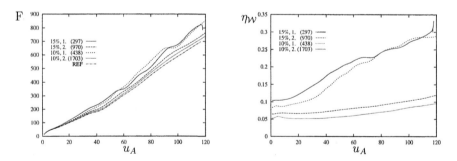

Figure 7.10: Applied load and relative error $\eta_\mathcal{W}$ without transfer.

As a comparison the same problem is computed with the strategy without transfer of the state variables. As initial meshes for the simulation an adaptive generated mesh is used which undergoes the prescribed tolerance in the first load step. With this approach unnecessary computations over all load steps can be avoided since the adaptive computation needs some iterations in remeshing to catch the single load. To satsify both tolerances over all load steps only one further remeshing is needed. In Fig. 7.10 the applied load and the relative error η_W are shown. About the same number of elements are needed as with the simulation with transfer of the state variables in the last adaptive mesh.

Pressure forming of a cylinder

The forming of a cylinder with pressure loading is now described. The cylinder of the previous example is taken as the structure to be deformed with the difference that the ends of the cylinder are free now. The outer part of the form in which the cylinder is cylindrical and is enlarged in the middle to an elliptic paraboloid is shown in Fig. 7.11, first picture. The simulations are performed again with the correction of the curvature in the reference configuration as in the previous problem (Fig. 7.7). Since large plastic deformations are expected in this problem the adaptive strategy without transfer of the state variables is applied. As relative tolerance 7% with respect to $\|.\|_\Psi$ is used. The first mesh is generated in the same manner as in the previous example by adaptive remeshing in the first load step until the tolerance is satisfied.

Figure 7.11: Press form, first mesh and second adapted mesh

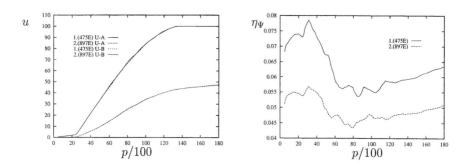

Figure 7.12: Displacements of points A, B and relative error η_Ψ.

Only one further adaptive refinement is needed to satisfy the tolerance. The adaptive discretizations in the final configuration are presented in Fig. 7.11 with their contact zones.

The relative error η_Ψ and the displacements of the points A in $\mathbf{E}_3$ and B in $\mathbf{E}_1$–directions are plotted in Fig. 7.12. The hinge in the displacements occurs with the beginning of the plastic deformation. The displacements at point A remain constant beyond a pressure of 1.27 when the point makes contact with the form. The highest stresses are found in the middle of the structure where the cylinder has expanded the most.

Deep drawing of a sheet

In the last shell example the deep drawing of a sheet is simulated with a rigid punch, blank holder and die as shown in Fig. 7.13 with the material data as shown. Again a quarter of the sheet is considered here due to symmetry. A perspective view of the tools described with several NURBS–free–form–surfaces shown in Fig. 7.13, and data for the definition of the master surfaces can be found in [9]. In the simulation the die and blank holder are positioned at a distance of $d = 0.1$ to each other. The punch is advanced after the first contact with the sheet by intervals of $\Delta u_s = 0.05$ until it has reached the final position $u_s = 50.0$. The adaptive simulations are performed for 15% and 10% tolerance with respect to $\|.\|_\Psi$. Again meshes fulfilling the prescribed tolerance in the first load step have been used as initial meshes.

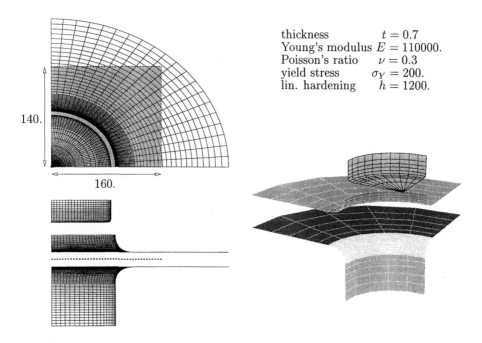

thickness $t = 0.7$
Young's modulus $E = 110000.$
Poisson's ratio $\nu = 0.3$
yield stress $\sigma_Y = 200.$
lin. hardening $h = 1200.$

Figure 7.13: Deep drawing of a sheet.

The sum over the vertical contact force and the relative error with respect to $\|.\|_\Psi$ are plotted in Fig. 7.14. The relative error decreases for higher load steps as in the previous example. The highest strains occur in the contact area of the die and at the border of

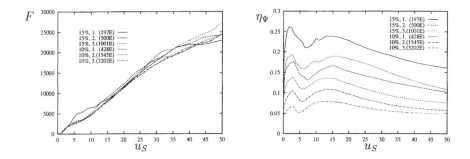

Figure 7.14: Vertical applied contact force and relative error η_Ψ.

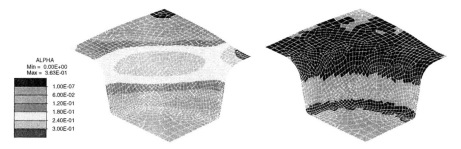

Figure 7.15: Plastic hardening variable α and contact zone of the final configuration of the second mesh of the adaptive computation with $tol = 10\%$.

the shorter edge of the structure where with $\alpha = 0.363$ the maximum value of the plastic hardening variable is reached, see Fig. 7.15. As in the previous example almost the whole structure is deformed by membrane forces in the final configuration; significant bending can be found only in the contact zone of the punch.

Hertzian problem

In this example we apply the error estimation procedures, global as well as local SPR based, to solve the well known 2D Hertzian problem of an elastic cylinder (Young's modulus $E = 7000$, Poisson's ratio $\nu = 0.3$) making contact with a planar rigid surface. The cylinder with a radius $r = 1$ and infinite length is loaded by an overall load of $F = 100$. Due to the symmetry we discretize one half of the problem using linear triangular elements. For this problem an exact solution for the contact pressure can be obtained analytically. Hence we compare the results computed by the adaptive methods directly with the exact solution. To avoid difficulties related to a point load in elasticity, the load is distributed over a small surface on top of the cylinder. In order to simulate the rigid surface we set Young's modulus to $E = 100000$ and Poisson's ratio to $\nu = 0.45$ in the finite element model. The initial mesh is depicted in Fig. 7.16. Within the duality based method we assume that the maximum contact pressure at the bottom of the cylinder is

Figure 7.16: Initial mesh

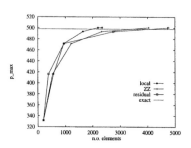

Figure 7.17: Convergence behaviour

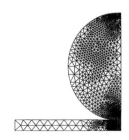

Figure 7.18: Global SPR

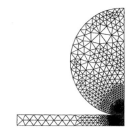

Figure 7.19: Duality based SPR

the quantity to be computed with best accuracy.

Figs. 7.18 and 7.19 show respectively the meshes optimized with respect to the local stresses and the global energy norm. While the local error control focuses on the point of interest, the global energy adaptive strategy leads also to refinements in the vicinity of load initiation. The maximum contact pressure obtained for the two different finite element models is compared with the analytical solution ($p_{max} = 495$) in Fig. 7.17. We observe a significantly higher effectivity of the mesh optimized with respect to the maximum contact pressure, which reaches the exact value with half the number of the elements required for the global method.

Example: Crossing tubes
In this example we consider two crossing rubber tubes ($E = 3000, \nu = 0.3$)
each with a length of $L = 100$mm and
radii of $r_o = 50$mm , $r_i = 30$mm. Due
to symmetry a one fourth model is used.
The mesh adaption is performed with respect to the global error energy norm and
with a prescribed relative error tolerance
of $\eta_{err} = 15$ %. Fig. 7.23 and Fig. 7.24
depict the mean stress field.

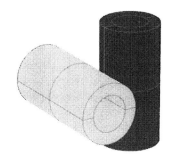

Figure 7.20: Initial geometry

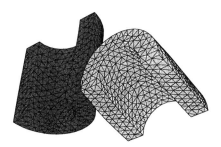

Figure 7.21: One fourth model,initial mesh

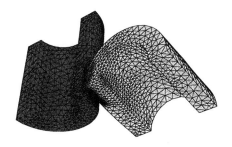

Figure 7.22: adapted mesh

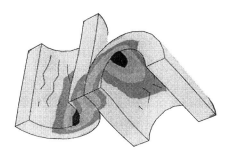

Figure 7.23: σ_I , initial mesh

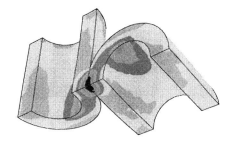

Figure 7.24: σ_I , adapted mesh

Example: Sliding sphere

This example is associated with large displacements and sliding. The initial geometry is shown in Fig.7.25 depicting a rubber sphere ($E = 3000, \nu = 0.3$) which slides between two rubber spheres ($E = 6000, \nu = 0.3$). Again the mesh adaption is performed with respect to the global error energy norm and with a prescribed relative error tolerance of $\eta_{err} = 15\%$. Fig. 7.28 and Fig. 7.29 depict the mean stress field.

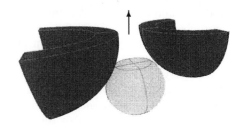

Figure 7.25: Initial geometry

Figure 7.26: One fourth model, initial mesh

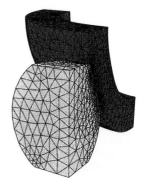

Figure 7.27: adapted mesh

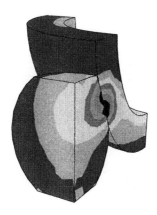

Figure 7.28: σ_I , initial mesh

Figure 7.29: σ_I , adapted mesh

REFERENCES

[1] I. Babuška and W. C. Rheinboldt, (1978). Error estimates for adaptive finite element computations, *SIAM J Numer. Analysis*, **15**, 736–754.

[2] I. Babuška, T. Strouboulis, C. S. Upadhyay, S. Gangaraj, and K., Copps, (1994). Validation of a posteriori error estimators by numerical approach, *Int. J. Numer. Meth. Engng.*, **37**, 1073–1123.

[3] I. Babuška and C. Schwab, (1993). A posteriori error estimation for hierarchic models of elliptic boundary value problems on thin domains , Institute for Physical Science and Technology, University of Maryland College Park, MD20742, USA, BN–1175 / CMC Rep. 94–06, Technical Note BN 1148.

[4] R. Becker, and R. Rannacher, (1996). A Feed-Back Approach to Error Control in Finite Element Methods: Basic Analysis and Examples, *EAST-WEST J. Numer. Math. 4*, 237–264.

[5] C. Carstensen, O. Scherf and P. Wriggers, (1997). Numerical Analysis for Contact of Elastic Bodies, *SIAM J. Sci. Comp.*.

[6] R. Eberlein, (1997). Finite–Elemente–Konzepte für Schalen mit großen elastischen und plastischen Verzerrungen, Dissertation, Technische Hochschule Darmstadt, Institut für Mechanik IV.

[7] C.-S. Han and P. Wriggers, (1998). A simple local a posteriori bending indicator for axisymmetric membrane and bending elements, *Engng. Computations*, **7**, 977–988.

[8] C.-S. Han and P. Wriggers, (1999). An h–adaptive method for elasto–plastic shell problems, accepted for publication in *Comput. Meth. Appl. Mech. Engrg*.

[9] Han, C.-S., (1999). Eine h-adaptive FE-Methode für elasto–plastische Schalenprobleme in unilateralem Kontakt, Dissertation to appear, Technische Hochschule Darmstadt, Institut für Mechanik IV.

[10] R. Hauptmann and K. Schweizerhof, (1998). A Systematic Developement of 'Solid–Shell' Element Formulation for Linear and Non–Linear Analysis Employing only Displacement Degrees of Freedom, *Int. J. Numer. Meth. Engng.*, **42**, 49–69.

[11] I. Hlaváček, J. Haslinger, J. Nečas and J. Lovíšek, (1988). J., *Solution of variational inequalities in mechanics*, Springer, New York.

[12] A. Ibrahimbegovic, (1994). Finite Elastoplastic Deformations of Space–Curved Membranes, *Comput. Meth. Appl. Mech. Engrg.*, **119**, 371–394.

[13] Ph. Jetteur, (1987). Improvement of the quadrilateral JET shell element for a particular class of shell problems, Ecole Polytechnique Federale de Lausanne, IREM internal report 87/1.

[14] Johnson, C., (1991). Adaptive finite element methods for the obstacle problem, Technical Report, Chalmers University of Technology, Göteborg.

[15] C. Johnson and P. Hansbo, (1992). Adaptive finite element methods in computational mechanics, *Comput. Meth. Appl. Mech. Engrg.*, **101**, 143–181.

[16] C. Johnson and P. Hansbo, (1992). Adaptive finite element methods for small strain elasto–plasticity. In D. Besdo and E. Stein (eds.). 'Finite inelastic deformations – Theory and Applications', Springer, Heidelberg.

[17] Kikuchi, N., and Oden, J. T., (1988). *Contact problems in elasticity: A study of variational inequalities and finite element methods*, SIAM, Philadelphia, 1988.

[18] C. Miehe, (1998). A theoretical and computational model for isotropic elastoplastic stress analysis in shells at large strains, *Comput. Meth. Appl. Mech. Engrg.*, **155**, 193–233.

[19] J. Tinsley Oden and J. R. Cho, (1996). Local *A Posteriori* Error Estimation of Hierarchical Models for Plate– and Shell–Like Structures, *Comput. Meth. Appl. Mech. Engrg.*

[20] S. I. Oh and S. Kobayashi, (1980). Finite Element Analysis of Plane–Strain Sheet Bending, Int. J. Mech. Sci, **22**, 583–594.

[21] E. Ramm, (1976). Geometrisch nichtlineare Elastostatik und Finite Elemente, Institut für Baustatik, Universität Stuttgart, Habilitation, Bericht Nr. 76–2.

[22] E. Rank and M. Schweingruber and M. Sommer, (1993). Adaptive mesh generation and transformation of triangular to quadrilateral meshes, Commun. Numer. Methods Eng., **9**, 121–129.

[23] W. C. Rheinboldt, (1985). Error Estimates for Nonlinear Finite Element Computations, *Comp. Struct.*, **20**,91–98.

[24] O. Scherf, (1997). Kontinuumsmechanische Modellierung nichtlinearer Kontaktprobleme und ihre numerische Analyse mit adaptiven Finite-Element-Methoden, Institute for Mechanics, University of Darmstadt.

[25] J.C. Simo and T.J.R. Hughes, (1998). *Computational Inelasticity*, Springer, Heidelberg.

[26] E. Stein and S. Ohnimus, (1995). Expansion method for the integrated solution– and model adaptivity within the FEM–analysis of plates and shells, EWCM to the honour of Alf Samuelsson and NSCM in Göteborg November 16–18.

[27] R. L. Taylor, (1988). Finite element analysis of linear shell problems. In J. R. Whiteman (ed.). 'The Mathematics of Finite Elements and Application', 1988, Academic Press, London, 191–204.

[28] P. Wriggers, O. Scherf and C. Carstensen, (1994). Adaptive techniques for the contact of elastic bodies, In *Recent Developments in Finite Element Analysis*, J.R. Hughes, E. Onate, O.C. Zienkiewicz (eds.), CIMNE, Barcelona.

[29] P. Wriggers and O. Scherf, (1995). Adaptive finite element methods for contact problems in plasticity, Proceedings of COMPLAS IV, Owen, D. R. J., Hinton, E., Oñate, E. (eds.), Pineridge Press, Swansea, 787-807.

[30] P. Wriggers and O. Scherf, (1995). An Adaptive Finite Element Algorithm for Contact Problems in Plasticity, Comp. Mech., **17**, 88–97.

[31] P. Wriggers and O. Scherf, (1994). An Adaptive Finite Element Technique for Nonlinear Contact Problems, Contact Mechanics, Proceedings of the 2nd Contact Mechanics International Symposium, Raous, M., Jean, M., Moreau, J. J. (eds.), Plenum Publishing, London.

[32] P. Wriggers, O. Scherf and C. Carstensen, (1995). An adaptive methods for frictionless contact problems, ZAMM,**75**,37-40.

[33] P. Wriggers and O. Scherf, (1997). Different a posteriori error estimators and indicators for contact problems, *Mathematics and Computer Modelling*, **28**,No.4-8,437-447.

[34] P. Wriggers and O. Scherf (1999). Adaptive Finite Element Techniques for Frictional Contact Problems Involving Large Elastic Strains, to appear in:*Comp. Mech.*

[35] Zienkiewicz, O.C., and J. Z. Zhu, (1997). A simple error estimator and adaptive procedure for practical engineering analysis, *Int. J. Num. Meth. Engng.*, **24**, 337–357.

[36] O. C. Zienkiewicz and R. L. Taylor (1988). *The Finite Element Method*, 4th ed., Vol. I, McGraw Hill, London.

[37] O. C. Zienkiewicz and J. Z. Zhu, (1993). The superconvergent patch recovery and a posteriori error estimates. Part 1: The recovery technique.*Int. J. Num. Meth. Engrg.*, **33**, 1331–1364.

8 HP-FINITE ELEMENT METHODS FOR HYPERBOLIC PROBLEMS

Endre Süli[a], Paul Houston[a] and Christoph Schwab[b]

[a]Oxford University Computing Laboratory, Wolfson Building, Parks Road, Oxford
OX1 3QD, United Kingdom[1]
[b]Seminar for Applied Mathematics, ETH Zürich, CH-8092 Zürich, Switzerland

ABSTRACT

This paper is devoted to the *a priori* and *a posteriori* error analysis of the *hp*-version of the discontinuous Galerkin finite element method for partial differential equations of hyperbolic and nearly-hyperbolic character. We consider second-order partial differential equations with nonnegative characteristic form, a large class of equations which includes convection-dominated diffusion problems, degenerate elliptic equations and second-order problems of mixed elliptic-hyperbolic-parabolic type. An *a priori* error bound is derived for the method in the so-called DG-norm which is optimal in terms of the mesh size h; the error bound is either 1 degree or 1/2 degree below optimal in terms of the polynomial degree p, depending on whether the problem is convection-dominated, or diffusion-dominated, respectively. In the case of a first-order hyperbolic equation the error bound is *hp*-optimal in the DG-norm. For first-order hyperbolic problems, we also discuss the *a posteriori* error analysis of the method and implement the resulting bounds into an *hp*-adaptive algorithm. The theoretical findings are illustrated by numerical experiments.

Key words. *hp*-finite element methods, hyperbolic problems, nonnegative characteristic form, *a priori* error analysis, *a posteriori* error analysis, adaptivity

8.1 INTRODUCTION

The discontinuous Galerkin finite element method (DGFEM) has a long and distinguished history. Its roots can be traced back to the work of Pian and collaborators [20] in the early 1960s on hybrid methods for elliptic problems (see also [19] for a historical survey); the mathematical analysis of hybrid methods was initiated by Babuška [4]. In 1971, Nitsche [18] considered an alternative scheme where the boundary multipliers present in

---

[1]Endre Süli and Paul Houston acknowledge the financial support of the EPSRC (Grant GR/K76221).

the hybrid formulation were eliminated in terms of normal fluxes and stabilisation terms were added to recover the optimal convergence rate. In a different context, discontinuous finite element methods were introduced by Reed and Hill [21], and Lesaint and Raviart [17] in order to overcome the stability limitations of conventional continuous finite element approximations to first-order hyperbolic problems. Although subsequently much of the research in the field of numerical analysis of partial differential equations concentrated on the development and the analysis of conforming finite element methods, in recent years there has been an upsurge of interest in discontinuous schemes. This paradigm shift was stimulated by several factors: the desire to handle, within the finite element framework, nonlinear hyperbolic problems (see [10] and [11]) which are known to exhibit discontinuous solutions even when the data are perfectly smooth; the need to treat convection-dominated diffusion problems without excessive numerical stabilisation; the computational convenience of discontinuous finite element methods due to a large degree of locality; and the necessity to accommodate high-order hp-adaptive finite element discretisations in a flexible manner (see [9]). The DGFEM can also be thought of as the higher-order extension of the classical cell centred finite volume method – a popular discretisation technique in the computational aerodynamics community.

In the present paper we develop the error analysis of the hp-DGFEM for partial differential equations of hyperbolic and nearly hyperbolic character. We begin by considering the *a priori* error analysis of the hp-DGFEM for second-order partial differential equations with nonnegative characteristic form; this represents a continuation of our earlier work [14] for first-order hyperbolic equations. In [14] an error bound, optimal both in terms of the local mesh size h and the local polynomial degree p, was derived for the hp-DGFEM supplemented by a streamline-diffusion type stabilisation involving a stabilisation parameter δ of size h/p. Here, we establish a similar result in the case of partial differential equations with nonnegative characteristic form; the resulting error bound is optimal in terms of powers of h, the part of the error bound which arises from the diffusion term is one power of p below the optimal rate, while the parts which stem from the advection and reaction terms are of optimal order in p. For convection-dominated diffusion equations, suboptimality in p is compensated by the fact that the leading term in the error bound is multiplied by a small number, proportional to the square root of the norm of the diffusion matrix. Indeed, in the case of a first-order hyperbolic equation, our error bound collapses to one that is hp-optimal. On the other hand, when the advective term is absent, the error bound is optimal in terms of powers of h and it is $1/2$ a power below optimal in terms of the polynomial degree p. The hp-DGFEM considered in this paper involves a discontinuity-penalisation device based on the ideas of Nitsche [18], Wheeler [26] and Arnold [3], albeit with a small but significant modification which permits us to pass to the hyperbolic limit with inactive discontinuity-penalisation. The error analysis of the hp-DGFEM discretisation considered here can also be viewed as an extension of the work of Baumann [6], Oden, Babuška and Baumann [19] and Riviere and Wheeler [22] in the reaction-diffusion case; also, they present an improvement over our earlier results presented in [25] where the error analysis of the hp-DGFEM was considered for partial differential equations with nonnegative characteristic form, albeit without streamline diffusion stabilisation. While in [25] the size of the discontinuity penalisation parameter was required to be $Const.p^2/h$, in the present paper it has been reduced to $Const/h$.

The second part of the paper is concerned with the *a posteriori* error analysis of the *hp*–DGFEM for hyperbolic problems. Here, we derive an *a posteriori* bound on the error for *hp*-DGFEM approximations of linear functionals of the analytical solution. The *a posteriori* error bound is based on an error representation formula which stems from a duality argument and the Galerkin orthogonality property of the *hp*-DGFEM. The error representation formula involves the computable finite element residual and the difference between the dual solution and its projection. We exemplify the relevance of the theoretical results by implementing the *a posteriori* error bound into an *hp*-adaptive algorithm for calculating the outflow normal flux of the solution to within a prescribed tolerance.

8.2 MODEL PROBLEM AND DISCRETISATION

Suppose that Ω is a bounded Lipschitz domain in $\mathbb{R}^d$, $d = 2, 3$, and consider the linear second-order partial differential equation

$$\mathcal{L}u \equiv -\sum_{i,j=1}^d \partial_j \left(a_{ij}(x)\,\partial_i u\right) + \sum_{i=1}^d b_i(x)\,\partial_i u + c(x)u = f(x) \;, \tag{8.1}$$

where f is a real-valued function belonging to $L^2(\Omega)$, and the real-valued coefficients a, b, c are such that:

$$\begin{aligned}
a(x) &= \{a_{ij}(x)\}_{i,j=1}^d \in L^\infty(\Omega)_{\mathrm{sym}}^{d\times d}, \\
b(x) &= \{b_i(x)\}_{i=1}^d \in W^{1,\infty}(\Omega)^d, \quad c(x) \in L^\infty(\Omega) \;.
\end{aligned} \tag{8.2}$$

It will be assumed throughout that the characteristic form associated with the principal part of the partial differential operator $\mathcal{L}$ is nonnegative; that is,

$$\boldsymbol{\xi}^T a(x)\,\boldsymbol{\xi} \geq 0 \quad \forall \boldsymbol{\xi} \in \mathbb{R}^d \text{ and a.e. } x \in \bar{\Omega} \;. \tag{8.3}$$

In order to ensure that the restriction of the matrix a to the boundary $\partial\Omega$ of Ω is well defined, we shall assume, for simplicity, that the entries of a are piecewise continuous on $\bar{\Omega}$. This assumption is sufficiently general to cover most cases of practical significance. Now let $\boldsymbol{\mu}(x) = \{\mu_i(x)\}_{i=1}^d$ denote the unit outward normal vector to $\Gamma = \partial\Omega$ at $x \in \Gamma$ and define the following subsets of Γ:

$$\Gamma_0 = \{x \in \Gamma : \boldsymbol{\mu}^T a(x)\boldsymbol{\mu} > 0\} \;,$$

$$\Gamma_- = \{x \in \Gamma \backslash \Gamma_0 : \boldsymbol{b} \cdot \boldsymbol{\mu} < 0\} \quad \text{and} \quad \Gamma_+ = \{x \in \Gamma \backslash \Gamma_0 : \boldsymbol{b} \cdot \boldsymbol{\mu} \geq 0\} \;.$$

The sets $\Gamma_\mp$ will be referred to as the inflow and outflow boundary, respectively. With these definitions we have that $\Gamma = \Gamma_0 \cup \Gamma_- \cup \Gamma_+$. We shall further decompose Γ_0 into two connected parts, Γ_D where a Dirichlet boundary condition is imposed and Γ_N where a Neumann condition is given, and we supplement the partial differential equation (8.1) with the following boundary conditions:

$$u = g_D \quad \text{on } \Gamma_D \cup \Gamma_- \quad \text{and} \quad \boldsymbol{\mu}^T a \nabla u = g_N \text{ on } \Gamma_N \;. \tag{8.4}$$

The boundary value problem (8.1), (8.4) includes a range of physically relevant instances, such as the mixed boundary value problem for an elliptic equation corresponding to the case when (8.3) holds with strict inequality, as well as the case of a linear transport problem associated with the choice of $a \equiv 0$ on $\bar{\Omega}$.

8.2.1 Finite element spaces

Suppose that $\mathcal{T}$ is a subdivision of Ω into open element domains κ such that $\bar{\Omega} = \cup_{\kappa \in \mathcal{T}} \bar{\kappa}$. Let us assume that the family of subdivisions $\mathcal{T}$ is shape-regular and that each $\kappa \in \mathcal{T}$ is a smooth bijective image of a fixed master element $\hat{\kappa}$, that is, $\kappa = \mathcal{F}_\kappa(\hat{\kappa})$ for all $\kappa \in \mathcal{T}$ where $\hat{\kappa}$ is either the open unit simplex or the open unit hypercube in $\mathbb{R}^d$. For an integer $r \geq 1$, we denote by $\mathcal{P}_r(\hat{\kappa})$ the set of polynomials of total degree $\leq r$ on $\hat{\kappa}$; when $\hat{\kappa}$ is the unit hypercube, we also consider $\mathcal{Q}_r(\hat{\kappa})$, the set of all tensor-product polynomials of degree $\leq r$ in each coordinate direction. The case of $r = 0$ can be easily incorporated into our analysis, but we have chosen to exclude it for simplicity of presentation so as to ensure that $1/r$ is meaningful for any polynomial degree under consideration. Next, to $\kappa \in \mathcal{T}$ we assign an integer $p_\kappa \geq 1$, collect the p_κ and $\mathcal{F}_\kappa$ in the vectors $\mathbf{p} = \{p_\kappa : \kappa \in \mathcal{T}\}$ and $\mathbf{F} = \{\mathcal{F}_\kappa : \kappa \in \mathcal{T}\}$, respectively, and consider the finite element space

$$S^{\mathbf{p}}(\Omega, \mathcal{T}, \mathbf{F}) = \{u \in L^2(\Omega) : u|_\kappa \circ \mathcal{F}_\kappa \in \mathcal{R}_{p_\kappa}(\hat{\kappa}) \quad \forall \kappa \in \mathcal{T}\} ,$$

where $\mathcal{R}_{p_\kappa}$ is either $\mathcal{P}_{p_\kappa}$ or $\mathcal{Q}_{p_\kappa}$. Assuming that $\mathcal{T}$ is a subdivision of Ω and $s > 0$, $H^s(\Omega, \mathcal{T})$ will denote the associated broken Sobolev space of index s.

8.2.2 The numerical method

Discretisation of the Low-Order Terms. Since the emphasis in this paper is on problems of hyperbolic and nearly-hyperbolic character, we begin by considering the *hp*-DGFEM approximation of the first-order partial differential operator $\mathcal{L}_b$ defined by

$$\mathcal{L}_b w = \mathbf{b} \cdot \nabla w + cw .$$

Assuming that κ is an element in the subdivision $\mathcal{T}$, we denote by $\partial \kappa$ the union of open faces of κ. This is non-standard notation in that $\partial \kappa$ is a subset of the boundary of κ; we have adopted it so as to ensure that the unit outward normal vector $\boldsymbol{\mu}(x)$ to $\partial \kappa$ at $x \in \partial \kappa$ is correctly defined. With these conventions, we define the inflow and outflow parts of $\partial \kappa$, respectively, by

$$\partial_- \kappa = \{x \in \partial \kappa : \mathbf{b}(x) \cdot \boldsymbol{\mu}(x) < 0\} , \qquad \partial_+ \kappa = \{x \in \partial \kappa : \mathbf{b}(x) \cdot \boldsymbol{\mu}(x) \geq 0\} . \tag{8.5}$$

For each $v \in H^1(\Omega, \mathcal{T})$ and any $\kappa \in \mathcal{T}$, we denote by v^+ the interior trace of v on $\partial \kappa$ (the trace taken from within κ). Let us consider an element κ such that the set $\partial_- \kappa \backslash \Gamma_-$ is nonempty; then for each $x \in \partial_- \kappa \backslash \Gamma_-$ (with the exception of a set of $(d-1)$-dimensional measure zero) there exists a unique element κ', depending on the choice of x, such that $x \in \partial_+ \kappa'$. If $\partial_- \kappa \backslash \Gamma_-$ is nonempty for some $\kappa \in \mathcal{T}$, then we can also define the outer trace v^- of v on $\partial_- \kappa \backslash \Gamma_-$ relative to κ as the inner trace v^+ relative to those elements κ' for which $\partial_+ \kappa'$ has intersection with $\partial_- \kappa \backslash \Gamma_-$ of positive $(d-1)$-dimensional measure. Furthermore, we introduce the oriented jump of v across $\partial_- \kappa \backslash \Gamma_-$: $\lfloor v \rfloor = v^+ - v^-$.

Given that $v, w \in H^1(\Omega, \mathcal{T})$, we define, as in [16], for example, the bilinear form

$$B_b(w, v) = \sum_{\kappa \in \mathcal{T}} \int_\kappa (\mathcal{L}_b w) v \, dx \tag{8.6}$$

$$- \sum_{\kappa \in \mathcal{T}} \int_{\partial_- \kappa \backslash \Gamma_-} (\mathbf{b} \cdot \boldsymbol{\mu}) \lfloor w \rfloor v^+ \, ds - \sum_{\kappa \in \mathcal{T}} \int_{\partial_- \kappa \cap \Gamma_-} (\mathbf{b} \cdot \boldsymbol{\mu}) w^+ v^+ \, ds ,$$

and the linear functional

$$\ell_b(v) = \sum_{\kappa \in \mathcal{T}} \int_\kappa fv \, dx - \sum_{\kappa \in \mathcal{T}} \int_{\partial_-\kappa \cap \Gamma_-} (\boldsymbol{b} \cdot \boldsymbol{\mu}) \, gv^+ \, ds \; . \tag{8.7}$$

Next, we focus on the discretisation of the leading term in the partial differential equation.

Discretisation of the Leading Term. Let us suppose that the elements in the subdivision have been numbered in a certain way, regardless of the direction of the advective velocity vector $\boldsymbol{b}$. We denote by $\mathcal{E}$ the set of element faces (edges for $d = 2$ or faces for $d = 3$) associated with the subdivision $\mathcal{T}$. Since hanging nodes are permitted in DGFEM, $\mathcal{E}$ will be understood to consist of the smallest faces in $\partial \kappa$. Also, let $\mathcal{E}_{\text{int}}$, resp. Γ_{int}, denote the set, resp. union, of all faces $e \in \mathcal{E}$ which do not lie on $\partial \Omega$. Given that $e \in \mathcal{E}_{\text{int}}$, there exist indices i and j such that $i > j$ and κ_i and κ_j share the interface e; we define the (numbering-dependent) jump of $v \in H^1(\Omega, \mathcal{T})$ across e and the mean value of v on e, respectively, by $[v] = v|_{\partial \kappa_i \cap e} - v|_{\partial \kappa_j \cap e}$ and $\langle v \rangle = \left(v|_{\partial \kappa_i \cap e} + v|_{\partial \kappa_j \cap e} \right) / 2$.

It is clear that, in general, $[v]$ will be distinct from $\lfloor v \rfloor$ in that the latter depends on the direction of the unit outward normal to an element boundary, while the former is only dependent on the element numbering; however, $|[v]| = |\lfloor v \rfloor|$. With each face $e \in \mathcal{E}_{\text{int}}$ we associate the normal vector $\boldsymbol{\nu}$ which points from κ_i to κ_j; on boundary faces we define $\boldsymbol{\nu} = \boldsymbol{\mu}$. Finally, we introduce, following [19], the bilinear form

$$B_a(w, v) = \sum_{\kappa \in \mathcal{T}} \int_\kappa a(x) \nabla w \cdot \nabla v \, dx + \int_{\Gamma_D} \{ w((a\nabla v) \cdot \boldsymbol{\nu}) - ((a\nabla w) \cdot \boldsymbol{\nu})v \} \, ds$$

$$+ \int_{\Gamma_{\text{int}}} \{ [w] \langle (a\nabla v) \cdot \boldsymbol{\nu} \rangle - \langle (a\nabla w) \cdot \boldsymbol{\nu} \rangle [v] \} \, ds \; , \tag{8.8}$$

associated with the principal part of the differential operator $\mathcal{L}$, and the linear functional

$$\ell_a(v) = \int_{\Gamma_D} g_D((a\nabla v) \cdot \boldsymbol{\nu}) \, ds + \int_{\Gamma_N} g_N v \, ds \; .$$

Discontinuity-Penalisation Term. Let $\bar{a} = ||a||_2$, with $|| \cdot ||_2$ denoting the matrix norm subordinate to the l^2 vector norm on $\mathbb{R}^d$, and let $\bar{a}_\kappa = \bar{a}|_\kappa$. To each e in $\mathcal{E}_{\text{int}}$ which is a common face of elements κ_i and κ_j in $\mathcal{T}$ we assign the nonnegative function $\langle \bar{a} \rangle_e = (\bar{a}_{\kappa_i}|_e + \bar{a}_{\kappa_j}|_e)/2$. Letting $\mathcal{E}_D$ denote the set of all faces contained in Γ_D, to each $e \in \mathcal{E}_D$ we assign the element $\kappa \in \mathcal{T}$ with that face and define $\langle \bar{a} \rangle_e = \bar{a}_\kappa|_e$. Consider the function σ defined on $\Gamma_{\text{int}} \cup \Gamma_D$ by $\sigma(x) = K \langle \bar{a} \rangle_e / |e|$ for $x \in e$ and $e \in \mathcal{E}_{\text{int}} \cup \mathcal{E}_D$, where $|e| = \text{meas}_{d-1}(e)$ and K is a positive constant (whose value is irrelevant for the present analytical study, so we put $K = 1$) and introduce

$$B_s(w, v) = \int_{\Gamma_D} \sigma wv \, ds + \int_{\Gamma_{\text{int}}} \sigma [w][v] \, ds \; , \quad \ell_s(v) = \int_{\Gamma_D} \sigma g_D v \, ds \; . \tag{8.9}$$

We highlight the fact that since the weight-function σ involves the norm of the matrix a, in the hyperbolic limit of $a \equiv 0$ the bilinear form $B_s(\cdot, \cdot)$ and the linear functional ℓ_s both vanish. This is a desirable property, since linear hyperbolic equations may possess solutions that are discontinuous across characteristic hypersurfaces, and penalising discontinuities across faces which belong to these would be unnatural.

Streamline-diffusion stabilisation.   Let $\delta \in H^1(\Omega, \mathcal{T})$ be a nonnegative function. In the present context δ will play the role of a *stabilisation parameter*; typically δ is chosen to be constant on each $\kappa \in \mathcal{T}$, although we shall not require this for now. We define the bilinear form and the linear functional, respectively,

$$B_\delta(w, v) = \sum_{\kappa \in \mathcal{T}} \int_\kappa \delta(\mathcal{L}w)(\boldsymbol{b} \cdot \nabla v) \, \mathrm{d}x \ , \qquad \ell_\delta(v) = \sum_{\kappa \in \mathcal{T}} \int_\kappa \delta f(\boldsymbol{b} \cdot \nabla v) \, \mathrm{d}x \ . \qquad (8.10)$$

The precise choice of the stabilisation parameter will be given in the next section.

Definition of the Method.   Finally, we define the bilinear form $B_{\mathrm{DG}}(\cdot, \cdot)$ and the linear functional $\ell_{\mathrm{DG}}(\cdot)$, respectively, by

$$
\begin{aligned}
B_{\mathrm{DG}}(w, v) &= B_a(w, v) + B_b(w, v) + B_s(w, v) + B_\delta(w, v) \ , \\
\ell_{\mathrm{DG}}(v) &= \ell_a(v) + \ell_b(v) + \ell_s(v) + \ell_\delta(v) \ .
\end{aligned}
$$

The hp-DGFEM approximation of (8.1), (8.4) is: find $u_{\mathrm{DG}} \in S^{\mathbf{P}}(\Omega, \mathcal{T}, \mathbf{F})$ such that

$$B_{\mathrm{DG}}(u_{\mathrm{DG}}, v) = \ell_{\mathrm{DG}}(v) \qquad \forall v \in S^{\mathbf{P}}(\Omega, \mathcal{T}, \mathbf{F}) \ . \qquad (8.11)$$

In the next section we state the key properties of this method. Before we do so, however, we note that in the definitions of the bilinear forms and linear functionals above and in the arguments which follow it has been tacitly assumed that $a \in C(\kappa)$ for each $\kappa \in \mathcal{T}$, that the fluxes $(a\nabla u) \cdot \boldsymbol{\nu}$ and $(\boldsymbol{b} \cdot \boldsymbol{\mu})u$ are continuous across element interfaces, and that u is continuous in an (open) neighbourhood of the subset of Ω where a is not identically equal to zero. If the problem under consideration violates these properties, the scheme and the analysis have to be modified accordingly.

8.3   ANALYTICAL RESULTS

Our first result concerns the positivity of the bilinear form $B_{\mathrm{DG}}(\cdot, \cdot)$ and the existence and uniqueness of a solution to (8.11). In order to prove it, we shall require the following *inverse inequality* (see [23]): there exists a positive constant C_{inv}, dependent only on the constant of the angle condition such that

$$\|\nabla \cdot W\|_{L^2(\kappa)} \le C_{\mathrm{inv}} \frac{p_\kappa^2}{h_\kappa} \|W\|_{L^2(\kappa)} \qquad (8.12)$$

for all $\kappa \in \mathcal{T}$ and all $W = (w_1, \dots, w_d) \in [S^{\mathbf{P}}(\Omega, \mathcal{T}, \mathbf{F})]^d$.

Theorem 8.1 *Suppose that, in addition to (8.2) and (8.3), there exists a positive constant γ_0 such that $\gamma \equiv c - \frac{1}{2}\nabla \cdot \boldsymbol{b} \ge \gamma_0$ on $\bar{\Omega}$. Let us also assume that*

$$0 \le \delta \le \frac{1}{2} \min\left(\frac{h_\kappa^2}{C_{inv}^2 p_\kappa^4 \bar{a}_\kappa}, \frac{\gamma}{\bar{c}_\kappa^2} \right) \qquad \forall \kappa \in \mathcal{T} \ , \qquad (8.13)$$

where $\bar{c}_\kappa = \|c\|_{L^\infty(\kappa)}$. *Then, assuming that the matrix a is piecewise constant on the partition in* $\mathcal{T}$

$$B_{\mathrm{DG}}(w,w) \geq |\|w\||^2_{\mathrm{DG}} \equiv D + \frac{1}{2}\sum_{\kappa\in\mathcal{T}} E_\kappa + \frac{1}{2}\sum_{\kappa\in\mathcal{T}} F_\kappa + \frac{1}{2}\sum_{\kappa\in\mathcal{T}} G_\kappa \ , \qquad (8.14)$$

where

$$D \equiv \int_{\Gamma_D} \sigma w^2\, ds + \int_{\Gamma_{\mathrm{int}}} \sigma[w]^2\, ds \ , \quad E_\kappa \equiv \|\sqrt{a}\nabla w\|^2_{L^2(\kappa)} + \|\sqrt{\gamma}w\|^2_{L^2(\kappa)} \ ,$$

$$F_\kappa \equiv \int_{\partial_-\kappa\cap\Gamma_-} |\boldsymbol{b}\cdot\boldsymbol{\mu}|w_+^2\, ds + \int_{\partial_+\kappa\cap\Gamma_+} |\boldsymbol{b}\cdot\boldsymbol{\mu}|w_+^2\, ds + \int_{\partial_-\kappa\backslash\Gamma_-} |\boldsymbol{b}\cdot\boldsymbol{\mu}|\lfloor w\rfloor^2\, ds \ ,$$

$$G_\kappa \equiv \|\sqrt{\delta}(\boldsymbol{b}\cdot\nabla w)\|^2_{L^2(\kappa)} \ ,$$

with $\sqrt{a}$ *denoting the (nonnegative) square-root of the matrix a, and* σ *as in the definition of the discontinuity-penalisation. Furthermore, the hp-DGFEM (8.11) has a unique solution* u_{DG} *in* $S^{\mathbf{P}}(\Omega,\mathcal{T},\mathbf{F})$.

PROOF: We begin by proving (8.14). First, we note that, trivially,

$$B_s(w,w) = \int_{\Gamma_D} \sigma w^2\, \mathrm{d}s + \int_{\Gamma_{\mathrm{int}}} \sigma[w]^2\, \mathrm{d}s \ .$$

Further, as $(\boldsymbol{b}\cdot\nabla w)w = \frac{1}{2}\boldsymbol{b}\cdot\nabla(w^2)$, after integration by parts we have that

$$B_b(w,w) = \frac{1}{2}\sum_{\kappa\in\mathcal{T}} F_\kappa + \sum_{\kappa\in\mathcal{T}}\int_\kappa |\sqrt{\gamma(x)}w(x)|^2\, \mathrm{d}x \ .$$

Next, we observe that

$$B_a(w,w) = \sum_{\kappa\in\mathcal{T}}\int_\kappa |\sqrt{a(x)}\nabla w(x)|^2\, \mathrm{d}x \ .$$

Finally, by the Cauchy-Schwarz inequality,

$$B_\delta(w,w) \geq \sum_{\kappa\in\mathcal{T}}\left[\frac{1}{2}\|\sqrt{\delta}(\boldsymbol{b}\cdot\nabla w)\|^2_{L^2(\kappa)} - \|\sqrt{\delta}(\nabla\cdot(a\nabla w))\|^2_{L^2(\kappa)} - \|\sqrt{\delta}cw\|^2_{L^2(\kappa)}\right] \ .$$

Noting (8.12) and (8.13), this implies that

$$B_\delta(w,w) \geq \frac{1}{2}\sum_{\kappa\in\mathcal{T}}\left(\|\sqrt{\delta}(\boldsymbol{b}\cdot\nabla w)\|^2_{L^2(\kappa)} - \|\sqrt{a}\nabla w\|^2_{L^2(\kappa)} - \|\sqrt{\gamma}w\|^2_{L^2(\kappa)}\right) \ .$$

Upon recalling the definition of the bilinear form $B_{\mathrm{DG}}(\cdot,\cdot)$, we arrive at (8.14).

To complete the proof of the theorem, we note that since $\gamma > 0$ on each element κ in the subdivision $\mathcal{T}$, then $B_{\mathrm{DG}}(w,w) > 0$ for all w in $S^{\mathbf{P}}(\Omega,\mathcal{T},\mathbf{F})\setminus\{0\}$, and hence we deduce the uniqueness of the solution u_{DG}. Further, since the linear space $S^{\mathbf{P}}(\Omega,\mathcal{T},\mathbf{F})$ is finite-dimensional, the existence of the solution to (8.11) follows from the fact that its homogeneous counterpart has the unique solution $u_{\mathrm{DG}} \equiv 0$. $\square$

Our second result provides a bound on the global error $e = u - u_{\mathrm{DG}}$. For simplicity, we shall assume that the entries of the matrix a are constant on each element $\kappa \in \mathcal{T}$ (with possible discontinuities across faces $e \in \mathcal{E}$). We quote the following result [5, 23].

Lemma 8.1 *Suppose that* $u \in H^{k_\kappa}(\kappa)$, $k_\kappa \geq 0$, $\kappa \in \mathcal{T}$. *Then, there exists* $\Pi_{hp}u$ *in the finite element space* $S^{\mathbf{p}}(\Omega, \mathcal{T}, \mathbf{F})$, *a constant* C *dependent on* k_κ *and the angle condition of* κ, *but independent of* u, $h_\kappa = diam(\kappa)$ *and* p_κ, *such that*

$$\|u - \Pi_{hp}u\|_{H^s(\kappa)} \leq C \frac{h_\kappa^{\tau_\kappa - s}}{p_\kappa^{k_\kappa - s}} \|u\|_{H^{k_\kappa}(\kappa)} \; , \tag{8.15}$$

where $0 \leq s \leq \tau_\kappa$ *and* $\tau_\kappa = \min(p_\kappa + 1, k_\kappa)$ *for* $\kappa \in \mathcal{T}$.

Our main result concerns the accuracy of the hp-DGFEM (8.11) and is stated in the next theorem. We recall the notation introduced earlier on: given that $\bar{a} = \|a\|_2$, with $\|\cdot\|_2$ denoting the matrix norm subordinate to the l^2 vector norm on $\mathbb{R}^d$, we let $\bar{a}_\kappa = \bar{a}|_\kappa$. Similarly, we define $\bar{b}_\kappa = \|\boldsymbol{b}\|_{L^\infty(\kappa)}$, $\bar{c}_\kappa = \|c\|_{L^\infty(\kappa)}$ and $\bar{\gamma}_\kappa = \|\gamma\|_{L^\infty(\kappa)}$.

Theorem 8.2 *In addition to the hypotheses of Theorem 8.1, let us assume that for any* $\kappa \in \mathcal{T}$ *such that* $\bar{b}_\kappa \neq 0$,

$$\delta(x) = \frac{1}{2} \min \left(\frac{h_\kappa^2}{C_{inv}^2 p_\kappa^4 \bar{a}_\kappa}, \frac{h_\kappa}{p_\kappa \bar{b}_\kappa}, \frac{\gamma}{\bar{c}_\kappa^2} \right), \qquad x \in \kappa \; . \tag{8.16}$$

Then, the solution $u_{\mathrm{DG}} \in S^{\mathbf{p}}(\Omega, \mathcal{T}, \mathbf{F})$ *of (8.11) obeys the error bound*

$$\||u - u_{\mathrm{DG}}\||_{\mathrm{DG}}^2 \leq C \sum_{\kappa : \bar{b}_\kappa \neq 0} \left(\bar{a}_\kappa \frac{h_\kappa^{2(\tau_\kappa - 1)}}{p_\kappa^{2(k_\kappa - 2)}} + \bar{b}_\kappa \frac{h_\kappa^{2(\tau_\kappa - 1/2)}}{p_\kappa^{2(k_\kappa - 1/2)}} + \bar{\gamma}_\kappa \frac{h_\kappa^{2\tau_\kappa}}{p_\kappa^{2k_\kappa}} \right) \|u\|_{H^{k_\kappa}(\kappa)}^2$$

$$+ C \sum_{\kappa : \bar{b}_\kappa = 0} \left(\bar{a}_\kappa \frac{h_\kappa^{2(\tau_\kappa - 1)}}{p_\kappa^{2(k_\kappa - 3/2)}} + \bar{\gamma}_\kappa \frac{h_\kappa^{2\tau_\kappa}}{p_\kappa^{2k_\kappa}} \right) \|u\|_{H^{k_\kappa}(\kappa)}^2 \; ,$$

where $\tau_\kappa = \min(p_\kappa + 1, k_\kappa)$, *and* $u \in H^{k_\kappa}(\kappa)$ *with* $k_\kappa \geq 2$ *when* $\bar{b}_\kappa \neq 0$, $k_\kappa > 3/2$ *when* $\bar{b}_\kappa = 0$, *for* $\kappa \in \mathcal{T}$.

PROOF: Let us decompose $e = u - u_{\mathrm{DG}}$ as $e = \eta + \xi$ where $\eta = u - \Pi_{hp}u$, $\xi = \Pi_{hp}u - u_{\mathrm{DG}}$, and Π_{hp} is as in Lemma 8.1. Then, by virtue of Theorem 8.1,

$$\||\xi\||_{\mathrm{DG}}^2 \leq B_{\mathrm{DG}}(\xi, \xi) = B_{\mathrm{DG}}(e - \eta, \xi) = -B_{\mathrm{DG}}(\eta, \xi) \; ,$$

where we have used the Galerkin orthogonality property $B_{\mathrm{DG}}(u - u_{\mathrm{DG}}, \xi) = 0$ which follows from (8.11) with $v = \xi$ and the definition of the boundary value problem (8.1), (8.4), given the assumed smoothness of u. Thus, we deduce that

$$\||\xi\||_{\mathrm{DG}}^2 \leq |B_a(\eta, \xi)| + |B_b(\eta, \xi)| + |B_s(\eta, \xi)| + |B_\delta(\eta, \xi)| \; . \tag{8.17}$$

Now, from (8.9) we have that

$$|B_s(\eta, \xi)| \leq \||\xi\||_{\mathrm{DG}} \left(\int_{\Gamma_D} \sigma |\eta|^2 \, ds + \int_{\Gamma_{\mathrm{int}}} \sigma [\eta]^2 \, ds \right)^{1/2} \; . \tag{8.18}$$

Next we consider $B_b(\eta, \xi)$. Upon integration by parts, we obtain

$$B_b(\eta, \xi) = \sum_\kappa \int_\kappa (c - \nabla \cdot \boldsymbol{b}) \eta \xi \, dx - \sum_{\kappa \in \mathcal{T}} \int_\kappa \eta (\boldsymbol{b} \cdot \nabla \xi) \, dx + \sum_{\kappa \in \mathcal{T}} \int_{\partial_+ \kappa \cap \Gamma_+} (\boldsymbol{b} \cdot \boldsymbol{\mu}) \eta^+ \xi^+ \, ds$$

$$+ \sum_{\kappa \in \mathcal{T}} \int_{\partial_+ \kappa \backslash \Gamma_+} (\boldsymbol{b} \cdot \boldsymbol{\mu}) \eta^+ \xi^+ \, ds + \sum_{\kappa \in \mathcal{T}} \int_{\partial_- \kappa \backslash \Gamma_-} (\boldsymbol{b} \cdot \boldsymbol{\mu}) \eta^- \xi^+ \, ds \ . \tag{8.19}$$

Denoting by $S_1, \ldots, S_5$ the five terms on the right-hand side of (8.19), we find, after shifting the 'indices' in the summation in S_4, that

$$|S_4 + S_5| \le \sum_{\kappa \in \mathcal{T}} \left(\int_{\partial_- \kappa \backslash \Gamma_-} |\boldsymbol{b} \cdot \boldsymbol{\mu}| |\eta^-|^2 \, ds \right)^{1/2} \left(\int_{\partial_- \kappa \backslash \Gamma_-} |\boldsymbol{b} \cdot \boldsymbol{\mu}| \lfloor \xi \rfloor^2 \, ds \right)^{1/2} \ .$$

Also, we note that elements $\kappa \in \mathcal{T}$ with $\bar{b}_\kappa = 0$ can be omitted from the summation in term S_2. Thus, after multiplying and dividing by $\sqrt{\gamma}$ and $\sqrt{\delta}$ under the integral signs in S_1 and S_2, respectively, (8.19) yields

$$|B_b(\eta, \xi)| \le C \||\xi\||_{\mathrm{DG}} \left(\sum_{\kappa : \bar{b}_\kappa \ne 0} \|(1 + \delta^{-1/2}) \eta\|^2_{L^2(\kappa)} + \sum_{\kappa \in \mathcal{T}} \int_{\partial_+ \kappa \cap \Gamma_+} |\boldsymbol{b} \cdot \boldsymbol{\mu}| |\eta^+|^2 \, ds \right.$$

$$\left. + \sum_{\kappa \in \mathcal{T}} \int_{\partial_- \kappa \backslash \Gamma_-} |\boldsymbol{b} \cdot \boldsymbol{\mu}| |\eta^-|^2 \, ds \right)^{1/2} , \tag{8.20}$$

where C is a generic positive constant, as in the statement of the theorem.

Next, we consider the term $B_a(\eta, \xi)$:

$$|B_a(\eta, \xi)| \le I + II + III \ ,$$

where

$$I \equiv \left| \sum_{\kappa \in \mathcal{T}} \int_\kappa a \nabla \eta \cdot \nabla \xi \, dx \right| \ , \quad II \equiv \left| \int_{\Gamma_{\mathrm{D}}} \{ \eta((a \nabla \xi) \cdot \boldsymbol{\nu}) - ((a \nabla \eta) \cdot \boldsymbol{\nu}) \xi \} \, ds \right| \ ,$$

$$III \equiv \left| \int_{\Gamma_{\mathrm{int}}} \{ [\eta] \langle (a \nabla \xi) \cdot \boldsymbol{\nu} \rangle - \langle (a \nabla \eta) \cdot \boldsymbol{\nu} \rangle [\xi] \} \, ds \right| \ .$$

Now, we have that

$$I^2 \le \||\xi\||^2_{\mathrm{DG}} \sum_{\kappa \in \mathcal{T}} \|\sqrt{a} \nabla \eta\|^2_{L^2(\kappa)} \ ,$$

$$II^2 \le C \||\xi\||^2_{\mathrm{DG}} \sum_{\kappa : \partial \kappa \cap \Gamma_{\mathrm{D}} \ne \emptyset} \left(\frac{\bar{a}_\kappa p_\kappa^2}{h_\kappa} \|\eta\|^2_{L^2(\partial \kappa \cap \Gamma_{\mathrm{D}})} + \bar{a}_\kappa h_\kappa \|\nabla \eta\|^2_{L^2(\partial \kappa \cap \Gamma_{\mathrm{D}})} \right) \ ,$$

$$III^2 \le C \||\xi\||^2_{\mathrm{DG}} \sum_{\kappa : \partial \kappa \cap \Gamma = \emptyset} \left(\frac{\bar{a}_\kappa p_\kappa^2}{h_\kappa} \|[\eta]\|^2_{L^2(\partial \kappa)} + \bar{a}_\kappa h_\kappa \|\nabla \eta\|^2_{L^2(\partial \kappa)} \right) \ .$$

Collecting the bounds on the terms I, II and III gives,

$$|B_a(\eta,\xi)| \leq C|||\xi|||_{\mathrm{DG}} \left(\sum_{\kappa \in \mathcal{T}} \|\sqrt{a}\nabla\eta\|^2_{L^2(\kappa)} \right.$$

$$+ \sum_{\kappa : \partial\kappa \cap \Gamma_{\mathrm{D}} \neq \emptyset} \left(\frac{\bar{a}_\kappa p_\kappa^2}{h_\kappa}\|\eta\|^2_{L^2(\partial\kappa \cap \Gamma_{\mathrm{D}})} + \bar{a}_\kappa h_\kappa \|\nabla\eta\|^2_{L^2(\partial\kappa \cap \Gamma_{\mathrm{D}})} \right)$$

$$\left. + \sum_{\kappa : \partial\kappa \cap \Gamma = \emptyset} \left(\frac{\bar{a}_\kappa p_\kappa^2}{h_\kappa}\|[\eta]\|^2_{L^2(\partial\kappa)} + \bar{a}_\kappa h_\kappa \|\nabla\eta\|^2_{L^2(\partial\kappa)} \right) \right)^{1/2}. \tag{8.21}$$

Finally, for $B_\delta(\eta,\xi)$ we have the bound

$$|B_\delta(\eta,\xi)| \leq |||\xi|||_{\mathrm{DG}} \left(\sum_{\kappa : \bar{b}_\kappa \neq 0} \|\sqrt{\delta}\mathcal{L}\eta\|^2_{L^2(\kappa)} \right)^{1/2}. \tag{8.22}$$

The required result now follows by noting that

$$|||u - u_{\mathrm{DG}}|||_{\mathrm{DG}} \leq |||\eta|||_{\mathrm{DG}} + |||\xi|||_{\mathrm{DG}},$$

inserting the estimates (8.18), (8.20), (8.21) and (8.22) into (8.17) to bound $|||u - u_{\mathrm{DG}}|||_{\mathrm{DG}}$ in terms of $|||\eta|||_{\mathrm{DG}}$ and other norms of η, and applying Lemma 8.1, together with the Trace Inequality to estimate norms over e and $\partial\kappa$ in terms of norms of η over κ, $\kappa \in \mathcal{T}$. $\square$

We note that in the purely hyperbolic case of $a \equiv 0$ the error bound in Theorem 8.2 collapses to the hp-optimal error bound $O(h^{\tau-1/2}/p^{k-1/2})$ in the DG-norm established in [14], which represents a generalisation of the optimal h-version bound for the DGFEM (see [16]) to the hp-version. In fact, for $a \equiv 0$ the error bound of Theorem 8.2 is by $1/2$ a p-order sharper than the corresponding estimate of Bey and Oden [8], except that there the streamline-diffusion parameter was $\delta = h/p^2$, while in the case of $a \equiv 0$ Theorem 8.2 corresponds to $\delta = h/p$. We note in this respect that the error bound of [8] may be reproduced even with $\delta = 0$, i.e. with less damping in the streamwise direction than required in [8]; see [25]. For further developments in this direction, we refer to [14, 25].

8.4 A POSTERIORI ERROR ANALYSIS

For the second half of this paper, we turn our attention to the subject of *a posteriori* error analysis of first-order hyperbolic problems, corresponding to $a_{ij} \equiv 0$ for $i, j = 1, \ldots, d$. In particular, using the approach in [7], we discuss the question of error estimation for linear functionals, such as the outflow flux and the local average of the solution. For simplicity, we restrict ourselves to the case when the streamline-diffusion stabilisation δ is set to zero; the case when $\delta > 0$ may be treated analogously, cf. [13]. Under these assumptions,

$$B_{\mathrm{DG}}(\cdot,\cdot) \equiv B_b(\cdot,\cdot) \quad \text{and} \quad \ell_{\mathrm{DG}}(\cdot) \equiv \ell_b(\cdot),$$

where $B_b(\cdot,\cdot)$ and $\ell_b(\cdot)$ are as defined in (8.6) and (8.7), respectively.

Given a linear functional $J(\cdot)$, our aim is to control the discretisation error between the true value $J(u)$, based on the analytical solution u to (8.1), and the actual computed value $J(u_{\mathrm{DG}})$. The proceeding error analysis is based on a hyperbolic duality argument; for full details and further numerical experiments, see [15]. To this end, we introduce the following *dual* or *adjoint* problem: find z in $H(\mathcal{L}^*, \Omega)$ such that

$$B_{\mathrm{DG}}(w, z) = J(w) \quad \forall w \in H(\mathcal{L}, \Omega) , \tag{8.23}$$

where $H(\mathcal{L}, \Omega)$, resp. $H(\mathcal{L}^*, \Omega)$, denotes the graph space of the first–order operator $\mathcal{L} \equiv \mathcal{L}_b$, resp. $\mathcal{L}^*$. Here, we assume that (8.23) has a unique solution; the validity of this assumption depends on the precise definition of the linear functional $J(\cdot)$ under consideration. In the case when $J(\cdot)$ represents the (weighted) normal flux through the outflow boundary Γ_+, i.e.

$$J(w) \equiv N_\psi(w) = \int_{\Gamma_+} (\boldsymbol{b} \cdot \boldsymbol{\mu}) \, w\psi \, \mathrm{d}s , \tag{8.24}$$

z is the (unique) solution to the following partial differential equation: find z in $H(\mathcal{L}^*, \Omega)$ such that

$$\mathcal{L}^* z \equiv -\nabla \cdot (\boldsymbol{b}z) + cz = 0 , \quad x \in \Omega , \qquad z = \psi , \quad x \in \Gamma_+ . \tag{8.25}$$

Other examples include the mean flow of the field u over the computational domain Ω or some compact subset of Ω.

For each element κ in the mesh $\mathcal{T}$, we define the *internal residual* $r_{h,p}$ and the *boundary residual* $r_{h,p}^-$ by

$$r_{h,p}|_\kappa = (f - \mathcal{L}u_{\mathrm{DG}})|_\kappa \quad \text{and} \quad r_{h,p}^-|_{\partial_-\kappa \cap \Gamma_-} = (g - u_{\mathrm{DG}}^+)|_{\partial_-\kappa \cap \Gamma_-} , \tag{8.26}$$

respectively. With this notation we have the following general result.

Theorem 8.3 *Let u and u_{DG} denote the solutions of (8.1) and (8.11), respectively, and suppose that the dual solution z satisfies (8.23). Then, the following a posteriori error bound holds:*

$$|J(u) - J(u_{\mathrm{DG}})| \le \epsilon(u_{\mathrm{DG}}, h, p, z, z_{h,p}) \equiv \sum_{\kappa \in \mathcal{T}} \eta_\kappa , \tag{8.27}$$

where

$$\eta_\kappa = |(r_{h,p}, z - z_{h,p})_\kappa + ((\boldsymbol{b} \cdot \boldsymbol{\mu})\lfloor u_{\mathrm{DG}} \rfloor, (z - z_{h,p})^+)_{\partial_-\kappa \backslash \Gamma_-} - ((\boldsymbol{b} \cdot \boldsymbol{\mu}) r_{h,p}^-, (z - z_{h,p})^+)_{\partial_-\kappa \cap \Gamma_-}|$$

and $z_{h,p}$ belongs to the finite element space $S^{\mathbf{P}}(\Omega, \mathcal{T}, \mathbf{F})$.

PROOF: Choosing $w = u - u_{\mathrm{DG}}$ in (8.23) and exploiting the Galerkin orthogonality property of the hp–DGFEM (cf. proof of Theorem 8.2), we deduce that

$$J(u) - J(u_{\mathrm{DG}}) = J(u - u_{\mathrm{DG}}) = B_{\mathrm{DG}}(u - u_{\mathrm{DG}}, z) = B_{\mathrm{DG}}(u - u_{\mathrm{DG}}, z - z_{h,p}) .$$

Recalling the definition of the bilinear form $B_{\mathrm{DG}}(\cdot,\cdot)$ and applying the triangle inequality gives the desired result.   □

While the residual terms $r_{h,p}$ and $r_{h,p}^-$ and the 'jump' term $\lfloor u_{\mathrm{DG}}\rfloor$ are easily evaluated once the numerical solution u_{DG} has been computed, the calculation of the corresponding 'weights' involving the dual solution z requires special care. As in [13], z will be estimated by numerically solving the dual problem (8.23); this will be discussed in detail in Section 8.5.1. First, however, in the case when the functional of interest $J(\cdot)$ is defined to be the mean flow of the field u over Ω, we derive an *a posteriori* bound on the error $u - u_{\mathrm{DG}}$ in negative Sobolev norms, cf. [13, 24], for example.

Theorem 8.4 *Let u and u_{DG} denote the solutions of (8.1) and (8.11), respectively. Then there exists a positive constant C, dependent only on the dimension d, the shape regularity of $\mathcal{T}$ and m, $m > 0$, such that*

$$\|u - u_{\mathrm{DG}}\|_{H^{-m}(\Omega)} \leq C \left[\left(\sum_{\kappa \in \mathcal{T}} \frac{h_\kappa^{2\tau_\kappa}}{p_\kappa^{2m}} \|r_{h,p}\|_{L_2(\kappa)}^2 \right)^{1/2} + \left(\sum_{\kappa \in \mathcal{T}} \frac{h_\kappa^{2\tau_\kappa-1}}{p_\kappa^{2m-1}} \|\lfloor u_{\mathrm{DG}}\rfloor\|_{\partial_-\kappa\backslash\Gamma_-}^2 \right)^{1/2} \right.$$
$$\left. + \left(\sum_{\kappa \in \mathcal{T}} \frac{h_\kappa^{2\tau_\kappa-1}}{p_\kappa^{2m-1}} \|r_{h,p}^-\|_{\partial_-\kappa\cap\Gamma_-}^2 \right)^{1/2} \right] ,$$

where $\tau_\kappa = \min(p_\kappa + 1, m)$ for all κ in $\mathcal{T}$.

PROOF: The proof is based on a duality argument using the Galerkin orthogonality of the hp–DGFEM, together with stability bounds for the dual problem, see [15].   □

We end this section by stating an *a priori* bound on the error in the computed functional in terms of Sobolev norms of the analytical solution u and the dual solution z which indicates the expected rate of convergence for $|J(u) - J(u_{\mathrm{DG}})|$ as the finite element space is enriched, i.e. as $h \to 0$ and $p \to \infty$. This will play a crucial role in the design of an hp–adaptive algorithm for automatically controlling the error in the computed functional, see Section 8.5.2 below. To this end, we assume for the moment that

$$b \in \left[S^1(\Omega, \mathcal{T}, \mathbf{F}) \cap C(\Omega) \right]^d , \qquad c \in S^0(\Omega, \mathcal{T}, \mathbf{F}) , \qquad f \in S^p(\Omega, \mathcal{T}, \mathbf{F}) . \tag{8.28}$$

Theorem 8.5 *Let u and u_{DG} denote the solutions of (8.1) and (8.11), respectively. Given that $u|_\kappa \in H^{k_\kappa}(\kappa)$, $k_\kappa \geq 1$, and $z|_\kappa \in H^{l_\kappa}(\kappa)$, $l_\kappa \geq 1$, for all κ in $\mathcal{T}$, we have*

$$|J(u) - J(u_{\mathrm{DG}})|^2 \leq C \sum_{\kappa \in \mathcal{T}} \frac{h_\kappa^{2\tau_\kappa-1}}{p_\kappa^{2k_\kappa-2}} \|u\|_{H^{k_\kappa}(\kappa)}^2 \cdot \sum_{\kappa \in \mathcal{T}} \frac{h_\kappa^{2\theta_\kappa-1}}{p_\kappa^{2l_\kappa-2}} \|z\|_{H^{l_\kappa}(\kappa)}^2 , \tag{8.29}$$

where $\tau_\kappa = \min(p_\kappa + 1, k_\kappa)$ and $\theta_\kappa = \min(p_\kappa + 1, l_\kappa)$ for all $\kappa \in \mathcal{T}$. Here, C is a positive constant, dependent only on d, the shape regularity of $\mathcal{T}$ and k_κ and l_κ, $\kappa \in \mathcal{T}$.

PROOF: See [15] for details.   □

For uniform orders, $p_\kappa = p$, $k_\kappa = k \geq 1$, $l_\kappa = l \geq 1$, and $h_\kappa = h$ for all κ in $\mathcal{T}$, we have

$$|J(u) - J(u_{\mathrm{DG}})| \leq C \frac{h^{\tau+\theta-1}}{p^{k+l-1}} p \, \|u\|_{H^k(\Omega)} \|z\|_{H^l(\Omega)} , \tag{8.30}$$

where $\tau = \min(p+1, k)$ and $\theta = \min(p+1, l)$. Here, the bound (8.30) is optimal in h and suboptimal in p by one order; in the case of fixed p, (8.30) reduces the optimal h–convergence error bound proved in [13] for a stabilised continuous approximation to u. From (8.30) we may deduce the following *a priori* error bound

$$\|u - u_{\text{DG}}\|_{H^{-m}(\Omega)} \leq C \frac{h^{\tau+\theta-1}}{p^{k+m-1}} p \, \|u\|_{H^k(\Omega)} \,, \tag{8.31}$$

where $\tau = \min(p+1, k)$ and $\theta = \min(p+1, m)$. In the presence of streamline–diffusion stabilisation, with stabilisation parameter $\delta = h/p$, the bounds (8.30) and (8.31) can be sharpened to ones that are simultaneously optimal in both h and p.

Finally, we note that the dependence of the constant C appearing in the *a priori* bound (8.29) on the regularity of the primal solution u and the dual solution z may be made explicit using the approximation results derived in [23]. In particular, this allows us to deduce that the error in the computed functional $J(\cdot)$ decays exponentially as $p \to \infty$ if *either* u or z are elementwise analytic, cf. [15]; this will be demonstrated in Section 8.5.3.

8.5 NUMERICAL IMPLEMENTATION

8.5.1 *Numerical approximation of the dual solution*

To ensure that the *a posteriori* error bound stated in Theorem 8.3 is fully computable, the dual solution z must be numerically approximated. In this section we describe a DGFEM for this purpose. As stated in Section 8.4, the particular form of the dual problem is dependent on the functional under consideration. For simplicity, let us suppose that $J(\cdot) = N_\psi(\cdot)$, i.e. J represents the outflow normal flux, cf. (8.24). In this case the dual solution z satisfies (8.25) for a given weight function ψ.

As in Section 8.2.1, we define $\tilde{S}^{\tilde{\mathbf{p}}}(\Omega, \tilde{\mathcal{T}}, \tilde{\mathbf{F}})$ to be the finite element space consisting of piecewise polynomials of degree $\tilde{\mathbf{p}}|_{\tilde{\kappa}} = \tilde{p}_{\tilde{\kappa}}$ on a mesh $\tilde{\mathcal{T}}$ consisting of shape regular elements $\tilde{\kappa}$ of size $\tilde{h}_{\tilde{\kappa}}$. With $\partial_+\tilde{\kappa}$ defined as in (8.5), we introduce the bilinear form and linear functional

$$
\begin{aligned}
\tilde{B}_{\text{DG}}(w, v) &= \sum_{\tilde{\kappa}\in\tilde{\mathcal{T}}} \int_{\tilde{\kappa}} \mathcal{L}^* w \, v \, \mathrm{d}x + \sum_{\tilde{\kappa}\in\tilde{\mathcal{T}}} \int_{\partial_+\tilde{\kappa}\backslash\Gamma_+} (\boldsymbol{b}\cdot\boldsymbol{\mu})\lfloor w\rfloor v^+ \, \mathrm{d}s \\
&\quad + \sum_{\tilde{\kappa}\in\tilde{\mathcal{T}}} \int_{\partial_+\tilde{\kappa}\cap\Gamma_+} (\boldsymbol{b}\cdot\boldsymbol{\mu})w^+ \, v^+ \, \mathrm{d}s \,, \\
\tilde{\ell}_{\text{DG}}(v) &= \sum_{\tilde{\kappa}\in\tilde{\mathcal{T}}} \int_{\partial_+\tilde{\kappa}\cap\Gamma_+} (\boldsymbol{b}\cdot\boldsymbol{\mu}) \, \psi v^+ \, \mathrm{d}s \,,
\end{aligned}
$$

respectively. The *hp*-DGFEM approximation of (8.25) is defined as follows: find $\tilde{z}_{\text{DG}} \in \tilde{S}^{\tilde{\mathbf{p}}}(\Omega, \tilde{\mathcal{T}}, \tilde{\mathbf{F}})$ such that

$$\tilde{B}_{\text{DG}}(\tilde{z}_{\text{DG}}, v) = \tilde{\ell}_{\text{DG}}(v) \quad \forall v \in \tilde{S}^{\tilde{\mathbf{p}}}(\Omega, \tilde{\mathcal{T}}, \tilde{\mathbf{F}}) \,. \tag{8.32}$$

8.5.2   Adaptive algorithm

For a user–defined tolerance TOL, we now consider the problem of designing the hp–finite element space $S^{\mathbf{P}}(\Omega, \mathcal{T}, \mathbf{F})$ such that

$$|J(u) - J(u_{\mathrm{DG}})| \leq \mathrm{TOL} \; , \tag{8.33}$$

subject to the constraint that the total number of degrees of freedom in $S^{\mathbf{P}}(\Omega, \mathcal{T}, \mathbf{F})$ is minimised. To ensure that (8.33) holds, we use the *a posteriori* error bound (8.27) to construct $S^{\mathbf{P}}(\Omega, \mathcal{T}, \mathbf{F})$ such that

$$\epsilon(u_{\mathrm{DG}}, h, p, z, z_{h,p}) \leq \mathrm{TOL} \; . \tag{8.34}$$

The stopping criterion (8.34) is enforced by equidistributing $\epsilon|_\kappa \equiv \eta_\kappa$ over the elements κ in the mesh $\mathcal{T}$. Thus, we insist that

$$\eta_\kappa \approx \mathrm{TOL}/N \tag{8.35}$$

holds for each κ in $\mathcal{T}$; here, N denotes the number of elements in the mesh $\mathcal{T}$.

Thereby, each of the elements in the mesh is flagged for either refinement or derefinement to ensure that the equidistribution principle (8.35) holds. Once an element κ has been flagged a decision must be made whether the local mesh size h_κ or the local degree of the approximating polynomial p_κ should be adjusted accordingly. Let us first deal with refinement, i.e. when the local error estimator η_κ is larger than the 'localised–tolerance' TOL/N. Clearly, if the error in the functional is locally 'smooth', then p–enrichment will be more effective than h–refinement, since the error will be expected to decay quickly within the current element κ as p_κ is increased. However, if the error in the functional has low regularity within the element κ, then h–refinement will be performed. Thus, regions in the computational domain where the error is locally non-smooth are isolated from smooth regions, thereby reducing the influence of singularities/shocks as well as making p–enrichment more effective.

To ensure that the desired level of accuracy is achieved efficiently, an automatic procedure for deciding when to h– or p–refine must be implemented. To this end, we first compute the local error indicator η_κ on each element κ in the mesh $\mathcal{T}$ using both a p_κ and a $p_\kappa - 1$ representation for u_{DG}. Thereby, assuming that $\eta_\kappa(p_\kappa - 1) \neq 0$, the perceived smoothness of the local error may be estimated using the ratio

$$\rho_\kappa = \eta_\kappa(p_\kappa)/\eta_\kappa(p_\kappa - 1) \; ; \tag{8.36}$$

here, we have written $\eta_\kappa(p_\kappa)$ to emphasise the dependence of the local error indicator η_κ on the local degree p_κ of the approximating polynomial, cf. Adjerid *et al.* [1] and Gui & Babuška [12], for example. If $\rho_\kappa \leq \gamma$, $0 < \gamma < 1$, the error is decreasing as the polynomial degree is increased, indicating that p–enrichment should be performed. On the other hand, $\rho_\kappa > \gamma$ means that the element κ should be locally subdivided. The number γ is referred to as the *type–parameter* [12]. Clearly, the choice of γ is critical to the success of this algorithm and will depend on the asymptotic behaviour of the quantity of interest. Instead of assigning an *ad hoc* value to the type parameter γ, we use ρ_κ together with

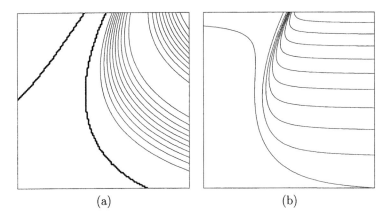

Figure 8.1: Piecewise bilinear interpolant on a 129×129 mesh of the analytical solution to: (a) Primal problem; (b) Dual Problem.

the *a priori* error bound (8.29) to directly estimate the local regularities k_κ and l_κ of the primal and dual solutions, respectively, on each element κ in $\mathcal{T}$. More precisely, motivated by (8.30), we assume that on a given element κ in $\mathcal{T}$

$$\eta_\kappa = \epsilon(u_{\mathrm{DG}}, h_\kappa, p_\kappa, z, z_{h,p})|_\kappa \approx C_\kappa\, p_\kappa^{-k_\kappa - l_\kappa + 1} \ .$$

Thus, we have that

$$k_\kappa + l_\kappa = \log(\rho_\kappa)/\log((p_\kappa - 1)/p_\kappa) + 1 \ .$$

Ideally, we would like to know k_κ and l_κ individually. The dual regularity l_κ may be estimated by calculating the $L^2(\kappa)$ norm of the error between the projection of $\tilde{z}_{\mathrm{DG}}$ onto the finite element spaces $S^{\mathbf{p}}(\Omega, \mathcal{T}, \mathbf{F})$ and $S^{\mathbf{p-1}}(\Omega, \mathcal{T}, \mathbf{F})$, together with the approximation result (8.15). Once, both k_κ and l_κ have been determined on element κ, then κ is p–enriched if either k_κ or l_κ is larger than $p_\kappa + 1$; otherwise the element is subdivided. For computational simplicity, only one hanging node is allowed on each side of a given element κ, though no restriction on the difference between the polynomial degrees on neighbouring elements is imposed. We note that this approach has been developed by Ainsworth and Senior [2] in the context of norm control for second–order elliptic problems.

On the other hand, if an element has been flagged for derefinement, then the strategy implemented here is to coarsen the mesh in smooth low–error–regions and decrease the degree of the approximating polynomial in non-smooth low–error–regions, cf. [1]. To this end, we again compute the local regularities k_κ and l_κ of the primal and dual solutions, respectively, on each element κ in $\mathcal{T}$ as described above. The element κ is then coarsened if either k_κ or l_κ is larger than $p_\kappa + 1$, otherwise the degree p_κ is reduced by one.

For the practical implementation of this adaptive algorithm, the dual solution z will be numerically approximated as outlined in Section 8.5.1. Here, we write $\hat{\epsilon}$ in lieu of $\epsilon(u_{\mathrm{DG}}, h, p, \tilde{z}_{\mathrm{DG}}, \tilde{z}_{h,p})$, where $\tilde{z}_{\mathrm{DG}}$ denotes the numerical approximation to z defined by (8.32) and $\tilde{z}_{h,p}$ denotes the L_2–projection of $\tilde{z}_{\mathrm{DG}}$ onto the finite element space $S^{\mathbf{p}}(\Omega, \mathcal{T}, \mathbf{F})$

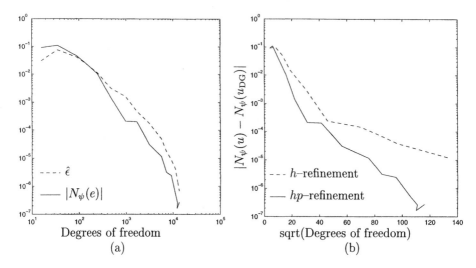

Figure 8.2: (a) $|N_\psi(u) - N_\psi(u_{\mathrm{DG}})|$ and $\hat{\epsilon}$ using hp–refinement; (b) True error in the functional using both h– and hp–refinement.

used to calculate u_{DG}. Furthermore, the finite element space $\tilde{S}^{\tilde{\mathbf{p}}}(\Omega, \tilde{\mathcal{T}}, \tilde{\mathbf{F}})$ used to approximate the dual solution z will be constructed adaptively at the same time as $S^{\mathbf{p}}(\Omega, \mathcal{T}, \mathbf{F})$. For this purpose, we define the following error indicator for the dual approximation

$$\eta_{-1,\tilde{\kappa}} = (\tilde{h}_{\tilde{\kappa}}/\tilde{p}_{\tilde{\kappa}}) \|\mathcal{L}^* \tilde{z}_{\mathrm{DG}}\|_{L_2(\tilde{\kappa})} + (\tilde{h}_{\tilde{\kappa}}/\tilde{p}_{\tilde{\kappa}})^{1/2} \left(\|\lfloor \tilde{z}_{\mathrm{DG}} \rfloor\|_{\partial_+ \tilde{\kappa} \backslash \Gamma_+} + \|\psi - \tilde{z}_{\mathrm{DG}}\|_{\partial_+ \tilde{\kappa} \cap \Gamma_+} \right) \, ,$$

which results from controlling the $H^{-1}(\Omega)$ norm of the error $z - \tilde{z}_{\mathrm{DG}}$, cf. Theorem 8.4. The hp–adaptive algorithm for the dual problem will be based on the *fixed fraction strategy*. Once the elements have been flagged for refinement/derefinement, $\tilde{h}_{\tilde{\kappa}}$ and $\tilde{p}_{\tilde{\kappa}}$ are altered accordingly by estimating the local regularity $\tilde{l}_{\tilde{\kappa}}$ of the dual solution on the dual mesh $\tilde{\mathcal{T}}$ as above by calculating $\eta_{-1,\tilde{\kappa}}$ using a $\tilde{p}_{\tilde{\kappa}}$ and $\tilde{p}_{\tilde{\kappa}} - 1$ representation of $\tilde{z}_{\mathrm{DG}}$, together with the *a priori* error bound (8.31).

8.5.3 Example

Here we consider a compressible hyperbolic problem subject to discontinuous inflow boundary condition, with $\mathbf{b} = (2y^2 - 4x + 1, 1 + y)$, $c = 0$ and $f = 0$. The characteristics enter the computational domain Ω from three sides of Γ, namely from $x = 0$, $y = 0$ and $x = 1$, and exit Ω through $y = 1$. Thus, we may prescribe

$$u(x,y) = \begin{cases} 0 & \text{for } x = 0 \ , \ 0.5 < y \leq 1 \ , \\ 1 & \text{for } x = 0 \ , \ 0 \leq y \leq 0.5 \ , \\ 1 & \text{for } 0 \leq x \leq 0.75 \ , \ y = 0 \ , \\ 0 & \text{for } 0.75 < x \leq 1 \ , \ y = 0 \ , \\ \sin^2(\pi y) & \text{for } x = 1 \ , \ 0 \leq y \leq 1 \ . \end{cases}$$

We define the weight function ψ in the functional $N_\psi(\cdot)$, cf. (8.24), by

$$\psi = 2 + \tanh((x - 1/2)/\varepsilon) \quad \text{for } 0 \leq x \leq 1 \ , \ y = 1 \ ,$$

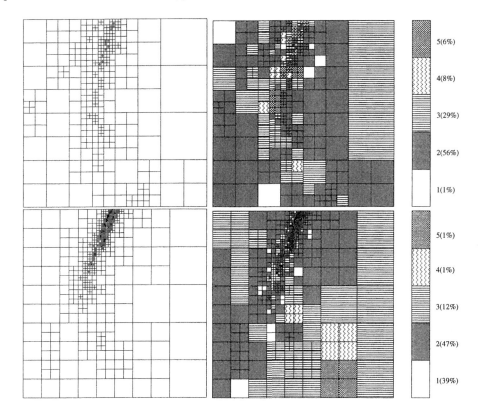

Figure 8.3: Mesh 9: Primal (top: 412 elements, 531 nodes and 5723 DOF) and Dual (bottom: 865 elements, 1064 nodes and 7037 DOF) $h-$ and $hp-$meshes

where $\varepsilon = 10^{-2}$. Thereby, the true value of the outward normal flux is $N_\psi(u) = 2.0115$. The analytical solutions to both the primal and dual problems are shown in Figure 8.1.

In Figure 8.2 we show the performance of the adaptive algorithm described in Section 8.5.2 for TOL $= 10^{-6}$; we note that this level of accuracy may be far beyond what is of practical importance, but is chosen to illustrate that the true error and the bound $\hat{\epsilon}$ exhibit the same asymptotic behaviour as the finite element space $S^\mathbf{p}(\Omega, \mathcal{T}, \mathbf{F})$ is enriched. In Figure 8.2(a) we plot the error in the computed functional $N_\psi(\cdot)$, together with the error bound $\hat{\epsilon}$. Here, we see that while on very coarse meshes $\hat{\epsilon}$ slightly underestimates the true error in the functional, as the finite element space is enriched the error bound over-estimates $|N_\psi(u) - N_\psi(u_{\mathrm{DG}})|$ by a consistent factor. Furthermore, in Figure 8.2(b) we compare the true error in the functional using both $h-$ and $hp-$adaptive refinement. We have plotted the error against the square–root of the number of degrees of freedom on a linear–log scale. While the error $|N_\psi(u) - N_\psi(u_{\mathrm{DG}})|$ using $h-$refinement 'tails–off' as $S^\mathbf{p}(\Omega, \mathcal{T}, \mathbf{F})$ is enriched, we see that after the initial transient, the error in the computed functional using $hp-$refinement becomes a straight line, thereby indicating exponential convergence. We note that the slight 'dip' and the subsequent rise in the true error in

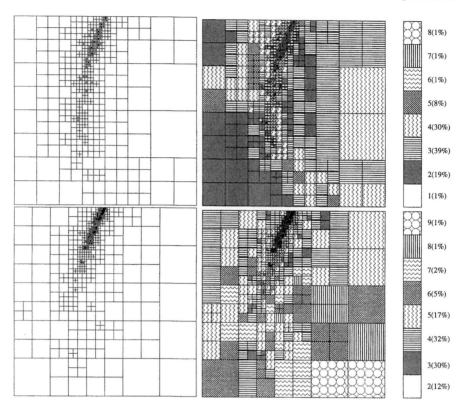

Figure 8.4: Mesh 16: Primal (top: 700 elements, 868 nodes and 13429 DOF) and Dual (bottom: 961 elements, 1210 nodes and 23601 DOF) h– and hp–meshes

the functional observed at the end of the hp–refinement algorithm, cf. Figure 8.2, is attributed to the fact that once the desired tolerance has almost been achieved, the last couple of iterations of the adaptive algorithm attempt to equidistribute the local error indicators η_κ over the elements κ in the computational mesh $\mathcal{T}$.

Finally in Figures 8.3 and 8.4 we show primal and dual meshes after 8 and 15 adaptive mesh refinements, respectively. For clarity, in each case we show the h–mesh alone, as well as the corresponding distribution of the polynomial degree and the percentage of elements with that degree. From Figure 8.3, we see that the elements in the primal mesh have been refined along the first discontinuity emanating from $(x, y) = (0.75, 0)$, since the dual solution has a layer in this region as well. In contrast, elements lying on the second discontinuity in the primal problem, which emanates from $(x, y) = (0, 0.5)$ have been less refined since the dual solution is smooth here. Furthermore, the mesh for the dual solution is concentrated within the steep layer in the weight function ψ; the inherent smoothing in the dual problem introduced by the compressible nature of $\boldsymbol{b}$ leads to p refinement in this layer as the flow moves away from Γ_+. The same behaviour is observed in Figure 8.4 for the primal and dual solutions.

ACKNOWLEDGEMENT

The authors wish to express their thanks to Mr. Mark Embree for writing the graphics packages used to display the *hp*–finite element meshes in this paper.

REFERENCES

[1] S. Adjerid, M. Aiffa and J.E. Flaherty (1998). Computational methods for singularly perturbed systems. In: J. Cronin and R.E. O'Malley, editors, *Singular Perturbation Concepts of Differential Equations*, AMS, Providence.

[2] M. Ainsworth and B. Senior (1998). An adaptive refinement strategy for *hp*–finite element computations. *Appl. Numer. Maths.* 26:165–178.

[3] D.N. Arnold (1982) An interior penalty finite element method with discontinuous elements. *SIAM J. Numer. Anal.*, 19:742–760.

[4] I. Babuška (1972) The finite element method with lagrangian multipliers. *Numer. Math.*, 20:179–192.

[5] I. Babuška and M. Suri (1987). The *hp*-Version of the Finite Element Method with quasiuniform meshes. *M²AN Mathematical Modelling and Numerical Analysis*, 21:199–238.

[6] C. Baumann (1997). An *hp*–Adaptive Discontinuous Galerkin FEM for Computational Fluid Dynamics. Doctoral Thesis. TICAM, University of Texas at Austin.

[7] R. Becker and R. Rannacher (1998). Weighted a posteriori error control in FE methods, ENUMATH-95, Paris, 18-22 Sept., 1995, in Proc. ENUMATH-97 (H.G. Bock, et al., eds.), pp. 621–637, World Scientific Publishing, Singapore.

[8] K.S. Bey and J.T. Oden (1996). *hp*-Version discontinuous Galerkin methods for hyperbolic conservation laws. *Comput. Meth. Appl. Mech. Engrg.*, 133:259–286.

[9] R. Biswas, K. Devine, and J.E. Flaherty (1994). Parallel adaptive finite element methods for conservation laws. *App. Numer. Math.*, 14:255–284.

[10] B. Cockburn, S. Hou, and C.-W. Shu (1990). TVB Runge-Kutta local projection discontinuous Galerkin finite elements for hyperbolic conservation laws. *Math. Comp.*, 54:545–581.

[11] B. Cockburn and C.-W. Shu (1998). The local discontinuous Galerkin method for time-dependent reaction-diffusion systems. *SIAM J. Numer. Anal.*, 35:2440–2463.

[12] W. Gui and I. Babuška (1986). The *h*, *p* and *h–p* versions of the finite element method in 1 Dimension. Part III. The adaptive *h–p* version. *Numer. Math.* 49:659–683.

[13] P. Houston, R. Rannacher and E. Süli (1999). A posteriori error analysis for stabilised finite element approximations of transport problems. Oxford University Computing Laboratory Technical Report NA-99/04, 1999 (submitted for publication).

[14] P. Houston, C. Schwab, and E. Süli (1998). Stabilised hp-finite element methods for first–order hyperbolic problems. Oxford University Computing Laboratory Technical Report NA-98/14 (submitted for publication).

[15] P. Houston and E. Süli (1999). hp-Adaptive discontinuous Galerkin finite element methods for first–order hyperbolic problems. In preparation.

[16] C. Johnson and J. Pitkäranta (1986). An analysis of the discontinuous Galerkin method for a scalar hyperbolic equation. Math. Comp. 46:1–26.

[17] P. Lesaint and P.-A. Raviart (1974). On a finite element method for solving the neutron transport equation. In: C.A. deBoor, (ed.), Mathematical Aspects of Finite Elements in Partial Differential Equations. Academic Press, New York, 89–145.

[18] J. Nitsche (1971). Über ein Variationsprinzip zur Lösung von Dirichlet Problemen bei Verwendung von Teilräumen, die keinen Randbedingungen unterworfen sind. Abh. Math. Sem. Univ. Hamburg, 36:9-15.

[19] J.T. Oden, I. Babuška, and C. Baumann (1998). A discontinuous hp-FEM for diffusion problems. J. Comp. Phys., 146:491–519.

[20] T.H.H. Pian (1965). Element stiffness matrices for boundary compatibility and for prescribed boundary stiffness. Proceedings of the Conference on Matrix Methods in Structural Mechanics, Wright–Patterson Air Force Base, AFFDL-TR-66-80, 457–477.

[21] W.H. Reed and T.R. Hill (1973). Triangular mesh methods for neutron transport equation. Los Alamos Scientific Laboratory report LA-UR-73-479, Los Alamos, NM.

[22] B. Riviere and M.-F. Wheeler (1999). Improved energy estimates for interior penalty, constrained and discontinuous Galerkin methods for elliptic problems. Part I. TICAM Technical Report, University of Texas at Austin.

[23] C. Schwab (1998). p- and hp-Finite Element Methods. Theory and Applications to Solid and Fluid Mechanics. Oxford University Press.

[24] E. Süli (1998). A posteriori error analysis and adaptivity for finite element approximations of hyperbolic problems. In: D. Kröner, (ed.), Theory and Numerics of Conservations Laws, 123–194, Springer, Heidelberg.

[25] E. Süli, C. Schwab, and P. Houston (1999). hp-DGFEM for partial differential equations with nonnegative characteristic form. In: B. Cockburn, G. Karniadakis, and C.-W. Shu, editors, Discontinuous Galerkin Finite Element Methods. Lecture Notes in Computational Science and Engineering. Springer-Verlag (to appear).

[26] M.F. Wheeler (1978). An elliptic collocation finite element method with interior penalties. SIAM J. Numer. Anal., 15:152–161.

9   WHAT DO WE WANT AND WHAT DO WE HAVE IN A POSTERIORI ESTIMATES IN THE FEM

Ivo Babuška[a], Theofanis Strouboulis[b], Dibyendu Datta[b], and Srihari Gangaraj[b]

[a]Texas Institute for Computational and Applied Mathematics,
The University of Texas at Austin, Austin, Texas 78712, U.S.A.

[b]Department of Aerospace Engineering,
Texas A& M University, College Station, Texas 77843, U.S.A.

ABSTRACT

This paper brings together some comments on major achievements in the theory and practice of *a posteriori* estimation for elliptic equations in the light of practical needs.

Key words. Guaranteed upper and lower estimates, bounds for outputs computed by the FEM, reliability interval for the computed outputs, guaranteed finite element computation.

9.1   INTRODUCTION

Since the first papers on *a posteriori* estimation and adaptive approaches [1] – [3] appeared some 20 years ago, significant progress has occured both in the FEM and its use in engineering pratice, as well as in the theory and practice of *a posteriori* estimation and its use in practical computation. Let us outline some significant aspects of the use of the FEM and *a posteriori* estimation in today's engineering practice.

1. Engineers often utilize commercial FE software as a "black box". Very often important studies are made by analysts with little understanding of the theory of the FEM, and lacking engineering experience which could be used to detect gross errors in the computed critical outputs. This practice is dangerous, because it can lead to accidents. Let us, for example, mention the Sleipner accident, where the error in the FE analysis employed led to the collapse of a drilling tower and a financial loss of the order of $ 700 million; see [4], [5].

2. Various software companies are beginning to implement *a posteriori* error estimation in their FE codes without explaining its precise meaning and range of applicability. This practice is dangerous because it can lead to misinterpretation of and unjustified confidence in the computed outputs.

3. Todays FE codes are capable of displaying the computed results using sophisticated 3-D color graphics. This can also be misleading because analysts and their managers seldom question the meaning of results displayed with fancy color graphics. As an example, let us mention the color display of the results of *a posteriori* estimation of the maximum stress in an L-shaped, and/or, cracked domain. This produces impressive color pictures with typically the red region indicating high error at the crack-tips and re-entrant corners; nevertheless such figures are meaningless because the exact value of the maximum error in the stress in the case that the domain has a re-entrant corner is infinite!

We will summarize below recent progress in quantitative *a posteriori* error estimation. The main result is that we can construct guaranteed intervals for the error in the computed outputs, for a large class of practical problems.

9.2   ERROR INDICATORS AND ESTIMATORS

The typical result in *a posteriori* error estimation is

$$\underline{\mathcal{E}} = \mathcal{C}_L \mathcal{E} \leq ||e_{FE}||_\mathcal{U} = ||u_{EX} - u_{FE}||_\mathcal{U} \leq \bar{\mathcal{E}} = \mathcal{C}_U \mathcal{E} \tag{9.1}$$

where u_{EX} is the exact solution, and u_{FE} its finite element approximation,

$$e_{FE} \overset{\text{def}}{=} u_{EX} - u_{FE} \tag{9.2}$$

is the error in u_{FE}, $|| \cdot ||_\mathcal{U}$ is the global energy-norm, $\mathcal{E}$ is the estimator for $||e_{FE}||_\mathcal{U}$. Typically $\mathcal{E}$ is obtained as a sum of the squares of element error indicators η_τ over the finite element mesh Δ, namely

$$\mathcal{E} = \sqrt{\sum_{\tau \in \Delta} (\eta_\tau)^2} \tag{9.3}$$

where

$$\eta_\tau = ||\tilde{u}_{FE} - u_{FE}||_{\mathcal{U}(\tau)} \tag{9.4}$$

and $\tilde{u}_{FE}$ is recovered from u_{FE}, either by solving local boundary-value problems (residual problems) with data obtained from u_{FE}, or by employing a local least-squares averaging. The two-sided estimate (9.1) expresses the equivalence $\mathcal{E} \approx ||e_{FE}||_\mathcal{U}$, and the constants $\mathcal{C}_L$ and $\mathcal{C}_U$ depend on the definition of the error indicators η_τ. Knowledge of the constants $\mathcal{C}_L$, and $\mathcal{C}_U$, allows us to compute, respectively, the lower estimator $\underline{\mathcal{E}} = \mathcal{C}_L \mathcal{E}$, and the upper estimator $\bar{\mathcal{E}} = \mathcal{C}_U \mathcal{E}$, and we have the two-sided estimate

$$\underline{\mathcal{E}} \leq ||e_{FE}||_\mathcal{U} \leq \bar{\mathcal{E}} \tag{9.5}$$

In general, it is rather difficult to obtain values for $\mathcal{C}_L$, $\mathcal{C}_U$ which lead to estimates with reasonable accuracy for practical computation. An exception is the subdomain residual estimator of [1] for which $\mathcal{C}_L$ and $\mathcal{C}_U$ depend only on the class of meshes employed and are easy to estimate; however this estimator has not been implemented in many codes. For the various element residual estimators given in [6] and [7], and for the very popular ZZ-SPR estimator [8], $\mathcal{C}_L$ or both constants depend on the class of meshes as well on the smoothness of the exact solution u_{EX} and their determination is difficult. One approach that has been employed (see [9]) to estimate $\mathcal{C}_L$, $\mathcal{C}_U$, has been to go through the proofs of equivalence between the various estimators and to estimate $\mathcal{C}_L$, $\mathcal{C}_U$, in terms of the various constants involved in the proof of equivalence between the various estimators. An example was given in [10], where $\mathcal{C}_U$ was estimated for the explicit residual estimator for certain meshes of linear triangles. However this estimate is rather poor and results in gross overestimation.

The reliability of an estimator is measured by its effectivity index $\kappa = \dfrac{\mathcal{E}}{||e_{FE}||_u}$ which satisfies

$$\frac{1}{\mathcal{C}_U} \leq \kappa = \frac{\mathcal{E}}{||e_{FE}||_u} \leq \frac{1}{\mathcal{C}_U} \tag{9.6}$$

If $\mathcal{C}_L$, $\mathcal{C}_U$, are known, we are mostly interested in the effectivity indices for the upper, and the lower estimator, $\bar{\mathcal{E}}$, and $\underline{\mathcal{E}}$ namely

$$1 \leq \bar{\kappa} = \frac{\bar{\mathcal{E}}}{||e_{FE}||_u} = \mathcal{C}_U \kappa \tag{9.7}$$

and

$$\underline{\kappa} = \frac{\underline{\mathcal{E}}}{||e_{FE}||_u} = \mathcal{C}_L \kappa \tag{9.8}$$

The main use of an upper estimator is for obtaining a safe point for stopping a calculation. An upper estimator which grossly overestimates can lead to expensive unnecessary overkill, especially in three dimensions. For example, the effectivity of the explicit upper estimator given in [10] can be close to 70 in many practical cases. The main use of the a lower estimator is for checking the effectivity of the upper estimator employed in the stopping criterion, and if it grossly underestimates it cannot serve the intended purpose.

In general the estimates that one gets for $\mathcal{C}_L$, $\mathcal{C}_U$, by going through the proofs of equivalence between the various estimators are too pessimistic to be of practical use, and they often employ assumptions about the mesh and the types of elements which do not hold for the meshes used in practical computation. Another approach for estimating $\mathcal{C}_L$, and $\mathcal{C}_U$, which was proposed in [12] – [16], is based on the splitting of the error into local and pollution errors (see [12] – [13]), and a local asymptotic analysis of the finite element solution (see [13] – [16]). The main idea is to employ the asymptotic analyses which are relevant in the cases of practical interest to determine the corresponding asymptotic values of $\mathcal{C}_L$ and $\mathcal{C}_U$ for the estimators of interest. For example, in the case that the finite element solution of plane elasticity problems in polygonal domains using quasi-uniform meshes of degree $1 \leq p \leq 3$, is of interest, it is sufficient to analyze $\mathcal{C}_L$, $\mathcal{C}_U$ for three types of asymptotics, in the interior of the domain, adjacent to a straight boundary, and at a corner point (see [14]). From these analyses the local asymptotic values $\mathcal{C}_L$, and $\mathcal{C}_U$ can be determined by solving generalized eigenvalue problems, see [13] – [16], taking into account all the possible connectivities of the mesh, the aspect ratio, the distortion of the elements, and the range of the coefficients (e.g. the Poisson's ratio, and/or the orthotropy). These can then be used to correct the estimators locally to obtain asymptotically upper and lower estimators. Note however that in this case we cannot talk about guaranteed upper and lower estimators.

Let us also comment on the meaning of the term indicator. Strictly speaking an indicator for an output of interest is the contribution of the element τ to the estimator for the output. Above we introduced $\mathcal{E}$ as an estimator for the global energy-norm of the error, $\|e_{FE}\|_\mathcal{U}$, and in this special case the indicator is determined as in (9.4). More generally, $\mathcal{E}_{\mathcal{F}(u)}$ could be an estimator for an output $\mathcal{F}(u)$, and in this case the error indicator is given by (see [17])

$$\eta_\tau = \sqrt{|\mathcal{B}_\tau(\tilde{u}_{FE} - u_{FE}, \tilde{G}_{FE} - G_{FE})|} \tag{9.9}$$

where $\mathcal{B}_\tau$ is the element inner product corresponding to the problem which is being solved, and G_{FE} is the finite element approximation of the Green's function G_{EX} which corresponds to the output $\mathcal{F}(u)$. The indicators given by (9.9) can then be used in mesh optimization algorithms to optimize the mesh and the distribution of the element degrees for the outputs of interest (see [13] and [17]). Another meaning of the word indicator is to denote some quantity which is used for local refinement and/or enrichment of the mesh. Let us mention, for example, the use of the modulus of the gradient of the finite element solution as an indicator for mesh refinement, namely

$$\eta_\tau = \sqrt{\int_\tau |\nabla u_{FE}|^2 \, d\tau} \tag{9.10}$$

Although the use of such an indicator leads to reasonable looking refinements of the mesh, the resulting meshes are often far from being optimal for any output, and hence they can be much more expensive than truly adaptive meshes which are obtained by employing a principle of minimization of the error in the outputs of interest (see [13] and [17]) and the appropriate indicators.

9.3   DESIRED PROPERTIES OF THE ESTIMATOR

In order for an estimator $\mathcal{E}$ to be suitable for guaranteed quantative computation it must have the following properties

(a) The estimator has to be a *guaranteed* upper estimator for the computed outputs which are of interest to the user of the program.

(b) All the constants involved in the definition of the estimator must be known for the most general types of meshes that may be used in the computation.

(c) The estimator must have a reasonable effectivity index, e.g. less than 1.5, for any mesh, however coarse this may be.

(d) A reliable lower estimator must be available for estimating the accuracy of the upper estimator, and must have a reasonable effectivity index, i.e. greater than 0.66.

(e) A relatively inexpensive procedure for improving the accuracy of the upper and the lower estimator must be available.

(f) The cost of the error estimator has to reasonable, e.g. comparable to the cost of computing the finite element solution.

Let us also give some comments related to the above desired properties for the estimator which we label using "ad.".

ad. (a) By a guaranteed estimator we mean that the estimator must be a guaranteed upper estimator for the exact error in the output. Often in the literature, the estimator is constructed to be upper for the error of the computed finite element solution with respect to an enriched/refined finite element solution and not the difference between the exact and the computed solution (see [11]). Such an estimator can often underestimate the exact error in the output of interest, and can be dangerous if it is used in the stopping criterion.

ad. (b) Sometimes the estimator is given with various constants specified only for certain meshes, such as, for example, the explicit upper estimator given in [10]. In practice the meshes are created by the available mesh generator and they may not satisfy the assumptions used in the estimation of the constants.

ad. (c) Many engineering computations employ one or at most two rather coarse meshes. It is therefore essential for the error estimator also to be reliable when the error is large, e.g. 40%, either because such an error may be acceptable to the user, or because there may be no resources to compute a better solution.

ad. (d) The cost of the estimator has to be properly measured to include its impact on the overall cost of the computation. For example, the explicit estimator given in [10] in many practical cases has effectivity index $\kappa \approx 70$, and hence if it is used in the stopping criterion it can lead to tremendous overkill which can be very expensive. The "cost" of an estimator which underestimates could be much higher, if it is misunderstood by the user as an upper one and its use leads to a costly accident, e.g. the Sleipner accident!

9.4   AN EXAMPLE OF AN ESTIMATOR WITH THE DESIRED PROPERTIES

We will now give an example of an error estimator which satisfies the requirements listed above in Section 9.3, namely:

1. It is a guaranteed upper estimator for the energy-norm of the exact error.

2. It can be supplemented by a lower estimator computed at negligible additional cost.

3. Both the lower and upper estimators can be improved by employing an inexpensive iteration.

4. It has relatively small total cost, e.g. 70% of the cost of computing the finite element solution.

Let us now outline the estimator and its properties in the case of the simplest two-dimensional model, Laplace's equation. A complete discussion may be found in [13], and also in [18] – [20].

Let Ω be the curvilinear polygonal domain shown in Fig 9.1a with boundary $\partial\Omega = \bar{\Gamma}_D \cup \bar{\Gamma}_N$, where $\Gamma_D = \Gamma_D^1 \cup \Gamma_D^2 \cup \Gamma_D^3$, is the Dirichlet boundary, and Γ_N the Neumann boundary, also as shown in Fig 9.1a. We will let u_{EX} denote the exact solution of the mixed boundary-value problem for the Laplacian.

Find

$$u_{EX} \in \mathcal{U}_{\bar{u}}(\Omega) = \left\{ v : \ \|v\|_u \overset{\text{def}}{=} \sqrt{\int_\Omega |\nabla v|^2 \, d\Omega} < \infty, \ \ v\big|_{\Gamma_D} = \bar{u} \right\} \qquad (9.11)$$

such that

$$\mathcal{B}_\Omega(u_{EX}, v) \overset{\text{def}}{=} \int_\Omega \nabla u_{EX} \cdot \nabla v \, d\Omega = 0 \qquad \forall v \in \mathcal{U}_0(\Omega) \qquad (9.12)$$

This is the variational formulation of the mixed boundary-value problem for the Laplacian with homogenous Neumann and non-homogenous Dirichlet boundary conditions

$$\Delta u\big|_\Omega \equiv 0, \qquad \frac{\partial u}{\partial n}\bigg|_{\Gamma_N} = 0, \qquad u\big|_{\Gamma_D} = \bar{u} \qquad (9.13)$$

Here we let $\bar{u} = 0$ on $\Gamma_D^1 \cup \Gamma_D^3$, and $\bar{u} = 1$ on Γ_D^2.

Letting T_h^Ω denote a regular mesh of curvilinear quadrilateral, e.g. the mesh shown in Fig 9.1b, we will define the finite element solution $u_{S_h^p}$, as the solution of the discrete variational problem:

Find

$$u_{S_h^p} \in S_{T_h,\bar{u}}^p(\Omega) = \left\{ v \in C^0(\Omega) : v\big|_\tau \cdot \mathbf{F}_\tau \in \hat{S}^p \ \ \forall \tau \in T_h^\Omega, \ v\big|_{\Gamma_D} = \bar{u} \right\} \qquad (9.14)$$

such that

$$\mathcal{B}_\Omega(u_{S_h^p}, v) = 0 \qquad \forall v \in S_{T_h,0}^p(\Omega) \qquad (9.15)$$

Here $\mathbf{F}_\tau$ denotes the mapping from the master square $\hat{\tau}$ to the curvilinear quadrilateral τ, and $\hat{S}^p$ is the bi-p master element space. Below we will use the notation S_h^p, instead of $S_{T_h}^p$. We will denote the error in $u_{S_h^p}$ by

$$e_u^{EX} \overset{\text{def}}{=} u_{EX} - u_{S_h^p} \qquad (9.16)$$

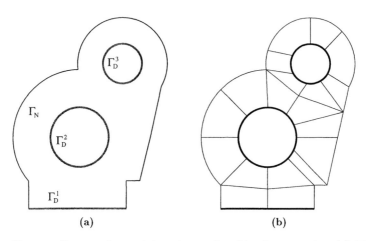

(a)   (b)

Figure 9.1: *The curvilinear polygonal domain employed in the examples*: (a) The domain Ω, with its boundary $\partial\Omega = (\bar\Gamma_D^1 \cup \bar\Gamma_D^2 \cup \bar\Gamma_D^3) \cup \bar\Gamma_N$ which consists of the Dirichlet and Neumann parts. (b) The coarse mesh of curvilinear quadrilaterals T_h^Ω.

We will also introduce the error with respect to an enriched and/or refined mesh finite element solution $u_{S_{h'}^{p'}} \in S_{T_h'}^{p'}$, namely

$$e_u^{S_{h'}^{p'}} \overset{\text{def}}{=} u_{S_{h'}^{p'}} - u_{S_h^p} \tag{9.17}$$

This the *truth-mesh* error employed in [11]. Here T_h' is a mesh obtained using nested refinement of Ω, or which coincides with T_h, and $p' > p$. We will call e_u^{EX}, the exact error to underline its difference from the truth-mesh error, $e_u^{S_{h'}^{p'}}$. Below we will omit the subscript u, from e_u^{EX}, and $e_u^{S_{h'}^{p'}}$, except in the latter part when we discuss the error in the outputs.

Substituting $u_{EX} = e^{EX} + u_{S_h^p}$ into (9.12), we get that e^{EX} is the exact solution of the *residual* equation: Find $e^{EX} \in \mathcal{U}_0(\Omega)$ such that

$$\mathcal{B}_\Omega(e^{EX}, v) = \mathcal{R}_{\Omega, u_{S_h^p}}(v) \overset{\text{def}}{=} \sum_{\tau \in T_h} \left(\int_\tau r_\tau \, v \, d\Omega + \sum_{\varepsilon \subseteq \partial\tau} \frac{1}{2} \int_\varepsilon J_\varepsilon \, v \, d\varepsilon \right) \qquad \forall v \in \mathcal{U}_0(\Omega) \quad (9.18)$$

where $r_\tau \overset{\text{def}}{=} \Delta u_{S_h^p}|_\tau$, is the interior residual and

$$J_{\varepsilon, u_{S_h^p}}(\mathbf{x}) \overset{\text{def}}{=} \begin{cases} \left(\nabla u_{S_h^p}|_{\tau_k} - \nabla u_{S_h^p}|_{\tau_j} \right)(\mathbf{x}) \cdot \mathbf{n}_\varepsilon(\mathbf{x}), & \mathbf{x} \in \varepsilon = \partial\tau_j \bigcap \partial\tau_k, \ j \neq k \\[2ex] 2\left(g(\mathbf{x}) - \nabla u_{S_h^p}(\mathbf{x}) \right) \cdot \mathbf{n}_{\Gamma_N}(\mathbf{x}), & \mathbf{x} \in \varepsilon \subseteq \Gamma_N \\[2ex] 0, & \mathbf{x} \in \varepsilon \subseteq \Gamma_D \end{cases}$$

$$(9.19)$$

is the jump of the normal derivative on the edge ε, where n_ε is the exterior unit normal on $\partial \tau_j$. Eq.(9.18) can also be understood as the definition of the residual $\mathcal{R}_\Omega$. Similarly the truth mesh error $e^{S^{p'}_{h'}} \in S^{p'}_{h',0}(\Omega)$, satisfies (9.18) for every $v \in S^{p'}_{h',0}(\Omega)$. We will employ a splitting of the residual $\mathcal{R}_\Omega$ into equilibrated element residuals

$$\mathcal{R}_\Omega(v) = \sum_{\tau \in T_h} \mathcal{R}_\tau^{EQ}(v|_\tau) \qquad \forall v \in \mathcal{U}_0(\Omega) \tag{9.20}$$

with

$$\mathcal{R}_\tau^{EQ}(v) \overset{\text{def}}{=} \int_\tau r_\tau \, v \, d\tau + \sum_{\varepsilon \subseteq \partial \tau} \int_\varepsilon (\frac{1}{2} J_\varepsilon + \text{sg}(\varepsilon, \tau) \, \theta_\varepsilon) \, v \, d\varepsilon = 0 \qquad \forall v \in S^p_{h,0}(\tau) \tag{9.21}$$

where $S^p_{h,0}(\tau)$ is the element space of functions obtained by restricting the functions from $S^p_{h,0}(\Omega)$ in τ. The idea of splitting the residual into equilibrated element residuals, and the construction of the θ_ε's was introduced in [21].

Let us now introduce $\hat{e}^{EX}$ and $\hat{e}^{S^{p'}_{h'}}$, respectively, the *exact* and the *computed* error indicator functions. The exact error indicator functions $\hat{e}^{EX}$, will be defined as the exact solutions of the element residual problems:

Find $\hat{e}^{EX} \in \mathcal{U}_0(\tau)$ such that

$$\mathcal{B}_\tau(\hat{e}_\tau^{EX}, v) = \mathcal{R}_\tau^{EQ}(v) \qquad \forall v \in \mathcal{U}_0(\tau) \tag{9.22}$$

where $\mathcal{U}_0(\tau)$ is defined analogously as $S^p_{h,0}(\tau)$. For the elements with $\partial \tau \cap \Gamma_D = \emptyset$, we will fix $\hat{e}^{EX}$ by imposing the condition

$$\int_\tau \hat{e}^{EX} \, d\tau = 0 \tag{9.23}$$

The computed indicator function $\hat{e}_\tau^{S^{p'}_{h'}} \in S^{p'}_{h',0}(\tau)$ is the solution of (9.22) – (9.23) with $\mathcal{U}_0(\tau)$ replaced by $S^{p'}_{h',0}(\tau)$. We will now define the exact element error indicator, and the exact estimator for the energy norm of the error

$$\eta_{\tau, u_{S^p_h}}^{EX} \overset{\text{def}}{=} \|\hat{e}_{\tau, u_{S^p_h}}^{EX}\|_{\mathcal{U}(\tau)}, \qquad \mathcal{E}_{u_{S^p_h}}^{EX} = \sqrt{\sum_{\tau \in T_h} (\eta_{\tau, u_{S^p_h}}^{EX})^2} \tag{9.24}$$

and the computed element error indicator, and the computed estimator

$$\eta_{\tau, u_{S^p_h}}^{S^{p'}_{h'}} \overset{\text{def}}{=} \|\hat{e}_{\tau, u_{S^p_h}}^{S^{p'}_{h'}}\|_{\mathcal{U}(\tau)}, \qquad \mathcal{E}_{u_{S^p_h}}^{S^{p'}_{h'}} = \sqrt{\sum_{\tau \in T_h} (\eta_{\tau, u_{S^p_h}}^{S^{p'}_{h'}})^2} \tag{9.25}$$

where below we will often ommit the subscript $u_{S^p_h}$ unless it is needed for clarity. We now have:

Theorem 9.1 *(Upper bounds)*

$$||e^{S_{h'}^{p'}}||_{\mathcal{U}} \le ||e^{EX}||_{\mathcal{U}} \le \mathcal{E}^{EX} \tag{9.26}$$

and

$$||e^{S_{h'}^{p'}}||_{\mathcal{U}} \le \mathcal{E}^{S_{h'}^{p'}} \le \mathcal{E}^{EX} \tag{9.27}$$

Proof: This follows directly from the definition of the energy norm by duality, the residual equation, and the definition of the indicator functions. □

Hence, the exact estimator $\mathcal{E}^{EX}$ is an upper estimator for $||e^{EX}||_{\mathcal{U}}$, and the computed estimator $\mathcal{E}^{S_{h'}^{p'}}$, is an upper estimator for $||e^{S_{h'}^{p'}}||_{\mathcal{U}}$. However, *we cannot say that the computed estimator $\mathcal{E}^{S_{h'}^{p'}}$ is an upper bound of $||e^{EX}||_{\mathcal{U}}$*. In fact it can easily be shown that this is not the case, in general, by a counterexample.

Counterexample Consider the example problem given above, and let T_h be the mesh shown in Fig 9.1b, and $u_{S_h^2}$ the corresponding finite element solution obtained using biquadratic elements. We then have

$$\mathcal{E}^{S_{h/8}^3} \approx 0.9||e^{u_{EX}-u_{S_h^2}}||_{\mathcal{U}} < ||e^{u_{EX}-u_{S_h^2}}||_{\mathcal{U}} \tag{9.28}$$

and hence, the computed estimator $\mathcal{E}^{S_{h/8}^3}$, underestimates the energy norm of the exact error $||e^{u_{EX}-u_{S_h^2}}||_{\mathcal{U}}$. Note also that

$$\mathcal{E}^{EX} = \lim_{n\to\infty, \, k\to\infty} \mathcal{E}^{S_{h/2^n}^{p+k}} \tag{9.29}$$

and for sufficiently high n and/or k, $\mathcal{E}^{S_{h/2^n}^{p+k}}$ is greater than $||e^{EX}||_{\mathcal{U}}$. However, in general, we do not know *a priori* what are the values of n, k for which this happens in the particular case of interest. □

The exact estimator $\mathcal{E}^{EX}$ is an upper estimator for $||e^{EX}||_{\mathcal{U}}$, but it cannot be computed. It is therefore necessary to construct a *computable* upper estimator. This can be done by employing the following lemmas.

Lemma 9.1

$$\mathcal{E}^{EX} = \sqrt{\left(\mathcal{E}^{S_{h'}^{p'}}\right)^2 + \sum_{\tau \in T_h^\Omega} ||\hat{e}_\tau^{EX} - \hat{e}_\tau^{S_{h'}^{p'}}||_{\mathcal{U}(\tau)}^2} \tag{9.30}$$

Proof: This follows directly from the orthogonalities

$$\mathcal{B}_\tau\left(\hat{e}_\tau^{EX} - \hat{e}_\tau^{S_{h'}^{p'}}, v\right) = 0 \qquad \forall v \in S_{h',0}^{p'}(\tau) \tag{9.31}$$

□

Lemma 9.2 *(Explicit upper estimator for the energy norm of the error in the error indicator)*

$$||\hat{e}_\tau^{EX} - \hat{e}_\tau^{S_h^{p+k}}||_{\mathcal{U}_\tau} \leq \mathcal{E}_{\hat{e}_\tau^{S_h^{p+k}}}^{EXPL} \tag{9.32}$$

where

$$\mathcal{E}_{\hat{e}_\tau^{S_h^{p+k}}}^{EXPL} \stackrel{\text{def}}{=} \mathcal{C}_1^\tau(p,k) \, ||r_\tau + \Delta \hat{e}_\tau^{S_h^{p+k}}||_{L^2(\tau)} + \sum_{\varepsilon \subset \partial \tau} \mathcal{C}_2^\varepsilon(p,k) \left\| J_\varepsilon^\tau - \frac{\partial}{\partial n_\varepsilon} \hat{e}_\tau^{S_h^{p+k}} \right\|_{L^2(\varepsilon)} \tag{9.33}$$

Proof: This follows directly from the residual equation for $\hat{e}_\tau^{EX} - \hat{e}_\tau^{S_h^{p+k}}$ and the Cauchy-Schwarz inequality.   □

Combining Lemma 9.1 and 9.2 we have:

Theorem 9.2 *(Computable upper estimator)*

$$||e^{EX}||_{\mathcal{U}} \leq \mathcal{E}^{U,S_h^{p+k}} \stackrel{\text{def}}{=} \sqrt{\left(\mathcal{E}^{S_h^{p+k}}\right)^2 + \sum_{\tau \in T_h} \left(\mathcal{E}_{\hat{e}_\tau^{S_h^{p+k}}}^{EXPL}\right)^2} \tag{9.34}$$

Proof: It follows directly from (9.30) – (9.33).   □

The constants $\mathcal{C}_1^\tau$, $\mathcal{C}_2^\tau$ can be estimated from

$$\mathcal{C}_1^\tau(p,k) \leq \sqrt{\frac{\max_\tau |\mathbf{J}_\tau|}{\mathcal{K}_\tau}} \, \hat{\mathcal{C}}_1(p,k), \qquad \mathcal{C}_2^\varepsilon(p,k) \leq \sqrt{\frac{\max_\varepsilon |\mathcal{J}_\varepsilon|}{\mathcal{K}_\tau}} \, \hat{\mathcal{C}}_2(p,k) \tag{9.35}$$

where $\mathbf{J}_\tau$, $\mathcal{J}_\varepsilon$ are, respectively, the Jacobians of the element and the edge mapping, and $\mathcal{K}_\tau = \min_\tau \left(|\mathbf{J}_\tau| \, \lambda_{\min}(\mathbf{J}_\tau^{-T}\mathbf{J}_\tau^{-1})\right)$, where $\lambda_{\min}(\mathbf{J}_\tau^{-T}\mathbf{J}_\tau^{-1})$ is the minimum eigenvalue of the matrix $\mathbf{J}_\tau^{-T}\mathbf{J}_\tau^{-1}$, and $\hat{\mathcal{C}}_1(p,k)$, $\hat{\mathcal{C}}_2(p,k)$ can be determined by solving a generalized eigenvalue problem in the master element; see [13], [19], [20].

A more general version of the Theorem $||e^{EX}||_{\mathcal{U}} \leq \mathcal{E}^{U,S_{h/2^n}^{p+k}}$, describes the construction of the computable upper estimator is given in [19] – [20].

Let us also derive a lower estimator for $||e^{EX}||_{\mathcal{U}}$, which uses the computed error indicator functions $\hat{e}_\tau^{S_{h'}^{p'}}$. We will use the following lemma:

Lemma 9.3 *For any $\tilde{e}^{S_{h'}^{p'}} \in S_{h',0}^{p'}(\Omega)$ we have*

$$||e^{EX}||_{\mathcal{U}}^2 = \left(\mathcal{E}^{S_{h'}^{p'}}\right)^2 - \sum_{\tau \in T_h^\Omega} ||\tilde{e}^{S_{h'}^{p'}} - \hat{e}_\tau^{S_{h'}^{p'}}||_{\mathcal{U}(\tau)}^2 + ||\tilde{e}^{S_{h'}^{p'}} - e^{EX}||_{\mathcal{U}(}^2 \tag{9.36}$$

Proof: This follows directly from the orthogonalities

$$\sum_{\tau \in T_h} \mathcal{B}_\tau(\hat{e}^{EX} - e^{EX}, v) = 0 \qquad \forall v \in \mathcal{U}_0(\Omega) \tag{9.37}$$

and

$$\sum_{\tau \in T_h} \mathcal{B}_\tau(\hat{e}_\tau^{EX} - \hat{e}_\tau^{S_{h'}^{p'}}, v) = 0 \qquad \forall v \in S_{h',0}^{p'}(\Omega) \tag{9.38}$$

$\square$

Using Lemma 9.3 we obtain:

Theorem 9.3 *(Lower estimator) For any $\tilde{e}^{S_{h'}^{p'}} \in S_{h',0}^{p'}(\Omega)$, we have*

$$\mathcal{E}^{L,S_{h'}^{p'}}(\tilde{e}^{S_{h'}^{p'}}) \stackrel{def}{=} \sqrt{\left(\mathcal{E}^{S_{h'}^{p'}}\right)^2 - \sum_{\tau \in T_h} ||\tilde{e}^{S_{h'}^{p'}} - \hat{e}_\tau^{S_{h'}^{p'}}||_{\mathcal{U}(\tau)}^2} \leq ||e^{EX}||_{\mathcal{U}} \tag{9.39}$$

Proof: Equation (9.39) follows immediately from (9.36) by omitting the last term. $\square$
Combining Theorems 9.3 and 9.4 we get the computable two-sided estimate

$$\mathcal{E}_{u_{S_h^p}}^{L,S_{h'}^{p'}}(\tilde{e}^{S_{h'}^{p'}}) \leq ||u_{EX} - u_{S_h^p}||_{\mathcal{U}} \leq \mathcal{E}_{u_{S_h^p}}^{U,S_{h'}^{p'}} \tag{9.40}$$

where we have used the subscript $u_{S_h^p}$ in the estimators $\mathcal{E}_{u_{S_h^p}}^{B,S_{h'}^{p'}}$, $B = L$ or R, to emphasise that these bounds are for the exact error in $u_{S_h^p}$. In the implementations $\tilde{e}^{S_{h'}^{p'}}$, was constructed by patching together the computed indicators $\hat{e}_\tau^{S_{h'}^{p'}}$, using nodal and edge averagings (see [13], [19], and [20]).
We can also use Lemma 9.3 to obtain improved upper and lower estimators. For this we will need the following result:

Lemma 9.4 *(Lower and upper estimators for any admissible function) Given any $\tilde{e}^{S_{h'}^{p'}} \in S_{h',0}^{p'}(\Omega)$, let $q_{\tilde{e}}^{S_h^p} \in S_{h,0}^p(\Omega)$ such that*

$$\mathcal{B}_\Omega(q_{\tilde{e}}^{S_h^p}, v) = -\mathcal{B}_\Omega(\tilde{e}^{S_{h'}^{p'}}, v) \qquad \forall v \in S_{h,0}^p(\Omega) \tag{9.41}$$

Then, for any $\tilde{\tilde{e}}^{S_{h'}^{p'}} \in S_{h',0}^{p'}(\Omega)$, we have

$$||q_{\tilde{e}}^{S_h^p}||_{\mathcal{U}}^2 + \left(\mathcal{E}_{\tilde{u}_{S_h^p}^*}^{L,S_{h'}^{p'}}(\tilde{\tilde{e}}^{S_{h'}^{p'}})\right)^2 \leq ||u_{EX} - \tilde{u}_{S_h^p}^*||_{\mathcal{U}}^2 \leq ||q_{\tilde{e}}^{S_h^p}||_{\mathcal{U}}^2 + \left(\mathcal{E}_{\tilde{u}_{S_h^p}^*}^{U,S_{h'}^{p'}}\right)^2 \tag{9.42}$$

where

$$\tilde{u}_{S_h^p}^* \stackrel{def}{=} u_{S_h^p} + \tilde{e}^{S_{h'}^{p'}} + q_{\tilde{e}}^{S_h^p} \tag{9.43}$$

is the improved finite element solution.

Proof: The proof follows from the observation that the error $u_{EX} - \tilde{u}^*_{S^p_h}$, satisfies the orthogonality

$$\mathcal{B}(u_{EX} - \tilde{u}^*_{S^p_h}, v) = 0 \qquad \forall v \in S^p_{h,0}(\Omega) \tag{9.44}$$

and hence Theorem 9.2 and 9.3 can be used to construct the two-sided estimate for $||u_{EX} - \tilde{u}^*_{S^p_h}||_{\mathcal{U}}$, and (9.42) follows from

$$||e^{EX} - \tilde{e}^{S^{p'}_{h'}}||^2_{\mathcal{U}} = ||q^{S^p_h}_{\tilde{e}}||^2_{\mathcal{U}} + ||u_{EX} - \tilde{u}^*_{S^p_h}||^2_{\mathcal{U}} \tag{9.45}$$

$\square$

Combining Lemmas 9.3 and 9.4 we get:

Theorem 9.4 *(Improved lower and upper estimators) For any* $\tilde{e}^{S^{p'}_{h'}}$, $\tilde{\tilde{e}}^{S^{p'}_{h'}} \in S^{p'}_{h',0}(\Omega)$ *we have*

$$\mathcal{E}^{L,S^{p'}_{h'}}_{u_{S^p_h}}(\tilde{e}^{S^{p'}_{h'}}) \leq \underline{\mathcal{E}}^{L,S^{p'}_{h'}}_{u_{S^p_h}}(\tilde{e}^{S^{p'}_{h'}}, \tilde{\tilde{e}}^{S^{p'}_{h'}}) \leq ||u_{EX} - u_{S^p_h}||_{\mathcal{U}} \leq \bar{\mathcal{E}}^{U,S^{p'}_{h'}}_{u_{S^p_h}}(\tilde{e}^{S^{p'}_{h'}}) \tag{9.46}$$

where

$$\underline{\mathcal{E}}^{L,S^{p'}_{h'}}_{u_{S^p_h}}(\tilde{e}^{S^{p'}_{h'}}, \tilde{\tilde{e}}^{S^{p'}_{h'}}) \overset{\text{def}}{=} \sqrt{\left(\mathcal{E}^{L,S^{p'}_{h'}}_{u_{S^p_h}}(\tilde{e}^{S^{p'}_{h'}})\right)^2 + ||q^{S^p_h}_{\tilde{e}}||^2_{\mathcal{U}} + \left(\mathcal{E}^{L,S^{p'}_{h'}}_{\tilde{u}^*_{S^p_h}}(\tilde{\tilde{e}}^{S^{p'}_{h'}})\right)^2} \tag{9.47}$$

and

$$\bar{\mathcal{E}}^{U,S^{p'}_{h'}}_{u_{S^p_h}}(\tilde{e}^{S^{p'}_{h'}}, \tilde{\tilde{e}}^{S^{p'}_{h'}}) \overset{\text{def}}{=} \sqrt{\left(\mathcal{E}^{L,S^{p'}_{h'}}_{u_{S^p_h}}(\tilde{e}^{S^{p'}_{h'}})\right)^2 + ||q^{S^p_h}_{\tilde{e}}||^2_{\mathcal{U}} + \left(\mathcal{E}^{U,S^{p'}_{h'}}_{\tilde{u}^*_{S^p_h}}\right)^2} \tag{9.48}$$

Proof: This follows from (9.36), (9.39), and (9.42). $\square$

The estimates (9.46) – (9.48) may also be employed recursively by constructing a $\tilde{\tilde{u}}^*_{S^p_h}$, etc; see also [13], and [18] – [20], and the following example.

Example A *Iterative improvement of the estimators:* We considered the mixed boundary-value problem for the Laplacian with non-homogenous Dirichlet boundary condition given above. For the finite element solution $u_{S^2_h}$, computed using the mesh T_h shown in Fig.9.1b, we constructed recursively the improved computable bounds $\mathcal{E}^{U,S^8_h}_{(m)}$, and $\mathcal{E}^{U,S^8_h}_{(m)}$, where m is the index indicating the number of recursions, where we let $\mathcal{E}^{B,S^8_h}_{(0)} \equiv \mathcal{E}^{B,S^8_h}$, $B = L$, or U. Fig. 9.2 gives the effectivity indices $\kappa^{B,(m)}_{S^8_h/S^2_h} \overset{\text{def}}{=} \mathcal{E}^{B,S^8_h}_{(m)}/||e^{EX}||_{\mathcal{U}}$, for the mth iterate of the $B = L$, and U estimator. Figure 9.2 also shows the ratio of the energy-norms of the truth-mesh error over the exact error, $||e^{S^8_h}||_{\mathcal{U}}/||e^{EX}||_{\mathcal{U}}$, which does not depend on the iteration, and $\mathcal{E}^{S^8_h}_{(m)}$, the upper estimator for the truth mesh error $||e^{S^8_h}||_{\mathcal{U}}$ is given by

$$\mathcal{E}^{S^8_h}_{(1)} = \sqrt{\left(\mathcal{E}^{L,S^8_h}\right)^2 + ||q^{S^p_h}_{\tilde{e}}||_{\mathcal{U}} + \left(\mathcal{E}^{S^8_h}_{\tilde{u}^*_{S^2_h}}\right)^2} \tag{9.49}$$

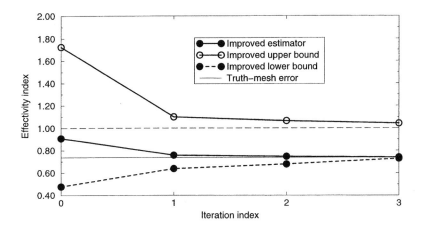

Figure 9.2: *Iterative improvement of the estimators*: Graphs of the effectivity indices $\kappa^{B,(m)}_{S_h^8/S_h^2}$, for $B = L, U$, $\kappa^{(m)}_{S_h^8/S_h^2}$, and the ratio $||e^{S_h^8}||_{\mathcal{U}}/||e^{EX}||_{\mathcal{U}}$ versus the iteration index (m). Note that the effectivity of the guaranteed bound $\mathcal{E}^{S_h^8,B}_{(m)}$ drastically improves in the first iteration. Let us emphasise that the relative error for this problem is $||e^{EX}||_{\mathcal{U}}/||u_{S_h^p}||_{\mathcal{U}} \approx 30\%$, and hence is not small.

and similarly for subsequent iterates and the effectivity indices $\kappa^{(m)}_{S_h^8/S_h^2} = \mathcal{E}^{S_h^8,(m)}/||e^{EX}||_{\mathcal{U}}$. From Fig. 9.2 it can be seen that the bounds drastically improve in the first iteration, and we have

$$\kappa^{U,(1)}_{S_h^8/S_h^2} \approx 1.1 \leq \frac{\mathcal{E}^{S_h^8,U}_{(1)}}{\mathcal{E}^{S_h^8,L}_{(1)}} \approx 1.9 \tag{9.50}$$

Thus, in this example, the first iteration is sufficient to guarantee that the effectivity index of $\mathcal{E}^{S_h^8,U}_{(1)}$, is less than two. From the graph it can also be seen that $\mathcal{E}^{S_h^8}_{(m)}$ (resp. $\mathcal{E}^{S_h^8,L}_{(m)}$) converges to the energy norm of the truth-mesh error $||e^{S_h^8}||_{\mathcal{U}}$ from above (resp. below), and that the upper estimator $\mathcal{E}^{S_h^8,U}_{(m)}$, converges to an upper bound of $||e^{EX}||_{\mathcal{U}}$

$$||e^{EX}||_{\mathcal{U}} < \lim_{m\to\infty} \mathcal{E}^{S_h^8,U}_{(m)} \tag{9.51}$$

Using the estimators for the energy norm of the error, we can now construct estimators for the error in desired outputs. Let $\mathcal{F} : \mathcal{U} \longmapsto \mathbb{R}$, be a bounded linear functional, e.g. $\mathcal{F}(u) = \frac{1}{|\omega|} \int_\omega \frac{\partial u}{\partial x_i} d\omega$, which corresponds to the average of the ith component of the gradient in a subdomain ω. Let $G^{\mathcal{F}}_{EX}$ denote the auxiliary function (Green's function) corresponding to $\mathcal{F}$, namely the solution of the problem:
Find $G^{\mathcal{F}}_{EX} \in \mathcal{U}_0(\Omega)$ such that

$$\mathcal{B}_\Omega(G^{\mathcal{F}}_{EX}, v) = \mathcal{F}(v) \qquad \forall v \in \mathcal{U}_0(\Omega) \tag{9.52}$$

We then have

$$\mathcal{F}(e_u^{EX}) = \mathcal{B}_\Omega(e_u^{EX}, e_G^{EX}) = \frac{1}{4}\left(\||s^{-1}e_u^{EX} + s\ e_G^{EX}\||_{\mathcal{U}}^2 - \||s^{-1}e_u^{EX} - s\ e_G^{EX}\||_{\mathcal{U}}^2\right) \qquad (9.53)$$

where $e_G^{EX} = G_{EX}^{\mathcal{F}} - G_{S_h^p}^{\mathcal{F}}$, and $G_{S_h^p}^{\mathcal{F}}$ is the finite element approximation of $G_{EX}^{\mathcal{F}}$. Using the fact that $s^{-1}e_u^{EX} \pm s\ e_G^{EX} = e_{s^{-1}u \pm s\ G}^{EX}$, we have

$$\mathcal{E}_{s^{-1}u_{S_h^p} \pm s\ G_{S_h^p}}^L \le \||s^{-1}e_u^{EX} + s\ e_G^{EX}\||_{u} \le \mathcal{E}_{s^{-1}u_{S_h^p} \pm s\ G_{S_h^p}}^U \qquad (9.54)$$

and using these bounds in (9.53) we obtain

$$\mathcal{E}_{(m),\mathcal{F}(u)}^{L,S_{h'}^{p'}} \le \mathcal{F}(e_u^{EX}) \le \mathcal{E}_{(m),\mathcal{F}(u)}^{U,S_{h'}^{p'}} \qquad (9.55)$$

where

$$\mathcal{E}_{\mathcal{F}(u)}^L = \frac{1}{4}\left((\mathcal{E}_{s^{-1}u_{S_h^p} + s\ G_{S_h^p}}^L)^2 - (\mathcal{E}_{s^{-1}u_{S_h^p} - s\ G_{S_h^p}}^U)^2\right) \qquad (9.56)$$

and is $\mathcal{E}_{\mathcal{F}(u)}^U$, is obtained from the right-hand side of (9.56) by interchanging L and U. Note that these bounds, of course, depend on $S_{h'}^{p'}$, m, and s, and that here we have for simplicity omitted to indicate this dependence in the notation. An optimal value of s, denoted by s_{opt}, may be obtained by minimizing the difference between the bounds, $\mathcal{E}_{\mathcal{F}(u)}^U - \mathcal{E}_{\mathcal{F}(u)}^L$. However, our numerical experience indicates that in practice very similar results may be obtained by employing s equal to one.

Let us now give an example.

Example B *Bounds for the error in the average derivatives:* Let us consider the example problem already employed in Example A, and let us be interested in the bounds for the average error in the derivative in the ith direction, $\frac{1}{\omega}\int_\omega \frac{\partial e^{EX}}{\partial x_i}d\omega$, for $\omega = \omega_1$, and $\omega = \omega_2$, shown in Fig. 9.3b. The regions ω_1, and ω_2 were chosen in the parts of the domain where the modulus of the gradient takes its largest, but finite, values, as can be seen in Fig. 9.3a. Fig. 9.3c shows the computed bound for the x-derivative in ω_1, and ω_2 as a function of the iteration index. Note that the accuracy of the zeroth iterate is unacceptable, while the accuracy of the second and subsequent iterates is excellent. □

9.5   CONCLUSIONS

We have addressed the problem of reliable quantitative estimation of the error in practical engineering computations. We have given an example of an estimator based on the computed solutions of element residual problems. The main conclusions are:

1. A guaranteed upper estimator for the energy norm of the exact error can be obtained by adding to the squares of the computed indicators an explicit estimate of their error.

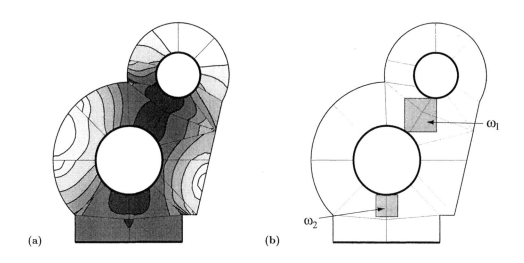

(a)   (b)

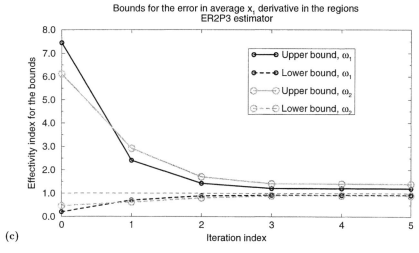

(c)

Figure 9.3: *Determination of bounds for the average error in the derivatives in a small region*: (a) The contours of the computed modulus of the gradient $|\nabla u_{S_h^p}^p|$. (b) The regions ω_1 and ω_2 for which the average error in the derivatives is computed. (c) Computed bounds for the error in the x-derivative in ω_1 and ω_2, versus the iteration index.

2. A lower estimator may be obtained by employing a continuous global approximation of the error constructed by patching together the element indicator functions via local averaging.

3. A recursive application of the upper and lower estimators leads to upper and lower bounds with improved accuracy.

4. Upper and lower estimators for the exact error in a linear functional can be constructed using the representation of the error in the functional as an inner-product of the error in the solution times the error in the Green's function.

5. The accuracy of the upper estimator can be monitored using the lower estimator.

6. The estimator presented is rather general and can be used with any mesh and also constructed for 3-D computations.

7. We have here addressed the guaranteed estimation of the error in the finite element solution. Our approach can also be applied to obtain guaranteed estimation of the error in recovered solutions, in a similar manner to the ZZ-SPR or other solutions constructed from local postprocessing of the finite element solution. This will be the subject of forthcoming work.

9.6   ACKNOWLEDGEMENT

This work was supported by the U.S.Army Research Office under Grant DAAL03-G-028, by the National Science Foundation under Grants MSS-9025110, DMS-91-20877, and DMS-95-0184, and by the Office of Naval Research under Grants N00014-96-1-0021, N00014-96-1-1015, N00014-99-1-0726, and N00014-90-J-1030. The support of Dr. Kailasam Iyer of the Army Research Office, and of Dr. Dick Lau, and Dr. Luise Couchman of the Office of Naval Research is greatly appreciated.

REFERENCES

[1] I. Babuška and W.C. Rheinboldt (1978). Error estimates for adaptive finite element computations. *SIAM J. Numer. Anal.*, 15:736–754.

[2] I. Babuška and W.C. Rheinboldt (1979). *A posteriori* error estimates for the finite element method. *Int. J. Numer. Meth. Engrg.*, 12:1597–1615.

[3] I. Babuška and W.C. Rheinboldt (1979). Adaptive approaches and reliability estimation in finite element analysis. *Comput. Meth. Appl. Mech. Engrg.*, 17/18:519–540.

[4] B. Jacobsen and F. Rosendahl (1994). The Sleipner platform accident. *Struct. Engrg. Internat.*, :190–193.

[5] W.K. Rettendal, O.T. Gudmestad and T. Aarum (1993). Design of concrete platforms after Sleipner A-1 sinking, in: S.K. Chakrabarti, C. Agee, H. Maeda,

A.N. Williams and D. Morrison, eds. *Offshore Technology Proc. 12th Int. Conf. on Offshore Mechanics and Arctic Engineering*, ASME, New York, 1:309–319.

[6] T. A. Westermann (1989). *A posteriori* estimation of errors in *h-p* finite element methods for linear elliptic boundary value problems. M. Sc. Thesis. *The University of Texas at Austin.*

[7] M. Ainsworth, and J. T. Oden (1993). A unified approach to *a posteriori* error estimator using element residual methods.

Numer. Math., 65:23–50.

[8] O. C. Zienkiewicz, and J. Z. Zhu (1992). The superconvergence patch recovery and *a posteriori* error estimates Part 1: The recovery technique Part 2: Error estimates and adaptivity. *Int. J. Numer. Meth. Engrg.*, 33:1333–1364, 1365–1382.

[9] R. Verfürth (1996). A Review of *a posteriori* Error Estimation and Adaptive Mesh-Refinement. *Wiley/Teubner*, New York.

[10] C. Carstensen and S.A. Funken (1997). Constants in Clement-interpolation error and residual based *a posteriori* estimates in finite element methods. *Christian Albrecht-Universität zu Kiel Bericht 97-11*, Mathematical Seminar (August 1997).

[11] M. Paraschivoiu, J. Peraire and A. T. Patera (1997). *A posteriori* finite element bounds for linear functional outputs of elliptic partial differential equations. *Comput. Meth. Appl. Mech. Engrg.*, 150:289–312.

[12] I. Babuška, T. Strouboulis, S.K. Gangaraj and C.S. Upadhyay (1997). Pollution error in the *h*-version of the finite element method and the local quality of the recovered derivatives. *Comput. Meth. Appl. Mech. Engrg.*, 140:1–37.

[13] I. Babuška and T. Strouboulis (2000). The finite element method and its reliability. *Oxford University Press*, (in preparation).

[14] I. Babuška, T. Strouboulis and C.S. Upadhyay (1994). A model study of the quality of *a posteriori* error estimators for linear elliptic problems. Error estimation in the interior of patchwise uniform grids of triangles. *Comput. Meth. Appl. Mech. Engrg.*, 114:307–378.

[15] I. Babuška, T. Strouboulis, C.S. Upadhyay, S.K. Gangaraj and K. Copps (1994). Validation of *a posteriori* error estimators by numerical approach, *Int. J. Numer. Meth. Engrg.*, 37:1073–1123.

[16] I. Babuška, T. Strouboulis and C.S. Upadhyay (1997). A model study of the quality of *a posteriori* error estimators for finite element solutions of linear elliptic problems with particular reference to the behaviour near the boundary. *Int. J. Numer. Meth. Engrg.*, 40:2521–2577.

[17] I. Babuška, T. Strouboulis, D.K. Datta, K. Copps and S.K. Gangaraj (1999). *A posteriori* estimation and adaptive control of the error in the quantity of interest: Part I : *A posteriori* estimation of the error in the Mises stress and the stress-intensity factors. *Comput. Meth. Appl. Mech. Engrg.*, (in press).

[18] I. Babuška, T. Strouboulis and S.K. Gangaraj (1999). Guaranteed computable bounds for the exact error in the finite element solution - Part I: One dimensional model problem. *Comput. Meth. Appl. Mech. Engrg.*, 176:51–79.

[19] T. Strouboulis, I. Babuška and S.K. Gangaraj (2000). Guaranteed computable bounds for the exact error in the finite element solution - Part II: Two dimensional model problem. *Comput. Meth. Appl. Mech. Engrg.*, (in press).

[20] S.K. Gangaraj (1999). *A posteriori* estimation of the error in the finite element solution by computation of the guaranteed upper and lower bounds. Ph.D. Dissertation. *Dept. of Aerospace Engineering, Texas A&M University, College Station, TX,*

[21] P. Ladeveze (1975). Comparison de modeles de milieux continus. These de Doctorat d'Etat es Science Mathematiques. *L'Universite P. et M. Curie, Paris.*

[22] I. Babuška, T. Strouboulis, S.K. Gangaraj, K. Copps and D.K. Datta (1998). Practical aspects of *a posteriori* estimation of the pollution error for reliable finite element analysis. *Computers & Structures*, 66:626–664.

[23] I. Babuška and A. Miller (1984). The post-processing approach in the finite element method Part 1: Calculation of displacements, stresses and other higher derivatives of the displacements. Part 2: The calculation of stress intensity factors. Part 3: *A posteriori* error estimates and adaptive mesh selection. *Int. J. Numer. Meth. Engrg.*, 20:1085–1109, 1111–1129, 2311–2325.

[24] I. Babuška and B. Szabo (1996). Trends and new problems in finite element methods. In J.R. Whiteman (ed.) *The Mathematics of Finite Elements and Applications - Highlights 1996*, Wiley, Chichester, 1–33.

[25] D.K. Datta (1997). *A posteriori* estimation and adaptive control of the error in the quantities of interest. M. S. Thesis. *Dept. of Aerospace Engineering, Texas A&M University, College Station, TX.*

[26] S. Prudhomme (1999). Adaptive control of error and stability of h-p approximations of the transient Navier-Stokes equations. Ph.D. Dissertation. *The University of Texas at Austin.*

10 SOLVING SHORT WAVE PROBLEMS USING SPECIAL FINITE ELEMENTS - TOWARDS AN ADAPTIVE APPROACH

Omar Laghrouche and Peter Bettess

School of Engineering, University of Durham,
Durham DH1 3LE, UK

ABSTRACT

Special finite elements capable of containing many wavelengths per nodal spacing are developed and used to solve problems such as the diffraction of a plane wave by a rigid body. The method used, called *the Partition of Unity Method*, is due to Melenk and Babuška. It consists of incorporating analytical information into a piecewise Galerkin approach. In the case of the Helmholtz problem, the wave potential in two dimensional space is approximated by systems of plane waves propagating in all directions. This leads to larger finite element matrices. However since the elements can contain many wavelengths, the dimension of the global system to be solved is greatly reduced. In a previous work, these special elements were developed by using the same approximating plane waves at all nodes. In this paper, the number and the directions of plane waves can vary from one node to another. It is shown that by taking only a few plane waves clustered around the main directions of propagation, the numerical results converge to the exact solution. This is a step towards developing self adaptive finite elements for short wave problems.

Key words. Short waves, finite elements, plane wave approximation, direction of propagation, wave diffraction, economic modelling

10.1 INTRODUCTION

This work aims at a major improvement in the ability to solve short wave diffraction problems, where the wavelength is a small fraction of the dimensions of the problem.

Such problems are of great economic importance in many fields, including acoustics, sonar and radar cross sections. The aim is to be able to predict patterns of wave diffraction, reflection and refraction, in closed spaces, and in unbounded domains, to any specified level of accuracy, irrespective of how short the wavelengths are, relative to the dimensions of the problem. No existing modelling methods are seen to be perfectly satisfactory. The economic impact of an effective and accurate solution to these very important problems would be enormous.

At the moment it is possible to predict the wave effects described above by using finite element or finite difference models. Boundary integrals may also be used, although they are generally limited to problems with constant wave speed. The difficulty with all these methods is that it is generally accepted that for results of 'engineering accuracy' of about 1%, around 10 nodes per wavelength are required. For the types of problem described above this leads to demands on the computer which are astronomically large. For example, consider a concert hall of say 20m by 20m by 20m, and a wavelength of about 3cm, for reasonably high frequency sounds of around 10kHz, the requirement would be for 600 by 600 by 600 elements, which is 216 million elements. Even with very powerful computers this is a huge problem. For the radar problem the wavelengths are down to millimetres and the size of the structure is even larger. Substantial efforts are being made in this field. Recent work by Morgan *et al.* [1–3] shows that significant problems can be modelled using explicit time stepping with a conventional finite element method which reduces storage demands. But the Morgan *et al.* analysis while very impressive uses millions of finite elements and is still for relatively long waves.

Another approach is to use ray tracing methods, which follow the wave past the object. These methods are numerically cheap and efficient. They deal with the short waves. Unfortunately they cannot cope with the effects of diffraction. They invariably predict 'sharp' edges to waves as they pass objects. Various 'fudges' can be adopted to attempt to patch in the diffraction effects, but this is generally unsatisfactory.

In recent years, an approach was developed which consists of incorporating the wave shape into the finite element shape function. This gives the possibility of lifting the ten nodes per wavelength restriction already mentioned. The difficulty with this approach in simple terms is that we do not know *a priori* the wave direction, or directions. There are two methods which have evolved for dealing with this, which will now be summarised. Basically in the first method we iterate to find the wave direction and in the second we retain many possible wave directions. The first method due to Astley and developed by Chadwick, Bettess and Laghrouche [4–7], guesses an initial wave direction, and then iterates to refine the wave direction, until convergence is achieved. The method was first applied in one dimension, where the wave direction is known. This demonstrated that many wavelengths could be modelled using a single finite element, which is one of the aims. In this form it is the wave envelope element proposed by Astley *et al.* [8–11]. However it has been extended to some relatively straightforward wave propagation problems in two dimensions, including plane waves in arbitrary directions, cylindrical waves and Hankel sources. This method leads to a relatively small set of equations to be solved. Also, the Astley type complex conjugate weighting is used, which makes the integration over the element domain with low order Gauss-Legendre rules straightforward. However, the method involves iterations which may not converge and thus depends upon the initial guess for wave direction, and has to be extended when waves propagating in several directions are present. The second method due to Melenk and Babuška [12–14], retains a separate wave potential for each possible wave direction, so that if waves were allowed at intervals of 10 degrees, there would be 36 complex variables at each node. (i.e. $360°/10° = 36$). In this case, a four node element might have a complex 144 by 144 matrix. In this method, no iterations are involved and the solution proceeds in one step. It has been applied successfully to simple rectilinear wave propagation problems, and many wavelengths could be modelled using a single finite element. But the integrations over the element domain are of complicated trigonometric functions and are carried out using high order Gauss-Legendre integration, typically 40 by 40 or higher.

Although both the above methods are promising, neither has as yet solved a really complicated wave problem involving multiple reflections and standing waves. It is possible that unforeseen drawbacks could be encountered in these situations. However, some method along these lines should be feasible. This is because it is known that a very fine mesh of finite elements would solve these problems, and yet such a mesh contains far too much information to describe the wave patterns. There simply has to be a better way of parametrising the short waves and solving for them. In a recent work, Laghrouche and Bettess [15] extended the method of Melenk and Babuška to deal with more realistic applications such as interference of plane waves with amphidromic points, the Hankel source and the diffraction of a plane wave by a circular cylinder. The wave number was increased while the mesh size was kept unchanged. The finite elements used contain many wavelengths in each direction and the element matrices were evaluated by using a high order Gauss-Legendre integration scheme. The directions of the approximating plane waves were evenly spaced and were the same through all the finite elements of the studied domain. This work has been extended. The directions are chosen not to be evenly spaced. They can also varies from one node to another. The number of directions also vary between nodes. The main objective is to show that one can cluster these directions around 'preferable' directions of wave propagation, increase their numbers in regions were many interferences occur, near diffracting objects, and decrease their number at the far field where waves radiate towards infinity. This improves the numerical efficiency, since few variables will be required, and is a step towards an adaptive method.

10.2 THEORY

10.2.1 *Plane wave diffracted by a rigid body*

An incident plane wave of potential ϕ_i encounters a rigid object of boundary Γ_1 (Figure 10.1), is modified and then radiates away to infinity. The total potential ϕ_t is then the summation of the incident potential and the diffracted potential denoted by ϕ_s. The studied region Ω is closed by the boundary Γ_2. The time independent potential ϕ_s satisfies the Helmholtz equation on the domain Ω

$$(\nabla^2 + k^2)\phi_s = 0 \tag{10.1}$$

where ϕ_s is a function of the cartesian coordinates x and y, ∇^2 is the Laplacian operator and k is the wave number. The time factor is removed by assuming a steady state problem.

On the scattering boundary Γ_1, the Neumann boundary condition gives

$$\frac{\partial \phi_s}{\partial n_1} = -\frac{\partial \phi_i}{\partial n_1} \qquad (10.2)$$

and on the exterior boundary Γ_2, the scattered potential is assumed to satisfy the following Robin boundary condition

$$\frac{\partial \phi_s}{\partial n_2} + ik\phi_s = g \qquad (10.3)$$

where g is the boundary condition.

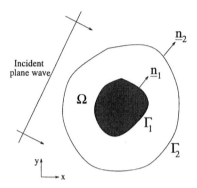

Figure 10.1: Scattering problem.

10.2.2   The residual scheme

The standard Galerkin finite element scheme is used to model the potential ϕ_s and so

$$\int_{\Omega} W(\nabla^2 \phi_s + k^2 \phi_s)d\Omega = 0 \qquad (10.4)$$

gives the governing finite element integral equation for the problem where W is the weighting function. Applying the divergence theorem gives an integral over the domain Ω and a line integral along the contours Γ_1 and Γ_2 involving the first derivatives of the unknown function ϕ_s and the weighting function W. Equation (10.4) becomes

$$\int_{\Omega}(-\boldsymbol{\nabla}W.\boldsymbol{\nabla}\phi_s + k^2 W\phi_s)d\Omega - \int_{\Gamma_1} W\boldsymbol{\nabla}\phi_s.\boldsymbol{n}_1 d\Gamma + \int_{\Gamma_2} W\boldsymbol{\nabla}\phi_s.\boldsymbol{n}_2 d\Gamma = 0 \qquad (10.5)$$

where $\boldsymbol{\nabla}$ is the gradient vector operator. By replacing the normal derivatives of the function ϕ_s by their expressions deduced from equations (10.2) and (10.3), this leads to the system

$$\int_{\Omega}(\boldsymbol{\nabla}W.\boldsymbol{\nabla}\phi_s - k^2 W\phi_s)d\Omega + ik\int_{\Gamma_2} W\phi_s d\Gamma = \int_{\Gamma_1} W\frac{\partial \phi_i}{\partial n_1}d\Gamma + \int_{\Gamma_2} Wg d\Gamma \qquad (10.6)$$

10.2.3   Waves in two space dimensions

The most general solution of the wave equation (10.1) in two space dimensions is given by

$$\phi_s(x,y) = \int_0^{2\pi} D(\theta) e^{i\boldsymbol{k}.\boldsymbol{r}} d\theta \tag{10.7}$$

where the position vector of a considered point is given by the radius vector

$$\boldsymbol{r} = x\boldsymbol{i} + y\boldsymbol{j} \tag{10.8}$$

where $\boldsymbol{i}$ and $\boldsymbol{j}$ are the unit vectors with respect to the x-axis and the y-axis, respectively. The wavenumber vector

$$\boldsymbol{k} = k\cos\theta\boldsymbol{i} + k\sin\theta\boldsymbol{j} \tag{10.9}$$

is given with respect to the direction of propagation θ to the x−axis. Finally, $D(\theta)$ is a distribution function of the direction of propagation. A combination of all such waves (10.7), propagating in all directions θ with variable amplitude and phase for each direction, forms a general solution to the Helmholtz equation in two-dimension space [16]. A different way of looking at this, is that a solution of the Helmholtz equation in two dimensions can be approximated by systems of plane waves propagating in different directions. Such approximating systems have been proven to be complete; the approximate solution converges to the exact solution as the number of the plane waves tends to infinity [17].

10.2.4   Finite element approximation

The domain Ω of Figure 10.1 is meshed into n-noded finite elements. The unknown function ϕ within each element is approximated using polynomial shape functions N_j and the nodal values ϕ_j of the potential as follows

$$\phi = \sum_{j=1}^{n} N_j \phi_j \tag{10.10}$$

By using the fact that a wave potential can be approximated by a combination of plane waves, one can write the potential at each node j of the finite element as follows

$$\phi_j = \sum_{l=1}^{m_j} A_j^l e^{ik(x\cos\theta_l + y\sin\theta_l)} \tag{10.11}$$

The m_j unknowns A_j^l, in a sense, represent the amplitudes of the plane waves

$$e^{ik(x\cos\theta_l + y\sin\theta_l)}, \quad l = 1, 2, ..., m_j,$$

at a node j. The number of the approximating plane waves and the chosen directions can vary from one node to another. These directions can be evenly spaced such that

$$\theta_l = l\frac{2\pi}{m_j} \qquad l = 1, 2, ..., m_j \tag{10.12}$$

or not evenly spaced such that they are clustered around certain directions of preference. In the case of the usual finite element piecewise approximation of expression (10.10), the element matrices would be of dimension n by n and the unknowns are the nodal values of the potential ϕ_j, $j = 1, 2, .., n$. However, the resulting element matrices given by the current formulation are of dimensions $\sum_{j=1}^{n} m_j \times \sum_{j=1}^{n} m_j$ and the unknowns of the problem are the amplitudes A_j^l, $j = 1, 2, .., n$ and $l = 1, 2, ..., m_j$. When calculating the element matrices, the following integrals

$$I_{jl} = \int_{-1}^{1} \int_{-1}^{1} f(\xi, \eta) e^{ik(x \cos \theta_j + y \sin \theta_j)} e^{ik(x \cos \theta_l + y \sin \theta_l)} d\xi d\eta \qquad (10.13)$$

are evaluated over each finite element domain by using high order Gauss-Legendre integration. The number of integration points is dependent on the nodal spacing in terms of the wavelength. The expression $f(\xi, \eta)$ involves the product of the polynomial shape functions, their derivatives, the determinant of the Jacobian and its inverse. More details about these elements can be found in [15]. A numerical integration scheme, which incorporates the knowledge of the plane wave functions, is being developed to evaluate such integrals in a better way. It already shows, for rectangular and triangular finite elements, large savings in number of operations and computation time [19].

10.3   APPLICATIONS

10.3.1   Plane wave diffracted by a rigid cylinder

A plane wave is incident horizontally upon a circular rigid cylinder (Figure 10.2). This classical problem has an analytical solution in terms of Hankel function series, due to Mac-Camy and Fuchs [18] for the case of a finite depth. The diffracted potential is expressed by

$$\phi_s = -\sum_{n=0}^{\infty} \epsilon_n i^n \frac{J_n'(ka)}{H_n'(ka)} H_n(kr) \cos n\alpha \qquad (10.14)$$

where r and α are the polar co-ordinates, a is the radius of the cylinder, $H_n(kr)$ and $J_n(kr)$ are, respectively, Hankel and Bessel functions of the first kind and of order n, ϵ_n is equal to 1 if $n = 0$ and equal to 2 for other values of n, and prime denotes differentiation with repect to the radius r. To solve this problem numerically using the finite elements described above, the mesh of Figure 10.6 is considered. In this the region around the cylinder extending to a radius of $R/a = 7$ is subdivided into 54 'quadratic' finite elements (Figure 10.6). The graphs on Figures 10.7 to 10.11 show the real part and the imaginary part of the diffracted potential around the cylinder for both the analytical solution and the numerical approximation. The number of approximating plane waves taken for each case and the number of integrating points are summarised in the following table

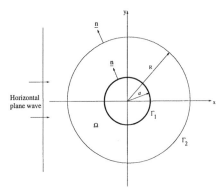

Figure 10.2: Horizontal plane wave diffracted by a rigid cylinder.

ka	$\sim$ number of plane waves	number of integration points
2π	30	24
4π	32	50
6π	34	74
8π	36	98
10π	36	120

In a previous work [15], the authors take a constant number of approximating plane waves and the same angles of expression (10.12) at all nodes of the meshed region. However, in this study, the number of the approximating plane waves shown in the table above (and preceded by $\sim$) is just an indication. There could be more or less plane waves attached to the considered node and the angles of these plane waves are different as well from one node to another. This is an important step towards implementing special finite elements where one can increase the number of approximating plane waves in regions of the studied domain with many wave interferences, and decrease their number at the far field where waves have mainly one direction. Dealing with the integration, it was shown that by increasing the size of a finite element in terms of the wavelength, one has to increase the number of integrating points [15]. In the case of $ka = 2\pi$, the finite elements on the exterior ring of the meshed region contain 2λ in the radial direction and 2.44λ in the angular direction, so 24 Gauss points were used to evaluate the element matrices. In the case of $ka = 10\pi$, the elements contain 10λ and 12.2λ in the radial and angular directions, respectively, and so 120 Gauss points were needed in the integration. In this application, it is obvious that the finite elements used to solve this problem of wave diffraction are very economical despite the fact that at this stage of the development, no special integration scheme is used. For comparison, in the case of $ka = 8\pi$, three different meshes can be considered for solving the diffraction problem of Figure 10.6. The first uses 10 nodal points per wavelength on the radius as well as on the exterior circular boundary. The generated mesh will contain 426160 nodes. In a second mesh, the computational region is supposed to be divided into finite elements with a constant area, with edges of a tenth of the wavelength in size. Such a mesh would be difficult to generate and would give a total of 241274 nodes. The last case is represented by the mesh of Figure 10.6 which contains 252 nodes. By taking 36 directions at each node, we will end up with 9072 degrees of freedom. Figure 10.3 shows the number of degrees of freedom corresponding to every case of mesh. The special finite elements of this study lead to a 98% reduction in comparison with the first case and to a 96% reduction in comparison with the second case and there is no need to update the mesh as a function of the wave number of the problem. It should be noted that the matrix bandwidth increases in the present scheme ,

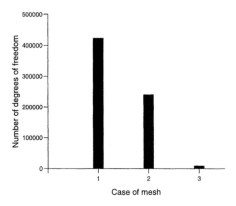

Figure 10.3: Number of degrees of freedom for different meshes.

and the total storage is therfore not so favourable. However there are reasonable grounds for expecting further reductions in the number of active variables. To take one obvious case of the 36 directions at nodes on the external boundary of the mesh, 18 are pointing inwards and cannot represent scattered waves. It should be possible to eliminate those.

10.3.2 Potential function with a local wave number

For a constant wave number k, the wave equation (10.1) is separable in the rectangular coordinates x and y. The potential function ϕ can be written as follows

$$\phi(x,y) = X(x)Y(y) \tag{10.15}$$

By substituting (10.15) into (10.1), we obtain

$$\frac{1}{X}\frac{d^2X}{dx^2} + \frac{1}{Y}\frac{d^2Y}{dy^2} + k^2 = 0 \tag{10.16}$$

Equation (10.16) can be satisfied for all values of x and y if each of the three terms of the equation is a constant and their sum is equal to zero. This gives

$$\frac{1}{X}\frac{d^2X}{dx^2} = -\alpha^2; \quad \frac{1}{Y}\frac{d^2Y}{dy^2} = -\beta^2; \quad \alpha^2 + \beta^2 = k^2 \tag{10.17}$$

Solutions of equations (10.17) are

$$\phi = e^{i\alpha(x\cos\gamma + y\sin\gamma)}e^{(-x\sin\gamma + y\cos\gamma)\sqrt{\alpha^2 - k^2}} \tag{10.18}$$

The potential of equation (10.18) is a solution of the Helmholtz equation [16], but one which does not look simply like a progressive wave, and which has a local wave number α and an angle of incidence γ. A rectangular mesh as shown on Figure 10.4 is considered and the problem consists of posing the given solution (10.18) on the boundaries of the rectangle and of recovering the correct solution in the interior of the mesh by considering systems of approximating plane waves with wave number k. First, the following parameters are

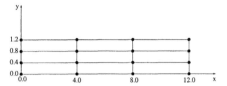

Figure 10.4: Potential function with a local wave number.

taken: $ka = 1.0$, $\alpha a = 1.5$ and $\gamma = 30°$ where a is equal to a unit of length. To solve the problem, 18 plane waves with angles evenly spaced are used in the approximating space. Then, the same problem is solved for $ka = 3.0$, $\alpha a = 5.0$ and $\gamma = 10°$ with 24 plane waves in the approximating space. Figures 10.12 and 10.13 show that the results for the analytical and the numerical solutions are in good agreement. To integrate the element matrices, 12 and 24 Gauss points were used respectively for the two cases. The current finite elements solve this problem of potential with a local wave α number by considering approximating plane waves with the global wave number k and this was not obvious to the authors initially.

10.3.3 Clustering the approximating plane waves

No algorithms have been developed in this area. The idea of the authors, which is speculative, is that some sort of p-type approximation for wave directions can be developed. The problem would first be solved with some standard number of wave directions at each node. From the resulting amplitudes, a new set of wave directions would be selected. For example, a polynomial in θ could be fitted through the calculated wave directions. New directions would be included at the maxima of this distribution. Wave directions with minimal or zero amplitudes would be eliminated. The procedure would continue until convergence had been obtained. Precise creteria, including error indicators for the selection of new directions, the elimination of old ones and the termination of the process, need to be developed. The authors are hopeful that existing concepts of h and p finite element adaptivity can be extended to solve these problems. Figure 10.5 gives a general idea of the way of implementing such an adaptive approach in order to make the method a powerful tool for dealing with short waves.

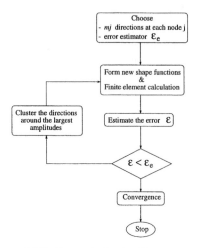

Figure 10.5: Suggested adaptive approach.

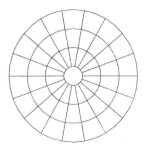

Figure 10.6: cylinder radius : $a = 1m$, radius of the meshed region $R/a = 7$

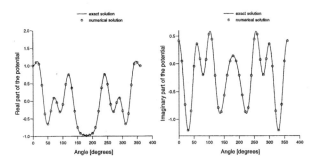

Figure 10.7: Diffracted potential around the cylinder, $ka = 2\pi$, $\lambda/a = 1.0$.

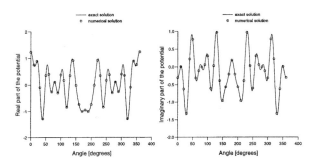

Figure 10.8: Diffracted potential around the cylinder, $ka = 4\pi$, $\lambda/a = 0.5$.

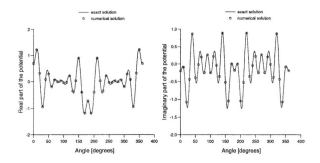

Figure 10.9: Diffracted potential around the cylinder, $ka = 6\pi$, $\lambda/a = 0.33$.

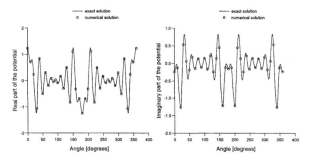

Figure 10.10: Diffracted potential around the cylinder, $ka = 8\pi$, $\lambda/a = 0.25$.

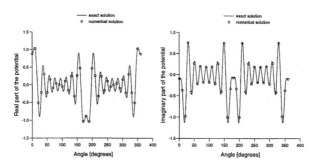

Figure 10.11: Diffracted potential around the cylinder, $ka = 10\pi$, $\lambda/a = 0.2$.

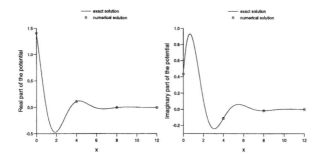

Figure 10.12: Potential function with a local wave number, $ka = 1.0, \alpha a = 1.5, \gamma = 30°$.

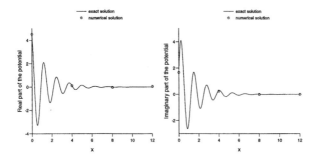

Figure 10.13: Potential function with a local wave number, $ka = 3.0, \alpha a = 5.0, \gamma = 10°$.

10.4 CONCLUSIONS

An extension of the Melenk and Babuška method has been presented, which demonstrates that it has the potential to be a very effective and powerful solution method for short wave problems. The examples shown demonstrate dramatic reductions in the total number of degrees of freedom for a given solution accuracy. The range of problems solved gives confidence in the approach. If the outstanding problems of integration of the element matrices, error indicators and adaptivity can be solved, the method may well become the first choice in short wave problems. Other concerns relating to the conditioning of the system matrix also need to be investigated. There seem to be genuine grounds for optimism that a method for short wave problems, which is almost independent of wavelength (and frequency) is possible. If this were obtained it would have a great impact in scientific fields from electromagnetism to geophysics. The techniques described may also have application in boundary element methods.

REFERENCES

[1] K. Morgan, P.J. Brookes, O. Hassan and N.P. Weatherill (1998). Parallel processing for the simulation of problems involving scattering of electromagnetic waves. *Comput. Meth. Appl. Mech. Engrg.*, 152:157-174.

[2] K. Morgan, O. Hassan and J. Peraire (1996). A time domain unstructured grid approach to the simulation of electromagnetic scattering in piecewise homogeneous media. *Comput. Meth. Appl. Mech. Engrg.*, 134:17-36.

[3] P.J. Brookes, K. Morgan, O. Hassan and N.P. Weatherill (1998). Extending the frequency range of electromagnetic scattering simulations. *6th ACME conference*, University of Exeter, Extended Abstracts, E. Maunder (Ed.), 11-14.

[4] P. Bettess (1987). A simple wave envelope element example. *Comm. Appl. numer. meth.*, 3:77-80.

[5] P. Bettess and E.A. Chadwick (1995). Wave envelope examples for progressive waves. *Int. J. Numer. Meth. Engng.*, 38:2487-2508.

[6] E.A. Chadwick and P. Bettess(1997). Modelling of progressive short waves using wave envelopes. *Int. J. Numer. Meth. Engng.*, 40:3229-3245.

[7] E.A. Chadwick, P. Bettess and O. Laghrouche (1999). Diffraction of short waves modelled using new mapped wave envelope finite and infinite elements. *Int. J. Numer. Meth. Engng.*, 45:335-354.

[8] R.J. Astley and W. Eversman (1981). A note on the utility of a wave envelope approach in finite element duct transmission studies. *Jour. Sound and Vib.*, 76:595-601.

[9] R.J.Astley (1983). Wave envelope and infinite elements for acoustical radiation. *Int. J. Numer. Meth. Fluids*, 3:507-526.

[10] R.J. Astley and W. Eversman (1983). Finite element formulation for acoustical radiation. *Jour. Sound and Vib.*, 88:47-64.

[11] R.J. Astley and W. Eversman (1984). Wave envelope and finite element schemes for fan noise radiation from turbofan inlets. *A.I.A.A. Journal*, 22:1719-1726.

[12] J.M. Melenk (1995). On generalized Finite Element Methods. *Ph.D. thesis*, University of Maryland, USA.

[13] J.M. Melenk and I. Babuška (1996). The Partition of Unity Finite Element Method. Basic Theory and Applications. *Comput. Meths. Appl. Mech. Engrg.*, 139:289-314.

[14] J.M. Melenk and I. Babuška (1997). The Partition of Unity Method. *Int. J. Numer. Meth. Engng.*, 40:727-758.

[15] O. Laghrouche and P. Bettess (1999). Short wave modelling using special finite elements. *Special Issue of Journal of Computational Acoustics*, (submitted)

[16] P.M. Morse and H. Feshbach (1953). Methods of Theoretical Physics. McGraw-Hill Book Company, Inc., USA.

[17] F.J. Sánchez-Sesma, I. Herrera and J. Avilés (1982). A boundary Method for elastic wave diffraction: application to scattering of SH waves by surface irregularities. *Bull. Seism. Soc. Am.*, 72:491-506.

[18] MacCamay and R.A. Fuchs (1954). Wave forces on piles. A diffraction theory. *US Army Coastal Engineering Research Center, Tech. Mem.* 69.

[19] P. Bettess, O. Laghrouche and B. Peseux (1999). A numerical integration scheme for special finite elements for Helmholtz equation. *(in progress)*

11 FINITE ELEMENT METHODS FOR FLUID-STRUCTURE VIBRATION PROBLEMS

Alfredo Bermúdez[a], Pablo Gamallo[a], Duarte Santamarina[a] and Rodolfo Rodríguez[b]

[a]Departamento de Matemática Aplicada,
Universidade de Santiago de Compostela, Spain.

[b]Departamento de Ingeniería Matemática,
Universidad de Concepción, Chile.

ABSTRACT

Several finite element methods for the numerical computation of fluid-structure vibration problems are considered. They are applied to two formulations based on different variables to describe the fluid: pressure and/or displacement potential in one case, and displacements in the other. While in the first formulation the problem is discretized by standard Lagrangian finite elements for both variables, in the second "face" Raviart-Thomas elements are used. In each case we consider both tetrahedral and hexahedral meshes. The numerical results allow us to compare the different methods in terms of error versus the number of degrees of freedom and computing time.

Key words. Fluid-structure, Elastoacoustics, Hydroelasticity, Finite Elements.

11.1 INTRODUCTION

This paper is devoted to a survey of a number of articles by the authors concerning finite element approximation of some fluid-structure vibration problems. More precisely we are interested in two particular fields: elastoacoustics and hydroelasticity. The former consists of the interaction of an acoustic fluid and an elastic solid, while the latter concerns the influence of an incompressible fluid on the vibrations of a structure. In the first case the fluid is subject to pressure waves and in the second to sloshing waves due to gravity acting on an open boundary. In both cases we are interested in harmonic vibrations and hence in modal analysis. Moreover we consider only bounded domains.

An overview of these kinds of problems and their numerical solution can be found in the book by Morand and Ohayon[15] where some of the formulations included in the present paper are also considered.

In any mathematical model of the above mentioned problems, displacements are typically used to describe the solid. However several possibilities exists for modelling the fluid.

In the case of elastoacoustics, the first choice is to use pressure, as it is usually done in pure acoustics. The inconvenience is that, after discretization, we obtain non-symmetric generalized eigenvalue problems ([16]), for which the computational solution involves considerable complications. To avoid this drawback the fluid has been described using alternative variables such as velocity potential ([12]), displacement potential or combinations of some of these; for example the pressure/potential formulation introduced in [14].

Since the solid is generally described in terms of displacements, choosing the same variable to describe the fluid presents important advantages. Firstly, both kinematic and kinetic conditions at the fluid-structure interface are very easily handled. Secondly the method can be extended to nonlinear situations (see [1]). Finally it leads to sparse symmetric matrices. However, in principle, some disadvantages also arise: vector fields have to be computed intead of scalar variables and, additionally, Lagrangian finite elements produce spurious eigenvalues (see [13]). The latter can be avoided (see [2, 6, 7]) by using "face elements", which are somehow similar to the "edge elements" currently used in electromagnetism.

The situation is quite similar in hydroelasticity where the use either of pressure or of displacement potential leads to non-symmetric generalized eigenvalues problems. However, either of these two fields can be eliminated and the corresponding terms replaced by added mass operators, leading to symmetric formulations (see for instance [15]).

An alternative approach to these added mass formulations can be obtained by using displacements to describe the fluid, as in elastoacoustics. However, in this case, pressure cannot be eliminated except if the incompressibility constraint is introduced as an essential condition in the functional space of the weak formulation. Then a conforming approximation demands the use of special divergence-free finite elements, even though in the two-dimensional case this can also be done by introducing a scalar stream function, which can be discretized by standard finite elements ([3, 4]). In the three-dimensional case a mixed displacement/pressure formulation can be used for the fluid, and this can be discretized by mixed face Raviart-Thomas finite elements.

11.2   MATHEMATICAL MODELLING

In this section we recall the equations for the harmonic vibrations of the coupling between a linear elastic structure and an inviscid fluid. Two cases are considered depending on whether the fluid is compressible or incompressible.

11.2.1   Elastoacoustics

The fluid is assumed to be compressible and barotropic, i.e. pressure depends only on density. We consider the problem of determining the small amplitude motions of this fluid contained into a linear isotropic elastic structure which obeys Hooke's law. Let Ω_F

and Ω_S be the domains occupied by fluid and solid, respectively. Let Γ_I be the interface between the two media and $\boldsymbol{\nu}$ its unit normal vector pointing outwards from Ω_F. The exterior boundary of the solid is the union of two parts, Γ_D and Γ_N: the structure is fixed on Γ_D and free on Γ_N. Finally let $\boldsymbol{n}$ be the unit outward normal vector on Γ_N (see Figure 11.1).

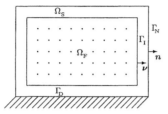

Figure 11.1: Elastoacoustic problem: vertical section of fluid and solid domains.

The governing equations in the frequency domain for free small amplitude motions of the coupled system are the following:

$$-\omega^2 \rho_F \boldsymbol{U} + \mathbf{grad}\, P = 0 \quad \text{in } \Omega_F, \tag{11.1}$$
$$\omega^2 \rho_S \boldsymbol{W} + \mathbf{div}\,(\boldsymbol{\sigma}(\boldsymbol{W})) = 0 \quad \text{in } \Omega_S, \tag{11.2}$$
$$P + \rho_F c^2 \,\mathrm{div}\,\boldsymbol{U} = 0 \quad \text{in } \Omega_F, \tag{11.3}$$
$$(\boldsymbol{U} - \boldsymbol{W}) \cdot \boldsymbol{\nu} = 0 \quad \text{on } \Gamma_I, \tag{11.4}$$
$$P\boldsymbol{\nu} + \boldsymbol{\sigma}(\boldsymbol{W})\boldsymbol{\nu} = 0 \quad \text{on } \Gamma_I, \tag{11.5}$$
$$\boldsymbol{W} = 0 \quad \text{on } \Gamma_D, \tag{11.6}$$
$$\boldsymbol{\sigma}(\boldsymbol{W})\boldsymbol{n} = 0 \quad \text{on } \Gamma_N, \tag{11.7}$$

where P is the amplitude of the fluid pressure, $\boldsymbol{U}$ and $\boldsymbol{W}$ are those of fluid and solid displacements, respectively, ω is the angular frequency, ρ_F and ρ_S the fluid and solid densities, respectively, c is the sound speed in the fluid, and $\boldsymbol{\sigma}$ the stress tensor in the solid, which relates to the linearized strain tensor $\varepsilon(\boldsymbol{W}) := \frac{1}{2}(\nabla \boldsymbol{W} + \nabla \boldsymbol{W}^t)$ by Hooke's law: $\boldsymbol{\sigma} = \lambda(\mathrm{tr}\,\varepsilon(\boldsymbol{W}))\mathbf{I} + \mu\varepsilon(\boldsymbol{W})$ (λ and μ being the Lamé coefficients).

11.2.2 Hydroelasticity

In this case the equations are exactly the same as for elastoacoustics, except for (11.3) which has to be replaced by the incompressibility condition:

$$\mathrm{div}\,\boldsymbol{U} = 0 \quad \text{in } \Omega_F. \tag{11.8}$$

Moreover, in many practical situations there is an open boundary Γ_0 (see Figure 11.2) subject to sloshing waves because of the effects of gravity. The corresponding boundary condition is:

$$P = \rho_F g \boldsymbol{U} \cdot \boldsymbol{\nu} \quad \text{on } \Gamma_0. \tag{11.9}$$

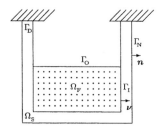

Figure 11.2: Hydroelasticity problem: vertical section of fluid and solid domains.

11.3   WEAK FORMULATIONS

In this section we recall different weak formulations for the above models. These are obtained by eliminating some of the unknowns describing the fluid.

11.3.1   Elastoacoustics

Pressure formulation

This was first considered in reference [16] (see also [17]) and can be obtained by eliminating the fluid displacement field from the above equations. Then the following weak formulation is obtained by standard procedures, i.e., multiplying by virtual displacement and pressure fields and then integrating by parts using a Green's formula:

Find $\omega \in \mathbb{R}$, $\boldsymbol{W} \in \mathrm{H}^1_{\Gamma_{\mathrm{D}}}(\Omega_{\mathrm{S}})^3$ and $P \in \mathrm{H}^1(\Omega_{\mathrm{F}})$, with $(\boldsymbol{W}, P) \neq 0$, such that

$$\int_{\Omega_{\mathrm{S}}} \boldsymbol{\sigma}(\boldsymbol{W}) : \boldsymbol{\varepsilon}(\boldsymbol{Z})\,dx + \frac{1}{\rho_{\mathrm{F}}} \int_{\Omega_{\mathrm{F}}} \mathbf{grad}\,P \cdot \mathbf{grad}\,Q\,dx - \int_{\Gamma_{\mathrm{I}}} P\,\boldsymbol{Z} \cdot \boldsymbol{\nu}\,d\Gamma$$

$$= \omega^2 \left(\int_{\Omega_{\mathrm{S}}} \rho_{\mathrm{S}} \boldsymbol{W} \cdot \boldsymbol{Z}\,dx + \frac{1}{\rho_{\mathrm{F}} c^2} \int_{\Omega_{\mathrm{F}}} PQ\,dx + \int_{\Gamma_{\mathrm{I}}} \boldsymbol{W} \cdot \boldsymbol{\nu}\,Q\,d\Gamma \right)$$

$$\forall \boldsymbol{Z} \in \mathrm{H}^1_{\Gamma_{\mathrm{D}}}(\Omega_{\mathrm{S}})^3, \quad \forall Q \in \mathrm{H}^1(\Omega_{\mathrm{F}}).$$

In the expression above, $\mathrm{H}^1_{\Gamma_{\mathrm{D}}}(\Omega_{\mathrm{S}})$ is the subspace of functions in $\mathrm{H}^1(\Omega_{\mathrm{S}})$ vanishing on Γ_{D}.

Pressure/potential formulation

The inconvenience of the pressure formulation lies in the fact that it is not symmetric. An alternative symmetric formulation can be obtained by introducing the displacement potential in the fluid as an additional unknown; that is a scalar field $\phi \in \mathrm{H}^1(\Omega_{\mathrm{F}})$ such that

$$\boldsymbol{U} = \mathbf{grad}\,\phi \qquad \text{in } \Omega_{\mathrm{F}},$$

as proposed in [14]. The corresponding weak formulation then reads as follows:

Find $\omega \in \mathbb{R}$, $\boldsymbol{W} \in H^1_{\Gamma_D}(\Omega_S)^3$, $P \in L^2(\Omega_F)$ and $\phi \in H^1(\Omega_F)$, with $(\boldsymbol{W}, P, \phi) \neq 0$, such that

$$\int_{\Omega_S} \boldsymbol{\sigma}(\boldsymbol{W}) : \boldsymbol{\varepsilon}(\boldsymbol{Z}) \, dx + \frac{1}{\rho_F c^2} \int_{\Omega_F} PQ \, dx = \omega^2 \left(\int_{\Omega_S} \rho_S \boldsymbol{W} \cdot \boldsymbol{Z} \, dx + \frac{1}{c^2} \int_{\Omega_F} Q\phi \, dx \right.$$

$$+ \frac{1}{c^2} \int_{\Omega_F} P\psi \, dx - \int_{\Omega_F} \rho_F \, \mathbf{grad} \, \phi \cdot \mathbf{grad} \, \psi \, dx + \int_{\Gamma_I} \rho_F \boldsymbol{W} \cdot \boldsymbol{\nu} \psi \, d\Gamma + \left. \int_{\Gamma_I} \rho_F \boldsymbol{Z} \cdot \boldsymbol{\nu} \phi \, d\Gamma \right)$$

$$\forall \boldsymbol{Z} \in H^1_{\Gamma_D}(\Omega_S)^3, \quad \forall Q \in L^2(\Omega_F), \quad \forall \psi \in H^1(\Omega_F).$$

The mathematical analysis of a numerical approximation of this problem is given in [9], where piecewise linear finite elements are used for the displacement potential and piecewise constant ones for pressure.

Displacement formulation

This is obtained by using equation (11.3) to eliminate the pressure in equations (11.1) and (11.5). Consider the space of displacements in each medium satisfying the kinematic constraint (11.4):

$$\mathcal{V} := \left\{ (\boldsymbol{U}, \boldsymbol{W}) \in H(\text{div}, \Omega_F) \times H^1_{\Gamma_D}(\Omega_S)^3 : \boldsymbol{U} \cdot \boldsymbol{\nu} = \boldsymbol{W} \cdot \boldsymbol{\nu} \text{ on } \Gamma_I \right\}.$$

The weak formulation of the displacement formulation looks very simple and compact. Moreover it is symmetric:

Find $\omega \in \mathbb{R}$ and $(\boldsymbol{U}, \boldsymbol{W}) \in \mathcal{V}$, with $(\boldsymbol{U}, \boldsymbol{W}) \neq 0$, such that

$$\int_{\Omega_S} \boldsymbol{\sigma}(\boldsymbol{W}) : \boldsymbol{\varepsilon}(\boldsymbol{Z}) \, dx + \int_{\Omega_F} \rho_F c^2 \, \text{div} \, \boldsymbol{U} \, \text{div} \, \boldsymbol{Y} \, dx = \omega^2 \left(\int_{\Omega_S} \rho_S \boldsymbol{W} \cdot \boldsymbol{Z} \, dx + \int_{\Omega_F} \rho_F \boldsymbol{U} \cdot \boldsymbol{Y} \, dx \right)$$

$$\forall (\boldsymbol{Y}, \boldsymbol{Z}) \in \mathcal{V}.$$

This formulation has been discretized by standard nodal finite elements in [13]. However, in such a case, spurious eigenvalues arise in addition to the physical ones. An alternative approach has been proposed in [7] and analyzed in [2] and [4]. This consists of using Raviart-Thomas finite elements, which are H(div)-conforming, to discretize the fluid displacements. The kinematic interface condition can be taken into account either in a weak sense by using a Lagrange multiplier or by eliminating the degrees of freedom of the fluid on the interface by static condensation. Non-compatible meshes on the fluid-solid interface can be used, which is very convenient for dealing with singularities arising from reentrant corners or dihedral angles in the solid case (see [6] and [5]).

11.3.2 Hydroelasticity

Added mass formulation

First of all, a new variable is introduced for the vertical displacement of the free surface:

$$\eta := \boldsymbol{U} \cdot \boldsymbol{\nu} \qquad \text{on } \Gamma_O. \tag{11.10}$$

Then, by applying the divergence operator to equation (11.1) and using the incompressibility condition (11.8), we can eliminate the fluid displacements $\boldsymbol{U}$ in equations (11.1)-(11.9). Thus we obtain:

$$-\Delta P = 0 \quad \text{in } \Omega_{\text{F}}, \tag{11.11}$$

$$\frac{\partial P}{\partial \nu} = \omega^2 \rho_{\text{F}} \eta \quad \text{on } \Gamma_{\text{O}}, \tag{11.12}$$

$$\frac{\partial P}{\partial \nu} = \omega^2 \rho_{\text{F}} \boldsymbol{W} \cdot \boldsymbol{\nu} \quad \text{on } \Gamma_{\text{I}}, \tag{11.13}$$

$$P = \rho_{\text{F}} g \eta \quad \text{on } \Gamma_{\text{O}}. \tag{11.14}$$

Now let us recall the *Neumann-to-Dirichlet operator* which is defined for any $\chi \in \text{L}^2(\partial\Omega_{\text{F}})$ such that $\int_{\partial\Omega_{\text{F}}} \chi \, d\Gamma = 0$ by $\mathcal{M}(\chi) := \phi|_{\partial\Omega_{\text{F}}}$, where ϕ is the unique solution in $\text{H}^1(\Omega_{\text{F}})$ of the Dirichlet problem:

$$\Delta \phi = 0 \quad \text{in } \Omega_{\text{F}},$$

$$\frac{\partial \phi}{\partial \nu} = \chi \quad \text{on } \partial\Omega_{\text{F}},$$

satisfying $\int_{\partial\Omega_{\text{F}}} \phi = 0$.

Let $\chi := (\boldsymbol{W} \cdot \boldsymbol{\nu}, \eta)$ denote the function in $\text{L}^2(\partial\Omega_{\text{F}})$ patchwise defined by $\chi = \boldsymbol{W} \cdot \boldsymbol{\nu}$ on Γ_{I} and $\chi = \eta$ on Γ_{O}. From the incompressibility condition (11.8), (11.4) and (11.10), such a function satisfies $\int_{\partial\Omega_{\text{F}}} \chi \, d\Gamma = 0$. The corresponding $\phi = \mathcal{M}(\boldsymbol{W} \cdot \boldsymbol{\nu}, \eta)$ is a potential for the fluid and the pressure is related to this by $P = \omega^2 \rho_{\text{F}}(\phi + C)$, for some constant C which can be obtained from condition (11.14). This constant can be eliminated from the weak formulation by using the space

$$\mathcal{W} := \left\{ (\boldsymbol{W}, \eta) \in \text{H}^1_{\Gamma_{\text{D}}}(\Omega_{\text{S}})^3 \times \text{L}^2(\Gamma_{\text{O}}) : \int_{\Gamma_{\text{I}}} \boldsymbol{W} \cdot \boldsymbol{\nu} \, d\Gamma + \int_{\Gamma_{\text{O}}} \eta \, d\Gamma = 0 \right\}.$$

Thus we obtain the following weak formulation:

Find $\omega \in \mathbb{R}$ and $(\boldsymbol{W}, \eta) \in \mathcal{W}$, with $(\boldsymbol{W}, \eta) \neq 0$, such that

$$\int_{\Omega_{\text{S}}} \boldsymbol{\sigma}(\boldsymbol{W}) : \boldsymbol{\varepsilon}(\psi) + \int_{\Gamma_{\text{O}}} \rho_{\text{F}} g \eta \xi \, d\Gamma$$

$$= \omega^2 \left(\int_{\Omega_{\text{S}}} \rho_{\text{S}} \boldsymbol{W} \cdot \psi + \int_{\Gamma_{\text{I}}} \rho_{\text{F}} \mathcal{M}(\boldsymbol{W} \cdot \boldsymbol{\nu}, \eta) \, \boldsymbol{Z} \cdot \boldsymbol{\nu} \, d\Gamma + \int_{\Gamma_{\text{O}}} \rho_{\text{F}} \mathcal{M}(\boldsymbol{W} \cdot \boldsymbol{\nu}, \eta) \xi \, d\Gamma \right)$$

$$\forall (\boldsymbol{Z}, \xi) \in \mathcal{W}.$$

A finite element discretization of this formulation has been proposed and analyzed in [10] and optimal error estimates have been obtained for the approximate eigenmodes.

Displacement/pressure formulation

In this case the space of kinematically admissible displacements/pressure fields is given by

$$\mathcal{X} = \{(\boldsymbol{U}, \boldsymbol{W}, P) \in \mathrm{H}^1_{\Gamma_D}(\Omega_S)^3 \times \mathrm{H}(\mathrm{div}, \Omega_F) \times \mathrm{L}^2(\Omega_F) : \boldsymbol{U} \cdot \nu = \boldsymbol{W} \cdot \nu \text{ on } \Gamma_I\},$$

and the weak formulation becomes:

Find $\omega \in \mathbb{R}$ and $(\boldsymbol{U}, \boldsymbol{W}, P) \in \mathcal{X}$, $(\boldsymbol{U}, \boldsymbol{W}, P) \neq 0$, such that

$$\int_{\Omega_S} \boldsymbol{\sigma}(\boldsymbol{W}) : \varepsilon(\boldsymbol{Z}) \, dx - \int_{\Omega_F} Q \, \mathrm{div} \, \boldsymbol{U} \, dx - \int_{\Omega_F} P \, \mathrm{div} \, \boldsymbol{Y} \, dx + \int_{\Gamma_O} \rho_F g \, \boldsymbol{U} \cdot \nu \, \boldsymbol{Y} \cdot \nu \, d\Gamma$$

$$= \omega^2 \left(\int_{\Omega_S} \rho_S \boldsymbol{W} \cdot \boldsymbol{Z} \, dx + \int_{\Omega_F} \rho_F \boldsymbol{U} \cdot \boldsymbol{Y} \, dx \right)$$

$$\forall (\boldsymbol{Y}, \boldsymbol{Z}, Q) \in \mathcal{X}.$$

In order to eliminate the pressure we should introduce the incompressibility condition in the space $\mathcal{X}$, which would necessitate use of divergence-free finite elements. However, in the 2D case we can take advantage of the existence of a scalar stream function. Indeed, since

$$\mathrm{div} \, \boldsymbol{U} = 0 \quad \Rightarrow \quad \exists \psi \in \mathrm{H}^1(\Omega_F) : \boldsymbol{U} = \mathbf{curl} \, \psi := \left(\frac{\partial \psi}{\partial x_2}, -\frac{\partial \psi}{\partial x_1} \right),$$

the pressure can be eliminated and $\boldsymbol{U}$ replaced by $\mathbf{curl} \, \psi$ in the variational equation above. The interface kinematic condition on Γ_I becomes $\mathbf{curl} \, \psi \cdot \nu = \boldsymbol{W} \cdot \nu$, which after discretization has a nonlocal character. This can be handled by using a Lagrange multiplier on the interface as proposed in [4], where optimal error estimates are obtained (see also [3, 8]).

11.4 NUMERICAL RESULTS

In this section we report numerical results corresponding to an elastoacoustic problem. For numerical experiments with the different formulations of the hydroelasticity problem see [3], [8] and [10]. We consider a test problem corresponding to water in a perfectly rigid cavity covered by a 3D clamped moderately thick steel plate (see Figure 11.3). The dimensions for the cavity are $4\,\mathrm{m} \times 6\,\mathrm{m} \times 1\,\mathrm{m}$ and for the plate $4\,\mathrm{m} \times 6\,\mathrm{m} \times 0.5\,\mathrm{m}$. The physical data are $\rho_F = 1000\,\mathrm{kg/m^3}$ and $c = 1430\,\mathrm{m/s}$ for water, and $\rho_S = 7700\,\mathrm{kg/m^3}$, Young's modulus $E = 1.44 \times 10^{11}\,\mathrm{Pa}$ and Poisson ratio $\nu = 0.35$ for steel. We have solved the problem in a quarter of the geometry to reduce the number of degrees of freedom, by imposing the symmetries of the different modes as constraints. As reference solution we have used that obtained by using a plate model (Reissner-Mindlin) for a fine mesh coupled with Raviart-Thomas elements for the fluid (see [11]).

This elastoacoustic problem has been solved with the following finite element discretizations of the fluid variables:

- pressure/potential formulation discretized with Lagrangian tetrahedral finite elements (degree 0 for pressure and degree 1 for potential);

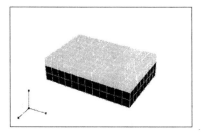

Figure 11.3: Geometry and hexahedral mesh ($h = 0.5$).

- pressure/potential formulation discretized with Lagrangian hexahedral finite elements (degree 0 for pressure and trilinear for potential);

- displacement formulation discretized with lowest degree Raviart-Thomas tetrahedral finite elements;

- displacement formulation discretized with lowest degree Raviart-Thomas hexahedral finite elements.

We have coupled these elements with linear tetrahedral (first and third cases) or trilinear hexahedral (second and fourth cases) for the solid.

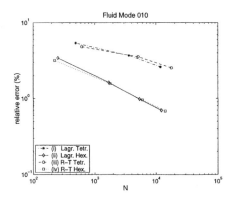

Figure 11.4: Error versus number of d.o.f (log-log scale).

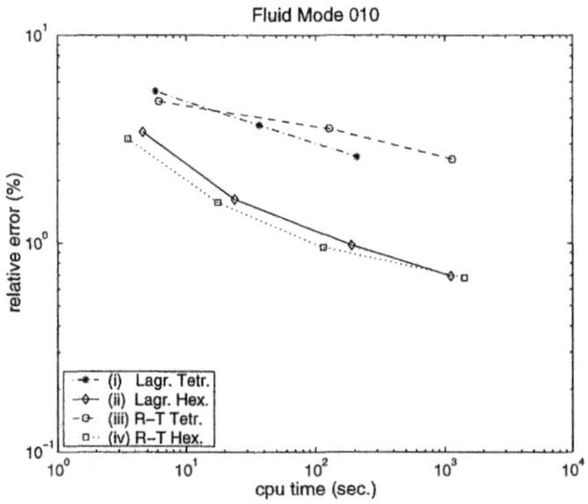

Figure 11.5: Error versus computer time (log-log scale).

Figures 11.4 to 11.7 show the error versus either the number N of degrees of freedom (d.o.f) or the computer time, required to calculate the ten eigenvalues nearest to a "shift" parameter by using Version 5.3 of the MATLAB eigensolver **eigs**. This eigensolver computes, by means of Arnoldi iterations, the eigenvalues and eigenvectors of a generalized eigenvalue problem of the form $\mathcal{A}x = \lambda \mathcal{B}x$ with $\mathcal{B}$ symmetric and positive definite. For the displacement formulation we obtain a discretized problem of this form by eliminating the degrees of freedom of the fluid on the interface in terms of those of the solid, by means of the kinematic constraint. So `eigs` can be directly applied. This is not the case, however, for the pressure/potential formulation and for this some previous transformations in the discrete problem have to be made (see [5]).

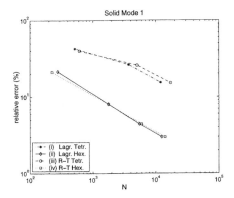

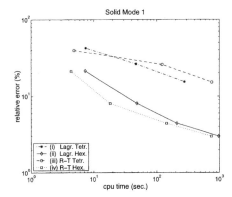

Figure 11.6: Error versus number of d.o.f (log-log scale).

Figure 11.7: Error versus computer time (log-log scale).

We observe from these figures that both formulations involve similar computational costs. However, in both cases, hexahedral elements are much less expensive than tetrahedral elements.

ACKNOWLEDGMENTS

The first three authors have been supported by research project PB97-0508, MEC (Spain) and the third one by grant FONDECYT 1.990.346 (Chile).

REFERENCES

[1] K. J. Bathe, C. Nitikitpaiboon, and X. Wang (1995). A mixed displacement-based finite element formulation for acoustic fluid-structure interaction. *Comp. Struct.*, 56:225-237.

[2] A. Bermúdez, R. Durán, M. A. Muschietti, R. Rodríguez, and J. Solomin (1995). Finite element vibration analysis of fluid-solid systems without spurious modes. *SIAM J. Numer. Anal.*, 32:1280-1295.

[3] A. Bermúdez, R. Durán, and R. Rodríguez (1997). Finite element solution of incompressible fluid-structure vibration problems. *Int. J. Numer. Meth. Engng.*, 40:1435-1448.

[4] A. Bermúdez, R. Durán, and R. Rodríguez (1998). Finite element analysis of compressible and incompressible fluid-solid systems. *Math. Comp.*, 32:1280-1295.

[5] A. Bermúdez, P. Gamallo, and R. Rodríguez (1999). An hexahedral face element for elastoacoustic vibration problems. *Technical Report 99-12,* Universidad de Concepción, Chile.

[6] A. Bermúdez, L. Hervella, and R. Rodríguez (1999). Finite element computation of three dimensional elastoacustic vibrations. *J. Sound Vibr.,* 219:277–304.

[7] A. Bermúdez and R. Rodríguez (1994). Finite element computation of the vibration modes of a fluid-solid system. *Comput. Meth. Appl. Mech. Engrg.,* 119:355-370.

[8] A. Bermúdez and R. Rodríguez (1999). Finite element analysis of sloshing and hydroelastic vibrations under gravity. *M²AN,* 33:305-327.

[9] A. Bermúdez and R. Rodríguez (1999). Analysis of a finite element method for pressure/potential formulation of elastoacoustic spectral problems. *Technical Report 98-25,* Universidad de Concepción, Chile.

[10] A. Bermúdez, R. Rodríguez, and D. Santamarina (1998). A finite element solution of an added mass formulation for coupled fluid-solid vibrations. *Technical Report 98-15,* Universidad de Concepción, Chile.

[11] R. Durán, L. Hervella-Nieto, E. Liberman, R. Rodríguez, and J. Solomin (1998). Finite element analysis of the vibration problem of a plate coupled with a fluid. *Technical Report 98-05,* Universidad de Concepción, Chile. (To appear in *Numer. Math.*).

[12] G. C. Everstine (1981). A symmetric potential formulation for fluid-structure interaction. *J. Sound Vibr.,* 79:157-160.

[13] L. Kiefling and G. C. Feng (1976). Fluid-structure finite element vibration analysis. *AIAA Journal,* 14:199-203.

[14] H. Morand and R. Ohayon (1979). Substructure variational analysis of the vibrations of coupled fluid-structure systems. Finite element results. *Int. J. Numer. Methods Engng.,* 14:741–755.

[15] H. Morand and R. Ohayon (1995). *Fluid Structure Interactions,* J. Wiley & Sons. Chichester.

[16] O. C. Zienkiewicz and R. E. Newton (1969). Coupled vibration of a structure submerged in a compressible fluid. *Proc. Int. Symp. on Finite Element Techniques,* Stuttgart, 1-15.

[17] O. C. Zienkiewicz and R. L. Taylor (1991). *The Finite Element Method,* Vol. 2. Mc Graw-Hill. London.

12 COUPLING DIFFERENT NUMERICAL ALGORITHMS FOR TWO PHASE FLUID FLOW

Malgorzata Peszyńska, Qin Lu and Mary F. Wheeler

Texas Institute for Computational and Applied Mathematics,
University of Texas, Austin, TX 78712 U.S.A.

ABSTRACT

We discuss the coupling of different numerical algorithms for the solution of two phase immiscible flow problems. The computational domain for the flow is split into several subdomains (blocks) on which different numerical algorithms or *models* are implemented. We discuss in particular two algorithms: a fully implicit and a sequential formulation. Both are based on a mixed finite element discretization in space and differ by the discretization in time. The codes for the two formulations are run on individual subdomains (blocks) and are coupled across the interface by a set of conditions imposing continuity of primary variables and of the component mass fluxes. The interface code is part of the multiblock multimodel framework which uses mortar spaces to handle nonmatching grids, and is capable of controlling different time stepping in different subdomains. We discuss numerical, mathematical and implementation issues involved in the coupling.

Key words. domain decomposition, multiphase flow, multiblock, mortar spaces, implicit solution, sequential solution, multimodel.

12.1 INTRODUCTION

It is generally believed that the coupling of different models or codes may be the only way to achieve progress in modelling and simulation of problems with complex geometry and physics. Currently many highly specialized algorithms and codes exist which can perform the local tasks in an already optimal or nearly optimal manner. The coupling of these specialized codes with applications to multiphase flow and subsurface modelling is the focus of this paper.

In industrial practice many codes have been coupled together in a loose fashion, for example by using interface values delivered by one code as boundary values for the next

Table 12.1: Computational cost (in seconds) of running a two phase quarter five–spot problem with n gridblocks using different codes.

problem size	n=500	n=4000	n=16000
Sequential	.040	.580	4.650
Implicit	.250	4.370	58.070
Black oil in two phase	2.510	41.190	338.000

time step of another code. Our multiblock multimodel framework allows for tight coupling. The interface values are the unknowns at every time step. Their values are sought iteratively with a domain decomposition procedure which stops when the conservation of the quantities in question has been satisfied to a given tolerance level. In our applications to multiphase flow in a subsurface, the quantities matched at an interface are phase pressures or other primary variables, and the conservation of mass across the interface is achieved by iterating the difference in the component fluxes to zero (or desired tolerance). More precisely, the matching condition is imposed in a weak sense.

In the traditional setting, if any part of the computational domain is occupied by n phases, then the n phase simulation code has to be run in the whole reservoir. However at a given time or at all times large parts of that reservoir may be occupied by fewer than n phases or components. For example, a black oil or compositional code (in petroleum industry this corresponds to a three or n phase code, respectively) may have to be run at higher elevations and around wells of an oil reservoir, while in the rest of the computational domain the flow can be simulated efficiently by a two phase code or even by a single phase code. In addition, these codes come in many flavors corresponding to different time and space discretization schemes, well models, solvers, etc. These different numerical algorithms may have optimal applications depending on the magnitude of flow rates, refinement needs, etc. However, without the multiblock multimodel approach, the use of a single complicated code is mandatory and it may be very costly (see Table 12.1 for timings). The multiblock multimodel allows for splitting of the computational domain into blocks in which different codes can be run efficiently.

In this paper we discuss the coupling of different numerical algorithms for the two phase code. Specifically, we consider two immiscible slightly compressible fluids, for example oil and water, and their pressures P_o, P_w, saturations S_o, S_w, concentrations N_o, N_w, and densities ρ_o, ρ_w, respectively, with $N_o = S_o\rho_o$ and $N_w = S_w\rho_w$, flowing in a three–dimensional porous medium (reservoir) with gravity G and depth $D(x)$. Note that phases here are equivalent to components. The porous medium is characterized by the porosity and permeability values $\phi(x)$, $K(x)$, as well as by the values of relative permeabilities and capillary pressure functions k_o, k_w, P_c which are rock– and fluid–specific and are functions of the water saturation S_w and may also be x–dependent. The flow is described by the classical equations of conservation of mass and momentum (Darcy's law) complemented by a set of algebraic constraints and constitutive equations (see [11]). The densities ρ_o, ρ_w are known functions of pressure of each phase with known compressibility constants c_w and c_o. The rock is assumed to be incompressible, but this constraint can easily be removed. We thus have:

$$\phi \frac{\partial N_o}{\partial t} - \nabla \cdot (\rho_o K \frac{k_o}{\mu_o}(\nabla P_o - \rho_o G \nabla D)) = 0, \tag{12.1}$$

$$\phi \frac{\partial N_w}{\partial t} - \nabla \cdot (\rho_w K \frac{k_w}{\mu_w}(\nabla P_w - \rho_w G \nabla D)) = 0, \tag{12.2}$$

$$S_w + S_o = 1, \tag{12.3}$$

$$P_o = P_w + P_c(S_w). \tag{12.4}$$

The above equations are complemented with a set of initial conditions and the no flow boundary conditions on the external boundary of the reservoir. The flow is driven by injection / production wells which are implemented in the numerical model using the *Peaceman model* (see [12]) and appear in the equations (12.1) and (12.2) as the right hand side terms $q_o(x)$, $q_w(x)$, respectively. Several numerical algorithms have been proposed (see [11]) and numerous codes in the petroleum industry, environment management, as well as in research labs, exist which employ now standard discretization techniques and are validated against field data and results.

While these codes could be understood as the single block approach, the domain decomposition or multiblock approach has been successfully applied to the multiphase flow, see [2, 16]. Several algorithms exist for coupling with grids matching or nonmatching across an interface. However, to our knowledge the coupling of different numerical algorithms is a new research direction in this domain of applications. In this paper we present results obtained as a joint effort of the research group at the Center for Subsurface Modelling at TICAM, UT Austin. The multiblock as well as the multiblock multimodel framework is a part of our in–house simulator framework IPARSv2 [14] (*Integrated Parallel Accurate Reservoir Simulator*). We would like to acknowledge several colleagues who were part of the project, most notably John Wheeler, Ivan Yotov, Manish Parashar, Steven Bryant and Srinivas Chippada.

The outline of the paper is as follows. In Section 12.2 we briefly formulate the multiblock multimodel technique including subdomain and interface algorithms. In Section 12.3 we present some numerical examples using which we discuss related mathematical, numerical and implementation issues. See [15] for more complex examples.

12.2 FORMULATION

In this section we briefly review the numerical algorithm used in the subdomains and on the interface.

12.2.1 Fully Implicit in Time Formulation

In the fully implicit formulation we choose as primary variables the pressure and the concentration of oil P_o, N_o. We first discuss semidiscretization in time. The discrete in time equations at the time t_{n+1} are obtained by the backward Euler formula and are solved for P_o^{n+1}, N_o^{n+1} which satisfy

$$\phi\frac{N_o^{n+1} - N_o^n}{\Delta t_{n+1}} - \nabla \cdot (\rho_o^{n+1}\frac{K}{\mu_o}k_o^{n+1}(\nabla P_o^{n+1} - \rho_o^{n+1}G\nabla D)) = 0, \qquad (12.5)$$

$$\phi\frac{N_w^{n+1} - N_w^n}{\Delta t_{n+1}} - \nabla \cdot (\rho_w^{n+1}\frac{K}{\mu_w}k_w^{n+1}(\nabla P_w^{n+1} - \rho_w^{n+1}G\nabla D)) = 0. \qquad (12.6)$$

The discretization in space is achieved through the use of expanded mixed finite element methods of lowest order Raviart-Thomas on a rectangular grid which by the appropriate quadrature reduce to cell centered finite differences (see [4, 3, 1]). The edge values are computed by upwinding. The resulting nonlinear system of equations is solved by the Newton method and the Jacobian equation is solved by one of the suite of iterative solvers capable of handling non–symmetric and non–positive systems arising from the Jacobian, for example GMRES. The Newton method stops when the residuals are less than a given tolerance ν.

The fully implicit formulation is known to be unconditionally stable and permits the use of large time steps which may vary adaptively while keeping the error in mass conservation to a minimum. The pitfall of the fully implicit method is the complexity of the Newtonian iteration which may be costly (see Table 12.1). The implementation of the wells in the implicit system allows perfect mass balances with small ν; however, it may critically affect the convergence of the Newtonian procedure. The implicit formulations are therefore applied in a limited number of simulations, in spite of the increase in the computational power and the wide spread of parallel computing. Other formulations known as *IMPES* (*implicit pressures, explicit saturations*) are attractive alternatives. The sequential formulation presented below is another example.

12.2.2 Sequential Formulation

The IPARSv2 two phase sequential model relies on the splitting of the model equations into the time lagged (formally) elliptic part and the parabolic–hyperbolic part. The splitting which we propose goes back to the paper of [6] where different time steps were proposed to be used for the pressure and for the concentration equations. Other formulations related to the one proposed here are those of *total pressure* [11] or *streamlines–streamtubes* methods [5]. Several others not mentioned here exist. The main idea behind the formulation is that even though the individual phase mobilities vary strongly with the water content function, the sum of them (the total mobility) remains close to a constant over many time steps. Therefore the pressure profiles remain stable even though water saturation varies strongly. Once the pressure profiles are known one solves for the saturation analytically (along streamlines) or numerically as we show below.

The primary variables in the sequential formulation are water pressure P_w and saturation S_w. The first equation defines the value of water pressure at the new time step P_w^{n+1} as a solution to the problem

$$\begin{aligned}
-\nabla \cdot \left(K\lambda_t^n(\nabla P_w^{n+1})\right) &= \nabla \cdot (K\lambda_o^n\nabla P_c(S_w^n)) \\
&\quad - \nabla \cdot (K(\lambda_o^n\rho_o^n + \lambda_w^n\rho_w^n)G\nabla D)
\end{aligned} \qquad (12.7)$$

where the oil, water, and total mobilities are $\lambda_o = \frac{k_o}{\mu_o}$, $\lambda_w = \frac{k_w}{\mu_w}$, and $\lambda_t = \lambda_o + \lambda_w$. Using the values of P_w^{n+1} the densities, the phase velocities u_o, u_w (defined as terms under $\nabla \cdot$ in (12.1)–(12.2)) and the total velocity $u_t = u_o + u_w$ are computed. Then the saturation equation is solved for S_w^{n+1}

$$\frac{\phi S_w^{n+1}}{\Delta t_{n+1}^S}\frac{\rho_w^{n+1}}{\rho_w^n} + \nabla \cdot [K\frac{\lambda_o^n \lambda_w^n}{\lambda_t^n}P_c'(S_w^n)\nabla S_w^{n+1}] = \frac{\phi}{\Delta t_{n+1}^S}S_w^n$$

$$- \nabla \cdot [\frac{\lambda_w^n}{\lambda_t^n}u_t^n] - \nabla \cdot [K\frac{\lambda_o^n \lambda_w^n}{\lambda_t^n}(\rho_w^n - \rho_o^n)G\nabla D)]. \qquad (12.8)$$

The time step for saturation may be much smaller than the pressure step or, alternatively, the pressure solution can be skipped and redone only every K saturation steps, with K as large as 10 or 60. The saturation time step is limited by the CFL–type stability condition $\frac{v}{\phi}\frac{\Delta t}{\Delta x} < 1$ on the time and spatial discretization steps $\Delta t, \Delta x$ in terms of the velocity v and porosity ϕ. Another limitation to the size of time step is a consequence of the presence of wells which are implemented using the Peaceman model. Since the densities as well as mobilities in the pressure equation are time-lagged (explicit) in the well terms, the (strict) material balances show discrepancy in mass conservation which can only be controlled by the time step size, as no parameter ν can be imposed.

The discretization in space is done analogously to what was described in the previous section. The set of two separate fully discrete equations (or more if multiple saturation steps are used) is solved each by a simple iterative linear solver like PCG for symmetric positive definite system. Since the system (12.7)–(12.8) is effectively a linearized version of (12.5)–(12.6), the computational cost per time step is much lower (see Table 12.1), but again, the time step may be severely restricted for reasons of accuracy and stability.

12.2.3 Interface coupling

In the previous sections we defined two alternative algorithms used to solve the same set of two phase flow equations. Below we describe the interface coupling of these two algorithms.

For simplicity we shall consider two blocks A and B with interface I. Wells are located in any part of $D = A \cup B \cup I$. The Neumann no–flow boundary conditions are imposed on ∂D. In the multiblock multimodel approach we use the implicit algorithm in the domain (block) A and the sequential algorithm in block B, and we seek the interface values of the primary unknowns on I such that the fluxes of the oil and water components match in a prescribed weak sense. This domain decomposition formulation for mixed methods stems from the classical paper [8] and was extended to the nonmatching grids with mortar spaces in [2] as well as implemented in our framework, see [15].

The choice of the primary unknowns on the interface is problem dependent. The physics of the flow imposes conditions on the interface which reflect the conservation of mass expressed by the matching of the component fluxes as well as the conservation of momentum that is equivalent to the equality of the pressures. For simplicity we assume below that the capillary pressure function is independent of the position x and then the matching of pressures is equivalent to the equality of (any pair of) saturations and concentrations. This assumption is true for a large class of problems. In such a case, the choice of primary variables influences and is motivated by the convergence properties, computational efficiency or coding convenience. In particular, in the tests presented below we use (P_o, N_o) as the interface primary variables. Note that this set matches the set of primary variables in the implicit formulation used in block A but is different from the one in block B.

We focus again on the semidiscrete in time coupling. The space discretization on the interface with the use of *mortar spaces* and the mathematical form of the matching conditions in the weak form have been described in [2, 16, 15]. The *mortar spaces* technique is capable of handling non–matching grids across the interface and uses suitable projections between the subdomain grids and the interface grids for both primary variables and fluxes. For notational convenience in the discussion below, depending on the context, we will understand as "values" the values of the primary unknowns (denoted by Λ) or the values of their projections into suitable spaces. Similar convention applies to the values of the fluxes of oil and water across I outward to subdomains $Flux_o, Flux_w$, respectively, and to their jump across I denoted below as $B(\Lambda)$. Furthermore, the interface problem $B(\Lambda) = 0$ is understood in a weak sense. See [2, 16] and references therein for details.

Interface values applied to the implicit equations in the block A.

Because of the choice of interface primary variables that we assumed above, the application of Dirichlet boundary values to the problem (12.5)–(12.6) is straightforward, and for a given current guess $\Lambda^{n+1} = (P_o^{*,n+1}, N_o^{*,n+1})$, we need to solve the problem (12.5)–(12.6) with

$$P_o^{n+1}|_I = P_o^{*,n+1} \tag{12.9}$$
$$N_o^{n+1}|_I = N_o^{*,n+1}. \tag{12.10}$$

Once the subdomain problem is solved, the normal fluxes of oil and water $Flux_o^A, Flux_w^A$ across I outward from A are computed.

Interface values applied to the sequential equations in the block B.

In the block B, one needs to find a map from the set of primary unknowns on the interface $\Lambda^{n+1} = (P_o^{*,n+1}, N_o^{*,n+1})$ to the set of primary unknowns in the subdomain for the sequential algorithm $(P_w^{*,n+1}, S_w^{*,n+1})$. There exists a direct algebraic relationship

$$P_w^* = P_o^* - P_c(S_w^*) \quad , \quad S_w^* = 1 - \frac{N_o^*}{\rho_o(P_o^*)} \tag{12.11}$$

which follows the capillary pressure relationship (12.4). However, note that the values Λ^{n+1} are imposed implicitly (in time) whereas the sequential solution uses time–lagged or explicit saturations and mobilities in the pressure equation. The direct application of (12.11) leads to inconsistency and failure of the interface algorithm. We propose instead to use the consistent formulae

$$P_w^{*,n+1} = P_o^{*,n+1} - P_c(S_w^{*,n}) \quad , \quad S_w^{*,n+1} = 1 - \frac{N_o^{*,n+1}}{\rho_o(P_o^{*,n+1})}. \tag{12.12}$$

The pressure equation (12.7) is then modified by the Dirichlet condition

$$P_w^{n+1}|_I = P_w^{*,n+1} \tag{12.13}$$

where the saturation values used for the mobilities are $S_w^{*,n}$. Next, the saturation equation (12.8) is solved and is complemented by

$$S_w^{n+1}|_I = S_w^{*,n+1}. \tag{12.14}$$

The computation of fluxes $Flux_o^B, Flux_w^B$ outward from B across I follows.

12.2.4 Solution of the interface problem.

The goal of the interface algorithm is to find, at every time step t_{n+1}, the interface values $\Lambda = \Lambda^{n+1} = (P_o^{*,n+1}, N_o^{*,n+1})$ so that

$$B(\Lambda) = |Flux_o^A - Flux_o^B| + |Flux_w^A - Flux_w^B| = 0$$

or, practically, $B(\Lambda) < \epsilon$, where ϵ is some prescribed tolerance and $|\cdot|$ is a suitable norm.

The problem $B(\Lambda) = 0$ can be solved by various solvers appropriate for general non-linear problems. In the results reported below we use the inexact Newton–Krylov method for which the Jacobian $B'(\Lambda)S$ is approximated by a finite difference and the equation to be solved in an interface Newton step is

$$B'(\Lambda)S \approx \frac{B(\Lambda + \sigma S) - B(\Lambda)}{\sigma}. = -B(\Lambda),$$

see [15]. Several parameters determine the efficiency and convergence of this technique; see [7, 9, 10]. For lack of space we only comment on the optimal choice of σ. For the multiblock implicit where all subdomain solvers are fully implicit the values of σ are controlled by the ν, and we used $\sigma \approx 10^{-8}$ or less with $\nu \approx 10^{-10}$. For multiblock sequential the optimal values, in the absence of ν, were rather large ($\approx 10^{-4}$). Therefore it is hard to choose a "perfect" σ for multiblock multimodel (some subdomain solvers are implicit, some are sequential), but in practice we use a large one $\sigma \approx 10^{-4}$.

12.3 COMPUTATIONAL EXAMPLES

12.3.1 Heterogeneous permeability example

In our first example we present the so–called quarter five–spot problem for a heterogeneous reservoir (Figure 12.1) with wells located in the lower left (water injection) and upper right (oil and water production) corners. The permeability was $K = 2md$ or $K = 200md$ as shown. We first solve the problem using a uniformly fine grid and a fully implicit code. Then we apply the multiblock multimodel strategy. The location of high permeability layers and wells suggests the use of a fine grid and an implicit algorithm in the high permeability zones and the use of a coarse grid and a sequential code in the remaining part of the reservoir where velocities are small. Further decomposition of the high permeability layers into regions around wells (implicit) and elsewhere (handled sequentially) is possible but not as desirable, because the high velocities in the sequential regions will limit the size of the time step. The gridding and the multimodel decomposition is shown in Figure 12.1. Figure 12.1 also shows the agreement of oil production rates calculated by the single block and by the multiblock multimodel codes, respectively.

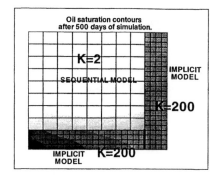 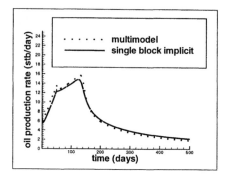

Figure 12.1: Grid and well rates for heterogeneous permeability example.

12.3.2 Numerical and mathematical issues

As was mentioned above, several factors determine the effectiveness of the interface algorithm for the multiblock multimodel problems. One of these is the difference in how the subdomain problems respond to the imposed boundary values. Consider a 1D (thin and long reservoir) with wells at opposite ends which has been split into three blocks so that I consists of the two disjoint parts. The flow is simulated by the multiblock implicit, multiblock sequential and multiblock multimodel codes. Figure 12.2 presents the values of the total jump $B(\Lambda)$ as a function of $\Lambda = (P_o^1, N_o^1)$ for the first time step of simulation. More precisely, at $t = t_1$ the waterflood front has not yet reached either part of the

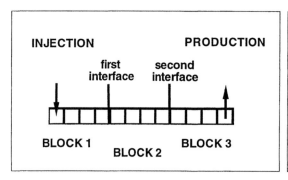 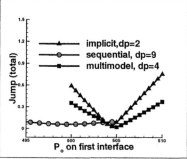

Figure 12.2: Qualitative behavior of $B(\Lambda)$ for 1D example.

interface I and for the purposes of this experiment we fix the value of N_o to be equal to the initial value. We let P_o imposed on the first interface vary and use the value of P_o on the second as $P_o + dp$ where the optimal value of dp is different for each experiment. Then the jump $B(\Lambda)$ is computed and plotted.

One can easily notice the qualitative and quantitative differences between $B(\Lambda)$ plotted for the three cases which are explained by the differences in the approximation properties of the implicit and the sequential formulations.

12.3.3   Implementation and parallel computation issues

In our last example we discuss the implementation issues. Aside from the coding effort spent on memory management, visualization, etc., the parallel issues are the most interesting to tackle, as the multimodel code by definition is really an MIMD (Multiple Instruction Multiple Data) code. Our implementation uses multiple MPI communicators [13]. Load balancing is an issue here. The traditional (non–multimodel) load balancing strategy is that the cells are divided more or less evenly between processors. Our experiments show that for optimal load balancing, one processor should never handle more than one model / code, if possible. Figure 12.3 shows the optimal load balancing decomposition and the speedup for the traditional and for the optimal load balancing strategies.

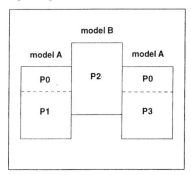 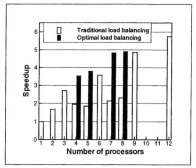

Figure 12.3: Load balancing issues in multimodel implementation.

12.4   CURRENT RESEARCH

The mortar spaces can be time dependent and chosen so as to weakly couple or strongly couple the subdomain problems. The interface solver type and convergence parameters may vary adaptively. This is a current research topic.

REFERENCES

[1] T. Arbogast, C. N. Dawson, P. T. Keenan, M. F. Wheeler and I. Yotov (1998). Enhanced cell-centered finite differences for elliptic equations on general geometry. *SIAM J. Sci. Comp.*, 19:404-425.

[2] T. Arbogast, L. C. Cowsar, M. F. Wheeler and I. Yotov. Mixed finite element methods on non-matching multiblock grids. *SIAM J. Num. Anal.*, to appear.

[3] T. Arbogast, M. F. Wheeler and I. Yotov (1997). Mixed finite elements for elliptic problems with tensor coefficients as cell-centered finite differences. *SIAM J. Numer. Anal.*, 34, 828-852.

[4] T. Arbogast, M. F. Wheeler and I. Yotov (1996). Logically rectangular mixed methods for flow in irregular, heterogeneous domains. *Computational Methods in Water Re-*

sources XI, A. A. Aldama and others editors, Computational Mech. Publ., Southampton, 621-628.

[5] R. P. Batycky, M. J. Blunt, M. R. Thiele (1996). A 3D Multi–Phase Streamline Simulator With Gravity and Changing Well Conditions. *17th Intl. Energy Agency Coll. Project on Enhanced oil Recovery*, Sydney, Australia, Sept. 29–Oct. 2

[6] J. Douglas,, R. E. Ewing, M. F. Wheeler (1983). A time-discretization procedure for a mixed finite element approximation of miscible displacement in porous media *R.A.I.R.O. Analyse Numerique, vol. 17, pp. 249–265.*

[7] S. C. Eigenstat and H. F. Walker (1994). Globally convergent inexact Newton method. *SIAM J. Sci. Optim.*, 4:393-422

[8] R. Glowinski and M. F. Wheeler (1988). Domain decomposition and mixed finite element methods for elliptic problems, *Domain Decomposition Methods for Partial Differential Equations*, SIAM, Philadelphia, 144-172.

[9] C. T. Kelly (1995). *Iterative methods for linear and nonlinear equations*. SIAM, Philadelphia.

[10] H. Klie (1996). Krylov-secant Methods for Solving Large Scale Systems of Coupled Nonlinear Parabolic Equations. *PhD thesis*, Rice University, Houston, Texas.

[11] D. W. Peaceman (1977). *Fundamentals of numerical reservoir simulation*. Elsevier, Amsterdam.

[12] D. W. Peaceman (1983). Interpretation of well-block pressure in numerical reservior simulation with non-square grid blocks and anisotropic permeability. *Tran. AIME*, Vol. 275.

[13] M. Snir, S. Otto, S. Huss-Lederman, D. Walker and J. Dongarra (1996). *MPI: the complete reference*. The MIT Press.

[14] P. Wang, I. Yotov, M. Wheeler, T. Arbogast, C. Dawson, M. Parashar and K. Sephernoori (1997). A New Generation EOS Compositional Reservoir Simulator: Part I – Formulation and Discretization. *SPE 37979, 1997 SPE Reservoir Simulation Symposium*, Dallas, Texas.

[15] M. F. Wheeler T. Arbogast, S. Bryant, J. Eaton, Q. Lu, M. Peszynska and I. Yotov (1999). A parallel multiblock/multidomain approach for reservoir simulation. *SPE 51884, 1999 SPE Symposium on Reservoir Simulation*, Houston, Texas.

[16] I. Yotov (1997). A mixed finite element discretization on non-matching multiblock grids for a degenerate parabolic equation arising in porous media flow. *East-West J. Numer. Math.*, 5:211-230

13 ANALYSIS AND NUMERICS OF STRONGLY DEGENERATE CONVECTION-DIFFUSION PROBLEMS MODELLING SEDIMENTATION- CONSOLIDATION PROCESSES

R. Bürger[a] and K. Hvistendahl Karlsen[b]

[a]Institute of Mathematics A, University of Stuttgart, Pfaffenwaldring 57, 70569 Stuttgart, Germany

[b]Dept. of Mathematics, University of Bergen, Johs. Brunsgt. 12, N-5008 Bergen, Norway, and RF-Rogaland Research, Thormølensgt. 55, N-5008 Bergen, Norway

ABSTRACT

In one space dimension, the phenomenological sedimentation-consolidation model reduces to an initial-boundary value problem (IBVP) for a nonlinear strongly degenerate convection-diffusion equation. Due to the mixed hyperbolic-parabolic nature of the model, its solutions are discontinuous and entropy solutions must be sought. In this contribution, we review recent existence and uniqueness results for this and a related IBVP, and present numerical methods that can be used to correctly simulate this model, i.e. conservative methods satisfying an entropy principle. Included in our discussion are finite difference methods and methods based on operator splitting, which are employed to simulate the settling of flocculated suspensions.

Key words. Degenerate convection-diffusion equation, operator splitting, front tracking, sedimentation-consolidation processes.

13.1 INTRODUCTION

In this contribution, we consider the quasilinear strongly degenerate parabolic equation

$$\partial_t u + \partial_x g(u, t) = \partial_x^2 A(u), \quad A(u) := \int_0^u a(s)\, ds, \quad a(u) \geq 0, \quad g(u, t) := q(t)u + f(u) \quad (13.1)$$

on a cylinder $Q_T := \Omega \times \mathcal{T}$, $\Omega := (0, 1)$, $\mathcal{T} := (0, T)$, $T > 0$. We allow that $a(u) = 0$ on an interval $[0, u_c]$, where equation (13.1) is then of parabolic type, and that $a(u)$ may be discontinuous at $u = u_c$. The flux density function $f(u)$ is (for simplicity) assumed to be piecewise differentiable with supp $f \subset [0, 1]$ and $f(u) \leq 0$, and $q(t)$ is a nonpositive piecewise differentiable Lipschitz continuous function. These assumptions are motivated by the model of sedimentation-consolidation processes of flocculated suspensions presented in [5, 6], to which we come back in § 13.4. Moreover, we require that $\|f'\|_\infty \leq \infty$, $\mathrm{TV}_\mathcal{T}(q) < \infty$ and $\mathrm{TV}_\mathcal{T}(q') < \infty$. We consider two different IBVPs. Problem A consists of equation (13.1) together with the initial and boundary conditions

$$u(x, 0) = u_0(x), \quad x \in \Omega; \quad u(1, t) = \varphi_1(t), \quad (f(u) - \partial_x A(u))(0, t) = 0, \quad t \in \mathcal{T}. \quad (13.2)$$

This problem has been studied previously by Bürger and Wendland [4]. The second IBVP, Problem B, is obtained from Problem A if the boundary condition (13.2) is replaced by

$$(g(u, t) - \partial_x A(u))(1, t) = \Psi(t), \quad t \in \mathcal{T}. \quad (13.3)$$

Let ω_ε be a standard C^∞ mollifier with supp $\omega_\varepsilon \subset (-\varepsilon, \varepsilon)$ and define $a_\varepsilon(u) := ((a + \varepsilon) * \omega_\varepsilon)(u)$ and $A_\varepsilon(u) := \int_0^u a_\varepsilon(s)\, ds$ for $\varepsilon > 0$. For Problem A, the assumptions on the initial and boundary data can be stated as

$$\varphi_1(t) \in [0, 1] \text{ for } t \in \overline{\mathcal{T}}, \quad \varphi_1 \text{ has a finite number of local extrema;} \quad (13.4)$$
$$u_0 \in \{u \in BV(\Omega) : u(x) \in [0, 1]; \exists M_0 > 0 : \forall \varepsilon > 0 : \mathrm{TV}_\Omega(\partial_x A_\varepsilon(u)) < M_0\}, \quad (13.5)$$

while for Problem B we require that (13.5) is valid and that either $\Psi \equiv 0$ or that there exist positive constants ξ and M_g such that $\xi a(u) - (q(t) + f'(u)) \geq M_g$ uniformly in ε.

Note that if $a(\cdot)$ is sufficiently smooth, then it is sufficient to require that $\mathrm{TV}_\Omega(\partial_x u_0)$ is finite. Multi-dimensional problems are treated in [2].

13.2 ENTROPY SOLUTIONS

It is well known that due both to the degeneracy of the diffusion coefficient $a(u)$ and to the nonlinearity of the flux density function $f(u)$, solutions of equation (13.1) are discontinuous and have to be considered as entropy solutions.

Definition 13.1 ([1]) *A function $u \in L^\infty(Q_T) \cap BV(Q_T)$ is an entropy solution of Problem A if the following conditions are satisfied:*

$$\partial_x A(u) \in L^2(Q_T); \quad (13.6)$$
$$\text{f. a. a. } t \in \mathcal{T}, \ \gamma_0(f(u) - \partial_x A(u)) = 0; \quad \text{f. a. a. } x \in \overline{\Omega}, \ \lim_{t \downarrow 0} u(x, t) = u_0(x), \quad (13.7)$$

and if $\forall \varphi \in C^{\infty}((0,1] \times \overline{\mathcal{T}})$, $\varphi \geq 0$, $\text{supp}\,\varphi \subset (0,1] \times \mathcal{T}$, $\forall k \in \mathbb{R}$:

$$\iint_{Q_T} \left\{ |u-k| \partial_t \varphi + \text{sgn}(u-k)[g(u,t) - g(k,t) - \partial_x A(u)] \partial_x \varphi \right\} dt dx$$

$$+ \int_0^T \left\{ -\text{sgn}(\varphi_1(t) - k)[g(\gamma_1 u, t) - g(k,t) - \gamma_1 \partial_x A(u)] \varphi(1,t) \right.$$

$$\left. + [\text{sgn}(\gamma_1 u - k) - \text{sgn}(\varphi_1(t) - k)][A(\gamma_1 u) - A(k)] \partial_x \varphi(1,t) \right\} dt \geq 0. \quad (13.8)$$

Definition 13.2 ([1]) *A function $u \in L^{\infty}(Q_T) \cap BV(Q_T)$ is an entropy solution of Problem B if (13.6) and (13.7) are valid, if for all $\varphi \in C_0^{\infty}(Q_T)$, $\varphi \geq 0$ and $k \in \mathbb{R}$ the inequality*

$$\iint_{Q_T} \left\{ |u-k| \partial_t \varphi + \text{sgn}(u-k)[g(u,t) - g(k,t) - \partial_x A(u)] \partial_x \varphi \right\} dt dx \geq 0 \quad (13.9)$$

holds, and if $\gamma_1(g(u,t) - \partial_x A(u)) = \Psi(t)$ for almost all $t \in \mathcal{T}$.

In these definitions, $\gamma_0 u := (\gamma u)(0,t)$ and $\gamma_1 u := (\gamma u)(1,t)$ denote the traces of u. Entropy inequalities like (13.8) go back to the pioneering papers of Kružkov [15] and Vol'pert [17] for first order equations and Vol'pert and Hudjaev [18] for second order equations.

We now briefly summarize some recent results on the existence and uniqueness of entropy solutions of Problems A and B, and state a new regularity result for the integrated diffusion coefficient for entropy solutions of Problem B. For details we refer to [1].

Theorem 13.1 ([1]) *Under the conditions stated in § 13.1, there exist entropy solutions to both problems A and B.*

Sketch of Proof. For both problems, existence of entropy solutions can be shown by the vanishing viscosity method. To this end, we consider the regularized parabolic IBVPs

$$\left. \begin{array}{l} \partial_t u^{\varepsilon} + \partial_x(q_{\varepsilon}(t)u^{\varepsilon} + f_{\varepsilon}(u^{\varepsilon})) = \partial_x^2 A_{\varepsilon}(u^{\varepsilon}), \ (x,t) \in Q_T; \ u^{\varepsilon}(x,0) = u_0^{\varepsilon}(x), \ x \in \Omega; \\ u^{\varepsilon}(1,t) = \varphi_1^{\varepsilon}(t), \ (f_{\varepsilon}(u^{\varepsilon}) - \partial_x A_{\varepsilon}(u^{\varepsilon}))(0,t) = 0, \ t \in (0,T] \end{array} \right\} (13.10)$$

$$\left. \begin{array}{l} \partial_t u^{\varepsilon} + \partial_x(q_{\varepsilon}(t)u^{\varepsilon} + f_{\varepsilon}(u^{\varepsilon})) = \partial_x^2 A_{\varepsilon}(u^{\varepsilon}), \ (x,t) \in Q_T; \ u^{\varepsilon}(x,0) = u_0^{\varepsilon}(x), \ x \in \Omega; \\ (g_{\varepsilon}(u^{\varepsilon},t) - \partial_x A_{\varepsilon}(u^{\varepsilon}))(1,t) = \Psi_{\varepsilon}(t), \ (f_{\varepsilon}(u^{\varepsilon}) - \partial_x A_{\varepsilon}(u^{\varepsilon}))(0,t) = 0, \ t \in (0,T] \end{array} \right\} (13.11)$$

where the functions q, f, u_0, φ_1 and Ψ have been replaced by particular smooth approximations for each problem that ensure compatibility conditions and existence of smooth solutions. It can then be shown that there exist constants M_1 to M_5 independent of ε such that the smooth solutions of Problem (13.10) satisfy

$$\|u^{\varepsilon}\|_{L^{\infty}(Q_T)} \leq M_1, \ \|\partial_x u^{\varepsilon}(\cdot,t)\|_{L^1(\Omega)} \leq M_2 \text{ for all } t \in \mathcal{T}, \ \|\partial_t u^{\varepsilon}\|_{L^1(Q_T)} \leq M_3, \quad (13.12)$$

while those of Problem (13.11) satisfy

$$\|u^{\varepsilon}\|_{L^{\infty}(Q_T)} \leq M_1, \ \|\partial_t u^{\varepsilon}(\cdot,t)\|_{L^1(\Omega)} \leq M_4 \text{ for all } t \in \mathcal{T}, \quad (13.13)$$

and, in the case where $\Psi \equiv 0$,

$$\|\partial_x u^\varepsilon(\cdot, t)\|_{L^1(\Omega)} \leq M_5 \text{ for all } t \in \mathcal{T} \tag{13.14}$$

and in the case where there exist constants $\xi, M_g > 0$ such that $\xi a(u) - (q(t) + f'(u)) \leq M_g$,

$$\|\partial_x u^\varepsilon\|_{L^1(Q_T)} \leq M_5. \tag{13.15}$$

Estimates (13.12) imply that the family $\{u^\varepsilon\}_{\varepsilon > 0}$ of solutions of Problem (13.10) is bounded in $W^{1,1}(Q_T) \subset BV(Q_T)$. Hence there exists a sequence $\varepsilon = \varepsilon_n \downarrow 0$ such that $\{u^{\varepsilon_n}\}$ converges in $L^1(Q_T)$ to a function $u \in L^\infty(Q_T) \cap BV(Q_T)$. The same is true for the family of solutions of Problem B^ε. To prove that u is an entropy solution of Problem A or B, it has to be shown that the diffusion function $A(u)$ has the required regularity. In both cases, it is fairly easy to show that $\|\partial_x A_\varepsilon(u^\varepsilon)\|_{L^2(Q_T)}$ is uniformly bounded independently of ε. Therefore, passing if necessary to a subsequence, $A_\varepsilon(u^\varepsilon) \to A(u)$ in $L^2(Q_T)$ and $\partial_x A_\varepsilon(u^\varepsilon) \to \partial_x A(u)$ weakly in $L^2(Q_T)$ as $\varepsilon \downarrow 0$. It is now easy to show that the limit function u satisfies the remaining parts of Definitions 13.1 and 13.2, respectively. $\square$

For the case of Problem B, the regularity result $\partial_x A(u) \in L^2(Q_T)$ can be considerably improved; namely, we have that $A(u)$ is Hölder continuous on $\overline{Q_T}$:

Theorem 13.2 ([1]) *Assume that $u^\varepsilon \to u$ a.e. on Q_T as $\varepsilon \downarrow 0$. Then there exists a subsequence $\varepsilon_n \downarrow 0$ such that $A(u^{\varepsilon_n}) \to A(u)$ uniformly on $\overline{Q_T}$ and $A(u) \in C^{1,1/2}(\overline{Q_T})$.*

Sketch of Proof. The proof is essentially based on the observation that if u^ε is a smooth solution of Problem B^ε, then the quantity $V^\varepsilon := -g_\varepsilon(u^\varepsilon, t) - a_\varepsilon(u^\varepsilon)\partial_x u^\varepsilon$ satisfies a linear parabolic IBVP with Dirichlet boundary data that are uniformly bounded in ε. From the maximum principle, we obtain that $\partial_x A_\varepsilon(u^\varepsilon)$ is uniformly bounded on $\overline{Q_T}$. This and estimates (13.13) to (13.15) allow the application of Kružkov's interpolation lemma [15, Lemma 5] to the linear IBVP. Hence there exists a constant M_7 such that

$$|A_\varepsilon(u^\varepsilon(x, t_2)) - A_\varepsilon(u^\varepsilon(x, t_1))| \leq M_7 \sqrt{|t_2 - t_1|}, \quad \forall (x, t_1), (x, t_2) \in \overline{Q_T}.$$

The Ascoli-Arzelà compactness theorem then yields the existence of a subsequence of $\{A(u^{\varepsilon_n})\}$ converging uniformly on $\overline{Q_T}$ to $A(u) \in C^{1,1/2}(\overline{Q_T})$. $\square$

Theorem 13.3 ([1]) *Let u and v be two entropy solutions either of Problem A or of Problem B with initial data u_0 and v_0, respectively. Then $\|u(\cdot, t) - v(\cdot, t)\|_{L^1(\Omega)} \leq \|u_0 - v_0\|_{L^1(\Omega)}$ is valid. In particular, both problems have at most one entropy solution.*

Sketch of Proof. The proof is based on the technique known as "doubling of the variables" introduced by Kružkov [15] as a tool for proving the L^1 contraction principle for entropy solutions of scalar conservation laws. This technique was recently extended by Carrillo [7] to a class of degenerate parabolic equations. This recent extension is adopted here to Problems A and B and leads to the inequality

$$\iint_{Q_T} \left\{ |u - v|\partial_t \varphi + \text{sgn}(u - v)[g(u, t) - g(v, t) - (\partial_x A(u) - \partial_x A(v))]\partial_x \varphi \right\} dt dx \geq 0,$$

valid for two entropy solutions u and v either of Problem A or of Problem B and for all test functions $\varphi \in C_0^\infty(Q_T)$, from which stability and uniqueness can be obtained in a standard fashion. $\square$

Remark 13.1 The proof of Theorem 13.3 (see [1]) is *not* based on a jump condition, in contrast to the uniqueness proof by Wu and Yin [20]. In fact, it is not clear whether a jump condition can be derived with integrated diffusion functions $A(u)$ that are only Lipschitz continuous. Moreover, it has been possible to derive jump conditions only in the 1-D case so far, while the new uniqueness proof can also be extended to multidimensions.

13.3 NUMERICAL METHODS

This section provides the necessary background for the development and application of numerical methods for mixed hyperbolic-parabolic problems.

13.3.1 Finite Difference Methods

To focus on the main ideas, we consider here the simplified problem

$$\partial_t u + \partial_x f(u) = \partial_x^2 A(u), \qquad u(x,0) = u_0(x), \tag{13.16}$$

where $(x,t) \in Q_T = \mathbb{R} \times (0,T)$ and $f = f(u)$, $A = \int^u a$, $a = a(u) \geq 0$, $u_0 = u_0(x)$ are sufficiently smooth functions. The difference methods described here can easily be modified to solve the full sedimentation-consolidation model. The material presented here is based on the series of papers by Evje and Karlsen [10–12], see also [3].

Selecting a mesh size $\Delta x > 0$, a time step $\Delta t > 0$, and an integer N so that $N\Delta t = T$, the value of the difference approximation at $(x_j, t_n) = (j\Delta x, n\Delta t)$ will be denoted by u_j^n. There are special difficulties associated with equation (13.1) which must be dealt with in developing numerical methods. For example, numerical methods based on naive finite difference replacements of the diffusion term may be adequate for smooth solutions but can give wrong results when discontinuities are present, see [11,12] for details. It turns out that it is preferable to use a conservative differencing of the second order term and upwind differencing of the convective flux and, i.e., a difference method of the form

$$\frac{u_j^{n+1} - u_j^n}{\Delta t} + \frac{F(u_j^n, u_{j+1}^n) - F(u_{j-1}^n, u_j^n)}{\Delta x} = \frac{A(u_{j-1}^n) - 2A(u_j^n) + A(u_{j+1}^n)}{(\Delta x)^2}, \tag{13.17}$$

where F is the upwind flux. For a monotone flux function f, the upwind flux is defined by $F(u_j^n, u_{j+1}^n) = f(u_j^n)$ if $f' > 0$ and $F(u_j^n, u_{j+1}^n) = f(u_{j+1}^n)$ if $f' < 0$. More generally, for a non-monotone flux function f, one needs the generalised upwind flux of Engquist and Osher defined by (see also see [11]) $F(u_j^n, u_{j+1}^n) = f^+(u_j^n) + f^-(u_{j+1}^n)$, where $f^+(u) = f(0) + \int_0^u \max(f'(s), 0)\, ds$ and $f^-(u) = \int_0^u \min(f'(s), 0)\, ds$. We assume that the following stability condition holds: $\max_u |f'(u)| \frac{\Delta t}{\Delta x} + 2\max_u |a(u)| \frac{\Delta t}{(\Delta x)^2} \leq 1$.

As is well known, upwind differencing stabilizes profiles which are liable to undergo sudden changes, i.e., discontinuities and other large gradient profiles. Therefore upwind differencing is perfectly suited to the treatment of discontinuities (and thus of the sedimentation model). Let u_Δ, $\Delta = (\Delta x, \Delta t)$, be the interpolant of degree one associated with the discrete data points $\{u_j^n\}$. Regarding the sequence $\{u_\Delta\}$, we have:

Theorem 13.4 ([11]) *The sequence $\{u_\Delta\}$ built from (13.17) converges in $L^1_{\text{loc}}(Q_T)$ to the unique entropy solution u of (13.16) as $\Delta \to 0$. Furthermore, $\{A(u_\Delta)\}$ converges uniformly on compact sets $\mathcal{K} \subset Q_T$ to $A(u) \in C^{1,1/2}(\bar{Q}_T)$ as $\Delta \to 0$.*

Sketch of Proof. An important part of the proof of this theorem is to establish the following three estimates for $\{u_j^n\}$: (a) a uniform L^∞ bound, (b) a uniform total variation bound, (c) L^1 Lipschitz continuity in the time variable, and the following two estimates for the discrete total flux $F(u^n; j) - \Delta_+ A(u_j^n)$: (d) a uniform L^∞ bound, (e) a uniform total variation bound. We refer to [11] for details concerning the derivation of these bounds. Then, using the three estimates (a)-(c), it is not difficult to show that there is a finite constant $C = C(T) > 0$ (independent of Δ) such that $\|u_\Delta\|_{L^\infty(Q_T)} + |u_\Delta|_{BV(Q_T)} \leq C$.

Hence, the sequence $\{u_\Delta\}$ is bounded in $BV(\mathcal{K})$ for any compact set $\mathcal{K} \subset Q_T$. It is thus possible to select a subsequence that converges in $L^1(\mathcal{K})$. Furthermore, using a standard diagonal process, we can construct a sequence that converges in $L^1_{\text{loc}}(Q_T)$ to a limit $u \in L^\infty(Q_T) \cap BV(Q_T)$. It is possible to use, among other things, estimates (d) and (e) to prove that $A(u_\Delta)$ is Hölder continuous on $\overline{Q}_T$ independently of Δ. Then by repeating the proof of the Ascoli-Arzela compactness theorem, we deduce the existence of a subsequence of $\{A(u_\Delta)\}$ converging uniformly to $A(u) \in C^{1,1/2}(\overline{\Pi}_T)$.

Finally, convergence of $\{u_\Delta\}$ to the correct physical solution of (13.16) follows from the cell entropy inequality ($k \in \mathbb{R}$)

$$\frac{|u_j^{n+1} - k| - |u_j^n - k|}{\Delta t} + \Delta_-\left(F(u^n \vee k; j) - F(u^n \wedge k; j) - \Delta_+|A(u_j^n) - A(k)|\right) \leq 0,$$

where $u \vee v = \max(u, v)$ and $u \wedge v = \min(u, v)$. This discrete entropy inequality is in turn an easy consequence of the monotonicity of the scheme. The reader is referred to [11] for further details on the convergence analysis.   □

Remark 13.2 In many applications it is desirable to avoid the explicit stability restriction associated with *(13.16)*. One way to overcome the restriction is of course to use an implicit version of *(13.17)*, see [12] for details. Moreover, the upwind method and all other monotone methods are at most first order accurate, giving poor accuracy in smooth regions. To overcome these problems, Evje and Karlsen [10] used the generalized MUSCL (Variable Extrapolation) idea of van Leer to formally upgrade the upwind method *(13.17)* to second order accuracy. Although more difficult than in the monotone case, it can be shown that the second order method also satisfies a discrete entropy condition and that it converges to the unique generalized solution of the problem, see [10] for details.

Finally, let us say a few words about the multi-dimensional case. For simplicity of notation, we consider only the two-dimensional problem

$$\partial_t u + \partial_x f(u) + \partial_y g(u) = \partial_x(d(u)\partial_x u) + \partial_y(d(u)\partial_y u), \quad u(x, y, 0) = u_0(x, y). \quad (13.18)$$

Let $u_{j,k}^n$ denote the finite difference approximation at $(x, y, t) = (j\Delta x, k\Delta y, n\Delta t)$. A conservative finite difference method for (13.18) takes the form

$$\frac{u_{j,k}^{n+1} - u_{j,k}^n}{\Delta t} + \Delta_{x,-}(F(u^n; j, k) - \Delta_{x,+}D(u_{j,k}^n)) + \Delta_{y,-}(G(u^n; j, k) - \Delta_{y,+}D(u_{j,k}^n)) = 0,$$
$$(13.19)$$

where $\Delta_{\ell,-}, \Delta_{\ell,+}$ are the backward and forward differences, respectively, in direction ℓ, for $\ell = x, y$, and F, G are convective numerical fluxes that are consistent with f, g, respectively. Roughly speaking, one can choose F, G to be any reasonable numerical flux for hyperbolic conservation laws. Let u_Δ, $\Delta = (\Delta x, \Delta y, \Delta t)$, be the interpolant of degree zero (piecewise constant) associated with the data points $\{u_{i,j}^n\}$. When (13.19) is monotone, one can establish the following convergence theorem:

Theorem 13.5 ([10, 11]) *Suppose that (13.19) is monotone. Then the sequence $\{u_\Delta\}$ built from (13.19) converges in $L^1_{\mathrm{loc}}(Q_T)$ to the unique entropy solution u of (13.16) as $\Delta \to 0$. Furthermore, $\{A(u_\Delta)\}$ converges to $A(u)$ weakly in $H^1(Q_T)$ as $\Delta \to 0$.*

13.3.2 Operator Splitting Methods

There are essentially two ways of constructing methods for solving convection-diffusion problems. One approach attempts to preserve some coupling between the two processes involved (convection and diffusion). The finite difference methods considered in the previous section try to follow this approach. Another approach is to split the convection-diffusion problem into a convection problem and a diffusion problem, which are then solved sequentially to approximate the exact solution of the model. The main attraction of splitting methods lies, of course, in the fact that one can employ the optimal existing methods for each subproblem. The splitting methods presented here are similar to the splitting methods that have been used over the years to simulate multi-phase flow in oil reservoirs. We refer to the lecture notes by Espedal and Karlsen [9] for an overview of this activity and an introduction to operator splitting methods in general. For simplicity of presentation, we restrict ourselves to multi-dimensional Cauchy problems of the form

$$\partial_t u + \nabla_x \cdot f(u) = \Delta_x A(u), \quad u(x,0) = u_0(x), \quad (x,t) \in Q_T = \mathbb{R}^d \times (0,T). \quad (13.20)$$

We emphasize that the numerical solution algorithms and their convergence analysis presented below carry over to more general convection-diffusion equations. To describe this operator splitting more precisely, we need the solution operator taking the initial data $v_0(x)$ to the entropy solution at time t of the hyperbolic problem

$$\partial_t v + \nabla_x \cdot f(v) = 0, \quad v(x,0) = v_0(x). \quad (13.21)$$

We denote this solution operator by $\mathcal{S}(t)$. Similarly, let $\mathcal{H}(t)$ be the solution operator (at time t) associated with the parabolic problem

$$\partial_t w = \Delta_x A(w), \quad w(x,0) = w_0(x). \quad (13.22)$$

Now choose a time step $\Delta t > 0$ and an integer N such that $N\Delta t = T$. Furthermore, let $t_n = n\Delta t$ for $n = 0, \ldots, N$ and $t_{n+1/2} = (n + \frac{1}{2})\Delta t$ for $n = 0, \ldots, N-1$. We then let the operator splitting solution $u_{\Delta t}$ be defined at the discrete times $t = t_n$ by

$$u_{\Delta t}(x, n\Delta t) = \left[\mathcal{H}(\Delta t) \circ \mathcal{S}(\Delta t) \right]^n u_0(x). \quad (13.23)$$

Of course, the ordering of the operators in (13.23) can be changed, and also the so-called Godunov formula (13.23) can be replaced by the more accurate Strang formula. Note that we have only defined $u_{\Delta t}$ at the discrete times t_n. Between two consecutive discrete times, we use a suitable time interpolant (see [13, 14]). Regarding $u_{\Delta t}$ we have:

Lemma 13.1 ([13, 14]) *The following a priori estimates hold: (a)* $\|u_{\Delta t}(\cdot, t)\|_{L^\infty} \leq$ $\|u_0\|_{L^\infty}$, *(b)* $|u_{\Delta t}(\cdot, t)|_{BV} \leq |u_0|_{BV}$, *(c)* $\|u_{\Delta t}(\cdot, t_2) - u_{\Delta t}(\cdot, t_1)\|_{L^1} \leq$ Const $\cdot |t_2 - t_1|^{1/2}$ *for all* $t_1, t_2 \geq 0$.

In view of estimates (a)-(c) in Lemma 13.1, there exists a subsequence $\{\Delta t_j\}$ and a limit function u such that $u_{\Delta t_j} \to u$ in $L^1_{\text{loc}}(Q_T)$ as $j \to \infty$. In addition, one can prove via an energy type argument that this limit satisfies $\nabla_x A(u) \in L^2(Q_T; \mathbb{R}^d)$ (see [13]). Finally, one can prove that $u_{\Delta t}$ satisfies a discrete entropy condition and consequently that the limit u satisfies the entropy condition (see [13]). Summing up, we have:

Theorem 13.6 ([13]) *The operator splitting solution* $u_{\Delta t}$ *converges in* $L^1_{\text{loc}}(Q_T)$ *to the unique entropy solution of the Cauchy problem* (13.20) *as* $\Delta t \to 0$.

So far we have assumed that the operators $\mathcal{S}^f(t)$ and $\mathcal{H}(t)$ determine exact solutions to their respective split problems and that discretization has been performed with respect to time only. In applications, the exact solution operators $\mathcal{S}^f(t)$ and $\mathcal{H}(t)$ are replaced by appropriate numerical approximations which involve discretization also with respect to space. For the split problem (13.21), one can choose from a diversity of methods for hyperbolic conservation laws. For the second split problem (13.22), one can also choose from a large collection of finite difference or element methods. Convergence results for fully discrete splitting methods can be found in, e.g., [13]. For a more complete overview of theoretical results for (fully discrete) operator splitting methods and references to papers dealing with such issues, we refer again to the lecture notes [9].

In what follows, we shall outline a fully discrete splitting method for the first sedimentation model (Problem A in §13.1), see [3] for a different one. This method has previously been employed by Bustos et al. [6] (see also [3]) and will be used in §13.4 below. This method splits the original Problem A into the second order problem

$$\partial_t w = \partial_x^2 A(w), \ (x, t) \in Q_T; \ w(x, 0) = w_0(x), \ z \in \Omega, \tag{13.24}$$

the linear convection problem

$$\partial_t u + q(t)\partial_x u = 0, \ (x, t) \in Q_T; \ u(x, 0) = u_0(x), \ x \in \Omega, \tag{13.25}$$

and the nonlinear hyperbolic IBVP

$$\left.\begin{array}{c} \partial_t v + \partial_x f(v) = 0, \ (x, t) \in Q_T; \ v(x, 0) = v_0(x), \ x \in \Omega; \\ (f(v) - a(v)\partial_x v)(0, t) = 0, \ v(1, t) = \varphi_1(t), \ t \in \mathcal{T}. \end{array}\right\} \tag{13.26}$$

Note that the ordering of the operators in (13.24)–(13.26) is different from the ordering used in (13.23). The splitting (13.24)–(13.26) can be analysed using the techniques of [13] together with an appropriate treatment of the boundary conditions (details will be presented in future work).

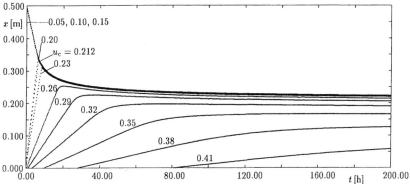

Figure 13.1: Numerical simulation of the settling of a flocculated suspension. The iso-concentration lines correspond to the annotated values.

13.4 APPLICATION TO THE SEDIMENTATION-CONSOLI-DATION MODEL

To illustrate the application to the sedimentation-consolidation model, we employ the splitting (13.24)–(13.26) to simulate the batch settling of an initially homogeneous suspension of concentration $u_0 = 0.18$ in a column of height $L = 0.5\,[\mathrm{m}]$. We use central differences to solve the second order problem (13.24), a first order upwind method to solve the linear convection problem (13.25), and, finally, a variant of Nessyahu and Tadmor's method [16] for the nonlinear convection problem (13.26), see [6] for details. Figure 13.1 shows the numerical solution calculated with $\Delta x/L = 1/400$, $\Delta t/\Delta z = 2000\,[\frac{\mathrm{s}}{\mathrm{m}}]$ and $\alpha = 1.3$, where α is the free parameter in Nessyahu and Tadmor's method [16]. The model functions (corresponding to a suspension of ground calcium carbonate in sea water, see [8]) are

$$f(u) = -1.87\times10^{-4}u(1-u)^{16.4}\,[\tfrac{\mathrm{m}}{\mathrm{s}}], \quad a(u) = \begin{cases} 0, & u \le u_{\mathrm{c}} := 0.212, \\ 0.0206u^6(1-u)^{16.4}\,[\tfrac{\mathrm{m}^2}{\mathrm{s}}], & u > u_{\mathrm{c}}. \end{cases}$$

REFERENCES

[1] R. Bürger, S. Evje and K.H. Karlsen (1999). On strongly degenerate convection-diffusion problems modeling sedimentation-consolidation processes. Applied Mathematics Report, Dept. of Mathematics, University of Bergen, Bergen, Norway.

[2] R. Bürger, S. Evje and K.H. Karlsen (1999). On strongly degenerate convection-diffusion problems in several space dimensions. In preparation.

[3] R. Bürger, S. Evje, K.H. Karlsen and K.-A. Lie (1999). Numerical methods for the simulation of the settling of flocculated suspensions. *Separ. Purif. Technol.*, to appear.

[4] R. Bürger and W.L. Wendland (1998). Existence, uniqueness and stability of generalized solutions of an initial-boundary value problem for a degenerating quasilinear parabolic equation. *J. Math. Anal. Appl.* 218:207–239.

[5] R. Bürger, W.L. Wendland and F. Concha (1999). Model equations for gravitational sedimentation-consolidation processes. *Z. Angew. Math. Mech.*, to appear.

[6] M.C. Bustos, F. Concha, R. Bürger and E.M. Tory, *Sedimentation and Thickening*, Kluwer Academic Publishers, Dordrecht, The Netherlands, to appear.

[7] J. Carrillo (1998). Solutions entropiques de problèmes non-linéaires dégénérés, *C. R. Acad. Sci. Paris Sér. I Math.* 327:155–160.

[8] F. Concha, R. Bürger and P. Garrido (1999). Optimization of thickener performance in the cement industry. In preparation.

[9] M.S. Espedal and K.H. Karlsen, Numerical solution of reservoir flow models based on large time step operator splitting algorithms, in A. Fasano and H. van Duijn (Eds.), Filtration in Porous Media and Industrial Applications, Lecture Notes in Mathematics, Springer Verlag, to appear.

[10] S. Evje and K.H. Karlsen (1999). Second order schemes for degenerate convection-diffusion equations. In preparation.

[11] S. Evje and K.H. Karlsen (1999). Monotone difference approximations of *BV* solutions to degenerate convection-diffusion equations. *SIAM J. Numer. Anal.*, to appear.

[12] S. Evje and K.H. Karlsen (1999). Degenerate convection-diffusion equations and implicit monotone difference schemes. In M. Fey and R. Jeltsch (Eds.), *Hyperbolic problems: theory, numerics, applications*, Int. Ser. of Numer. Math., Birkhäuser Verlag, Basel, 129:285–294.

[13] S. Evje and K.H. Karlsen (1999). Viscous splitting approximation of mixed hyperbolic-parabolic convection-diffusion equations. *Numer. Math.* 83:107–137.

[14] K.H. Karlsen and N.H. Risebro (1997). An operator splitting method for nonlinear convection-diffusion equations. *Numer. Math.* 77:365–382.

[15] S.N. Kružkov (1970). First order quasilinear equations in several independent variables. *Math. USSR Sb.* 10:217–243.

[16] H. Nessyahu and E. Tadmor (1990). Non-oscillatory central differencing for hyperbolic conservation laws *J. Comp. Phys.* 87:408–463.

[17] A.I. Vol'pert (1967). The spaces *BV* and quasilinear equations. *Math. USSR Sb.* 2:225–267.

[18] A.I. Vol'pert and S.I. Hudjaev (1969). Cauchy's problem for degenerate second order parabolic equations, *Math. USSR Sb.* 7:365–387.

[19] Z. Wu (1983). A boundary value problem for quasilinear degenerate parabolic equations, MRC Technical Summary Report #2484, University of Wisconsin.

[20] Z. Wu and J. Yin (1989). Some properties of functions in BV_x and their applications to the uniqueness of solutions for degenerate quasilinear parabolic equations, *Northeastern Math. J.* 5:395–422.

14 SOME EXTENSIONS OF THE LOCAL DISCONTINUOUS GALERKIN METHOD FOR CONVECTION-DIFFUSION EQUATIONS IN MULTIDIMENSIONS

Bernardo Cockburn[a] and Clint Dawson[b]

[a]Department of Mathematics,
University of Minnesota, Minneapolis, MN 55455 U.S.A.

[b]Texas Institute for Computational and Applied Mathematics,
The University of Texas, Austin, TX 78712 U.S.A.

ABSTRACT

The local discontinuous Galerkin method has been developed recently by Cockburn and Shu for convection-dominated convection-diffusion equations. In this paper, we extend the method to multidimensional equations with non-periodic boundary conditions, and with a positive semi-definite diffusion coefficient which may depend on space and time. Stability and *a priori* error estimates are derived.

Key words. discontinuous Galerkin method, convection-diffusion equation

14.1 INTRODUCTION

The local discontinuous Galerkin (LDG) method was developed recently by Cockburn and Shu [11] for convection-dominated convection-diffusion equations. This method is a generalization of the method introduced by Bassi and Rebay [2] for the compressible Navier-Stokes equations which is in turn an extension of the so-called Runge-Kutta discontinuous Galerkin methods for conservation laws developed by Cockburn and Shu in a series of papers [6,8–10,12]; see the review on the development of discontinuous Galerkin methods by Cockburn, Karnidakis and Shu [7]. The LDG method is similar in nature to the upwind-mixed or Godunov-mixed methods developed by Dawson *et al* [13–16]. Both methods can be viewed in an operator splitting context, where a high resolution upwind method is used for advection, combined with some type of finite element method for diffusion. In the case of the LDG method, a discontinuous Galerkin approach is used for diffusion, and in the case of the upwind-mixed method, a mixed finite element method approximates diffusion.

Both of these methods have the nice properties that they are based on conserving mass locally over each element and they approximate sharp fronts accurately and with minimal oscillation; these methods are easily extendible to nonlinear convection-diffusion systems. Moreover, the LDG method can easily be extended to higher-order polynomials, can be defined on any grid including non-conforming or non-matching grids, and also easily allows one to vary the degree of the approximating space from one element to the next.

In [11], the LDG method is described and analyzed for convection-diffusion equations with periodic boundary conditions. It is proven that, for general regular triangulations and approximate solutions using polynomials of degree k in each element, the upper bound of the 'energy norm' is of order h^k for the general convection-diffusion case and of order $h^{k+1/2}$ in the purely hyperbolic case; h stands for the mesh parameter. In the purely parabolic case, for uniform Cartesian grids and approximate solutions using tensor products of polynomials of degree k, the rate of convergence is of order h^{k+1} or h^k for k odd or even, respectively; this result holds for the special choice of numerical fluxes used by Bassi and Rebay [2]. Castillo [5] considered the one-dimensional bounded domain case and identified a special numerical flux for which the improved rate of convergence of order $k + 1$ is obtained for arbitrary meshes when the approximate solution uses polynomials of degree k.

In this paper, we extend the work done in [11] in two ways: (i) we consider non-periodic boundary conditions in a multi-dimensional setting, and (ii) we consider the case in which the diffusion/dispersion tensor depends on (x, t). Thus, we focus on the following standard transport equation,

$$\phi\, c_t + \nabla \cdot (uc - D\nabla c) = \phi\, f, \quad (x, t) \in \Omega \times (0, T], \tag{14.1}$$

where Ω is a bounded domain in $\mathbb{R}^d$, $d = 1$, 2, or 3, and $T > 0$. In porous media applications, see, for example, [3], $c(x, t)$ represents the concentration of some chemical component, $\phi(x)$ is the porosity of the medium and may also include adsorption effects, $u(x, t)$ is the Darcy velocity, $D(x, t)$ is a diffusion/dispersion tensor assumed to be symmetric and at worst positive semi-definite, and $f = f(x, t)$ is a source term. We will assume that positive constants ϕ_*, ϕ^* exist such that

$$\phi_* \leq \phi(x) \leq \phi^*, \tag{14.2}$$

and that the Darcy velocity u satisfies the continuity equation

$$\nabla \cdot u = 0. \tag{14.3}$$

The initial condition for the concentration is the following:

$$c(x, 0) = c^0(x), \quad \text{on } \Omega. \tag{14.4}$$

For boundary conditions, let n denote the unit outward normal to $\Gamma \equiv \partial\Omega$. We write $\Gamma = \overline{\Gamma}_I \cup \overline{\Gamma}_O$, where

$$\Gamma_I = \{x \in \partial\Omega : u \cdot n < 0\}, \tag{14.5}$$

and

$$\Gamma_O = \{x \in \partial\Omega : u \cdot n \geq 0\}. \tag{14.6}$$

On these boundaries we assume that the following conditions hold:

$$(uc - D\nabla c) \cdot n = uc_I \cdot n, \quad \text{on } \Gamma_I, \tag{14.7}$$

where c_I is specified, and

$$(D\nabla c) \cdot n = 0, \quad \text{on } \Gamma_O. \tag{14.8}$$

We assume that the coefficients, initial and boundary data, and domain Ω are sufficiently smooth so that a unique solution c exists for the problem above.

The paper is outlined as follows. In the next section, we define the LDG method in continuous time. In Section 3, we analyze the stability of this method and in Section 4, we derive an *a priori* error estimate. We end in Section 5 with some concluding remarks.

14.2 METHOD FORMULATION

The weak formulation. Let us motivate the weak formulation we will use to define the LDG method. We start by rewriting (14.1) in the following mixed form:

$$\phi\, c_t + \nabla \cdot (uc + z) = \phi\, f, \tag{14.9}$$

$$\tilde{z} = -\nabla c, \tag{14.10}$$

and

$$z = D\tilde{z}. \tag{14.11}$$

and by rewriting the boundary conditions accordingly, that is,

$$(uc + z) \cdot n = (uc_I) \cdot n, \text{ on } \Gamma_I \tag{14.12}$$

and

$$z \cdot n = 0, \text{ on } \Gamma_O. \tag{14.13}$$

Note that the introduction of the auxiliary variables $\tilde{z}$ and z allows us to treat tensors D that are not invertible; this is not possible to do in the standard mixed approach. This notion was introduced in [1] for the mixed finite element method applied to elliptic equations.

Next, to obtain our weak formulation, we simply multiply the above equations by test functions and integrate on each finite element. To describe this procedure, we need to introduce some notation. Let $\{\mathcal{T}_h\}_{h>0}$ denote a family of finite element partitions of Ω such that no element Ω_e crosses the boundaries of Γ_I or Γ_O, where h_e is the element diameter and h the maximal element diameter; each element Ω_e has a Lipschitz boundary $\partial\Omega_e$. For any function w in $W_e = H^1(\Omega_e)$, we denote its trace on $\partial\Omega_e$ by w^-. Finally, we denote the $L^2(\tilde{\Omega})$ inner product by $(\cdot,\cdot)_{\tilde{\Omega}}$, where we omit $\tilde{\Omega}$ if $\tilde{\Omega} = \Omega$. To distinguish integration over domains $\tilde{\Omega} \in \mathbb{R}^{d-1}$, e.g., surfaces or lines, we will use the notation $\langle\cdot,\cdot\rangle_{\tilde{\Omega}}$; we denote by n_e the unit outward normal vector to $\partial\Omega_e$, with $n_e = n$ on Γ. With this notation, and assuming that c, $\tilde{z}$, and z are smooth enough, we obtain the following weak formulation of equations (14.9)-(14.13):

$$(\phi\, c_t, w)_{\Omega_e} - (u\, c + z, \nabla w)_{\Omega_e} + \langle(c\,u + z)\cdot n_e, w^-\rangle_{\partial\Omega_e/\Gamma}$$
$$+ \langle c\,u\cdot n, w^-\rangle_{\partial\Omega_e\cap\Gamma_O} = -\langle c_I\, u\cdot n, w^-\rangle_{\partial\Omega_e\cap\Gamma_I} + (\phi f, w)_{\Omega_e}, \quad w \in W_e, \tag{14.14}$$

$$(\tilde{z}, v)_{\Omega_e} - (c, \nabla\cdot v)_{\Omega_e} + \langle c, v^-\cdot n_e\rangle_{\partial\Omega_e} = 0, \quad v \in (W_e)^d, \tag{14.15}$$

and

$$(z, \tilde{v})_{\Omega_e} = (D\tilde{z}, \tilde{v})_{\Omega_e}, \quad \tilde{v} \in (W_e)^d. \tag{14.16}$$

The LDG method. Let $W_{h,e} \subset W_e$ denote the set of all polynomials of degree at most k_e defined on Ω_e. On Ω_e, we approximate $c(\cdot,t)$ by $C(\cdot,t) \in W_{h,e}$, $z(\cdot,t)$ by $Z(\cdot,t) \in (W_{h,e})^d$, and $\tilde{z}(\cdot,t)$ by $\tilde{Z}(\cdot,t) \in (W_{h,e})^d$. To define C, Z, and $\tilde{Z}$, we simply have to use the weak formulation (14.14)-(14.16). However, we see that we have terms involving c and z on $\partial\Omega_e$. Since C and Z are discontinuous across these boundaries, we must define how we approximate these terms. For $x \in \partial\Omega_e$ we define

$$w^-(x) = \lim_{s\to 0^-} w(x + s n_e), \qquad w^+(x) = \lim_{s\to 0^+} w(x + s n_e), \qquad \overline{w} = \frac{1}{2}(w^+ + w^-).$$

We approximate the value of C on $\partial\Omega_e/\Gamma$ by the so-called "upwind value" of C which is defined as follows:

$$C^u = \begin{cases} C^-, & u\cdot n_e \geq 0, \\ C^+, & u\cdot n_e < 0. \end{cases} \tag{14.17}$$

The value of Z on $\partial\Omega_e/\Gamma$ is approximated by $\overline{Z}$. Finally, the value of C on $\partial\Omega_e \cap \Gamma_O$ is simply approximated by C^-. This takes care of the integrals of the equation (14.14). The approximation of the value of C on $\partial\Omega_e$ in equation (14.15) is taken to be equal to $\overline{C}$ on $\partial\Omega_e/\Gamma$ and equal to C^- on $\partial\Omega_e \cap \Gamma$.

We are now ready to formulate the LDG method. At $t = 0$ we define $C(\cdot,0) \equiv C^0 \in W_{h,e}$ by

$$(\phi(C^0 - c^0), w)_{\Omega_e} = 0, \quad w \in W_{h,e}. \tag{14.18}$$

For each $t > 0$, $C(\cdot,t)$, $Z(\cdot,t)$ and $\tilde{Z}(\cdot,t)$ are determined by the following equations:

$$(\phi\, C_t, w)_{\Omega_e} - (uC + Z, \nabla w)_{\Omega_e} + \langle(C^u\, u + \overline{Z})\cdot n_e, w^-\rangle_{\partial\Omega_e/\Gamma}$$
$$+ \langle C^-u\cdot n, w^-\rangle_{\partial\Omega_e\cap\Gamma_O} = -\langle c_I u\cdot n, w^-\rangle_{\partial\Omega_e\cap\Gamma_I} + (\phi f, w)_{\Omega_e}, \quad w \in W_{h,e}, \tag{14.19}$$

$$(\tilde{Z}, v)_{\Omega_e} - (C, \nabla \cdot v)_{\Omega_e} + \langle \overline{C}, v^- \cdot n_e \rangle_{\partial \Omega_e / \Gamma} + \langle C^-, v^- \cdot n \rangle_{\partial \Omega_e \cap \Gamma} = 0 \quad v \in (W_{h,e})^d, \quad (14.20)$$

and

$$(D\tilde{Z}, \tilde{v})_{\Omega_e} - (Z, \tilde{v})_{\Omega_e} = 0, \quad \tilde{v} \in (W_{h,e})^d. \tag{14.21}$$

Note that by (14.20), $\tilde{Z}$ can be eliminated *locally* in terms of C, and by (14.21), Z can be expressed *locally* in terms of $\tilde{Z}$. These relations can be substituted into (14.19), giving an equation for C alone. The stencil for any element Ω_e involves the neighbors of Ω_e, that is, those elements which share an edge with Ω_e, and the neighbors of the neighbors. Thus, on a conforming triangular mesh, for example, the unknowns in Ω_e are related to unknowns in at most nine neighboring elements.

14.3 STABILITY ANALYSIS

To develop an *a priori* error estimate for the scheme derived above, we need to examine its stability. The stability proof below also demonstrates the existence and uniqueness of the numerical solution. We proceed in three steps.

Step 1. First, we rewrite the LDG method (14.18)–(14.21) in compact form. To do that, we add all the left hand sides of equations (14.19)–(14.21), add over all the elements of the partition of Ω, integrate in time from 0 to T and call the result $\mathcal{B}(C, Z, \tilde{Z}; w, v, \tilde{v})$. Then equations (14.19)–(14.21) can be rewritten in compact form as follows:

$$\mathcal{B}(C, Z, \tilde{Z}; w, v, \tilde{v}) = - \int_0^T \langle c_I\, u \cdot n_e, w^- \rangle_{\Gamma_I}\, dt + \int_0^T (\phi\, f, w)\, dt, \tag{14.22}$$

and they have to hold for each test function $(w, v, \tilde{v})$ in $\mathcal{C}^0(0, T; W_h)$, where, of course, $W_h = \{(w, v, \tilde{v}) : \text{On each element } \Omega_e \in \mathcal{T}_h, (w, v, \tilde{v}) \in W_{h,e} \times (W_{h,e})^d \times (W_{h,e})^d\}$.

Now, to obtain our stability result we simply have to set $w = C$, $v = Z$, and $\tilde{v} = \tilde{Z}$ and perform some simple manipulations.

Step 2. By construction, we have that $B(C, Z, \tilde{Z}; C, Z, \tilde{Z}) = \Theta_1 + \Theta_2 + \Theta_3$, where

$$\Theta_1 = \int_0^T \sum_e \left[(\phi\, C_t, C)_{\Omega_e} + (D\tilde{Z}, \tilde{Z})_{\Omega_e} \right] dt,$$

$$\Theta_2 = \int_0^T \sum_e \left[-(C\, u, \nabla C)_{\Omega_e} + \langle C^u\, u \cdot n_e, C^- \rangle_{\partial \Omega_e / \Gamma} + \langle C^-\, u \cdot n_e, C^- \rangle_{\partial \Omega_e \cap \Gamma_O} \right] dt,$$

and

$$\Theta_3 = \int_0^T \sum_e \left[-(Z, \nabla C)_{\Omega_e} - (C, \nabla \cdot Z)_{\Omega_e} + \langle \overline{Z} \cdot n_e, C^- \rangle_{\partial \Omega_e / \Gamma} \right.$$
$$\left. + \langle \overline{C}, Z^- \cdot n_e \rangle_{\partial \Omega_e / \Gamma} + \langle C^-, Z^- \cdot n_e \rangle_{\partial \Omega_e \cap \Gamma} \right] dt.$$

It is very simple to realize that

$$\Theta_1 = \frac{1}{2} \| \phi^{1/2}\, C(T) \|^2 + \int_0^T \| D^{1/2}\, \tilde{Z} \|^2\, dt - \frac{1}{2} \| \phi^{1/2}\, C(0) \|^2 \tag{14.23}$$

where $\| f \|^2 = \int_\Omega f^2(x)\, dx$.

To deal with Θ_2 is not that simple. Integrating by parts and taking into account the continuity equation (14.3), we get

$$
\begin{aligned}
\Theta_2 &= \int_0^T \sum_e \Bigl[-\frac{1}{2}\langle C^- \, u \cdot n_e, C^- \rangle_{\partial\Omega_e/\Gamma} - \frac{1}{2}\langle C^- \, u \cdot n_e, C^- \rangle_{\partial\Omega\cap\Gamma_I} \\
&\qquad\qquad + \langle C^u \, u \cdot n_e, C^- \rangle_{\partial\Omega_e/\Gamma} + \frac{1}{2}\langle C^- \, u \cdot n_e, C^- \rangle_{\partial\Omega_e\cap\Gamma_O} \Bigr]\, dt \\
&= \int_0^T \Bigl[\frac{1}{2}\langle |\, u \cdot n\,|, (C^-)^2 \rangle_\Gamma + \sum_e \langle\, (C^u - \frac{1}{2}C^-)\, u \cdot n_e, C^- \rangle_{\partial\Omega_e/\Gamma} \Bigr]\, dt.
\end{aligned}
$$

Next, we would like to rewrite the sum over the elements e in terms of the sum over the set of *interior* edges $\{\, \gamma_l \,\}_l$. To do that, we need to introduce some notation. Let γ_l be an interior edge belonging to $\partial\Omega_e$ then, we set

$$
[F] = (-F^+ + F^-)\, n_e.
$$

Note that this quantity is well defined independently of the set Ω_e we take as a reference. We can now write that

$$
\begin{aligned}
\sum_e \langle\, (C^u - \frac{1}{2}C^-)\, u \cdot n_e, C^- \rangle_{\partial\Omega_e/\Gamma} &= \sum_l \langle u \cdot (C^u\,[C] - \frac{1}{2}[C^2]), 1 \rangle_{\gamma_l} \\
&= \sum_l \langle u \cdot (C^u\,[C] - \overline{C}\,[C]), 1 \rangle_{\gamma_l} \\
&= \sum_l \langle u \cdot [C], C^u - \overline{C} \rangle_{\gamma_l}, \\
&= \frac{1}{2}\sum_l \langle |\, u \cdot n_l\,|, [C]^2 \rangle_{\gamma_l},
\end{aligned}
$$

since

$$
(C^u - \overline{C})\, u \cdot [C] = \frac{1}{2}[C]^2 \, |\, u \cdot n_l\,|,
$$

where n_l denotes any unit normal to γ_l.

Hence, we get

$$
\Theta_2 = \frac{1}{2}\int_0^T \Bigl[\langle C^- \, |\, u \cdot n\,|, C^- \rangle_\Gamma + \sum_l \langle |\, u \cdot n_l\,|, [C]^2 \rangle_{\gamma_l} \Bigr]\, dt. \qquad (14.24)
$$

Let us now treat Θ_3. Integrating by parts and taking into account the definition of $\overline{Z}$ and $\overline{C}$, we get

$$
\begin{aligned}
\Theta_3 &= \int_0^T \sum_e \Bigl[\langle \overline{Z} \cdot n_e, C^- \rangle_{\partial\Omega_e/\Gamma} + \langle \overline{C}, Z^- \cdot n_e \rangle_{\partial\Omega_e/\Gamma} - \langle C^-, Z^- \cdot n_e \rangle_{\partial\Omega_e/\Gamma} \Bigr]\, dt \\
&= \int_0^T \sum_e \Bigl[\langle Z^+ \cdot n_e, C^- \rangle_{\partial\Omega_e/\Gamma} + \langle C^+, Z^- \cdot n_e \rangle_{\partial\Omega_e/\Gamma} \Bigr]\, dt \\
&= 0. \qquad\qquad\qquad\qquad\qquad\qquad\qquad\qquad\qquad\qquad\qquad\qquad\qquad (14.25)
\end{aligned}
$$

Since $B(C, Z, \tilde{Z}; C, Z, \tilde{Z}) = \Theta_1 + \Theta_2 + \Theta_3$, using the identities (14.23),(14.24), and(14.25), we get that

$$
\begin{aligned}
B(C, Z, \tilde{Z}; C, Z, \tilde{Z}) &= \frac{1}{2} \| \phi^{1/2} C(T) \|^2 + \int_0^T \| D^{1/2} \tilde{Z} \|^2 \, dt - \frac{1}{2} \| \phi^{1/2} C(0) \|^2 \\
&\quad + \frac{1}{2} \int_0^T \left[\langle | u \cdot n |, (C^-)^2 \rangle_\Gamma + \sum_l \langle | u \cdot n_l |, [C]^2 \rangle_{\gamma_l} \right] dt.(14.26)
\end{aligned}
$$

Step 3. Inserting the above identity into (14.22) with $(w, v, \tilde{v}) = (C, Z, \tilde{Z})$ we get, after some simple manipulations,

$$
\| (C, \tilde{Z}) \|^2 \leq \| \phi^{1/2} C(0) \|^2 + 2 \int_0^T \langle | u \cdot n |, (c_I)^2 \rangle_{\Gamma_I} \, dt + 2 \int_0^T \| \phi^{1/2} f \| \, \| \phi^{1/2} C \| \, dt.
$$

where

$$
\begin{aligned}
\| (C, \tilde{Z}) \|^2 &\equiv \| \phi^{1/2} C(T) \|^2 + \int_0^T \| D^{1/2} \tilde{Z} \|^2 \, dt \\
&\quad + \frac{1}{2} \int_0^T \left[\langle | u \cdot n |, (C^-)^2 \rangle_\Gamma + \sum_l \langle | u \cdot n_l |, [C]^2 \rangle_{\gamma_l} \right] dt. \tag{14.27}
\end{aligned}
$$

Finally, since (14.18) implies that

$$
\| \phi^{1/2} C(0) \| \leq \| \phi^{1/2} c^0 \|,
$$

we obtain our stability result stated below by a simple application of the following Lemma.

Lemma 14.1 *Suppose that for all $T > 0$ we have*

$$
\chi^2(T) + R(T) \leq A(T) + 2 \int_0^T B(t) \chi(t) \, dt,
$$

where R, A, and B are nonnegative functions. Then

$$
\sqrt{\chi^2(T) + R(T)} \leq \sup_{0 \leq t \leq T} A^{1/2}(t) + \int_0^T B(t) \, dt.
$$

Proposition 14.1 *[Stability of the LDG method] The scheme (14.18)-(14.21) satisfies*

$$
\| (C, \tilde{Z}) \| \leq \left\{ \| \phi^{1/2} c^0 \|^2 + 2 \int_0^T \| c_I | u \cdot n |^{1/2} \|_{\Gamma_I}^2 \, dt \right\}^{1/2} + \int_0^T \| \phi^{1/2} f \|_{\Omega_e} \, dt.
$$

Note that from the above result, the existence and uniqueness of the approximate solution (C, Z) follows. Since the problem is linear and finite dimensional, existence and uniqueness are equivalent. If we assume two solutions (C_1, Z_1) and (C_2, Z_2) exist and subtract them, then the difference satisfies the inequality above with $c_I = f = c^0 = 0$. Thus $(C_1, Z_1) = (C_2, Z_2)$.

14.4   AN *A PRIORI* ERROR ESTIMATE

We now develop an *a priori* error estimate for the scheme. As in the previous section, we proceed in several steps.

Step 1. We start by introducing some notation. Let $\Pi c(\cdot, t) \in W_{h,e}$ denote the $L^2(\Omega_e)$ projection of $c(\cdot, t)$:

$$(\Pi c - c, w)_{\Omega_e} = 0, \quad w \in W_{h,e}. \tag{14.28}$$

Similarly, let $\Pi \tilde{z}(\cdot, t) \in (W_{h,e})^d$ denote the L^2 projection of $\tilde{z}$ and $\Pi z(\cdot, t) \in (W_{h,e})^d$ the L^2 projection of z. Define Πc^u analogous to C^u. Set $\psi_c = C - \Pi c$, $\tilde{\psi}_z = \tilde{Z} - \Pi \tilde{z}$, $\psi_z = Z - \pi z$, $\theta_c = c - \Pi c$, $\tilde{\theta}_z = \tilde{z} - \Pi \tilde{z}$ and $\theta_z = z - \Pi z$. Our goal is to obtain an estimate of $(\psi_c, \psi_z, \tilde{\psi}_z)$ in terms of $(\theta_c, \theta_z, \tilde{\theta}_z)$.

To do that, we note that by the construction of the form B (14.22), we have

$$B(C, Z, \tilde{Z}; w, v, \tilde{v}) = B(c, z, \tilde{z}; w, v, \tilde{v}),$$

where $(c, z, \tilde{z})$ is the exact solution. This implies that

$$B(\psi_c, \psi_z, \tilde{\psi}_z; w, v, \tilde{v}) = B(\theta_c, \theta_z, \tilde{\theta}_z; w, v, \tilde{v}),$$

and, for $(w, v, \tilde{v}) = (\psi_c, \psi_z, \tilde{\psi}_z)$,

$$B(\psi_c, \psi_z, \tilde{\psi}_z; \psi_c, \psi_z, \tilde{\psi}_z) = B(\theta_c, \theta_z, \tilde{\theta}_z; \psi_c, \psi_z, \tilde{\psi}_z). \tag{14.29}$$

Since the left hand side of this equality is given by (14.26), we only need to obtain an upper bound of its right hand side to obtain the desired error estimate.

Step 2. Let us get the estimate of the right hand side of (14.29). By the definition of the form B, we get

$$
\begin{aligned}
& B(\theta_c, \theta_z, \tilde{\theta}_z; \psi_c, \psi_z, \tilde{\psi}_z) \\
& = \int_0^T \Bigg\{ \sum_e \Big[(\phi\,(\theta_c)_t, \psi_c)_{\Omega_e} - (u\theta_c, \nabla\psi_c)_{\Omega_e} - (D\tilde{\theta}_z, \tilde{\psi}_z)_{\Omega_e} \Big] \\
& \quad - \sum_l \Big[\langle (c - \Pi c^u) u \cdot n_l, [\psi_c] \rangle_{\gamma_l} + \langle (z - \overline{\Pi z}) \cdot n_l, [\psi_c] \rangle_{\gamma_l} + \langle (c - \overline{\Pi c}, [\psi_z] \cdot n_l) \rangle_{\gamma_l} \Big] \\
& \quad + \langle (c - \Pi c^-) u \cdot n, \psi_c^- \rangle_{\Gamma_O} + \langle c - \Pi c^-, \psi_z^- \cdot n \rangle_{\Gamma} \Bigg\}. \\
& = T_1 + \ldots + T_8. \tag{14.30}
\end{aligned}
$$

We now examine the terms T_1 through T_8. In the following, K will denote a generic positive constant and $\Pi_0 g$ will denote the projection of $g \in L^2(\Omega)$ into piecewise constants, that is,

$$\sum_e \int_{\Omega_e} (g - \Pi_0 g)\,dx = 0. \tag{14.31}$$

Consider

$$T_1 = \int_0^T \sum_e (\phi\,(\theta_c)_t, \psi_c)_{\Omega_e}.$$

Note that if ϕ were constant, this term would be zero, since Πc is the L^2 projection of c into W_h. Thus

$$
\begin{aligned}
(\phi\,(\theta_c)_t, \psi_c)_{\Omega_e} &= ((\phi - \Pi_0\phi)(\theta_c)_t, \psi_c)_{\Omega_e} \\
&= ((\phi - \Pi_0\phi)\phi^{-1/2}(\theta_c)_t, \phi^{1/2}\,\psi_c)_{\Omega_e} \\
&\leq (\phi_*)^{-1/2}\|\phi - \Pi_0\phi\|_{L^\infty(\Omega_e)}\|\,(\theta_c)_t\|_{\Omega_e}\|\phi^{1/2}\,\psi_c\|_{\Omega_e} \\
&\leq (\phi_*)^{-1/2}\,h_e\|\phi\|_{W^1_\infty(\Omega)}\|\,(\theta_c)_t\|_{\Omega_e}\,\|\phi^{1/2}\,\psi_c\|_{\Omega_e},
\end{aligned}
$$

for $\phi \in W^1_\infty(\Omega)$. Hence, after a simple application of Cauchy-Schwarz inequality, we get

$$\int_0^T \sum_e (\phi\,(\theta_c)_t, \psi_c)_{\Omega_e} dt \leq K_1 \int_0^T \left\{ \sum_e h_e^2\,\|(\theta_c)_t\|_{\Omega_e}^2 \right\}^{1/2} \|\phi^{1/2}\,\psi_c\|\,dt, \qquad (14.32)$$

where

$$K_1 = (\phi_*)^{-1/2}\|\phi\|_{W^1_\infty(\Omega)}. \qquad (14.33)$$

Next, consider

$$
\begin{aligned}
T_2 &= -\int_0^T \sum_e (u\theta_c, \nabla\psi_c)_{\Omega_e}\,dt \\
&= \int_0^T \sum_e ((u - \Pi_0 u)\theta_c, \nabla\psi_c)_{\Omega_e}\,dt,
\end{aligned}
$$

since each component of $\nabla\psi_c \in W_h$, where $\Pi_0 u$ represents the projection of each component of u into piecewise constants. Thus, for $u \in (W^1_\infty(\Omega))^d$,

$$T_2 \leq \|u\|_{(W^1_\infty(\Omega))^d} \int_0^T \sum_e h_e\|\theta_c\|_{\Omega_e}\,\|\nabla\psi_c\|_{\Omega_e}\,dt.$$

Assuming h_e is sufficiently small, and using an inverse estimate ([4],Lemma 4.5.3), we find

$$T_2 \leq K_i\,\|u\|_{(W^1_\infty(\Omega))^d} \int_0^T \sum_e \|\theta_c\|_{\Omega_e}\,\|\psi_c\|_{\Omega_e}\,dt.$$

Finally, using (14.2), we get

$$T_2 \leq K_2 \int_0^T \left\{ \sum_e \|\theta_c\|_{\Omega_e}^2\,dt \right\}^{1/2} \|\phi^{1/2}\psi_c\|\,dt \qquad (14.34)$$

where

$$K_2 = K_i \|u\|_{(W^1_\infty)^d} (\phi_*)^{-1/2}. \tag{14.35}$$

The estimates of T_3, T_4, and T_7 are straightforward:

$$
\begin{aligned}
T_3 &= -\int_0^T \sum_e (D\tilde{\theta}_z, \tilde{\psi}_z)_{\Omega_e} \, dt \\
&\leq 2 \int_0^T \sum_e \|D^{1/2}\tilde{\theta}_z\|^2_{\Omega_e} \, dt + \frac{1}{8} \int_0^T \sum_e \|D^{1/2}\tilde{\psi}_z\|^2_{\Omega_e} \, dt, \tag{14.36}
\end{aligned}
$$

$$
\begin{aligned}
T_4 &= -\int_0^T \sum_l \langle (u \cdot n_l)(c - \Pi c^u), [\psi_c] \rangle_{\gamma_l} \, dt \\
&\leq \int_0^T \sum_l \| \, |u \cdot n_l|^{1/2}(c - \Pi c^u)\|^2_{\gamma_l} \, dt + \frac{1}{4} \int_0^T \sum_l \| \, |u \cdot n_l|^{1/2}[\psi_c]\|^2_{\gamma_l} \, dt, \tag{14.37}
\end{aligned}
$$

and

$$
\begin{aligned}
T_7 &= \int_0^T \langle (c - \Pi c^-)u \cdot n, \psi_c^- \rangle_{\Gamma_O} \, dt \\
&\leq \| \, |u \cdot n|^{1/2}(c - \Pi c^-)\|^2_{\Gamma_O} + \frac{1}{4} \int_0^T \| \, |u \cdot n|^{1/2}\psi_c^-\|^2_{\Gamma_O} \, dt. \tag{14.38}
\end{aligned}
$$

The estimates of the remaining terms are more delicate. Let us start by considering T_5:

$$
\begin{aligned}
T_5 &= \int_0^T \sum_l \langle (z - \overline{\Pi z}) \cdot n_l, [\psi_c] \rangle_{\gamma_l} \, dt \\
&= \int_0^T \left\{ \sum_l \rho_l \|(z - \overline{\Pi z})\|^2_{\gamma_l} \right\}^{1/2} \left\{ \sum_l \rho_l^{-1} \|[\psi_c]\|^2_{\gamma_l} \right\}^{1/2} \, dt, \tag{14.39}
\end{aligned}
$$

where ρ_l is a positive number to be determined. Let $\{\Omega_{e_l}\}$ denote the set of elements whose boundaries intersect γ_l. Since Ω_e has a Lipschitz boundary, by ([4],Theorem 1.6.6),

$$
\begin{aligned}
\|[\psi_c]\|_{\gamma_l} &\leq K_{tr} \sum_{e_l} \|\psi_c\|^{1/2}_{L^2(\Omega_{e_l})} \|\psi_c\|^{1/2}_{H^1(\Omega_{e_l})} \\
&\leq K_{tr} K_i \sum_{e_l} h_{e_l}^{-1/2} \|\psi_c\|_{L^2(\Omega_{e_l})} \\
&\leq K_{tr} K_i h_l^{-1/2} \sum_{e_l} \|\psi_c\|_{L^2(\Omega_{e_l})} \\
&\leq K_{tr} K_i h_l^{-1/2} M_l^{1/2} \left\{ \sum_{e_l} \|\psi_c\|^2_{L^2(\Omega_{e_l})} \right\}^{1/2},
\end{aligned}
$$

$$\tag{14.40}$$

where $h_l = \min_{e_l} h_{e_l}$ and M_l is the maximum number of elements whose boundary intersects γ_l (in a conforming mesh, $M_l = 2$). Hence,

$$\sum_l \rho_l^{-1} \|\psi_c\|_{L^2(\Omega_{e_l})}^2 \leq \sum_l \rho_l^{-1} K_{tr}^2 K_i^2 h_l^{-1} M_l \sum_{e_l} \|\psi_c\|_{L^2(\Omega_{e_l})}^2$$

$$\leq \sum_l \sum_{e_l} \|\psi_c\|_{L^2(\Omega_{e_l})}^2, \tag{14.41}$$

if we take $\rho_l^{-1} K_{tr}^2 K_i^2 h_l^{-1} M_l = 1$. Finally,

$$\sum_l \rho_l^{-1} \|\psi_c\|_{L^2(\Omega_{e_l})}^2 \leq N \|\psi_c\|^2 \leq N \phi_*^{-1} \|\phi^{1/2} \psi_c\|^2, \tag{14.42}$$

where N is the maximum number of internal edge segments that any element intersects. Inserting this inequality in the bound for T_5, we get

$$T_5 \leq K_5 \int_0^T \left\{ \sum_l h_l^{-1} \|(z - \overline{\Pi z})\|_{\gamma_l}^2 \right\}^{1/2} \|\phi^{1/2} \psi_c\| \, dt, \tag{14.43}$$

where

$$K_5 = \left\{ K_{tr} K_i N \phi_*^{-1} \sup_l M_l \right\}^{1/2}.$$

In a similar manner,

$$T_6 = \int_0^T \sum_l \langle c - \overline{\Pi c}, [\psi_z] \cdot n_l \rangle_{\gamma_l} \, dt$$

$$\leq K_6 \int_0^T \left\{ \sum_l h_l^{-1} \|(c - \overline{\Pi c})\|_{\gamma_l}^2 \right\}^{1/2} \|\psi_z\| \, dt, \tag{14.44}$$

where

$$K_6 = \left\{ K_{tr} K_i N \sup_l M_l \right\}^{1/2}.$$

By (14.16) and (14.21),

$$\|\psi_z\|_{\Omega_e}^2 = (D\tilde{\psi}_z, \psi_z)_{\Omega_e} - (D\tilde{\theta}_z, \psi_z)_{\Omega_e},$$

and so,

$$\|\psi_z\|_{\Omega_e} \leq K_D \|D^{1/2}\tilde{\psi}_z\|_{\Omega_e} + \|D\tilde{\theta}_z\|_{\Omega_e},$$

Substituting into (14.44) and performing a few simple manipulations, we get

$$T_6 \leq K_7 \int_0^T \sum_l h_l^{-1} \|(c - \overline{\Pi c})\|_{\gamma_l}^2 \, dt + \frac{1}{8} \int_0^T \|D^{1/2}\tilde{\psi}_z\|^2 \, dt$$

$$+ K_6 \int_0^T \left\{ \sum_l h_l^{-1} \|(c - \overline{\Pi c})\|_{\gamma_l}^2 \right\}^{1/2} \left\{ \sum_e \|D\tilde{\theta}_z\|_{\Omega_e}^2 \right\}^{1/2} \, dt, \tag{14.45}$$

where

$$K_7 = 2 K_6^2 K_D^2.$$

Finally, we treat the last term

$$T_8 = \int_0^T \langle c - \Pi c^-, \psi_z^- \cdot n \rangle_\Gamma \, dt \qquad (14.46)$$

in a similar way. Let $\{\Omega_{e_\Gamma}\}$ denote the set of elements whose boundaries intersect Γ. Then we obtain

$$
T_8 \;\leq\; K_7 \int_0^T \sum_{e_\Gamma} h_{e_\Gamma}^{-1} \|(c - \overline{\Pi c})\|_{e_\Gamma \cap \Gamma}^2 \, dt + \frac{1}{8} \int_0^T \| D^{1/2} \tilde{\psi}_z \|^2 \, dt
$$

$$
+ K_6 \int_0^T \left\{ \sum_{e_\Gamma} h_{e_\Gamma}^{-1} \|(c - \overline{\Pi c})\|_{e_\Gamma \cap \Gamma}^2 \right\}^{1/2} \left\{ \sum_{e_\Gamma} \| D\tilde{\theta}_z \|_{\Omega_{e_\Gamma}}^2 \right\}^{1/2} dt, \quad (14.47)
$$

where $\{\Omega_{e_\Gamma}\}$ denotes the set of elements whose boundaries intersect Γ.

Thus, substituting the estimates (14.32), (14.34), (14.36), (14.37), (14.38), (14.43), (14.45) and (14.47), in the expression for $B(\theta_c, \theta_z, \tilde{\theta}_z; \psi_c, \psi_z, \tilde{\psi}_z)$, (14.30), we find

$$
B(\theta_c, \theta_z, \tilde{\theta}_z; \psi_c, \psi_z, \tilde{\psi}_z) \leq T_1 + T_2 + \int_0^T T_3 \, \| \phi^{1/2} \, \psi_c \|^2 \, dt,
$$

where

$$
T_1 \;=\; 2 \int_0^T \sum_e \| D^{1/2} \tilde{\theta}_z \|_{\Omega_e}^2 \, dt
$$

$$
+ \int_0^T \sum_l \| \, |u \cdot n_l|^{1/2} (c - \Pi c^u) \|_{\gamma_l}^2 \, dt + \int_0^T \| \, |u \cdot n|^{1/2} (c - \Pi c^-) \|_{\Gamma_o}^2 \, dt
$$

$$
+ K_6 \int_0^T \left\{ \sum_l h_l^{-1} \| c - \overline{\Pi c} \|_{\gamma_l}^2 \right\}^{1/2} \left\{ \sum_e \| D\tilde{\theta}_z \|_{\Omega_e}^2 \right\}^{1/2} dt
$$

$$
+ K_6 \int_0^T \left\{ \sum_{e_\Gamma} h_{e_\Gamma}^{-1} \| c - \overline{\Pi c} \|_{e_\Gamma \cap \Gamma}^2 \right\}^{1/2} \left\{ \sum_{e_\Gamma} \| D\tilde{\theta}_z \|_{\Omega_{e_\Gamma}}^2 \right\}^{1/2} dt, \quad (14.48)
$$

$$
T_2 \;=\; \frac{3}{8} \int_0^T \sum_e \| D^{1/2} \tilde{\psi}_z \|_{\Omega_e}^2 \, dt
$$

$$
+ \frac{1}{4} \int_0^T \sum_l \| \, |u \cdot n_l|^{1/2} [\psi_c] \|_{\gamma_l}^2 \, dt + \frac{1}{4} \int_0^T \| \, |u \cdot n|^{1/2} [\psi_c] \|_{\Gamma_o}^2 \, dt,
$$

and

$$
T_3 \;=\; K_1 \left\{ \sum_e h_e^2 \|(\theta_c)_t\|_{\Omega_e}^2 \right\}^{1/2} + K_2 \left\{ \sum_e \|\theta_c\|_{\Omega_e}^2 \right\}^{1/2} + K_5 \left\{ \sum_l h_l^{-2} \| z - \overline{\Pi z} \|_{\gamma_l}^2 \right\}^{1/2} \quad (14.49)
$$

Step 3. Now, we insert the above estimates in the identity (14.29) and use the identity (14.26) to obtain, after a few algebraic manipulations,

$$\| (\psi_c, \tilde{\psi}_z) \|^2 \;\leq\; A(T) + 2 \int_0^T B(t) \, \| \phi^{1/2} \psi_c \| \, dt,$$

where $\| (\cdot, \cdot) \|$ is defined by (14.27), $B = \mathcal{T}_3$, see (14.49), and $A(T) = \| \phi^{1/2} \psi_c(0) \|^2 + 2\mathcal{T}_1$, see (14.48). Now, a simple application of Lemma 14.1 gives the following result.

Theorem 14.1 *[First error estimate]. The scheme (14.18)-(14.21) satisfies the following error estimate*

$$\| (\psi_c, \tilde{\psi}_z) \| \leq A^{1/2}(T) + \int_0^T B(t) \, dt.$$

From this result, by the triangle inequality and some simple approximation results, we get the following estimate.

Theorem 14.2 *[Second error estimate]. The scheme (14.18)-(14.21) satisfies the following error estimate*

$$\| (c - C, \tilde{z} - \tilde{Z}) \| \leq \| (\theta_c, \tilde{\theta}_z) \| + A^{1/2}(T) + \int_0^T B(t) \, dt \leq K \, h^k,$$

if the exact solution is sufficiently smooth, where $k = \min_e k_e$.

14.5   CONCLUDING REMARKS

The stability estimate in Proposition 14.1 can be easily be obtained for nonlinear convection-diffusion equations. The error estimates in Theorems 14.1 and 14.2 can also be obtained for numerical fluxes more general than the ones presented in this paper, for more general boundary conditions, and for operators Π that are not necessarily the L^2-projection operators.

REFERENCES

[1] T. Arbogast, M.F. Wheeler and I. Yotov (1997). Mixed finite elements for elliptic problems with tensor coefficients as cell-centered finite differences. *SIAM J. Num. Anal.*, 34:828–852.

[2] F. Bassi and S. Rebay (1997). A high-order accurate discontinuous finite element method for the numerical solution of the compressible Navier-Stokes equations. *J. Comput. Phys.*, 131:267–279.

[3] J. Bear (1972). *Dynamics of Fluids in Porous Media*. Dover, New York.

[4] S. Brenner and L. R. Scott (1994). *The Mathematical Theory of Finite Element Methods*. Springer Verlag, New York.

[5] P. Castillo (to appear). An optimal error estimate for the local discontinuous Galerkin method. In B. Cockburn, G.E. Karniadakis and C.-W. Shu, editors, *First International Symposium on Discontinuous Galerkin Methods, Lecture Notes in Computational Science and Engineering*. Springer-Verlag, Berlin.

[6] B. Cockburn, S. Hou and C. W. Shu (1990). TVB Runge-Kutta local projection discontinuous galerkin finite element method for conservations laws IV: The multidimensional case. *Math. Comp.*, 54:545–581.

[7] B. Cockburn, G. E. Karniadakis and C.-W. Shu (to appear). The development of discontinuous Galerkin methods. In B. Cockburn, G.E. Karniadakis and C.-W. Shu, editors, *First International Symposium on Discontinuous Galerkin Methods, Lecture Notes in Computational Science and Engineering*. Springer-Verlag, Berlin.

[8] B. Cockburn, S. Y. Lin and C. W. Shu (1989). TVB Runge-Kutta local projection discontinuous Galerkin finite element method for conservations laws III: One dimensional systems. *J. Comput. Phys.*, 84:90–113.

[9] B. Cockburn and C. W. Shu (1989). TVB Runge-Kutta local projection discontinuous Galerkin finite element method for scalar conservations laws II: General framework. *Math. Comp.*, 52:411–435.

[10] B. Cockburn and C. W. Shu (1991). The Runge-Kutta local projection P^1-discontinuous Galerkin method for scalar conservations laws. M^2AN, 25:337–361.

[11] B. Cockburn and C. W. Shu (1998). The local discontinuous Galerkin method for time dependent convection-diffusion systems. *SIAM J. Numer. Anal.*, 35:2440–2463.

[12] B. Cockburn and C. W. Shu (1998). TVB Runge-Kutta discontinuous Galerkin finite element method for conservations laws V: Multidimensional systems. *J. Comput. Physics*, 141:199–224.

[13] C. Dawson and V. Aizinger. Upwind mixed methods for transport equations. Computational Geosciences, to appear.

[14] C. N. Dawson (1990). Godunov-mixed methods for immiscible displacement. *Int. J. Numer. Meth. Fluids*, 11:835–847.

[15] C. N. Dawson (1991). Godunov-mixed methods for advective flow problems in one space dimension. *SIAM J. Num. Anal.*, 28:1282–1309.

[16] C. N. Dawson (1993). Godunov-mixed methods for advection-diffusion equations in multidimensions. *SIAM J. Num. Anal.*, 30:1315–1332.

15 SCIENTIFIC COMPUTING TOOLS FOR 3D MAGNETIC FIELD PROBLEMS

Michael Kuhn, Ulrich Langer and Joachim Schöberl

SFB "Numerical and Symbolic Scientific Computing",
Johannes Kepler University Linz, A–4040 Linz, Austria

ABSTRACT

Three dimensional magnetic field problems are challenging, not only because of interesting applications in industry but also from a mathematical point of view. In order to handle technical magnetic field problems efficiently, it is not sufficient to have a fast Maxwell solver, but also a geometry modeller, an advanced 3D mesh generator, mesh handling and refinement methods, parallelization tools, and postprocessing tools including advanced visualization techniques are required. We will present here pre- and postprocessing tools, especially produced for adaptive multilevel methods used in the solver. In the magnetostatic case, the Maxwell solver is based on special mixed variational formulations of the Maxwell equations. In combination with an adaptive nested multilevel preconditioned iteration approach, we obtain an optimal solver with respect to the complexity. This is confirmed by the results of numerical experiments presented. We also propose a concept for coupling finite elements with boundary elements. Coupled finite and boundary element schemes are especially suited for problems where it is necessary to compute the exterior magnetic field.

Key words. Maxwell's equations, finite element method, boundary element method, multigrid preconditioners, domain decomposition, scientific computing tools, geometry modelling, mesh generation, visualization, parallelization

15.1 INTRODUCTION

Usually, technical 3D electromagnetic field problems are characterized by complicated interface geometries, possibly with moving parts (e.g. rotating parts), large jumps in the coefficients, nonlinearities, singularities, and the necessity of calculating the exterior magnetic field. In practice, the aim of the simulations is often the optimization of the magnetic device concerned. Coupled field problems in which the electromagnetic fields interact with other fields, e.g. with mechanical fields, lead to problems of high computational complexity. In order to handle such kinds of technical electromagnetic field problems, one needs not only a fast Maxwell solver and optimizer, but also a geometry modeller, an advanced 3D mesh generator, mesh handling and refinement methods, parallelization tools, and postprocessing tools including advanced visualization techniques are required.

We will present such pre- and postprocessing tools, especially modified for the adaptive multilevel methods used in the solver and the optimizer. Of course, the heart of our object oriented 3D field problem solver environment FEPP is an adaptive multilevel solver. In the magnetostatic case, the Maxwell solver is based on special mixed variational formulations of the Maxwell equations in $\boldsymbol{H}_0(\mathbf{curl}) \times \boldsymbol{H}_0^1(\Omega)$ and their discretization by the lowest order Nédélec and Lagrange finite elements, respectively. This mixed finite element scheme can be reduced to a symmetric and positive definite (spd) scheme for defining the vector potentials introduced for the $\boldsymbol{B}$-field. A clever combination of a multigrid regularisator with an adaptive nested multilevel preconditioned iteration approach for the spd system leads to an optimal solver with respect to the complexity. This is confirmed by the results of numerical experiments for both academic problems and real life applications. Let us mention that multilevel preconditioners for Maxwell's equations have also been studied in [1, 15]. In [8], hp-adaptive finite element methods were studied for Maxwell's equations.

Similar to the 2D case [7, 16, 21], we also propose a domain decomposition concept for coupling finite elements with boundary elements that leads to a symmetric, but indefinite, scheme. Coupled finite and boundary element schemes are especially suited for problems where it is necessary to take into account the exterior magnetic field. However, as in 2D, it may be advantageous to use the boundary element technique also in other air subdomains, e.g. in order to separate moving parts. In the finite element subdomains we introduce some vector potentials for the $\boldsymbol{B}$-field whereas scalar potentials for the $\boldsymbol{H}$-field are used in the boundary element subdomains. Another approach to the coupled finite and boundary element approximation of the magnetostatic Maxwell equations was studied in [2].

For the parallelization, two different strategies have been developed. The first approach uses a thread-based implementation that is especially suited for shared memory parallel computers such as the ORIGIN 2000. It is highly efficient if small numbers of processors are used. The second concept is based on distributed data algorithms and has been developed for massively parallel computers and workstation (PC) clusters.
The rest of the paper is organized as follows. In Section 15.2, we recall the Maxwell equations on which the numerical magnetic field simulations are based. Section 15.3 is devoted to the numerical solution of 3D magnetostatic field problems. In Section 15.4, we present the scientific computing tools that are necessary for treating more advanced 3D field problems. Section 15.5 contains some results of our numerical experiments. Finally, we add some concluding remarks in Section 15.6.

More information on the problem solver environment FEPP and on the presented results can be obtained from the homepage http://www.sfb013.uni-linz.ac.at of the special research project (SFB) "Numerical and Symbolic Scientific Computing" that is supported by the Austrian Science Foundation (FWF) under the grant SFB F013.

15.2 THE MAXWELL EQUATIONS

The Maxwell equations

$$\mathbf{curl}(H) = J + \frac{\partial D}{\partial t}, \ \ \text{div}(B) = 0, \ \ \mathbf{curl}(E) = -\frac{\partial B}{\partial t}, \ \ \text{div}(D) = \varrho \qquad (15.1)$$

couple the magnetic fields H (magnetic field intensity) and B (magnetic flux density) with the electric fields E (electric field intensity), D (electric flux density), J (electric current density) and ϱ (electric charge density). The partial differential equations (15.1) must be completed by constitutive relationships, initial and boundary conditions. In the case of isotropic media, the constitutive relationships have the form

$$J = \sigma E, \ \ B = \mu H, \ \ D = \varepsilon E, \qquad (15.2)$$

where σ, μ, ε denote the conductivity, the permeability, and the permittivity, respectively. In more general cases including anisotropic and inhomogeneous materials, these coefficients are tensors, where the single components depend on the locus (inhomogeneous materials) as well as on certain field quantities (nonlinear case). If moving bodies are involved, one has to add a term of the form $\mathbf{curl}(v \times B)$ to the right hand side of the third equation in (15.1), where v is the velocity [19, 23]. We come back to this case at the end of this paper in connection with coupled magnetomechanical problems. In the following, we restrict our considerations to the linear stationary case. Then, the equations describing the electric and magnetic fields can be separated. The magnetostatic equations, in which we are interested from now on, can be rewritten as

$$\mathbf{curl}(H) = J, \ \ H = \nu B, \ \ \text{div}(B) = 0, \qquad (15.3)$$

where $\nu = 1/\mu$ and the current density J is supposed to be given. If permanent magnets are present, J has to be replaced by $J + \mathbf{curl}(H_0)$, where H_0 is the initial (permanent) magnetic field intensity. Furthermore, we note that the current density J is divergence-free by physical reasons, i.e.,

$$\text{div}(J) = 0. \qquad (15.4)$$

Theoretically, the computational domain Ω coincides with the space $\mathbb{R}^3$ in any case. The behaviour of the magnetic field at infinity is described by radiation conditions. In practice, one may often simplify the problem by considering a bounded, simply connected computational domain $\Omega \subset \mathbb{R}^3$ with Lipschitz boundary $\Gamma = \partial\Omega$ and by replacing the radiation condition by the boundary condition

$$B \cdot n = 0 \ \ \text{on } \partial\Omega, \qquad (15.5)$$

where n stands for the unit outward normal with respect to $\partial\Omega$. Of course, this modification causes some modelling error. However, if the sources are completely surrounded by iron or, more generally, if Ω is chosen large enough, this error can be neglected. In Subsection 15.3.4, we will use the Boundary Element Method in order to avoid the restriction of the problem to bounded domains and thus the definition of artificial boundary

conditions. Instead, the physical radiation conditions are modelled exactly.
Introducing some vector potential u for the B-field $B = \mathbf{curl}(u)$ [9], we can easily derive
the canonical variational formulation

$$\text{Find } u \in H_0(\mathbf{curl}, \Omega) : \int_\Omega \nu \, \mathbf{curl}(u) \cdot \mathbf{curl}(v) \, dx = \int_\Omega J \cdot v \, dx \quad \forall \, v \in H_0(\mathbf{curl}, \Omega)$$

$$(15.6)$$

from the magnetostatic equations (15.3) in Ω and the boundary conditions (15.5) on Γ,
where $H_0(\mathbf{curl}, \Omega) := \{v \in L^2(\Omega)^3 \mid \mathbf{curl}(v) \in L^2(\Omega)^3, \, v \times n = 0 \text{ on } \Gamma\}$. Condition
(15.4) ensures the solvability of the variational problem (15.6). However, the solution is
not unique. Indeed, if one adds to some solution u of (15.6) a gradient-field $\nabla\varphi \in L^2(\Omega)^3$
with $\nabla\varphi \times n = 0$, one obtains again a solution.

From now one we allow the coefficient $\nu = 1/\mu$ to be piecewise constant. We mention
that on the interfaces the normal component of the B-field and the tangential component
of the H-field are continuous.

15.3   3D MAGNETOSTATIC FIELD PROBLEMS

15.3.1   Mixed FE Discretization

In the previous section we observed that the solution of the canonical variational formulation (15.6) is not unique. In order to obtain some variational formulation that has a
unique solution, we must take into account some gauging condition for the vector potential
u. The Coulomb gauging

$$\text{div}(u) = 0. \tag{15.7}$$

aims at divergence-free vector potentials. Treating the weak formulation of the Coulomb
gauging (15.7) as an equality constraint, we arrive at the following mixed variational
formulation that is fundamental for our approach to the numerical solution of the magne-
tostatic Maxwell equations (15.3): Find $(u, p) \in X \times M := H_0(\mathbf{curl}, \Omega) \times H_0^1(\Omega)$ such
that

$$a(u, v) + b(v, p) = \langle f, v \rangle \quad \forall \, v \in H_0(\mathbf{curl}, \Omega), \tag{15.8}$$
$$b(u, q) = 0 \qquad \forall \, q \in H_0^1(\Omega), \tag{15.9}$$

where $a(u, v) := \int_\Omega \nu \, \mathbf{curl}(u) \cdot \mathbf{curl}(v) \, dx$, $b(v, p) := \int_\Omega v \cdot \nabla p \, dx$, and $\langle f, v \rangle := \int_\Omega J \cdot v \, dx$. Now it is not difficult to conclude from the Brezzi-Babuška theory that the mixed
variational problem (15.8) - (15.9) has an unique solution. Moreover, choosing $v = \nabla p \in H_0(\mathbf{curl}, \Omega)$ in (15.8), we immediately observe from (15.4) and the Friedrichs inequality
that

$$p = 0. \tag{15.10}$$

This simple observation is crucial for our approach. Indeed, adding to the second equation
of our mixed variational problem (15.8) - (15.9) an arbitrary symmetric and positive

definite (spd) bilinear form $c(\cdot, \cdot) : M \times M \to \mathbb{R}^1$, we arrive at the equivalent mixed variational problem: Find $(\boldsymbol{u}, p) \in \boldsymbol{H}_0 \, (\mathbf{curl}, \Omega) \times H_0^1(\Omega)$ such that

$$a(\boldsymbol{u}, \boldsymbol{v}) + b(\boldsymbol{v}, p) \;=\; \langle \boldsymbol{f}, \boldsymbol{v} \rangle \quad \forall \boldsymbol{v} \in \boldsymbol{H}_0 \, (\mathbf{curl}, \Omega), \tag{15.11}$$

$$b(\boldsymbol{u}, q) - c(p, q) \;=\; 0 \qquad \forall q \in H_0^1(\Omega). \tag{15.12}$$

Let now be $X_h := \mathcal{N}_h^1 \subset X$ and $M_h := \mathcal{S}_h^1 \subset X$ the lowest order edge element space (see [25]) and the space of piecewise linear nodal elements on a shape-regular tetrahedral triangularization of Ω, respectively [6], where h is some discretization parameter. Then the mixed finite element approximation to the regularized mixed variational problem (15.11) - (15.12) reads as follows: Find $(\boldsymbol{u}_h, p_h) \in X_h \times M_h$ such that

$$a(\boldsymbol{u}_h, \boldsymbol{v}_h) + b(\boldsymbol{v}_h, p_h) \;=\; \langle \boldsymbol{f}, \boldsymbol{v}_h \rangle \quad \forall \boldsymbol{v}_h \in X_h, \tag{15.13}$$

$$b(\boldsymbol{u}_h, q_h) - c(p_h, q_h) \;=\; 0 \qquad \forall q_h \in M_h. \tag{15.14}$$

Again, using the Brezzi-Babuška theory, we conclude that the mixed finite element scheme (15.13) - (15.14) has an unique solution. From the same theory, we can derive the standard *a priori* error estimates (see, e.g., [4]). Similarly to the continuous case, we observe that $p_h = 0$ since $q_h \in M_h$ implies $\nabla q_h \in X_h$.

Let us choose the standard finite element basis in the spaces X_h and M_h. Then the mixed finite element scheme (15.13) - (15.14) is equivalent to the regular, symmetric, but indefinite system

$$\begin{pmatrix} A & B^T \\ B & -C \end{pmatrix} \begin{pmatrix} \boldsymbol{U}_h \\ \boldsymbol{P}_h \end{pmatrix} = \begin{pmatrix} \boldsymbol{F}_h \\ \boldsymbol{0} \end{pmatrix} \tag{15.15}$$

of linear finite element equations, where the matrices A, B, C and the first component $\boldsymbol{F}_h$ of the right-hand side are derived from the bilinear forms $a(\cdot, \cdot)$, $b(\cdot, \cdot)$, $c(\cdot, \cdot)$, and the linear form $\langle \boldsymbol{f}, \cdot \rangle$, respectively. $\boldsymbol{U}_h$ and $\boldsymbol{P}_h$ are respectively vectors of the nodal values of $\boldsymbol{u}_h$ and p_h.

15.3.2 SPD Reduction

Eliminating $\boldsymbol{P}_h = C^{-1} B \boldsymbol{U}_h$ from the second equation in (15.15) and inserting it into the first equation, we obtain the spd Schur complement system

$$G \boldsymbol{U}_h := (A + B^T C^{-1} B) \boldsymbol{U}_h = \boldsymbol{F}_h. \tag{15.16}$$

Let $\tilde{C}$ be some spd matrix that is spectrally equivalent to C (briefly, $\tilde{C} \approx C$), i.e. there are some positive constants c_1 and c_2, independent of h, such that

$$c_1(\tilde{C} \boldsymbol{Q}_h, \boldsymbol{Q}_h) \le (C \boldsymbol{Q}_h, \boldsymbol{Q}_h) \le c_2(\tilde{C} \boldsymbol{Q}_h, \boldsymbol{Q}_h) \quad \forall \boldsymbol{Q}_h \in \mathbb{R}^{m_h}, \tag{15.17}$$

where $m_h := \dim(M_h)$. Then the original Schur complement system (15.16) is equivalent to the modified Schur complement system

$$\tilde{G} \boldsymbol{U}_h := (A + B^T \tilde{C}^{-1} B) \boldsymbol{U}_h = \boldsymbol{F}_h. \tag{15.18}$$

Instead of solving the symmetric, but indefinite system (15.15), we solve the spd modified Schur complement system (15.18). We recall that the second component $\boldsymbol{P}_h$ is equal to $\boldsymbol{0}$. Furthermore, we can play around with the regularisor $\tilde{C}$ in order to improve the properties of the system matrix $\tilde{G}$ with respect to the solver we are going to construct now.

15.3.3   Solver

Let us choose the spd bilinear form

$$c(p, q) := \frac{1}{\nu_{\min}} \int_\Omega \nabla p \, \nabla q \, dx, \tag{15.19}$$

corresponding to the Laplace operator scaled by $1/\nu_{\min}$ (other choices for the $c(\cdot, \cdot)$ are discussed in [20]), and let us consider a spd preconditioner C_H for the spd matrix $\boldsymbol{H} := A + M$, i.e. there are some positive constants c_3 and c_4, independent of h, such that

$$c_3(C_H \boldsymbol{V}_h, \boldsymbol{V}_h) \leq (\boldsymbol{H} \boldsymbol{V}_h, \boldsymbol{V}_h) \leq c_4(C_H \boldsymbol{V}_h, \boldsymbol{V}_h) \quad \forall \boldsymbol{V}_h \in \mathbb{R}^{n_h}, \tag{15.20}$$

where $n_h := \dim(X_h)$ and M is the mass matrix in X_h defined by

$$(M \boldsymbol{U}_h, \boldsymbol{V}_h) := \nu_{\min} \int_\Omega \boldsymbol{u}_h \cdots \boldsymbol{v}_h \, dx.$$

The discrete LBB–condition and the spectral equivalence inequalities (15.20) imply that $C_H \approx H \approx G \approx \tilde{G}$ (see [20] for the detailed proof). Once a good preconditioner C_H and an appropriate regularisor $\tilde{C}$ are available, we can solve the modified Schur complement system (15.18) by the Preconditioned Conjugate Gradient (PCG) method. In practice, we choose the multigrid preconditioner

$$C_H := H(I - M_H)^{-1} \tag{15.21}$$

and the multigrid regularisor

$$\tilde{C} := C_C := C(I - M_C)^{-1}, \tag{15.22}$$

where M_H and M_C are the corresponding multigrid iteration operators with respect to $\boldsymbol{H}$ and C. Choosing appropriate symmetric multigrid cycles, we can now conclude from the results of [1, 12, 15] that the PCG method is asymptotically optimal with respect to the operation count and to the memory demand [17]. This is confirmed by the numerical results presented in Section 15.5.

15.3.4   Symmetric Coupling of FEM and BEM

If one is interested in the exterior magnetic field, then the coupling of the FEM with the BEM is certainly the natural technique for handling this problem. For simplicity of presentation, let us consider the case where the magnetic sources and the ferromagnetic materials are located in some bounded and simply connected Lipschitz domain Ω_F where we will use the FEM for approximating the magnetic field. Thus, we suppose that in the exterior BEM subdomain $\Omega_B := (\bar{\Omega}_F)^c$ the electric current density vanishes, i.e., $\boldsymbol{J} = \boldsymbol{0}$, and $\nu = \nu_B = 1/\mu_0$ (air). We can again introduce the vector potential $\boldsymbol{u}$ for the $\boldsymbol{B}$–field $\boldsymbol{B} = \mathbf{curl}(\boldsymbol{u})$ in Ω_F. Furthermore, we now assume that Ω_F is chosen such that Ω_B is also simply connected. Then in the exterior domain Ω_B, the $\boldsymbol{H}$–field can be represented as a gradient field of some scalar potential φ, i.e. $\boldsymbol{H} = \mathrm{grad}(\varphi)$ in Ω_B.

Therefore, in the exterior subdomain Ω_B, the magnetostatic Maxwell equations (15.3) are essentially reduced to the scaled Laplace equation for the scalar potential φ. The Cauchy data for the solution of this equation are related by Calderon's integral equation

$$\begin{pmatrix} \varphi \\ \lambda \end{pmatrix} = \mathcal{C} \begin{pmatrix} \varphi \\ \lambda \end{pmatrix} := \begin{pmatrix} \frac{1}{2}\mathcal{I} + \mathcal{K} & -\nu\mathcal{V} \\ -\frac{1}{\nu}\mathcal{D} & \frac{1}{2}\mathcal{I} - \mathcal{K}^* \end{pmatrix} \begin{pmatrix} \varphi \\ \lambda \end{pmatrix} \tag{15.23}$$

on the interface $\Gamma := \partial\Omega_F = \partial\Omega_B$, where φ denotes the trace of the scalar potential on Γ, $\lambda = \frac{1}{\nu}\frac{\partial\varphi}{\partial n} = B \cdot n$ on Γ, $n :=$ outer unit normal to Ω_F, $\mathcal{V} :=$ single layer potential operator on Γ, $\mathcal{K} :=$ double layer potential operator on Γ, $\mathcal{D} :=$ hypersingular operator on Γ.

Using now Coulumb's gauging condition (15.7) in Ω_F and Cauchy's representation formula (15.23) of the Cauchy data together with the interface condition predicting the continuity of the tangential part $H \times n$ of the H–field and of the normal component $B \cdot n$ of the B–field, we arrive at the mixed coupled FE-BE variational formulation: Find $(\boldsymbol{u}, \varphi, p) \in V := U \times \Phi \times P$ such that:

$$a(\boldsymbol{u}, \varphi; \boldsymbol{v}, \psi) + b(\boldsymbol{v}, p) = \langle \boldsymbol{f}, \boldsymbol{v} \rangle \quad \forall (\boldsymbol{v}, \psi) \in U \times \Phi, \tag{15.24}$$

$$b(\boldsymbol{u}, q) - c(p, q) = 0 \quad \forall q \in P, \tag{15.25}$$

where $U := \boldsymbol{H}(\mathrm{curl}, \Omega_F)$, $\Phi := H_*^{1/2}(\Gamma)$, and $P := H_*^1(\Omega_F)$. The bilinear forms are defined by the identities

$$a(\boldsymbol{u}, \varphi; \boldsymbol{v}, \psi) := \int_{\Omega_F} \nu \, \mathbf{curl}(\boldsymbol{u}) \cdot \mathbf{curl}(\boldsymbol{v}) \, dx$$

$$+ \langle \nu_B \mathcal{V}(\mathbf{curl}(\boldsymbol{u}) \cdot \boldsymbol{n}), \mathbf{curl}(\boldsymbol{v}) \cdot \boldsymbol{n} \rangle_\Gamma - \langle (\frac{1}{2}\mathcal{I} + \mathcal{K})\varphi, \mathbf{curl}(\boldsymbol{v}) \cdot \boldsymbol{n} \rangle_\Gamma$$

$$+ \langle \frac{1}{\nu_B}\mathcal{D}\varphi, \psi \rangle_\Gamma + \langle (\frac{1}{2}\mathcal{I} + \mathcal{K}^*)(\mathbf{curl}(\boldsymbol{u}) \cdot n), \psi \rangle_\Gamma,$$

$$b(\boldsymbol{u}, q) := \int_{\Omega_F} \boldsymbol{u} \cdot \nabla q \, dx, \quad c(p, q) := \int_{\Omega_F} \nabla p \cdot \nabla q \, dx, \quad \langle \boldsymbol{f}, \boldsymbol{v} \rangle := \int_{\Omega_F} \boldsymbol{J} \cdot \boldsymbol{v} \, dx.$$

The subscript "$*$" means that the function of the corresponding space should be L_2–orthogonal to the constant functions. We mention that $H_*^1(\Omega_F)$ can be changed to $H^1(\Omega_F)$ if the above bilinear form $c(\cdot, \cdot)$ is replaced by the bilinear form

$$c(p, q) := \int_\Omega (\alpha \nabla p \cdot \nabla q + \beta \, p \, q) \, dx, \tag{15.26}$$

that may be easier to handle in practice, where α and β are appropriately chosen, positive scaling constants. Again, in both cases, one can show existence and uniqueness of the solution. Moreover, $p = 0$ if $\int_{\Omega_F} \boldsymbol{J} \cdot \nabla q \, dx = 0 \quad \forall q \in H_*^1(\Omega_F)$ (see [20] for the proof).

Formally, the mixed domain and boundary integral variational formulation (15.24)-

(15.25) is equivalent the problem

$$
\begin{aligned}
\mathbf{curl}\,(\nu\,\mathbf{curl}(\boldsymbol{u})) &= \boldsymbol{J} &&\text{in}\quad \Omega_F \quad (\text{PDE}),\\
\operatorname{div}(\boldsymbol{u}) &= 0 &&\text{in}\quad \Omega_F \quad (\text{gauging condition}),\\
\operatorname{div}(\mu_B\,\nabla\varphi) &= 0 &&\text{in}\quad \Omega_B \quad (\text{PDE}),\\
\varphi &= O(1/|x|) &&\text{for}\quad |x|\to\infty \quad (\text{radiation condition}),\\
\nabla\varphi\times\boldsymbol{n} &= \nu_F(\mathbf{curl}(\boldsymbol{u})\times\boldsymbol{n}) &&(\text{interface condition: } \boldsymbol{H}\times\boldsymbol{n}\text{ continuous}),\\
\lambda := \mu_B(\nabla\varphi\cdot\boldsymbol{n}) &= \mathbf{curl}(\boldsymbol{u})\cdot\boldsymbol{n} &&(\text{interface condition: } \boldsymbol{B}\cdot\boldsymbol{n}\text{ continuous}),
\end{aligned}
$$

that corresponds to the magnetostatic Maxwell equations in $\mathbb{R}^3$.

Choosing the finite (boundary) element subspaces $U_h := \mathcal{N}_h^1 \subset U$, $\Phi_h := \mathcal{S}_h^1 \subset \Phi$ and $P_h := \mathcal{S}_h^1 \subset P$ in analogy to Subsection 15.3.1, we derive from (15.24) the symmetric coupled FE-BE Galerkin scheme: Find $(\boldsymbol{u}_h, \varphi_h, p_h) \in V_h := U_h \times \Phi_h \times P_h$ such that

$$
\begin{aligned}
a(\boldsymbol{u}_h, \varphi_h; v_h, \psi_h) + b(\boldsymbol{v}_h, p_h) &= \langle \boldsymbol{f}, v_h \rangle && \forall\, (\boldsymbol{v}_h, \psi_h) \in U_h \times \Phi_h, & (15.27)\\
b(\boldsymbol{u}_h, q_h) - c(p_h, q_h) &= 0 && \forall\, q_h \in P_h, & (15.28)
\end{aligned}
$$

that is again equivalent to the following symmetric, but indefinite system of coupled FE-BE equations

$$
\begin{pmatrix} A_h & K_h^T & B_h^T \\ K_h & -D_h & 0 \\ B_h & 0 & -C_h \end{pmatrix}
\begin{pmatrix} \boldsymbol{U}_h \\ \varphi_h \\ \boldsymbol{P}_h \end{pmatrix}
=
\begin{pmatrix} \boldsymbol{F}_h \\ 0 \\ 0 \end{pmatrix},
\tag{15.29}
$$

where $A_h = A_h^F + A_h^B$ consists of the contributions from the first two terms of the bilinear form $a(\cdot,\cdot)$. Eliminating again $\boldsymbol{P}_h = C^{-1}B\boldsymbol{U}_h$ from the third equation in (15.29) and inserting it into the first equation, we obtain the Schur complement system

$$
\begin{pmatrix} \tilde{A}_h & K_h^T \\ K_h & -D_h \end{pmatrix}
\begin{pmatrix} \boldsymbol{U}_h \\ \varphi_h \end{pmatrix}
=
\begin{pmatrix} \boldsymbol{F}_h \\ 0 \end{pmatrix},
$$

where $\tilde{A}_h = A_h^F + A_h^B + B_h^T C_h^{-1} B_h$. In contrast to the finite element case considered in Subsection 15.3.2, here the Schur complement system remains symmetric and indefinite. Similar to the 2D case discussed in [5, 21, 24], we can now construct efficient solvers on the basis of the Bramble-Pasciak transformation [3, 30].

15.4   SCIENTIFIC COMPUTING TOOLS

In this section we discuss the scientific computing tools that are required for the solution of general field problems in 3D. We apply these tools to the simulation of Maxwell's equations introduced in Section 15.3.

The interplay of these computing tools is illustrated in Figure 15.1. Clearly, the central module is the simulation code. Usually, this is the most CPU-time consuming part of the whole process, and it is a topic of intensive research. The input to the simulation module is the description of the problem which is to be solved. This includes the computational

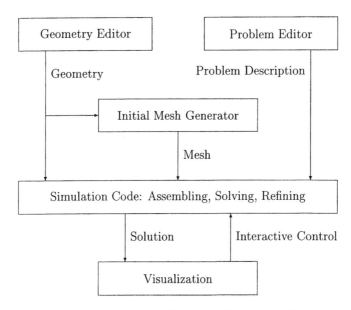

Figure 15.1: Interplay of Scientific Computing Tools.

domain, the underlying equations, the boundary conditions etc. These input data have to be described by proper data structures. Therefore, the user should have access to tools for an interactive generation of the geometry as well as the problem description. These tools are definitely different for scientific computing and for commercial applications. For scientific processes, the tool may be just based on a text editor. In addition to the small amount of data for geometry and equations, a mesh is needed for the simulation. Since the mesh data are much more complex, an automatic tool for mesh generation is needed. The simulation produces a nearly arbitrarily large amount of data. For the output of the information, advanced visualization tools are necessary.

We will describe the tools developed within the research project "Numerical and Symbolic Scientific Computing", namely the mesh generator NETGEN, the simulation code FEPP and the visualization module VIPP. The data structures for the geometry representation and the problem description are also described by means of a specific example.

15.4.1 Geometry Description

There are many different techniques for describing the geometry of domains. Their advantages and disadvantages depend on the specific application we are dealing with. For example, a surface representation is useful for visualization of virtual reality environments, but for simulation purposes solid representations are necessary. Surfaces can be described very easily by parametric classes, like a cylinder with two points on the axis and a radius, or by approximation with spline patches. While the spline approximation is very flexible

```
solid kernel =
                orthobrick (-8, -6, -1; 8, 6, 1)
        and not orthobrick (-6, -4, -2; -1, 4, 2)
        and not orthobrick ( 1, -4, -2;  6, 4, 2);

solid coil1 =
                cylinder (-7, -3, 0;-7, 3, 0; 3)
        and not cylinder (-7, -3, 0;-7, 3, 0; 2)
        and plane (0, -3, 0; 0, -1, 0)
        and plane (0,  3, 0; 0,  1, 0);

solid coil2 =
                cylinder ( 0, -3, 0; 0, 3, 0; 3)
        and not cylinder ( 0, -3, 0; 0, 3, 0; 2)
        and plane (0, -3, 0; 0, -1, 0)
        and plane (0,  3, 0; 0,  1, 0);

solid coil3 =
                cylinder ( 7, -3, 0; 7, 3, 0; 3)
        and not cylinder ( 7, -3, 0; 7, 3, 0; 2)
        and plane (0, -3, 0; 0, -1, 0)
        and plane (0,  3, 0; 0,  1, 0);

solid box = orthobrick (-12, -8, -5; 12, 8, 5);

solid air = box and not (kernel or coil1 or coil2 or coil3);
```

Figure 15.2: Geometry description for the transformer.

for describing complicated geometries like car bodies, human bodies or geodetic data, it requires special tools for the generation of the geometry description itself.

Many realistic geometries like the transformer shown below can be described very easily by a Constructive Solid Geometry (CSG) model. An open number of parametric classes forms primitives. These primitives are combined by the Eulerian operations *union*, *intersection* and *complement* to form more complicated solids. It is of moderate effort to define such geometries by means of a text editor.

Figure 15.2 contains the whole description of the geometry of the transformer. It contains five subdomains, namely the kernel, three coils and the air domain around these parts. The outer boundary is given by an iron casing of high conductivity that motivates the boundary condition (15.5). The file is more or less self-explaining. Primitives are, e.g., cylinders of infinite length defined by two points on the axis and the radius, or a brick in normal direction defined by a minimal and a maximal point. For example, one coil is defined by the outer cylinder, from which an inner cylinder is cut off, and by two planes which restrict the infinite cylinder to a finite length. The geometry is drawn in Figure 15.3. Fast geometry visualization is done by NETGEN on the basis of (nonconforming) triangle approximations.

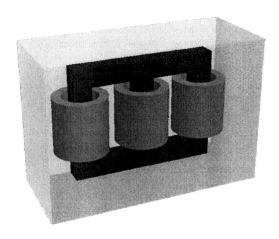

Figure 15.3: Geometry visualization of the transformer.

15.4.2 The 3D Mesh Generator NETGEN

A mesh generator is required to generate a triangulation based on the geometry description. We now explain the principal parts of the fully automatic tetrahedral mesher NETGEN [27]. Figure 15.4 shows the sequence of steps performed by NETGEN. The principle is bottom up, from 0-dimensional objects (points) to 3-dimensional objects (tetrahedra). First, one has to detect all corner points (and other special points) of the geometry. These points are defined by nonlinear equations. The algorithm used here is a combination of a globally convergent bisection method and a fast, locally convergent Newton method. Starting from the corner points, the edges are identified by curve following methods.

After handling the edges, the surfaces are meshed. This is done by a version of the advancing front method. Starting from the boundary of a surface in 3D, triangle by triangle is cut off to reduce the remaining part of the surface. The principle is shown in Figure 15.5. The rules defining how to generate triangles for a given boundary configuration are described by data structures instead of hard-coded instructions. This design decision keeps the code relatively short and maintainable.

The last part is the volume mesh generation. Originally, NETGEN used the same advancing front method for volume meshing as for surface meshing. Together with a back-tracking strategy the algorithm is very robust. The drawback of this approach is that it is very time consuming. Now, we use a combination of the Delaunay algorithm with the rule based advancing front method. The Delaunay algorithm is very fast, but often fails to generate the whole volume mesh compatible with the boundary mesh. Then, a small number of local zones are left. These are filled out by the original, robust rule based advancing front method.

From the geometry description of the transformer given in Figure 15.2, NETGEN generated a coarse mesh with 2465 tetrahedra as shown in Figure 15.6.

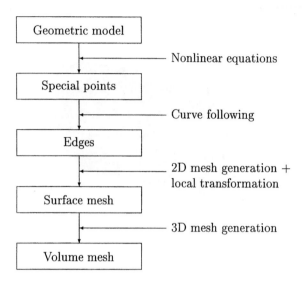

Figure 15.4: Subproblems in mesh generation.

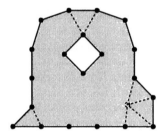

Figure 15.5: Advancing Front Method: cutting off triangles.

15.4.3 Problem and Solver Description

The simulation module FEPP is designed to handle different types of equations with one code - even within one run. This is achieved by a problem description language (see Figure 15.7 for our model problem). It enables us to define an arbitrary number of finite element spaces, gridfunctions, bilinear forms or linear forms. By certain flags the objects can be adapted to the problem. For Maxwell equations we use edge elements. This is indicated by the -edge flag in the definition of the finite element space called -vedge.

The bilinear form is built up from a number of predefined forms, e.g., the token `rotrotedge` defines the form

$$\int_\Omega \nu(x)\,\mathbf{curl}(u)\cdot\mathbf{curl}(v)\,dx.$$

The coefficient $\nu(x)$ can be given as constant, as piecewise constant, by a generic formula

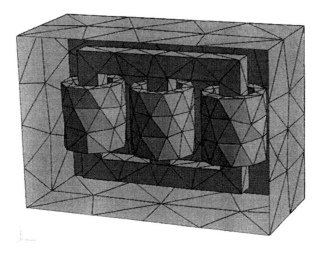

Figure 15.6: Initial mesh of the transformer.

or as grid function being the solution of another problem. The linear form is similarly defined. In our application the matrix H corresponds to the bilinear form (ah) whereas the modified Schur complement matrix $\tilde{G}$ corresponds to the bilinear form (ae) which is basically realized by some special form denoted by the token `rotrotedgestab`.

Next, we have to define a preconditioner. We want to use the multigrid preconditioner described in Section 15.5. The preconditioner needs the bilinear form to work with. It can be tuned for the problem by a list of parameters. Clearly, the type of smoothing and prolongation operators have to be chosen to meet the properties of the underlying problem. The multigrid cycle, the number of smoothing steps and the coarse grid solver can be adapted to the specific example.

The problem to be solved is a boundary value problem. This needs the bilinear form, the linear form, and the gridfunction for the solution. These objects are internally represented by matrices and vectors. The solver for symmetric problems is a preconditioned conjugate gradient method. We use the multigrid preconditioner (ce) defined in Figure 15.7 (cf. also Section 15.3.3).

Next, the type of post-processing is specified. The function `calcflux` computes the natural fluxes for the problem at hand. For the Maxwell equations it computes the induction field B. Finally, the error estimator and element selection strategy for the refinement of the grid is defined. Since the fluxes are already computed, the Zienkiewicz-Zhu error estimator `estzz` is very cheap to compute (see, e.g., [29]).

```
loadgrid  trafo

// edge element space:
define space vedge -edge -domains=[ 1, 2, 3, 4, 5 ]

// define forms and gridfunctions:
define bilinearform ae -spc=vedge
define bilinearform ah -spc=vedge
define linearform fe -spc=vedge
define gridfunction ue -spc=vedge
define gridfunction b -dim=3 -elconst

// build up forms:
add ah rotrotedge ( ([ 1e-3, 1, 1, 1, 1 ])) -quad=3
add ah massedge (1e-3) -quad=2
add ah robin (1e8) -quad=5 -bound=1
add ae rotrotedgestab ( ([ 1e-3, 1, 1, 1, 1 ]), (0), (1), (1000) ) -quad=3
add ae robin (1e8) -quad=5 -bound=1

add fe sourceedge1 ( (x3), 0, (-x1-7) ) -dom=2 -quad=1
add fe sourceedge1 ((-x3), 0,    (x1) ) -dom=3 -quad=1
add fe sourceedge1 ( (x3), 0, (-x1+7) ) -dom=4 -quad=1

// define multigrid preconditioner:
define preconditioner ce -name=mg -mat=ah -prol=21 -smoother=3 -cycle=1

// solve boundary value problem:
define problem bvp -mat=ae -rhs=fe -sol=ue -pre=ce

// define prostprocessing (fluxes):
define problem calcflux -mat=ae -sol=ue -flux=b

// ZZ - error estimator:
define gridfunction elerr -elconst
define problem estzz -mat=ae -flux=b -sol=ue  -elerr=elerr
define problem badelementmarker -elerr=elerr -fac=0.1 -minlevel=1
```

Figure 15.7: Problem description for the transformer.

15.4.4   The Adaptive Multilevel (Parallel) Solver FEPP

The numerical simulations are carried out using the object oriented C++ code *FEPP* [28]. The required input including the mesh, geometry and problem description has been described already. Starting with a coarse mesh, a hierarchy of refined meshes is produced based on error estimators until a desired accuracy is reached. The hierarchy of those meshes is essential for the application of efficient multigrid and multilevel preconditioners which guarantee optimal complexity of the algorithms.

After starting *FEPP* the problem description file is processed starting with loading the mesh. Then, the objects defined in the file are created. The command

```
>assemble
```

causes the assembling of all objects on the current grid. In our case, this includes the allocation of the solution vector, the allocation of the vectors and matrices corresponding to linear and bilinear forms, followed by the assembling of the integrators given by the **add** operation. The assembling of the problem **bvp** involves immediately the iterative solution of the finite element equations. The post-processing includes the calculation of fluxes according to the components of the bilinear form as well as the error estimator giving the refinement information, i.e., elements with large local error are marked for refinement.

The user can proceed interactively with the command

```
>refine
```

which causes the refinement of the grid and the assembling of all objects on the new grid. After 5 refinement steps, the grid shown in Figure 15.8 is obtained. Alternatively, the command

```
>draw3d
```

allows an interactive visualization of the current data including meshes and data of grid functions. The concept of the visualization is described in the next subsection.

15.4.5   Scientific Visualization

The development of our own visualization tool *VIPP* [22] was driven by the goal of interactive visualization during adaptive multilevel computations and by the requirements caused by very large data sets. In particular, since modern numerical algorithms as described in this paper allow us to execute 3D computations with several 10^5 unknowns within minutes even on notebook computers, the performance of interactive visualization including rotating objects and zooming is an important issue. So, any system based on file transfers is prohibitive for online visualization.

Instead, interactive visualization has to be considered as a part of the simulation process. In our case, a shared data concept was chosen. Namely, *VIPP* accesses the data owned by *FEPP* directly. The interface between *VIPP* and *FEPP* is realized by general *FeppDataAccess*–functions. Moreover, the visualization can make use of the hierarchical data structures supplied by *FEPP*. That means, during CPU intensive visualization tasks, e.g., slicing through objects, smaller data sets corresponding to coarser meshes are used.

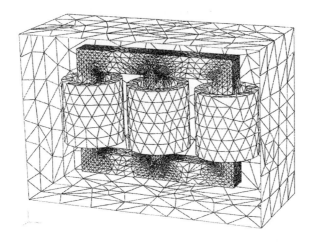

Figure 15.8: Level 6 mesh of the transformer.

The finest resolution (corresponding to the access to data of the finest grid) is restored as soon as the user stops interactive tasks.

The tool *VIPP* [22] allows us to visualize surface and volume meshes, including shrunk meshes. Slicing planes can be defined and are drawn continuously according to a given point and a normal direction. Scalar gridfunctions (solutions) can be visualized by colors, iso-lines or iso-surfaces.

A 3D vector valued solution (**B**–field) is shown in Figure 15.9 as a vector plot.

15.5 NUMERICAL RESULTS

In this section, we present numerical results for our model problem. Both, geometry and problem data have already been given. The magnetic field which is to be computed is generated by tangential currents within the three coils. The currents are now normalized to 1 in contrast to the description file given above. The iron core has a permeabilty of 1000.

As proposed we solve the modified Schur complement system (15.18). The matrix $\tilde{C}$ is implicitly defined by (15.22). The action of $\tilde{C}^{-1}$ corresponds to a multigrid V–cycle with 2 Gauss–Seidel pre- and post-smoothing steps. As a solver for system (15.18) we apply the PCG method with the preconditioner C_H as given in (15.21). The action of C_H^{-1} corresponds to a multigrid V–cycle using a block–Gauss–Seidel smoother with 1 pre- and post-smoothing step [1]. In particular those edges adjacent to one vertex belong to a common block.

We start the calculations on the coarsest mesh with 2805 unknowns (one unknown per

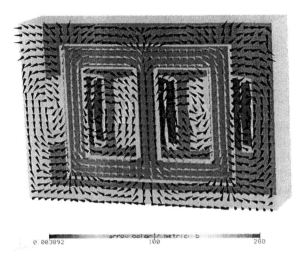

Figure 15.9: Solution (B–field) in a vector plot.

edge). Finer meshes are obtained by adaptive refinement leading to 258867 unknowns after 5 refinement steps. The solution (B–field) is shown in Figure 15.9. As expected, areas of rapid changes in the B-field along edges correspond to areas of mesh refinement (cf. Figure 15.8) according to the error estimator.

Table 15.1 shows the results obtained on a SGI Octane workstation with a RS 10000 processor, 250 MHz and 2GB memory. The CPU times are given in seconds. Column 4 (acc. time) shows the overall (accumulated) time including mesh refinement, assembling and solution, whereas the last column shows partial times (per level) required for the assembling of linear and bilinear forms (assemble) and the PCG solver (solve). The number of iterations and times are given for a relative accuracy of 10^{-6} required by the PCG. The constant numbers indicate that the preconditioners are robust with respect to

Table 15.1: Numerical results for the transformer.

Level	Unknowns	Iter.	acc. time [sec]	level times (assemble/solve)
1	2805	7	5.7	1.4/4.0
2	6211	8	12.8	3.1/2.2
3	10995	9	27.3	5.2/5.9
4	32865	10	73.2	15.4/19.7
5	81614	10	193.4	38.4/54.5
6	258867	11	614.4	135.5/194.7

the discretization parameter h. Since the CPU time grows from level to level with almost the same factor as the number of unknowns, the algorithm presented in this paper is of the same asymptotically optimal complexity as was theoretically predicted.

15.6 CONCLUDING REMARKS

In the second part of this paper we have presented scientific computing tools that are necessary for studying numerically 3D field problems of technical complexity by means of advanced adaptive multilevel algorithms. We have just demonstrated the interplay of these tools for the simulation of 3D linear magnetostatic field problems. In the first part, we have developed a robust, asymptotically optimal, multigrid preconditioned CG solver for regularized finite element schemes approximating the 3D magnetostatic Maxwell equations. These numerical techniques can be extended to more advanced electromagnetical field problems including nonlinear problems [13, 14], time-harmonic problems, transient problems and coupled magnetomechanical problems. For instance, in [26], these adaptive multigrid techniques have been successfully extended to 3D transient coupled magnetomechanical problems. The numerical simulation of the acceleration of a magnetic excited aluminum plate was in quite good agreement with the corresponding measured quantities. Moreover, our multigrid solver was 30 times faster than some commercial ICCG solver.

Furthermore, we have proposed a concept for constructing and solving symmetric coupled FE-BE schemes approximating the magnetostatic Maxwell equations in the whole $\mathbb{R}^3$. For the sake of simplicity, we have only considered the case of two subdomains where the BEM takes cares of a precise approximation of the magnetic field in the exterior air subdomain and the FEM is used in the bounded complement. As in 2D, this technique can be used to construct nonoverlapping domain decomposition methods. In dependence of the data, one can use the BEM in some subdomains whereas the FEM is applied in the other subdomains. We mention that on the interfaces between two BEM subdomains the additional unknown $\lambda := \boldsymbol{B} \cdot \boldsymbol{n}$ occurs (see [20] for details). These domain decomposition schemes are especially suited for parallel computing. Following the parallelization concepts proposed in [24], [21] and [10], we have developed a parallel version of our multigrid package FEPP under MPI. For this purpose, at least, one further scientific computing tool that handles some automatic domain decomposition and data distribution is required. We use a modified recursive spectral bisection technique that was before developed in 2D [11, 18, 20].

REFERENCES

[1] D. Arnold, R. Falk and R. Winther (1999). Multigrid in H(div) and $\boldsymbol{H}$(curl). *Numer. Math.* (to appear).

[2] A. Bossavit (1991). Mixed methods and the marriage between mixed finite elements and boundarys elements. *Numer. Meth. for Partial Differential Equations*, 7:347–362.

[3] J. H. Bramble and J. E. Pasciak (1988). A preconditioning technique for indefinite

systems resulting from mixed approximations of elliptic problems. *Math. Comp.*, 50(181):1–17.

[4] F. Brezzi and M. Fortin (1991). *Mixed and Hybrid Finite Element Methods*, volume 15 of *Springer Series in Computational Mathematics*. Springer-Verlag.

[5] C. Carstensen, M. Kuhn and U. Langer (1998). Fast parallel solvers for symmetric boundary element domain decomposition equations. *Numer. Math.*, 79:321–347.

[6] P. Ciarlet (1978). *The Finite Element Method for Elliptic Problems*. North-Holland Publishing Company, Amsterdam-New York-Oxford.

[7] M. Costabel (1987). Symmetric methods for the coupling of finite elements and boundary elements. In C. A. Brebbia, W. L. Wendland and G. Kuhn, eds., *Boundary Elements IX*, pages 411–420. Springer–Verlag.

[8] L. Demkowicz and L. Vardaptyan (1998). Modeling of electromagnetic absorption/scattering using hp-adaptive finite elements. *Comput. Meth. Appl. Mech. Engrg.*, 152:103–124.

[9] V. Girault and P.A. Raviart (1986). *Finite Element Methods for Navier-Stokes Equations*. Springer-Verlag.

[10] G. Haase (1999). *Parallelisierung numerischer Algorithmen für partielle Differentialgleichung*. Teubner Verlag, Stuttgart.

[11] G. Haase and M. Kuhn (1999). Preprocessing for 2d FE-BE domain decomposition methods. *Computing and Visualization in Science.* (to appear).

[12] W. Hackbusch (1985). *Multi–Grid Methods and Applications*, volume 4 of *Springer Series in Computational Mathematics*. Springer–Verlag, Berlin.

[13] B. Heise (1994). Analysis of a fully discrete finite element method for a nonlinear magnetic field problem. *SIAM J. Numer. Anal.*, 31(3):745–759.

[14] B. Heise, M. Kuhn and U. Langer (1997). A mixed variational formulation for 3D magnetostatics in the space $H(\mathrm{rot}) \cap H(\mathrm{div})$. Technical Report 97-1, Johannes Kepler University Linz, Institute of Mathematics.

[15] R. Hiptmair (1999). Multigrid methods for Maxwell's equations. *SIAM J. Numer. Anal.*, 36:204–225.

[16] G. C. Hsiao and W. Wendland (1991). Domain decomposition in boundary element methods. In *Proc. of IV Int. Symposium on Domain Decomposition Methods, (R. Glowinski, Y. A. Kuznetsov, G. Meurant, J. Périaux, O. B. Widlund eds.), Moscow, May 1990*, pages 41–49. SIAM Publ., Philadelphia.

[17] M. Jung, U. Langer, A. Meyer, W. Queck and M. Schneider (1989). Multigrid preconditioners and their applications. In G. Telschow, editor, *Third Multigrid Seminar, Biesenthal 1988*, pages 11–52. Karl–Weierstrass–Institut, Berlin. Report R–MATH–03/89.

[18] R. Koppler, G. Kurka and J. Volkert (1997). Exdasy - an user-friendly and extentable data distribution system. *Lecture Notes in Computer Science*, 1300:118–127.

[19] A. Kost (1994). *Numerische Methoden in der Berechnung elektromagnetischer Felder.* Springer–Lehrbuch. Springer Verlag, Berlin, Heidelberg.

[20] M. Kuhn (1998). *Efficient Parallel Numerical Simulation of Magnetic Field Problem.* Ph.D. thesis, Johannes Kepler University Linz.

[21] M. Kuhn and U. Langer (1997). Adaptive domain decomposition methods in FEM and BEM. In J.R. Whiteman, ed., *The Mathematics of Finite Elements and Applications*, pages 103–122. John Wiley & Sons.

[22] G. Kurka (1998). A fast visualization system for adaptive grids. SFB Report 98-11, Johannes Kepler University Linz, SFB "Numerical and Symbolic Scientific Computing".

[23] St. Kurz, J. Fetzer, G. Lehner and W.M. Rucker (1998). A novel formulation for 3d eddy current problems with moving bodies using a Lagrangian description and BEM-FEM coupling. *IEEE Transaction on Magnetics*, 34:3068–3073.

[24] U. Langer (1994). Parallel iterative solution of symmetric coupled BE/FE–equations via domain decomposition. *Contemp. Math.*, 157:335–344.

[25] J. Nédélec (1986). A new family of mixed finite elements in R^3. *Numer. Math.*, 50:57–81.

[26] M. Schinnerl, J. Schöberl, M. Kaltenbacher, U. Langer and R. Lerch (1999). Multigrid methods for the fast numerical simulation of coupled magnetomechanical systems. *ZAMM.* (to appear).

[27] J. Schöberl (1997). NETGEN - An advancing front 2D/3D-mesh generator based on abstract rules. *Computing and Visualization in Science*, 1:41–52.

[28] J. Schöberl (1997). Object-oriented Finite Element Code *FEPP*. See also: http://www.sfb013.uni-linz.ac.at.

[29] R. Verfürth (1996). *A Review of A Posteriori Error Estimation and Adaptive Mesh–Refinement Techniques.* Wiley-Teubner Series in Advances in Numerical Mathematics. Wiley and Teubner, Chichester, Stuttgart.

[30] W. Zulehner (1998). Analysis of iterative methods for saddle point problems: A unified approach. Institutsbericht 538, Johannes Kepler University Linz, Department of Numerical Analysis.

16 DUALITY BASED DOMAIN DECOMPOSITION WITH ADAPTIVE NATURAL COARSE GRID PROJECTORS FOR CONTACT PROBLEMS

Zdeněk Dostál[a], Francisco A. M. Gomes Neto[b] and Sandra A. Santos[b]

[a]FEI VŠB-Technical University Ostrava, Ostrava, CZ 70833, Czech Republic

[b]IMECC – UNICAMP, University of Campinas, CP 6065, 13081–970 Campinas SP, Brazil

ABSTRACT

An efficient non-overlapping domain decomposition algorithm of the Neumann-Neumann type for solving both coercive and semicoercive contact problems is presented. The discretized problem is first turned by the duality theory of convex programming into a quadratic programming problem with bound and equality constraints. The latter is further modified by means of orthogonal projectors to the natural coarse space introduced by Farhat and Roux in the framework of their FETI method. The resulting problem is then solved by an augmented Lagrangian type algorithm with an outer loop for the Lagrange multipliers for the equality constraints and an inner loop for the solution of the bound constrained quadratic programming problems. The algorithm exploits the efficient quadratic programming algorithms proposed earlier for effective identification of the contact interface and preserves numerical and parallel scalability of the FETI method in the solution of auxiliary linear problems. Further improvement may be achieved by using projectors that are adapted to the current face and standard preconditioners.

Key words. Contact problem, domain decomposition, natural coarse grid

[16]This research has been supported by FAPESP (grant 97/12676-4),CNPq, FINEP, FAEP–UNICAMP, by grants GA ČR No. 201/97/0421 and 101/98/0535, and by project CEZ:J17/98:272401 MŠMT ČR.

16.1   INTRODUCTION

Duality based domain decomposition methods have proved to be practical and efficient tools for the parallel solution of large elliptic boundary value problems [15, 24]. It has been observed that these methods may be even more successful for variational inequalities, including those that describe the equilibrium of a system of elastic bodies in frictionless unilateral contact, since duality not only reduces the dimension and improves conditioning of the original problem, but also removes singularities and reduces the inequalities to nonnegativity bounds on variables [5]. Moreover, relatively expensive preparation of the solution of subproblems may be reused during the contact interface iterations, and the spatial domain of contact problems may be naturally decomposed by identifying the subdomains with particular bodies. In this case, duality yields a quadratic programming (QP) problem in unknown Lagrange multipliers which can be interpreted as contact forces.

In this paper, we review our work related to development of effective algorithms for the solution of multibody contact problems, which blends progress in duality based domain decomposition methods with our results on solving QP problems. For the sake of simplicity, we consider only frictionless problems of linear elasticity, but the results may be exploited for the solution of more general problems [14].

We start our exposition by describing the decomposition of a system of elastic bodies in frictionless contact into subdomains and giving the discretized conditions of equilibrium of the system as an indefinite QP problem in nodal displacements with a block diagonal stiffness matrix and general equality and inequality constraints. Then we show that the difficulties arising from general inequality constraints and semidefiniteness of the problem in displacements may be reduced by application of the duality theory [5, 10, 12]. The matrix of the dual quadratic form turns out to be positive definite with a spectrum that is favorably distributed for application of conjugate gradient based methods.

The dual formulation is modified in Section 16.3 in order to redistribute the spectrum of the Hessian of the augmented Lagrangian so that it is favorably distributed for the conjugate gradient iterations, even with a large penalty parameter that is necessary to achieve a sufficient penalization effect. The main tools are the projectors to the natural coarse space [17, 24], results of Axelsson [1, 2] on the rate of convergence of the conjugate gradient method with a gap in the spectrum, our recent results on distribution of the spectrum of penalized matrices [7], and optional preconditioning [23, 22].

The algorithms for the solution of the modified problem are reviewed in Section 16.4. We have adapted the augmented Lagrangian type algorithm proposed by Conn, Gould and Toint [4] for the solution of more general problems, but we use the special structure of our problem to improve the performance of the algorithm. The modifications exploit projectors and adaptive precision control of the solution of auxiliary problems, using mostly the results of Friedlander, Martínez and the authors [19, 6, 11, 8]. Theoretical results on convergence of the algorithm are reported.

Results of numerical experiments that demonstrate the power of our algorithms are given in Section 16.5. Finally, in Section 16.6, some comments and conclusions are presented.

16.2 EQUILIBRIUM OF A SYSTEM OF BODIES IN CONTACT

Consider a system of elastic bodies decomposed into s subdomains each of which occupies, in a reference configuration, a domain Ω^p in $\mathbb{R}^d$, $d = 2, 3$ with sufficiently smooth boundary Γ^p as in Figure 16.1. Suppose that each boundary Γ^p consists of four disjoint parts $\Gamma^p_U, \Gamma^p_F, \Gamma^p_E$ and Γ^p_C, $\Gamma^p = \Gamma^p_U \cup \Gamma^p_F \cup \Gamma^p_E \cup \Gamma^p_C$, and that the displacements $\mathbf{U}^p : \Gamma^p_U \to \mathbb{R}^d$ and forces $\mathbf{F}^p : \Gamma^p_F \to \mathbb{R}^d$ are given. The part Γ^p_C denotes the part of Γ^p that may make unilateral contact with some other subregion, and Γ^p_E denotes the part of Γ^p that is "glued" to other subdomains. In particular, we shall denote by Γ^{pq}_C the part of Γ^p that can be, in the solution, in contact with the body Ω^q, and we shall denote by Γ^{pq}_E the part of Γ^p that is glued to Ω^q. Obviously $\Gamma^{pq}_E = \Gamma^{qp}_E$ and $\Gamma^{pq}_C = \Gamma^{qp}_C$. Let us recall that the gluing conditions require continuity of displacements and of their derivatives across all Γ^p_E.

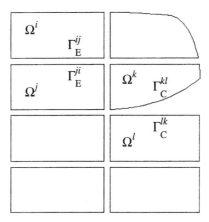

Figure 16.1: Contact problem with optional decomposition

We shall look for the displacements that satisfy the conditions of equilibrium in the set $\mathbf{K} = \mathbf{K}^E \cap \mathbf{K}^I$ of all kinematically admissible displacements $\mathbf{v}$ of the Sobolev space

$$\mathcal{V} = H^1(\Omega^1)^d \times \ldots \times H^1(\Omega^s)^d, \tag{16.1}$$

where $\mathcal{K}^E = \{\mathbf{v} \in \mathcal{V} : \mathbf{v}^p = \mathbf{U} \text{ on } \Gamma^p_U \text{ and } \mathbf{v}^p(\mathbf{x}) = \mathbf{v}^q(\mathbf{x}), \mathbf{x} \in \Gamma^{pq}_E\}$ comprises elements of $\mathcal{V}$ that are continuous across all Γ^{pq}_E and $\mathbf{K}^I$ comprises those that satisfy the noninterpenetration conditions [20, 21]. The displacement $\mathbf{u} \in \mathbf{K}$ of the system of bodies in equilibrium then minimizes the energy functional J so that

$$J(\mathbf{u}) \leq J(\mathbf{v}) \text{ for any } \mathbf{v} \in \mathbf{K}. \tag{16.2}$$

Conditions that guarantee existence and uniqueness of the solution may be expressed in terms of coercivity of J and may be found, for example, in [20, 21].

The linearization and finite element discretization of $\Omega = \Omega^1 \cup \ldots \cup \Omega^s$ with suitable numbering of nodes results in the QP problem

$$\frac{1}{2}\boldsymbol{u}^T K \boldsymbol{u} - \boldsymbol{f}^T \boldsymbol{u} \to \min \quad \text{subject to} \quad B_I \boldsymbol{u} \le \boldsymbol{c} \quad \text{and} \quad B_E \boldsymbol{u} = \boldsymbol{0} \qquad (16.3)$$

with a symmetric block-diagonal matrix $K = \mathrm{diag}(K_1, \ldots, K_s)$ of order n, $\boldsymbol{f} \in \mathbb{R}^n$, an $m \times n$ full rank matrix B_I, $\boldsymbol{c} \in \mathbb{R}^m$, and an $l \times n$ full rank matrix B_E. The diagonal blocks K_p that correspond to subdomains Ω^p are positive definite or semidefinite sparse matrices. Moreover, we assume that the nodes are numbered in such a way that the K_p are banded matrices that can be effectively decomposed by a variant of the Cholesky factorization [16]. The vector $\boldsymbol{f}$ describes the nodal forces arising from the volume forces or some other tractions, the matrix B_I and the vector $\boldsymbol{c}$ describe the linearized incremental non-interpenetration conditions, and B_E is used to enforce the continuity of the displacements across the auxiliary interfaces Γ_E^p. More details may be found in [10].

Even though (16.3) is a standard convex QP problem, it is not suitable for numerical solution. The reasons are that K is typically ill-conditioned or singular, and that the feasible set is so complex that projections onto it can hardly be effectively computed, so that it would be very difficult to achieve fast identification of the contact interface and fast solution of auxiliary linear problems. The complications mentioned above may be essentially reduced by applying the duality theory of convex programming (e.g. Dostál et al. [5, 10]). Since the case with regular K was described in [10], we shall assume that the matrix K has a nontrivial null space that defines the natural coarse grid [24].

The Lagrangian associated with problem (16.3) is

$$L(\boldsymbol{u}, \boldsymbol{\lambda}_I, \boldsymbol{\lambda}_E) = \frac{1}{2}\boldsymbol{u}^T K \boldsymbol{u} - \boldsymbol{f}^T \boldsymbol{u} + \boldsymbol{\lambda}_I^T (B_I \boldsymbol{u} - \boldsymbol{c}) + \boldsymbol{\lambda}_E^T B_E \boldsymbol{u}, \qquad (16.4)$$

where $\boldsymbol{\lambda}_I$ and $\boldsymbol{\lambda}_E$ are the Lagrange multipliers associated with the inequalities and equalities, respectively. Introducing notation

$$\boldsymbol{\lambda} = \begin{bmatrix} \boldsymbol{\lambda}_I \\ \boldsymbol{\lambda}_E \end{bmatrix}, \quad B = \begin{bmatrix} B_I \\ B_E \end{bmatrix} \quad \text{and} \quad \hat{\boldsymbol{c}} = \begin{bmatrix} \boldsymbol{c} \\ \boldsymbol{0} \end{bmatrix},$$

we can write the Lagrangian briefly as

$$L(\boldsymbol{u}, \boldsymbol{\lambda}) = \frac{1}{2}\boldsymbol{u}^T K \boldsymbol{u} - \boldsymbol{f}^T \boldsymbol{u} + \boldsymbol{\lambda}^T (B\boldsymbol{x} - \hat{\boldsymbol{c}}).$$

It is well known that (16.3) is equivalent to the problem of finding $(\hat{\boldsymbol{u}}, \hat{\boldsymbol{\lambda}})$ so that

$$L(\hat{\boldsymbol{u}}, \hat{\boldsymbol{\lambda}}) = \sup_{\boldsymbol{\lambda}_I \ge 0} \inf_{\boldsymbol{u}} L(\boldsymbol{u}, \boldsymbol{\lambda}). \qquad (16.5)$$

If we eliminate $\boldsymbol{u}$ from (16.5), we shall get the minimization problem

$$\min \Theta(\boldsymbol{\lambda}) \quad \text{s.t.} \quad \boldsymbol{\lambda}_I \ge 0 \quad \text{and} \quad R^T(\boldsymbol{f} - B^T \boldsymbol{\lambda}) = 0, \qquad (16.6)$$

where R denotes a matrix whose columns span the null space of K, $K^\dagger$ denotes any matrix that satisfies $KK^\dagger K = K$, and

$$\Theta(\boldsymbol{\lambda}) = \frac{1}{2}\boldsymbol{\lambda}^T B K^\dagger B^T \boldsymbol{\lambda} - \boldsymbol{\lambda}^T (B K^\dagger \boldsymbol{f} - \hat{\boldsymbol{c}}). \qquad (16.7)$$

Farhat and Roux [15] proposed to use for $K^\dagger$ the left generalized inverse that satisfies $K_p^\dagger = K_p^{-1}$ whenever K_p is non-singular and $K^\dagger = \text{diag}(K_1^\dagger, \ldots, K_s^\dagger)$. The most important fact is that the product of such a $K^\dagger$ with a vector may be carried out effectively [16]. Once the solution $\boldsymbol{\lambda}$ of (16.6) is obtained, the vector $\boldsymbol{u}$ that solves (16.5) can be evaluated by means of explicit formula that may be found in [5, 10].

We assume that the matrix $R^T B^T$ is a full rank matrix, so that the Hessian of Θ is positive definite. Moreover, the Hessian is closely related to that of the basic FETI method by Farhat and Roux [18, 15], so that its spectrum is relatively favorably distributed for the application of the conjugate gradient method.

16.3 PROJECTORS AND AUGMENTED LAGRANGIAN

Even though problem (16.6) is much more suitable for computations than (16.3) and was used for efficient solution of contact problems [10], further improvement may be achieved by adapting simple observations and results of Farhat, Mandel and Roux [17]. We shall formulate a problem that is equivalent to (16.6), but for which the augmented Lagrangian has such a spectral distribution that it is possible to give an in a sense optimal estimate of the rate of convergence of unconstrained minimization by the conjugate gradient method.

Let us denote $F = BK^\dagger B^T$, $\widetilde{\boldsymbol{d}} = BK^\dagger \boldsymbol{f}$, $\widetilde{G} = R^T B^T$, $\widetilde{\boldsymbol{e}} = R^T \boldsymbol{f}$, and let T denote a regular matrix such that the matrix $G = T\widetilde{G}$ has orthonormal rows. After denoting $\boldsymbol{e} = T\widetilde{\boldsymbol{e}}$, problem (16.6) reads

$$\min \ \frac{1}{2}\boldsymbol{\lambda}^T F\boldsymbol{\lambda} - \boldsymbol{\lambda}^T \widetilde{\boldsymbol{d}} \quad \text{s.t} \quad \boldsymbol{\lambda}_I \geq 0 \quad \text{and} \quad G\boldsymbol{\lambda} = \boldsymbol{e}. \tag{16.8}$$

Using the same reasoning as in the the linear case [24], the minimization problem (16.8) may be transformed to minimization on the subset of a vector space by means of $\overline{\boldsymbol{\lambda}}$ that satisfies $G\overline{\boldsymbol{\lambda}} = \boldsymbol{e}$. After denoting $\boldsymbol{d} = \widetilde{\boldsymbol{d}} - F\overline{\boldsymbol{\lambda}}$, the modified problem reads

$$\min \ \frac{1}{2}\boldsymbol{\lambda}^T F\boldsymbol{\lambda} - \boldsymbol{d}^T \boldsymbol{\lambda} \quad \text{s.t} \quad G\boldsymbol{\lambda} = 0 \quad \text{and} \quad \boldsymbol{\lambda}_I \geq -\overline{\boldsymbol{\lambda}}_I. \tag{16.9}$$

To assess our progress, let us compare the distribution of the spectrum of the Hessians $H_1 = F + \rho \widetilde{G}^T \widetilde{G}$ and $H_2 = F + \rho G^T G$ of the augmented Lagrangians for problems (16.6) and (16.9), respectively. Let us assume that the eigenvalues of F are in the interval $[a, b]$ and that the nonzero eigenvalues of $\widetilde{G}^T \widetilde{G}$ are in $[\gamma, \delta]$. For each square matrix A, let $\sigma(A)$ denote its spectrum. Using the analysis of [8, 7], it follows that

$$\sigma(H_1) \subseteq [a, b] \cup [a + \rho\gamma, b + \rho\delta] \quad \text{and} \quad \sigma(H_2) \subseteq [a, b] \cup [a + \rho, b + \rho].$$

If ρ is sufficiently large and $\gamma < \delta$, then the spectrum of H_1 is distributed in two intervals with the larger one on the right and the analysis of Axelsson [1] shows that the rate of convergence of the conjugate gradients for minimization of the quadratic function with Hessian H_1 depends on the penalization parameter ρ. However, the spectrum of H_2 is distributed in two intervals of the same length, so that the rate of convergence may be expressed by means of the effective condition number $\overline{\kappa}(H_2) = 4b/a$ and does not depend on the penalization parameter ρ [1].

Further improvement is based on observation that the augmented Lagrangian for (16.9) may be decomposed by the orthogonal projectors $Q = G^T G$ and $P = I - Q$ on the image space of G^T and on the kernel of G, respectively. Indeed, problem (16.9) is equivalent to

$$\min \quad \frac{1}{2}\boldsymbol{\lambda}^T PFP\boldsymbol{\lambda} - \boldsymbol{\lambda}^T Pd \quad \text{s.t} \quad G\boldsymbol{\lambda} = 0 \quad \text{and} \quad \lambda_I \geq -\bar{\lambda}_I \tag{16.10}$$

and the Hessian $H_3 = PFP + \rho Q$ of the augmented Lagrangian

$$L(\boldsymbol{\lambda}, \boldsymbol{\mu}, \rho) = \frac{1}{2}\boldsymbol{\lambda}^T (PFP + \rho Q)\boldsymbol{\lambda} - \boldsymbol{\lambda}^T Pd + \boldsymbol{\mu}^T G\boldsymbol{\lambda} \tag{16.11}$$

is decomposed by projectors P and Q whose image spaces are invariant subspaces of H_3. If $[a_P, b_P]$ denotes the interval that contains the non-zero eigenvalues of PFP, it follows that the eigenvalues of H_3 satisfy

$$\sigma(H_3) \subseteq [a_P, b_P] \cup \{\rho\} \quad \text{and} \quad [a_P, b_P] \subseteq [a, b] \tag{16.12}$$

so that the number k of conjugate gradient iterations that reduce the gradient of the augmented Lagrangian (16.11) for (16.10) by ϵ satisfies [2]

$$k \leq \frac{1}{2}\text{int}\left(\sqrt{\frac{b_P}{a_P}}\ln\left(\frac{2}{\epsilon}\right) + 3\right). \tag{16.13}$$

Moreover, analysis of the FETI method by Farhat, Mandel and Roux [17] implies that, for the regular decomposition,

$$\frac{b_P}{a_P} \leq \text{const}\,\frac{H}{h}, \tag{16.14}$$

where h and H are the mesh and subdomain diameters, respectively. Examining (16.13) and (16.14), we conclude that the rate of convergence for unconstrained minimization of the augmented Lagrangian (16.11) does not depend on either the penalization parameter ρ or the discretization parameter h provided the ratio H/h is kept bounded by a constant.

Our final step is based on the observation that we can further improve the rate of convergence of the conjugate gradient iterations if we shall use adaptive projectors P_J and Q_J to the intersections of the ranges of the projectors P and Q with the faces $\mathbf{W}_J = \{u : u_J = 0\}$ defined by active sets $J \subseteq I$, respectively. For example, to implement the projector Q_J, replace the submatrix G_J of G by zeros, orthonormalize the rows of the modified matrix and use the resulting matrix $\widehat{G}_J$ to define the adaptive projector $Q_J = \widehat{G}_J^T \widehat{G}_J$. Since the range of P_J is contained in the range of P and the range of Q_J is contained in the range of Q, we can assume improved convergence. Still further improvement may be achieved by adapting the standard FETI preconditioners [23, 22]. For example, the lumped preconditioner should be used in the form

$$C^{-1} = P_J BKB^T P_J + Q_J$$

to preserve both the preconditioning and penalization effects.

16.4   SOLUTION OF BOUND AND EQUALITY CONSTRAINED QUADRATIC PROGRAMMING PROBLEMS

The algorithm that we propose here may be considered a variant of the augmented Lagrangian type algorithm proposed by Conn, Gould and Toint [4] for identification of stationary points of more general problems. To solve our problem, the algorithm approximates new Lagrange multipliers for the equality constraints of (16.10) in the outer loop while QP problems with simple bounds are solved in the inner loop. However, the algorithm that we describe here is modified in order to exploit the specific structure of our problem. The most important of such modifications consists in including the adaptive precision control of auxiliary problems in Step 1 [11, 8]. The algorithm treats each type of constraint separately, so that efficient algorithms using projections and adaptive precision control [19, 6] may be used for the bound constrained QP problems.

To simplify our notation, let us denote $F_P = PFP$ so that the augmented Lagrangian for problem (16.10) and its gradient are given by

$$L(\boldsymbol{\lambda}, \boldsymbol{\mu}, \rho) = \frac{1}{2}\boldsymbol{\lambda}^T F_P \boldsymbol{\lambda} - \boldsymbol{\lambda}^T P \boldsymbol{d} + \boldsymbol{\mu}^T G \boldsymbol{\lambda} + \frac{1}{2}\rho ||Q\boldsymbol{\lambda}||^2$$

and

$$g(\boldsymbol{\lambda}, \boldsymbol{\mu}, \rho) = F_P \boldsymbol{\lambda} - P \boldsymbol{d} + G^T(\boldsymbol{\mu} + \rho G \boldsymbol{\lambda}),$$

respectively. The *projected gradient* $g^P = g^P(\boldsymbol{\lambda}, \boldsymbol{\mu}, \rho)$ of L at $\boldsymbol{\lambda}$ is then given entry-wise by

$$g_i^P = g_i \text{ for } \boldsymbol{\lambda}_i > -\overline{\boldsymbol{\lambda}}_i \text{ or } i \notin I \text{ and } g_i^P = g_i^- \text{ for } \boldsymbol{\lambda}_i = -\overline{\boldsymbol{\lambda}}_i \text{ and } i \in I$$

with $g_i^- = \min(g_i, 0)$, where I is the set of indices of constrained entries of $\boldsymbol{\lambda}$.

All of the parameters that must be defined prior to the application of the algorithm are listed in Step 0. Typical values of these parameters for our problems are given in brackets.

Algorithm 16.1 *(Simple bound and equality constraints)*

  Step 0.   { *Initialization of parameters*}
        Set $0 < \alpha < 1$ $[\alpha = 0.1]$, $1 < \beta$ $[\beta = 10]$, $\rho_0 > 0$ $[\rho_0 = 10^4]$, $\eta_0 > 0$ $[\eta_0 = 0.1]$,
        $M > 0$ $[M = 10^4]$, $\boldsymbol{\mu}^0$ $[\boldsymbol{\mu}^0 = 0]$ *and* $k = 0$.

  Step 1.   *Find* $\boldsymbol{\lambda}^k$ *so that* $||g^P(\boldsymbol{\lambda}^k, \boldsymbol{\mu}^k, \rho_k)|| \leq M||G\boldsymbol{\lambda}^k||$.

  Step 2.   *If* $||g^P(\boldsymbol{\lambda}^k, \boldsymbol{\mu}^k, \rho_k)||$ *and* $||G\boldsymbol{\lambda}^k||$ *are sufficiently small, then stop.*

  Step 3.   *If* $||G\boldsymbol{\lambda}^k|| \leq \eta_k$

  Step 3a.   *then* $\boldsymbol{\mu}^{k+1} = \boldsymbol{\mu}^k + \rho_k G\boldsymbol{\lambda}^k$, $\rho_{k+1} = \rho_k$, $\eta_{k+1} = \alpha\eta_k$

  Step 3b.   *else* $\rho_{k+1} = \beta\rho_k$, $\eta_{k+1} = \eta_k$

      end if.

Step 4.   Increase k and return to Step 1.

An implementation of Step 1 is carried out by the minimization of the augmented Lagrangian L subject to $\boldsymbol{\lambda}_I \geq 0$ by means of the algorithm that we shall describe later.

The algorithm has been proved [8] to converge for any set of parameters that satisfy the prescribed relations. Moreover, it has been proved that the asymptotic rate of convergence is the same as for the algorithm with exact solution of the auxiliary QP problems (i.e. $M = 0$) and that the penalty parameter is uniformly bounded. These results, with the above discussion on elimination of the negative effect of penalization, give theoretical support to Algorithm 16.1.

Let us describe the implementation of Step 1 of Algorithm 16.1, assuming that $\boldsymbol{\mu}$ and ρ are fixed and denoting $\theta(\boldsymbol{\lambda}) = L(\boldsymbol{\lambda}, \boldsymbol{\mu}, \rho)$. Let us recall that the set of indices of the dual variables λ_i is decomposed into two disjoint sets I and E with I denoting the indices of the constrained entries of $\boldsymbol{\lambda}$, and let us denote by $\mathcal{A}(\boldsymbol{\lambda})$ and $\mathcal{F}(\boldsymbol{\lambda})$ the *active set* and *free set* of indices of $\boldsymbol{\lambda}$, respectively, i.e.

$$\mathcal{A}(\boldsymbol{\lambda}) = \{i \in I : \lambda_i = -\overline{\lambda}_i\} \quad \text{and} \quad \mathcal{F}(\boldsymbol{\lambda}) = \{i : \lambda_i > -\overline{\lambda}_i \text{ or } i \in E\}. \tag{16.15}$$

The *chopped gradient* $\boldsymbol{g}^C$ and the *inner gradient* $\boldsymbol{g}^I$ of $\theta(\boldsymbol{\lambda})$ are defined by

$$g_i^I = g_i \text{ for } i \in \mathcal{F}(\boldsymbol{\lambda}) \text{ and } g_i^I = 0 \text{ for } i \in \mathcal{A}(\boldsymbol{\lambda}) \tag{16.16}$$

$$g_i^C = 0 \text{ for } i \in \mathcal{F}(\boldsymbol{\lambda}) \text{ and } g_i^C = g_i^- \text{ for } i \in \mathcal{A}(\boldsymbol{\lambda}). \tag{16.17}$$

Hence the Karush-Kuhn-Tucker conditions for the solution of the problem to find

$$\min \ \theta(\boldsymbol{\lambda}) \quad \text{s.t.} \quad \boldsymbol{\lambda}_I \geq -\overline{\boldsymbol{\lambda}}_I \tag{16.18}$$

are satisfied iff the projected gradient $\boldsymbol{g}^P = \boldsymbol{g}^I + \boldsymbol{g}^C$ vanishes.

An efficient algorithm for the solution of convex QP problems with simple bounds has been proposed independently by Friedlander and Martínez [19, 3] and Dostál [6]. The algorithm uses projections to avoid the lower bound on the number of iterates when many active constraints of the solution are missing in the initial active set and the precision control that speeds up releasing constraints from the active set and safeguards the algorithm from oscillations often attributed to the active set strategy.

If the inequality $\|\boldsymbol{g}^C(\boldsymbol{y}^i)\| \leq \Gamma \|\boldsymbol{g}^I(\boldsymbol{y}^i)\|$ holds for $\Gamma > 0$, then we call $\boldsymbol{y}^i$ *proportional* [6]. The algorithm explores the face $W_J = \{\boldsymbol{y} : y_i = -\overline{\lambda}_i \text{ for } i \in J\}$ with a given active set $J \subseteq I$ as long as the iterates are proportional. If $\boldsymbol{y}^i$ is not proportional, we generate $\boldsymbol{y}^{i+1}$ by means of the descent direction $\boldsymbol{d}^i = -\boldsymbol{g}^C(\boldsymbol{y}^i)$ in a step that we call *proportioning*, and then we continue exploring the new face defined by $J = \mathcal{A}(\boldsymbol{y}^{i+1})$. The class of algorithms driven by proportioning may be defined as follows.

Algorithm 16.2 *(General Proportioning Scheme - GPS)*
Let a feasible $\boldsymbol{\lambda}^0$ and $\Gamma > 0$ [$\Gamma = 1$] be given. For $i > 0$, choose $\boldsymbol{\lambda}^{i+1}$ by the following rules:
(i) If $\boldsymbol{\lambda}^i$ is not proportional, define $\boldsymbol{\lambda}^{i+1}$ by proportioning.
(ii) If $\boldsymbol{\lambda}^i$ is proportional, choose $\boldsymbol{\lambda}^{i+1}$ feasible so that $\theta(\boldsymbol{\lambda}^{i+1}) \leq \theta(\boldsymbol{\lambda}^i)$ and $\boldsymbol{\lambda}^{i+1}$ satisfies at least one of the conditions: $\mathcal{A}(\boldsymbol{\lambda}^i) \subset \mathcal{A}(\boldsymbol{\lambda}^{i+1})$, $\boldsymbol{\lambda}^{i+1}$ is not proportional, or $\boldsymbol{\lambda}^{i+1}$ minimizes θ subject to $\boldsymbol{\lambda} \in W_J, J = \mathcal{A}(\boldsymbol{\lambda}^i)$.

The set relation $\subset$ is used in the strict sense. Basic theoretical results have been proved in [19, 6, 3]. In particular, it has been shown that $\boldsymbol{\lambda}^k$ generated by Algorithm GPS with given $\boldsymbol{\lambda}^0$ and $\Gamma > 0$ converges to the solution of (16.18) and that the solution is reached in a finite number of steps provided Γ is sufficiently large or the problem is not dual degenerate.

Step (ii) of Algorithm GPS may be implemented by means of the conjugate gradient method. The most simple implementation of this step starts from $\boldsymbol{y}^0 = \boldsymbol{\lambda}^k$ and generates the conjugate gradient iterations $\boldsymbol{y}^1, \boldsymbol{y}^2, \ldots$ for $\min\{\theta(\boldsymbol{y}) : \boldsymbol{y} \in \mathcal{W}_J, \ J = \mathcal{A}(\boldsymbol{y}^0)\}$ until $\boldsymbol{y}^i$ is found that is not feasible or not proportional or minimizes $\theta(\boldsymbol{\lambda})$ subject to $\boldsymbol{\lambda}_I \geq -\overline{\boldsymbol{\lambda}}_I$. If $\boldsymbol{y}^i$ is feasible, then we put $\boldsymbol{\lambda}^{k+1} = \boldsymbol{y}^i$, otherwise $\boldsymbol{y}^i = \boldsymbol{y}^{i-1} - \alpha^i \boldsymbol{p}^i$ is not feasible and we can find $\tilde{\alpha}^i$ so that $\boldsymbol{\lambda}^{k+1} = \boldsymbol{y}^i - \tilde{\alpha}^i \boldsymbol{p}^i$ is feasible and $\mathcal{A}(\boldsymbol{\lambda}^k) \not\subset \mathcal{A}(\boldsymbol{\lambda}^{k+1})$. We shall call the resulting algorithm *feasible proportioning* [6]. An obvious drawback of the feasible proportioning algorithm is that it is usually unable to add more than one index to the active set in one iteration. A simple but efficient alternative is to replace the feasibility condition by $\theta(P\boldsymbol{y}^{i+1}) \leq \theta(P\boldsymbol{y}^i)$, where $P\boldsymbol{y}$ denotes the projection on the set $\Omega = \{\boldsymbol{y} : y_i \geq -\overline{\lambda}_i \text{ for } i \in I\}$. If the conjugate gradient iterations are interrupted by condition $\theta(P\boldsymbol{y}^{i+1}) > \theta(P\boldsymbol{y}^i)$, then a new iteration is defined by $\boldsymbol{\lambda}^{k+1} = P\boldsymbol{y}^i$. This modification of the feasible proportioning algorithm is called *monotone proportioning* [6].

The algorithm uses the single parameter Γ. We believe that a good choice is $\Gamma \approx 1$, as it seems reasonable to change the face when the error in the chopped gradient dominates that in the free gradient as the conjugate gradient method typically reduces only the latter.

The performance of the algorithm depends essentially on the rate of convergence of the conjugate gradient method that minimizes θ in faces. In our case, the optimality results (16.13) and (16.14) suggest that the examination of faces can be carried out efficiently.

16.5 NUMERICAL EXPERIMENTS

In this section, we report some results of numerical experiments to illustrate the practical behavior of various implementations of our algorithm. We have implemented Algorithm 16.1 to solve the basic dual problem (16.6) so that we can plug in the orthonormalization of the constraints (16.8) or the projectors to the natural coarse space (16.10). We have implemented our algorithm also with the modified lumped preconditioner.

In particular, we have solved the model problem resulting from the finite difference discretization of the following continuous problem:

$$\text{Minimize} \quad q(u_1, u_2) = \sum_{i=1}^{2} \left(\int_{\Omega^i} |\nabla u_i|^2 \, d\Omega - \int_{\Omega^i} f u_i \, d\Omega \right)$$
$$\text{subject to} \quad u_1(0, y) \equiv 0 \text{ and } u_1(1, y) \leq u_2(1, y) \text{ for } y \in [0, 1],$$

where $\Omega^1 = (0, 1) \times (0, 1)$, $\Omega^2 = (1, 2) \times (0, 1)$, $f(x, y) = -5$ for $(x, y) \in (0, 1) \times [0.75, 1)$, $f(x, y) = 0$ for $(x, y) \in (0, 1) \times (0, 0.75)$, $f(x, y) = -1$ for $(x, y) \in (1, 2) \times (0, 0.25)$ and $f(x, y) = 0$ for $(x, y) \in (1, 2) \times (0.25, 1)$. This problem is semicoercive due to the lack of Dirichlet data on the boundary of Ω^2.

The solution of the model problem may be interpreted as the displacement of two membranes under the traction f. The left membrane is fixed on the left and the left edge of the right membrane is not allowed to penetrate below the edge of the left membrane. The solution is unique because the right membrane is pressed down. More details about this model problem including some other results may be found in [13].

The model problem was discretized by regular grids defined by the stepsize $h = 1/n$ with $n + 1$ nodes in each direction per subdomain Ω^i, $i = 1, 2$. Each subdomain Ω^i was decomposed into $m \times m$ identical squares with dimensions $H = 1/m$.

The model problem was solved with the stopping criterium

$$\|g^P(\lambda, \mu, 0)\| \leq 10^{-4}\|d\| \quad \text{and} \quad \|G\lambda\| \leq 10^{-4}\|f\|.$$

The results confirmed high numerical scalability of the algorithm with projector to the natural coarse grid. If we kept $H/h = 32$, the number of the conjugate gradient iterations that were necessary to solve the problem ranged from 32 to 107 for the discretizations characterized by n ranging from $n = 64$ to $n = 512$. The finest discretization comprised 557040 primal variables and 31728 dual variables with 513 nodes on the free boundary. The number of iterations dropped by some 20% by application of the lumped preconditioners, so that it ranged from 23 to 81. The experiments were run on a SUN Sparc Ultra1 computer, under SunOS 5.5.1, using the f77 (version 4.0) FORTRAN compiler and double precision. The auxiliary problems were solved by QUACON, a routine developed in the Institute of Mathematics, Statistics and Scientific Computation at Unicamp [3]. The solution of the largest problem with more than half million primal variables required 590.5 and 476.2 seconds for unpreconditioned and preconditioned implementation, respectively. More details may be found in [12, 13].

To test performance of our algorithm on a more realistic problem, we have also considered a 3D contact problem of [9]. The problem comprises three elastic blocks with boundary conditions defined by prescribed zero normal displacements on the vertical boundaries and on the bottom of the model with exception of the boundaries of the excavation in the bottom block. The problem was discretized by the finite element method so that the resulting discrete problem comprised 6419 nodal variables and 382 dual variables on the contact interface. The problem was solved to the same precision as above to comply with our earlier experiments with the basic algorithm without the natural coarse space preconditioning We have not used any secondary decomposition so that our coarse grid was formed just by the two dimensional null space of the stiffness matrix that is generated by independent vertical movements of the upper two blocks. In spite of this, the number of the conjugate gradient iterations dropped from 296 reported in [9] to 211 showing that realistic medium size semicoercive contact problems may be solved efficiently even without optional decomposition of bodies.

16.6   COMMENTS AND CONCLUSIONS

We have described a new domain decomposition algorithm for the solution of coercive and semicoercive frictionless contact problems of elasticity. The method combines a variant of the FETI method with projectors to the natural coarse grid and recently developed algorithms for the solution of special quadratic programming problems. A new feature of these algorithms is the adaptive control of precision of the solution of auxiliary problems with effective usage of the projections to the natural coarse grid.

The implementation of this approach deals separately with each body or subdomain, so that it is suitable for parallelization. Theoretical results are presented that guarantee convergence and high numerical scalability of the algorithm. In particular, the algorithm is shown to be in a sense optimal with respect to both penalization and discretization parameters. First numerical experiments are in agreement with the theory and give further evidence that the algorithms presented are efficient. We believe that the performance of the algorithms may be further improved by adapting the standard regular preconditioners to the unconstrained minimization in faces.

REFERENCES

[1] O. Axelsson (1976). A class of iterative methods for finite element equations. *Comput. Meth. Appl. Mech. and Engrg.*, 9:127–137.

[2] O. Axelsson (1995). *Iterative Solution Methods.* Cambridge Univ. Press, Cambridge.

[3] R. H. Bielschowski, A. Friedlander, F. A. M. Gomes, J. M. Martínez and M. Raydan (1997). An adaptive algorithm for bound constrained quadratic minimization. *Investigación Operativa*, 7:67–102.

[4] A. R. Conn, N. I. M. Gould and Ph. L. Toint (1991). A globally convergent augmented lagrangian algorithm for optimization with general constraints and simple bounds. *SIAM J. Numer. Anal.*, 28:545–572.

[5] Zdeněk Dostál (1995). Duality based domain decomposition with proportioning for the solution of free boundary problems. *J. Comp. Appl. Math.*, 63:203–208.

[6] Zdeněk Dostál (1997). Box constrained quadratic programming with proportioning and projections. *SIAM J.Opt.*, 7:871–887.

[7] Zdeněk Dostál (1999). On preconditioning and penalized matrices. *Num. Lin. Alg. Appl.*, 6:109–114.

[8] Zdeněk Dostál, Ana Friedlander and Sandra A. Santos (October 1996). Augmented lagrangians with adaptive precision control for quadratic programming with simple bounds and equality constraints. Technical Report RP 74/96, IMECC-UNICAMP, University of Campinas.

[9] Zdeněk Dostál, Ana Friedlander and Sandra A. Santos (1997). Analysis of block structures by augmented lagrangians with adaptive precision control. In Z. Rakowski, editor, *Proc. of Geomechanics'96*, pages 175–180. A. A. Balkema, Rotterdam.

[10] Zdeněk Dostál, Ana Friedlander and Sandra A. Santos (1998). Solution of coercive and semicoercive contact problems by feti domain decomposition. *Contemporary Math.*, 218:82–93.

[11] Zdeněk Dostál, Ana Friedlander and Sandra A. Santos (1999). Augmented lagrangians with adaptive precision control for quadratic programming with equality constraints. *Comp. Opt. Appl.*, 14:37–53.

[12] Zdeněk Dostál, Francisco A. M. Gomes Neto and Sandra A. Santos. Solution of contact problems by feti domain decomposition with natural coarse-space projections. accepted for publication in Comput. Meth. Appl. Mech. Engrg.

[13] Zdeněk Dostál, Francisco A. M. Gomes Neto and Sandra A. Santos (October 1998). Duality-based domain decomposition with natural coarse-space for variational inequalities. Technical Report RP 63/98, IMECC-UNICAMP, University of Campinas.

[14] Zdeněk Dostál and Vít Vondrák (1997). Duality based solution of contact problems with Coulomb friction. *Arch. Mech.*, 49:453–460.

[15] C. Farhat, P. Chen and F.-X. Roux (1993). The dual schur complement method with well posed local neumann problems. *SIAM J. Sci. Stat. Comput.*, 14:752–759.

[16] C. Farhat and M. Gérardin (1997). On the general solution by a direct method of a large scale singular system of linear equations: application to the analysis of floating structures. *Int. J. Num. Met. Eng.*, 41:675–696.

[17] C. Farhat, J. Mandel and F.-X. Roux (1994). Optimal convergence properties of the feti domain decomposition method. *Comput. Meth. Appl. Mech. Engrg.*, 115:365–385.

[18] C. Farhat and F.-X. Roux (1992). An unconventional domain decomposition method for an efficient parallel solution of large-scale finite element systems. *SIAM J. Sci. Stat. Comput.*, 13:379–396.

[19] A. Friedlander and J. M. Martínez (1994). On the maximization of concave quadratic function with box constraints. *SIAM J. Opt.*, 4:177–192.

[20] I. Hlaváček, J. Haslinger, J. Nečas and J. Lovíšek (1988). *Solution of Variational Inequalities in Mechanics*. Springer Verlag, Berlin.

[21] N. Kikuchi and J. T. Oden (1988). *Contact Problems in Elasticity*. SIAM, Philadelphia.

[22] A. Klawonn and O. B. Widlund (1999). A domain decomposition method with lagrange multipliers for linear elasticity. Technical Report No. 780, Courant Institute, New York University.

[23] J. Mandel and R. Tezaur (1996). Convergence of a substructuring method with lagrange multipliers. *Numer. Math.*, 73:473–487.

[24] F.-X. Roux and C. Farhat (1998). Parallel implementation of direct solution strategies for the coarse grid solvers in 2-level feti method. *Contemporary Math.*, 218:158–173.

17 A MULTI-WELL PROBLEM FOR PHASE TRANSFORMATIONS

Mieczysław S. Kuczma

Institute of Structural Engineering, Poznan University of Technology
60-965 Poznan, Poland

ABSTRACT

The work[1] deals with the formulation and solution of the boundary value problem for a body made of a multiphase thermoelastic material undergoing stress-induced coherent martensitic phase transformations. The proposed approach rests on the minimization of elastic energy and the second principle of thermodynamics. The effective free energy density of the multiphase system is a result of the homogenization of the piecewise quadratic potential adopted. The deformation process is formulated as an evolution variational inequality, which is finally solved as a sequence of linear complementarity problems by an iterative algorithm. The existence and uniqueness of a solution to the problem is shown. The results of numerical simulation of the tensile test on a two-phase strip are in good qualitative agreement with experimental observations.

Key words. variational inequality, linear complementarity problem, finite element method, martensitic phase transformation, shape memory alloy, pseudoelasticity.

17.1 INTRODUCTION

We are concerned with the deformation process of a solid body composed of a thermo-elastic material undergoing stress-induced coherent martensitic phase transformations. The martensitic transformation is a first-order solid-to-solid phase change occurring in various crystalline solids, including pseudoelastic *shape memory alloys* (SMA). This intriguing material response is attributed to discontinuous changes in the crystal lattice of the high temperature phase, *austenite*, which possesses a greater symmetry, and that of the low temperature phase, *martensite*, which may exist in many variants. A characteristic feature of the first-order transformation is that it takes place provided some barrier in energy is overcome. So, although the resulting microstructure of SMA is usually reversible

---

[1]Dedicated to Professor Andrzej Gawęcki on the occasion of his 60th birthday.

under loading/unloading cycles, the deformation process is accompanied by dissipation of energy.

The mathematical structure of the problem under consideration resembles that of a plastic flow problem [14, 15]. For further details of the concepts used here the reader may consult [13], for other numerical methods used in phase transformation problems [17, 19].

17.2   MATHEMATICAL MODEL

In this section we discuss briefly a variational inequality formulation for the quasi-static evolution of a pseudoelastic solid undergoing martensitic phase transformations. Our approach is based on the minimization of the elastic energy [1, 5, 9, 21, 22, 3, 4, 20] and the second principle of thermodynamics supplemented with the postulate of realizability [16]. Let $\Omega \subset \mathbb{R}^d$ with $d = 1, 2, 3$ be an open region occupied by the body in its undeformed state at a temperature $\theta_0 > A_f^0$. Further, suppose that the material may appear in $N + 1$ preferred strain states: the parent phase (austenite), indexed with $i = N + 1 \equiv a$, and the N-variants of martensite. At a material point $\boldsymbol{x} \in \Omega$ we postulate the Helmholtz free energy W_i in the form

$$W_i(\boldsymbol{E}, \theta) = \frac{1}{2}(\boldsymbol{E} - \boldsymbol{D}_i) \cdot \mathbb{A}_i[\boldsymbol{E} - \boldsymbol{D}_i] + \varpi_i(\theta), \qquad 1 \leq i \leq N + 1,$$

where $\mathbb{A}_i$ is the isotropic elasticity tensor for i-th phase (variant), assumed the same for all phases, $\mathbb{A}_i = \mathbb{A}$, and $\boldsymbol{E} = \frac{1}{2}(\nabla\boldsymbol{u} + (\nabla\boldsymbol{u})^{\mathrm{T}})$ is the tensor of (small) strains, $\boldsymbol{u}$ being a continuous displacement vector field on Ω. By $\boldsymbol{D}_i$ we denote the transformation strain (domain) in i-th phase [2]. (Note that we may take the transformation strain of austenite equal to zero, $\boldsymbol{D}_a = \boldsymbol{0}$.) The function $\varpi_i(\theta)$ depends on temperature θ, which is treated here as a parameter, and it may be assumed that $\varpi_i(\theta) = C_v(\theta - \theta_0) - C_v\theta \ln(\theta/\theta_0) + e_i^0 - \theta s_i^0$ where e_i^0, s_i^0 are the energy and entropy constants of i-th phase, C_v the common specific heat. The free energy function of the material is a multi-well functional which is piecewise quadratic

$$W(\boldsymbol{E}) = \min_{1 \leq i \leq N+1} \{W_i(\boldsymbol{E})\}. \tag{17.1}$$

The function $W(\boldsymbol{E})$ is not quasiconvex and it is possible to find a boundary value problem for which there is no minimizer of the associated energy potential. This mathematical property of $W(\boldsymbol{E})$ reflects the tendency of the material system to form a finer and finer microstructure. Very often we are, however, interested in the description of the overall, macroscopic properties of media with fine microstructure, and then the techniques of homogenization allow one to surmount this deficiency and provide the desired results. Convexification and quasiconvexification are among these techniques. Denoting by c_i the volume fraction of i-th phase we can define the quasiconvexified (or, relaxed at fixed volume fractions) free energy of the mixture, cf. Kohn [9], as the following expression

$$\widetilde{W}(\boldsymbol{E}, \boldsymbol{c}) = \frac{1}{2}(\boldsymbol{E} - \boldsymbol{D}(\boldsymbol{c})) \cdot \mathbb{A}[\boldsymbol{E} - \boldsymbol{D}(\boldsymbol{c})] + \sum_{i=1}^{N+1} c_i \varpi_i + \frac{1}{2}\sum_{i=1}^{N+1}\sum_{j=1}^{N+1} c_i c_j B_{ij}. \tag{17.2}$$

where the effective transformation strain $D(c)$ is the convex combination,

$$D(c) \equiv \sum_{i=1}^{N+1} c_i D_i, \quad \text{with } c_i \geq 0, \ \sum_{i=1}^{N+1} c_i = 1, \tag{17.3}$$

that is an element of the convex hull spanned by the set of transformation strains D_i, $D(c) \in \mathrm{co}\{D_1, D_2, \dots, D_{N+1}\}$. Moreover B_{ij} are material constants, being entries of the positive semi-definite matrix B with zero diagonal elements. In a special case of two-phase material the homogenized energy function, $\widetilde{W}$, is illustrated in Fig. 17.1.

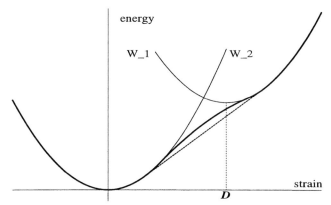

Figure 17.1: Homogenized free energy of a two phase system with parabolic energies W_1 and W_2, and transformation strains $D_1 \equiv D$ of martensite, and $D_2 = 0$ of austenite. The bold line corresponds to the quasiconvexification whereas the dotted bold line to the convexification of W.

We control the evolution of volume fractions c_i by making use of the second law of thermodynamics which requires that the (mechanical) dissipation should be non-negative,

$$\mathcal{D} = T \cdot \dot{E} - (\dot{\widetilde{W}} + s\dot{\theta}) \geq 0. \tag{17.4}$$

Under the assumption of constrained equilibrium and the relations for the stress tensor $T = \partial \widetilde{W}/\partial E = \mathbb{A}[E(u) - D(c)]$ and entropy $s \equiv -\partial \widetilde{W}/\partial \theta$, we may reduce (17.4) to

$$\mathcal{D} = -\frac{\partial \widetilde{W}}{\partial c} \cdot \dot{c} \equiv X \cdot \dot{c} = \sum_{m=1}^{N} X_m \dot{c}_m \geq 0. \tag{17.5}$$

where the N-tuple quantity X is the driving force of phase transformations, cf. [16, 23],

$$X_m = D_m \cdot \mathbb{A}[E] - \sum_{i=1}^{N} (D_m \cdot \mathbb{A}[D_i] + B_{mi}^{\star}) c_i - (\varpi_m - \varpi_a) - B_{am}. \tag{17.6}$$

with $c_a, a = N + 1$, being eliminated by the condition (17.3)$_3$, $B_{mi}^{\star} = B_{mi} - B_{ai} - B_{am}$.

For each variant of martensite let us define the phase transformation functions

$$F_m^+ = k_m^+ - X_m \geq 0, \qquad F_m^- = X_m - k_m^- \geq 0 \qquad \text{for} \quad 1 \leq m \leq N, \tag{17.7}$$

the threshold functions

$$
\begin{aligned}
k_m^+ &= \max\{L_m(c_m - p_m), 0\} \geq 0, \qquad \text{with} \quad p_m = 0, \\
k_m^- &= \min\{L_m(c_m - p_m), 0\} \leq 0, \qquad \text{with} \quad p_m = 1,
\end{aligned} \tag{17.8}
$$

and decompose the rate of volume fraction $\dot{c}_m$ into the positive and the negative parts

$$\dot{c}_m^+ = \max\{\dot{c}_m, 0\} \geq 0, \qquad \dot{c}_m^- = \max\{-\dot{c}_m, 0\} \geq 0. \tag{17.9}$$

The material constants $L_m > 0$ have the meaning of the energy dissipated during the forward and reverse transformation of the unit volume of m-th variant of martensite.

Accounting for the dissipation inequality (17.5) we stipulate the following phase transformation rules for the multi-phase mixture, requiring for each component X_m ($1 \leq m \leq N$) of the driving force that

$$
\begin{aligned}
&\text{if} \qquad X_m(\boldsymbol{c}) = k_m^+(c_m) \qquad \text{then} \quad \dot{c}_m \geq 0 \\
&\text{if} \qquad X_m(\boldsymbol{c}) = k_m^-(c_m) \qquad \text{then} \quad \dot{c}_m \leq 0 \\
&\text{if} \quad k_m^-(c_m) < X(\boldsymbol{c}) < k_m^+(c_m) \quad \text{then} \quad \dot{c}_m = 0
\end{aligned} \tag{17.10}
$$

which can be expressed as the following system of variational inequalities, cf. [14],

$$c_m \geq 0, \ \sum_{m=1}^{N} c_m \leq 1, \qquad \sum_{m=1}^{N} F_m^{\pm}(\boldsymbol{c}) \cdot (y_m^{\pm} - \dot{c}_m^{\pm}) \geq 0 \quad \text{for all } y_m^{\pm} \geq 0. \tag{17.11}$$

With $\boldsymbol{f}$ being a body force per unit volume, the equilibrium equations are

$$\text{div}\,\mathbb{A}[\boldsymbol{E}(\boldsymbol{u}) - \boldsymbol{D}(\boldsymbol{c})] + \boldsymbol{f} = \boldsymbol{0}. \tag{17.12}$$

Relations (17.10), or (17.11), and (17.12) describe completely the deformation process at a material point $\boldsymbol{x} \in \Omega$. Next a global formulation of the incremental boundary value problem is given.

17.2.1   Weak formulation

We integrate the rate problem (17.11), (17.12) by an implicit time integration method, imposing conditions (17.11) and (17.12) at selected (process) times $t_n \in [0, T]$, with $n = 1, 2, \ldots$, and $T < \infty$. Using the notations $\boldsymbol{u}_n \equiv \boldsymbol{u}(\cdot, t_n)$, $\boldsymbol{c}_n \equiv \boldsymbol{c}(\cdot, t_n)$ for displacements and volume fractions at time $t = t_n$ and the symbol Δ for finite increments, we define

$$
\begin{aligned}
\Delta \boldsymbol{u}_n &\equiv \boldsymbol{u}_n - \boldsymbol{u}_{n-1} \\
\Delta \boldsymbol{c}_n &\equiv \boldsymbol{c}_n - \boldsymbol{c}_{n-1}
\end{aligned}
$$

and we split the function Δc_n into its positive and negative parts,

$$\Delta c_n = \Delta c_n^+ - \Delta c_n^-.$$

Let $V(t_n)$ designate the set of kinematically admissible increments of displacements of the body Ω at time $t = t_n$, $V(t_n) = \{v \in H^1(\Omega, \mathbb{R}^d) | \ v(x) = w(x, t_n) \ \text{for a.e.} \ x \in \partial\Omega_u\}$ where $H^1(\Omega, \mathbb{R}^d)$ is a usual Hilbert space of vector-valued functions defined on Ω. By $\partial\Omega_u$ we denote a non-empty part of the boundary $\partial\Omega$ where displacements w are prescribed (at time t_n). The conditions $(17.11)_{1,2}$ dictate the following constraint sets on the increments of volume fraction, Δc_n, Δc_n^+, and Δc_n^-,

$$
\begin{aligned}
K(z) &= \left\{ w \in L^2(\Omega, \mathbb{R}^N) \ | \ |w_i| \leq 1, \ 0 \leq \textstyle\sum_{i=1}^{N}(z_i + w_i) \leq 1, \ z \in Z \right\} \\
K_+(z) &= \left\{ w \in L^2(\Omega, \mathbb{R}^N) \ | \ w_i \geq 0, \ \textstyle\sum_{i=1}^{N}(z_i + w_i) \leq 1, \ z \in Z \right\} \\
K_-(z) &= \left\{ w \in L^2(\Omega, \mathbb{R}^N) \ | \ w_i \geq 0, \ z_i - w_i \geq 0, \ z \in Z \right\} \\
Z &= \left\{ z \in L^2(\Omega, \mathbb{R}^N) \ | \ z_i \geq 0, \ \textstyle\sum_{i=1}^{N} z_i \leq 1 \right\}.
\end{aligned}
\tag{17.13}
$$

Now we are in a position to formulate a typical time-step of the incremental boundary value problem defined by (17.12) and (17.11) as the variational inequality.

Find $(\Delta u_n, \Delta c_n) \in V(t_n) \times K(c_{n-1})$ such that

$$a(\Delta u_n, v) - g(\Delta c_n, v) \ = \ f_{n,n-1}(v) \qquad \forall v \in V$$

$$\mp g(z_\pm - \Delta c_n^\pm, \Delta u_n) \pm h(\Delta c_n, z_\pm - \Delta c_n^\pm) \ \geq \ \mp b_{n-1}^\pm (z_\pm - \Delta c_n^\pm) \ \ \forall z_\pm \in K_\pm(c_{n-1})$$
$$\tag{17.14}$$

The bilinear forms are defined as follows

$$
\begin{aligned}
a(w_n, v) &= \int_\Omega \mathbb{A}[\nabla w_n] \cdot \nabla v \ dx \\
g(w_n, v) &= \sum_{m=1}^{N} \int_\Omega w_{m,n} \mathbb{A}[D_m] \cdot \nabla v \ dx \\
h(w_n, v) &= \sum_{m=1}^{N} \sum_{i=1}^{N} \int_\Omega (D_m \cdot \mathbb{A}[D_i] + B_{mi}^\star + \delta_{mi} L_m) w_{m,n} v_i \ dx
\end{aligned}
\tag{17.15}
$$

where δ_{ij} is the Kronecker delta, while the linear forms

$$
\begin{aligned}
f_{n,n-1}(v) &= \int_\Omega \Delta f_n \cdot v \ dx + (\text{terms on } \partial\Omega)_{n,n-1} \\
b_{n-1}^\pm(w) &= \sum_{m=1}^{N} \int_\Omega [B_{am} + (\varpi_m - \varpi_a) + L_m (c_{m,n-1} - p_{m,n-1})^\pm] w_m \ dx \quad (17.16) \\
&\quad \mp g(w, u_{n-1}) \pm h(c_{n-1}, w).
\end{aligned}
$$

17.3   EXISTENCE AND UNIQUENESS

We shall show that a bilinear form A associated with the variational inequality (17.14) is continuous and coercive on the product space $\bar{V} = V \times \Lambda$ with $\Lambda = L^2(\Omega, \mathbb{R}^N)$, under some assumptions to be presented here. Let $\bar{K} = V \times K$, with K of (17.13), stand for the convex set of constraints, $\bar{K} \subset \bar{V}$. The elements $\bar{u}, \bar{v} \in \bar{V}$ represent the pairs

$$\bar{u} = (\boldsymbol{u}, \boldsymbol{c}), \qquad \bar{v} = (\boldsymbol{v}, \boldsymbol{z}).$$

We consider finite increments of displacements $\Delta \boldsymbol{u}_n$ and volume fractions $\Delta \boldsymbol{c}_n$, but for the sake of simplicity in notation we designate here:

$$\boldsymbol{u} \equiv \Delta \boldsymbol{u}_n, \qquad \boldsymbol{c} \equiv \Delta \boldsymbol{c}_n, \qquad \langle l, \bar{v} \rangle \equiv f_{n,n-1}(\boldsymbol{v}) - b_{n-1}^+(\boldsymbol{z}) + b_{n-1}^-(\boldsymbol{z}). \tag{17.17}$$

The bilinear form $A : \bar{V} \times \bar{V} \longrightarrow \mathbb{R}$ is defined by

$$\begin{aligned} A(\bar{u}, \bar{v} - \bar{u}) &\equiv a(\boldsymbol{u}, \boldsymbol{v} - \boldsymbol{u}) - g(\boldsymbol{c}, \boldsymbol{v} - \boldsymbol{u}) - g(\boldsymbol{z} - (\boldsymbol{c}^+ - \boldsymbol{c}^-), \boldsymbol{u}) + \\ &\quad h(\boldsymbol{c}, \boldsymbol{z} - (\boldsymbol{c}^+ - \boldsymbol{c}^-)) \end{aligned} \tag{17.18}$$

and the problem (17.14) can be rewritten as the variational inequality:

Find $\bar{u} \in \bar{K}$ such that

$$A(\bar{u}, \bar{v} - \bar{u}) \geq \langle l, \bar{v} - \bar{u} \rangle \quad \text{for all } \bar{v} \in \bar{K}. \tag{17.19}$$

Regarding the material we assume that the elasticity tensor $\mathbb{A}$ possesses the usual properties of symmetry (used already in the previous section), and is uniformly point-wise stable with bounded entries, i.e., there exist constants $k_A, k' > 0$ such that for all symmetric 2nd order tensors $\boldsymbol{S}$,

$$\boldsymbol{S} \cdot \mathbb{A}[\boldsymbol{S}] \geq k_A |\boldsymbol{S}|^2, \qquad \max_{i,j,k,l} \|A_{ijkl}\|_\infty < k' < \infty. \tag{17.20}$$

The critical role in the question of uniqueness will be played by the symmetric matrix $\boldsymbol{B}^L$ whose elements depend upon the material properties $B_{ij}^\star$ and L_i,

$$B_{ij}^L = B_{ij}^\star + \delta_{ij} L_i, \quad 1 \leq i, j \leq N. \tag{17.21}$$

We assume that $\boldsymbol{B}^L$ is positive definite, i.e. there exists a constant $H_p > 0$ such that

$$\boldsymbol{y} \cdot \boldsymbol{B}^L \boldsymbol{y} \geq H_p |\boldsymbol{y}|^2 \quad \text{for all } \boldsymbol{y} \in \mathbb{R}^N. \tag{17.22}$$

Furthermore, it is natural to assume that each $\boldsymbol{D}_m$ is bounded, so for the average phase transformation strain tensor $\boldsymbol{D}(\boldsymbol{z})$ we have the estimation,

$$\boldsymbol{D}(\boldsymbol{z}) \cdot \boldsymbol{D}(\boldsymbol{z}) \leq k_D |\boldsymbol{z}|^2 \qquad k_D < \infty. \tag{17.23}$$

The required properties of the bilinear form A are formulated in two lemmas. The proof of the first is clear, in the latter we use the idea of Jiang [7] and a modification of Korn's inequality due to Kikuchi and Oden [8], which says that for a sufficiently smooth region Ω there exists a positive constant $k_K > 0$ such that

$$\int_\Omega \boldsymbol{E}(\boldsymbol{v}) \cdot \boldsymbol{E}(\boldsymbol{v}) \, \mathrm{d}x \geq k_K \|\boldsymbol{v}\|_V^2 \qquad \text{for every } \boldsymbol{v} \in V. \tag{17.24}$$

Lemma 17.1 *The bilinear form A is continuous on $\bar{V}$, i.e., there exists a positive constant k such that*

$$|A(\bar{u}, \bar{v})| \leq k \, \|\bar{u}\|_{\bar{V}} \, \|\bar{v}\|_{\bar{V}} \quad \text{for all} \quad \bar{u}, \bar{v} \in \bar{V}. \tag{17.25}$$

Lemma 17.2 *Let $\boldsymbol{B}^L$ be positive definite. The bilinear form A is coercive on $\bar{V}$, i.e., there exists a constant $k_0 > 0$ such that*

$$A(\bar{v}, \bar{v}) \geq k_0 \, \|\bar{v}\|_{\bar{V}}^2 \quad \text{for all} \quad \bar{v} \in \bar{V}. \tag{17.26}$$

Proof. Let $\beta = H_p/(H_p + 2k_A k_D)$. We have,

$$A(\bar{v}, \bar{v}) = \int_{\Omega} \left\{ (\boldsymbol{E}(\boldsymbol{v}) - \boldsymbol{D}(\boldsymbol{z})) \cdot \mathbb{A} \left[\boldsymbol{E}(\boldsymbol{v}) - \boldsymbol{D}(\boldsymbol{z}) \right] + \boldsymbol{z} \cdot \boldsymbol{B}^L \boldsymbol{z} \right\} \, \mathrm{d}x$$

$$\geq \int_{\Omega} \left\{ k_A (\boldsymbol{E}(\boldsymbol{v}) - \boldsymbol{D}(\boldsymbol{z})) \cdot (\boldsymbol{E}(\boldsymbol{v}) - \boldsymbol{D}(\boldsymbol{z})) + \boldsymbol{z} \cdot \boldsymbol{B}^L \boldsymbol{z} \right\} \, \mathrm{d}x$$

$$\geq \int_{\Omega} \left\{ k_A \beta \, |\boldsymbol{E}(\boldsymbol{v})|^2 - \frac{k_A \beta}{1 - \beta} |\boldsymbol{D}(\boldsymbol{z})|^2 + H_p |\boldsymbol{z}|^2 \right\} \, \mathrm{d}x +$$

$$\int_{\Omega} k_A \left| \sqrt{1 - \beta} \boldsymbol{E}(\boldsymbol{v}) - \frac{1}{\sqrt{1 - \beta}} \boldsymbol{D}(\boldsymbol{z}) \right|^2 \, \mathrm{d}x$$

$$\geq \int_{\Omega} k_A \beta \, |\boldsymbol{E}(\boldsymbol{v})|^2 \, \mathrm{d}x + \int_{\Omega} \left(H_p - \frac{k_A \beta k_D}{1 - \beta} \right) |\boldsymbol{z}|^2 \, \mathrm{d}x$$

$$\geq k_0 \left(\|\boldsymbol{v}\|_V^2 + \|\boldsymbol{z}\|_{\Lambda}^2 \right) = k_0 \, \|\bar{v}\|_{\bar{V}}^2, \tag{17.27}$$

with

$$k_0 = \min \left\{ \frac{k_A H_p k_K}{H_p + 2k_A k_D}, \frac{1}{2} H_p \right\}.$$

In the light of these lemmas the proof of the following theorem on existence and uniqueness is a standard result in the theory of variational inequalities [18].

Theorem 17.1 *Let the elasticity tensor $\mathbb{A}$ satisfy conditions (17.20), the matrix $\boldsymbol{B}^L$ be positive definite, and transformation strain tensors $\boldsymbol{D}_m$ satisfy condition (17.23). There exists a unique solution of the problem (17.19). The solution depends continuously on data $l \in \bar{V}^*$.*

17.4 NUMERICAL ALGORITHM AND EXAMPLE

To illustrate some of the effects covered by the proposed model we have simulated numerically the basic, uniaxial tensile test on a two-phase material strip, as a plane stress displacement-driven problem. For the solution of the finite dimensional counterpart of the system (17.14) we have developed two algorithms. The first algorithm [11] is based on the classical idea of pivoting. The other algorithm, which was used for solving the present example and will be described elsewhere, is an adaptation of the two-step iterative scheme (SSORP plus PCG) due to Kočvara and Zowe [10] to the box constraints (17.13).

The calculated strip and the imposed boundary conditions (17.28) are schematically displayed in Fig. 17.3a. The problem we chose is motivated by the laboratory experiments of Ichinose *et al.* [6]. Allowing the right-hand side of the strip to move transversely, we have applied the following boundary conditions:

on the left-hand side of the strip $\quad \begin{cases} u(0, y) = 0 & 0 \le y \le b, \\ v(0, b/2) = 0, \end{cases}$ (17.28)

on the right-hand side of the strip $\quad \{ \ u(a, y) = w(t) \quad 0 \le y \le b.$

The loading program $w(t)$, t being a process time, is a linear function increasing from zero to the scaled maximum value of $w(t_E)/a = 0.05159$ corresponding to point E in Fig. 17.3a. Further, we assumed that the thickness of the strip changes linearly along the x-axis with characteristic values of 1.00 mm in the middle ($x = 12.00$ mm) and of 0.99 mm at the ends ($x = 0.00$ and $x = 20.00$), while being constant along the y-direction. The variable thickness introduced makes the problem nonhomogeneous, and is to reflect in some sense the influence of temperature induced in the course of the austenite-martensite phase transformation, [24, 12]. Due the lack of a complete set of material data we have used the following material parameters [13]: Young's modulus $E = 10000.00$ MPa, $B_{12} = B_{21} = 0.60$ J/m^3, $L_1 = 1.2012$ J/m^3, so $B_{11}^L = 0.0012$, Poisson's ratio $\nu = 0.30$, difference in energies at the stress-free state, $\varpi_1 - \varpi_2 = 3.756$ J/m^3. The transformation strain tensor adopted is now

$$D = \begin{bmatrix} 0.04 & 0.06 \\ 0.06 & 0.04 \end{bmatrix}.$$

For the field of displacements $\boldsymbol{u}(\cdot, t) \equiv (u(\cdot, t),\, v(\cdot, t))$ we have used a 6-node triangle

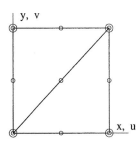

Figure 17.2: One cell of the finite element mesh in Fig. 17.3b: $\circ$ = node of the mesh of displacements (u, v), $\bigcirc$ = node of the mesh of volume fraction c.

finite element with quadratic shape functions, whilst for the volume fraction $c(\cdot, t)$ a 3-node linear triangle, see Fig. 17.2. The strip was divided into $(12 \times 100) \times 2$ finite elements with 5025 nodes of displacement field and 1313 nodes of volume fraction, so the number of unknowns of the problem was 12 676. The calculations of the deformation process were performed with the initial mesh of the undeformed strip, while Fig. 17.3b shows just by means of the mesh the characteristic deformation stages of the strip.

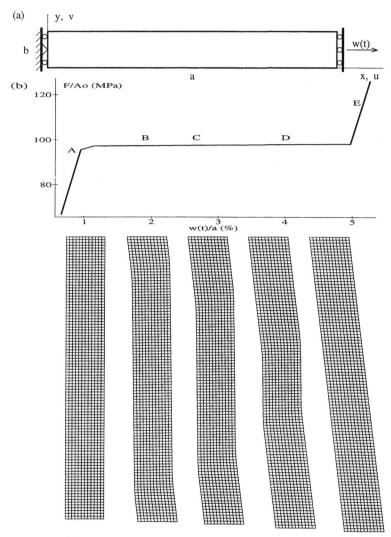

Figure 17.3: (a) The strip made of a material with two preferred states, of length $a = 20.00$ mm and width $b = 2.50$ mm, under uniaxial tension $w(t)$ and imposed boundary conditions. (b) The diagram of scaled force (F/A_0) vs. scaled elongation $(w(t)/a)$, and the shape of the deformed strip at selected states: A = (0.950, 95.479), B = (1.907, 97.370), C = (2.672, 97.557), D = (4.011, 97.885), E = (5.159, 116.454).

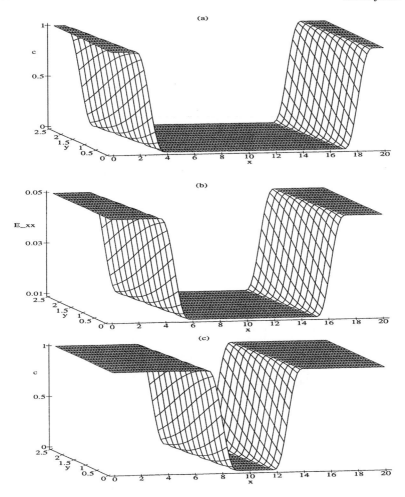

Figure 17.4: The distribution of volume fraction c at state B, (a), and state D, (c), and that of strain E_{xx} at state C, (b), of Fig. 17.3b.

In its initial state the strip was in the pure austenite phase, and elastically deformed the strip remains in this state at point A of Fig. 17.3b. Then the phase transformation starts at the left and right ends of the strip, leading finally to the state E of pure martensite in the whole strip. In the intermediate stages of the process the phase transformation fronts move from the ends to the middle, cf. states B, C, and D in Figs. 17.3b and 17.4. During the process there are moving regions of the strip in which the material is in the pure austenite phase ($c = 0$) or martensite phase ($c = 1$), and the transition zone where there is a mixture of both phases ($0 < c < 1$). Since the total strain consists of elastic strain and phase transformation strain, the normal component E_{xx} is non-zero also in the middle part of the strip at state C where $c = 0$, see Fig. 17.4b (the distribution of c

at state C is similar to that of E_{xx}). Note that even though the average strain at point D, $w/a = 0.04011$, is greater than the transformation strain component $D_{xx} = 0.04$, the phase transformation has not taken place in the whole strip at this level of elongation. The reason is the two-dimensional stress-strain state of the material.

17.5   CONCLUSIONS

The results obtained confirm the dominant shear character of the martensitic phase transformation, and are in good qualitative agreement with the experimental findings of Ichinose *et al.* [6]. On the other hand with reference to the numerical results presented in [13], our results show how different may be the deformation mode of the strip if boundary conditions are changed and/or some inhomogeneity is present. The proposed variational inequality formulation of the phase transformation process seems to be a very promising approach for dealing with the associated hysteretic boundary value problems.

ACKNOWLEDGEMENTS

The author would like to thank Professors V. Levitas, A. Mielke and E. Stein for valuable discussions. The work was supported by the Volkswagen Foundation under grant I/70284 and by the KBN grant to the Poznan University of Technology (DPB 11-669/99).

REFERENCES

[1] J.M. Ball and R.D. James (1987). Fine phase mixtures as minimizers of energy. *Arch. Rational Mech. Anal.*, 100(1):13–52.

[2] K. Bhattacharya (1993). Comparison of the geometrically nonlinear and linear theories of martensitic transformation. *Continuum Mech. and Themodyn.*, 5:205–242.

[3] M. Fremond (1996). *Shape memory alloy. A thermomechanical macroscopic study*, volume 351 of *CISM*, chapter 1, pages 1–67. Springer, Wien, New York.

[4] E. Fried and M. E. Gurtin (1996). Semi-quadratic variational problems for multiphase equilibria. *Quart. Appl. Math.*, 1:73–84.

[5] Y. Huo and I. Müller (1993). Nonequlibrium thermodynamics of pseudoelasticity. *Continuum Mech. and Thermodyn.*, 5:163–204.

[6] S. Ichinose, Y. Funatsu and K. Otsuka (1985). Type II deformation twinning in γ'_1 martensite in a Cu-Al-Ni alloy. *Acta metall.*, 33(9):1613–1620.

[7] L. Jiang (1984). On an elastic-plastic problem. *Jour. Diff. Eqns.*, 51:97–115.

[8] N. Kikuchi and J. T. Oden (1988). *Contact Problems in Elasticity: A Study of Variational Inequalities and Finite Elements Methods.* SIAM, Philadelphia, PA.

[9] R.V. Kohn (1991). The relaxation of a double-well energy. *Continuum Mech. Thermodyn.*, 3:193–236.

[10] M. Kočvara and J. Zowe (1994). An iterative two-step algorithm for linear complementarity problems. *Numer. Math.*, 68:95–106.

[11] M.S. Kuczma (1999). A viscoelastic-plastic model for skeletal structural systems with clearances. *Comp. Assist. Mech. Engng. Sci.*, 6(1):83–106.

[12] M.S. Kuczma, V.I. Levitas, A. Mielke and E. Stein (May 1997). Nonisothermal hysteresis loops in pseudoelasticity. In Proceedings of the XIII Conference on *Computer Methods in Mechanics*, pages 711–718. Poznan.

[13] M.S. Kuczma, A. Mielke and E. Stein (1999). Modelling of hysteresis in two-phase systems. *Arch. Mech.* (in press).

[14] M.S. Kuczma and E. Stein (1994). On nonconvex problems in the theory of plasticity. *Arch. Mech.*, 46(4):505–529.

[15] M.S. Kuczma and J.R. Whiteman (1995). Variational inequality formulation for flow theory plasticity. *Int. J. Engng Sci.*, 33(8):1153–1169.

[16] V. I. Levitas (1995). The postulate of realizability: Formulation and applications to post-bifurcation bahavior and phase transitions in elastoplastic materials. Part I and II. *Int. J. Engng. Sci.*, 33:921–971.

[17] V. I. Levitas, A.V. Idesman and E. Stein (1998). Finite element simulation of martensitic phase transformations in elastoplastic materials. *Int. J. Solids Structures*, 35:855–887.

[18] J.L. Lions and G. Stampacchia (1967). Variational inequalities. *Comm. Pure Appl. Math.*, XX:493–519.

[19] F. Marketz and F.D. Fischer (1994). Micromechanical modelling of stress-assisted martensitic transformation. *Modelling Simul. Mater. Sci. Eng.*, 2:1017–1046.

[20] A. Mielke, F. Theil and V.I. Levitas (September 1998). Mathematical formulation of quasistatic phase transformations with friction using an extremum principle. IfAM-Preprint A8, University of Hannover, Institute for Applied Mathematics.

[21] A.C. Pipkin (1991). Elastic materials with two prefferred states. *Quart. J. Mech Appl. Math.*, 44:1–15.

[22] B. Raniecki and Ch. Lexcellent (1998). Thermodynamics of isotropic pseudoelasticity in shape memory alloys. *Eur. J. Mech., A/Solids*, 17(2):185–205.

[23] B. Raniecki and K. Tanaka (1994). On the thermodynamic driving force for coherent phase transformations. *Int. J. Engng Sci.*, 32(12):1845–1858.

[24] J.A. Shaw and S. Kiriakides (1997). On the nucleation and propagation of phase transformation fronts in a NiTi alloy. *Acta mater.*, 45(2):683–700.

18 ADVANCED BOUNDARY ELEMENT ALGORITHMS

Christian Lage [2] and Christoph Schwab

[1] Seminar für Angewandte Mathematik,
ETH Zürich, CH-8092 Zürich, Switzerland

ABSTRACT

We review recent algorithmic developments in the boundary element method (BEM) for large scale engineering calculations. Two classes of algorithms, the clustering and the wavelet-based schemes are compared. Both have $O(N(\log N)^a)$ complexity with some small $a \geq 0$ and allow in-core simulations with up to $N = O(10^6)$ DOF on the boundary on serial workstations. Clustering appears more robust for complex surfaces.

Key words. Boundary element method, cluster method, wavelet method.

18.1 BOUNDARY ELEMENT METHOD

Let $\Omega \subset \mathbb{R}^3$ be a bounded polyhedron with N_0 straight faces Γ_k and boundary $\Gamma = \partial\Omega$. To present our algorithms, we consider the model problem

$$\Delta U = 0 \text{ in } \Omega, \quad U = f \text{ on } \Gamma . \tag{18.1}$$

Using the fundamental solution $e(x, y)$, (18.1) can be reduced in various ways to a boundary integral equation (BIE) for an unknown density u on Γ [2, 19, 21, 28, 43]:

$$Au = \lambda u + Ku = g \text{ on } \Gamma . \tag{18.2}$$

Here g is related to f, $\lambda = 0$ in formulations leading to equations of the first kind, and $\lambda \neq 0$ in indirect methods, based on layer potentials. The integral operator K is given by

$$(Ku)(x) = \int_{y \in \Gamma} k(x, y)\, u(y) ds_y, \quad x \in \Gamma , \tag{18.3}$$

---

[2]Present address: Coyote Systems, 2740 Van Ness Ave #210, San Francisco, CA 94109, USA

[1]This work was supported in part under the TMR network "Multiscale Methods in Numerical Analysis" of the EC by the Swiss National Fund

where the kernel $k(x, y)$ is a normal/ tangential derivative of $e(x, y)$ and satisfies for some $s \geq 0$

$$|D_x^\alpha D_y^\beta \, k(x, y)| \leq \frac{C(\alpha, \beta)}{|x - y|^{s + |\alpha| + |\beta|}}, \quad \forall \alpha, \, \beta \in \mathbb{N}_0^3. \tag{18.4}$$

Typically $k(x, y)$ is analytic if $x \neq y$ and singular at $x = y$. In most applications, $s = 1, 2$ or 3 and the "integral" in (18.3) has to be understood as finite part integral if $s \geq 2$ (see [41], [17] for more on such integrals). Many other, more general problems admit a BIE reformulation (18.2); for a reference, see e.g. [2, 21, 43] (in the case of vector-valued U in (18.1), the BIE (18.2) will consist of a matrix of integral operators with entries satisfying (18.3), (18.4)).

Boundary Element Methods (BEM) are discretizations which reduce (18.2) to the linear system

$$\boldsymbol{Au} = \lambda \boldsymbol{Mu} + \boldsymbol{Ku} = \boldsymbol{g} \tag{18.5}$$

for the unknown solution vector $\boldsymbol{u}$. Here $\boldsymbol{M} \in \mathbb{R}^{N \times N}$ is a "mass" matrix, and $\boldsymbol{K} \in \mathbb{R}^{N \times N}$ is the stiffness-matrix corresponding to K in (18.3). In BEM, it is obtained by

$$\boldsymbol{K} = \{K_{\lambda \lambda'}\}_{\lambda, \lambda' \in \Lambda}, \ K_{\lambda \lambda'} = (\mathcal{X}_\lambda(\mathcal{Y}_{\lambda'} \, k))_{\lambda, \lambda' \in \Lambda} \tag{18.6}$$

where Λ is some index set of cardinality $N = |\Lambda|$ and $\mathcal{X}_\lambda$, $\mathcal{Y}_\lambda$ are linear functionals on a suitable function space V on Γ in which (18.2) is well-posed and uniquely solvable. For example, let $\{x_\lambda\}_{\lambda \in \Lambda}$ be a set of collocation points on Γ and let $V^N \subset V$ with $V^N = \text{span}\{\varphi_\lambda : \lambda \in \Lambda\}$ be a finite element space on Γ. One may take

$$\mathcal{X}_\lambda \varphi := \varphi(x_\lambda), \quad \mathcal{Y}_{\lambda'} \varphi; = \varphi(x_{\lambda'}) \quad \text{Nyström Method,} \tag{18.7a}$$

$$\mathcal{X}_\lambda \varphi := \varphi(x_\lambda), \quad \mathcal{Y}_{\lambda'} \varphi := \langle \varphi, \varphi_{\lambda'} \rangle \quad \text{Collocation Method,} \tag{18.7b}$$

$$\mathcal{X}_\lambda \varphi := \langle \varphi, \varphi_\lambda \rangle, \quad \mathcal{Y}_{\lambda'} \varphi := \langle \varphi, \varphi_{\lambda'} \rangle \quad \text{Galerkin Method} \tag{18.7c}$$

where $\langle \cdot, \cdot \rangle$ is a suitable duality pairing on Γ. (18.7a) is not possible in general if $k(x, y)$ is singular for $x = y$, (18.7b), (18.7c) require special numerical integrations which are by now completely understood ([17, 38] for (18.7b) and [1, 30, 27, 37] for (18.7c), see also [15] for the farfield integrals).

Most of the classical BEM algorithms in engineering require manipulation of the full matrix $\boldsymbol{A}$ leading to operations of $O(N^2)$ complexity. The performance of such BEM algorithms is hence inferior to e.g. FE-multigrid methods applied directly to the PDE (18.1). Nevertheless, in recent years a large body of literature on the *formulation* of BIEs has appeared, in particular the derivation of well-posed variational first-kind BIEs ([21, 43, 28]), which includes mathematical convergence analysis of the discretizations.

We conclude that *for the viability of the BEM in industrial problems it is mandatory to reduce the complexity of* $\boldsymbol{u} \longmapsto \boldsymbol{Au}$ from $O(N^2)$ to (essentially) $O(N)$. This reduction should a) apply for general surfaces Γ of possibly complicated shape, b) be noticeable for N in the practical range of unknowns and c) be possible for all kernels $k(x, y)$ arising in applications.

We present in the following two classes of BE-Algorithms, referred to as *clustering-*, resp. *wavelet-algorithms*, to achieve these goals. Both replace $\boldsymbol{A}$ in (18.5) by approximations $\widetilde{\boldsymbol{A}}$ which can be manipulated in $O(N)$ complexity, while the error $\|\boldsymbol{A} - \widetilde{\boldsymbol{A}}\|$ is small.

Unless stated otherwise, we assume in the following that $\{V^N\}$ is a finite element space of piecewise polynomials on a quasiuniform mesh $\mathcal{M}$ on Γ of shape regular triangles of diameter $h = O(N^{-\frac{1}{2}})$.

18.2 CLUSTER METHODS

The cluster BEM contains panel-clustering [18] as well as multipole techniques [16] as special cases. The main idea is the approximation of $\boldsymbol{K}$ by

$$\widetilde{\boldsymbol{K}} = \boldsymbol{N} + \sum_{\sigma, \tau} \boldsymbol{X}_\sigma^T \, \boldsymbol{F}_{\sigma\tau} \, \boldsymbol{Y}_\tau \tag{18.8}$$

with a sparse *near-field matrix* $\boldsymbol{N} \in \mathbb{R}^{N \times N}$ and *far-field matrices* $\boldsymbol{X}_\sigma \in \mathbb{R}^{M \times N}$, $\boldsymbol{Y}_\tau \in \mathbb{R}^{M \times N}$ and $\boldsymbol{F}_{\sigma\tau} \in \mathbb{R}^{M \times M}$. The matrices $\boldsymbol{X}_\sigma$, $\boldsymbol{F}_{\sigma\tau}$, $\boldsymbol{Y}_\tau$ are never formed explicitly and $\boldsymbol{u} \longmapsto \widetilde{\boldsymbol{K}} \boldsymbol{u}$ is realized as

$$\widetilde{\boldsymbol{K}} \boldsymbol{u} = \boldsymbol{N} \boldsymbol{u} + \sum_{\sigma, \tau} \boldsymbol{X}_\sigma \left(\boldsymbol{F}_{\sigma\tau}(\boldsymbol{Y}_\tau \, \boldsymbol{u}) \right) , \tag{18.9}$$

without direct access to entries of $\boldsymbol{K}$. The amount of storage used to keep the necessary information of the far field is of order $O(M'N)$ with $M' \ll N$ which reduces the $O(N^2)$ complexity. The basic assumption on the kernel $k(x, y)$ underlying cluster methods is its local approximability by a degenerate kernel. Let $D \subset \mathbb{R}^3$ be any domain containing Γ.

Assumption 18.1 *Let* $0 \le \eta < 1$, $k \colon D \times D \to \mathbb{C}$ *a kernel function and* $\mathcal{I}$ *an index set. Then for all* $x_0, y_0 \in D$, $x_0 \ne y_0$, *and* $m \in \mathbb{N}_0$ *there exists an approximation* $\tilde{k}$ *of the form*

$$k(x, y) \sim \tilde{k}(x, y; x_0, y_0, m) := \sum_{(\mu, \nu) \in \mathcal{I}_m} \kappa_{(\mu, \nu)}^m(x_0, y_0) \, X_\mu(x; x_0) \, Y_\nu(y; y_0), \quad \mathcal{I}_m \subset \mathcal{I} \times \mathcal{I} \tag{18.10}$$

such that for all $x, y \in D$ *satisfying*

$$|y - y_0| + |x - x_0| \le \eta |y_0 - x_0| \tag{18.11}$$

the error is bounded by

$$|k(x, y) - \tilde{k}(x, y; x_0, y_0, m)| \le C \, e^{-C(\eta)m} |y - x|^{-s} \tag{18.12}$$

with $C(\eta) > 0$ *a decreasing function and* C *a constant, both independent of* m.

Examples of kernel approximations $\tilde{k}$ follow:

18.2.1   Taylor expansion ([18, 22, 35])

Let the kernel function k depend only on the difference of its arguments:

$$k(x, y) = k(y - x) .$$ (18.13)

We expand $k(y - x)$ formally into a Taylor series centered at $y_0 - x_0$ with $x_0, y_0 \in \mathbb{R}^d$:

$$
\begin{aligned}
k(y - x) &= \sum_{\mu \in \mathbb{N}_0^d} \frac{1}{\mu!} (D^\mu k)(y_0 - x_0)(y - x - y_0 + x_0)^\mu \\
&= \sum_{(\nu,\mu) \in \mathbb{N}_0^d \times \mathbb{N}_0^d} (D^{\mu+\nu} k)(y_0 - x_0) \frac{(x_0 - x)^\mu}{\mu!} \frac{(y - y_0)^\nu}{\nu!}
\end{aligned}
$$

This motivates an approximation (18.10) described by

$$
\begin{aligned}
\mathcal{I} &:= \mathbb{N}_0^d, \quad \mathcal{I}_m := \{(\mu, \nu) \in \mathcal{I} \times \mathcal{I} : |\mu + \nu| < m\}, \\
\kappa_{(\mu,\nu)}^m(x_0, y_0) &:= (D^{\mu+\nu} k)(y_0 - x_0), \\
X_\mu(x, x_0) &:= (x_0 - x)^\mu / \mu!, \quad Y_\nu(y, y_0) := (y - y_0)^\nu / \nu! .
\end{aligned}
$$ (18.14)

In [18, 19, 20, 22] the error bound (18.12) has been verified for several kernel functions. For example, (18.12) holds for the fundamental solution of Laplace's equation in $\mathbb{R}^3$,

$$k(x, y) = \frac{1}{4\pi} |y - x|^{-1}, \quad x, y \in \mathbb{R}^3, \ x \neq y$$ (18.15)

with $C = 1$ and $C(\eta) = \log \frac{1}{\eta}$.

18.2.2   Multipole expansion [16, 34]

The multipole expansions are a special case of (18.10). They are kernel specific in that the expansion coefficients must be evaluated analytically (but once and for all) for the kernel of interest. For the ubiquitous Coulomb-potential $k(x, y) = |x - y|^{-1}$, we have (18.10) with

$$\mathcal{J}_m := \{\mu \in \mathbb{N}_0 \times \mathbb{Z} : |\mu_2| \le \mu_1, \ \mu_1 < m\}, \quad \mathcal{I}_m := \{(\mu, \nu) \in (\mathcal{J}_m)^2 : \mu_1 + \nu_1 < m\}$$ (18.16)

$$\kappa_{(\mu,\nu)}^m(x_0, y_0) := \kappa_{\mu+\nu}(x_0, y_0) := \frac{1}{C_{\mu_1+\nu_1}^{\mu_2+\nu_2} |y_0 - x_0|^{\mu_1+\nu_1+1}} Y_{\mu_1+\nu_1}^{\mu_2+\nu_2} \left(\frac{y_0 - x_0}{|y_0 - x_0|} \right)$$ (18.17)

$$X_\mu(x; x_0) := C_{\mu_1}^{\mu_2} |x - x_0|^{\mu_1} Y_{\mu_1}^{-\mu_2} \left(\frac{x - x_0}{|x - x_0|} \right), \quad Y_\nu(y; y_0) := X_\nu(-y; -y_0)$$ (18.18)

with

$$C_\ell^p := \frac{i^{|p|}}{\sqrt{(\ell - p)!(\ell + p)!}}, \quad Y_\ell^p(x) := P_\ell^{|p|}(\cos \theta) e^{ip\phi}$$ (18.19)

for $x = (\cos\phi\sin\theta,\ \sin\theta,\ \cos\theta)^T \in \boldsymbol{S}_2$. The functions X_μ and Y_ν are solid spherical harmonics of positive degree whereas the expansion coefficients $\kappa^m_{(\mu,\nu)}$ are homogeneous harmonic polynomials of negative degree. *Note that the multipole expansion is nothing else but an efficient representation of the Taylor expansion* of $|y-x|^{-1}$. While for arbitrary kernel functions k, the index set of a truncated Taylor expansion contains $O(m^3)$ indices, only $O(m^2)$ coefficients must be stored to evaluate the Taylor expansion of $|y-x|^{-1}$ using the multipole ansatz according to (18.16)–(18.18). For the Helmholtz Equation, X_μ, Y_ν should depend on the wavenumber [14, 4], for elasticity [20] gives a multipole expansion and [13] shows how to obtain FMM for isotropic elasticity from FMM for the Laplacean.

18.2.3 Clustering Error

To employ a kernel approximation (18.10) for the construction of $\widetilde{\boldsymbol{K}}$ in (18.8), we partition the set $D \times D$. Let $\mathcal{P}(D)$ denote the set of all subsets of D, let $\check{r}_A, \check{c}_A$ be the Chebyšev radius and center, respectively, and 1_A the characteristic function of a set $A \subset \mathbb{R}^3$.

Definition 18.1 *Subsets σ, τ of D are called clusters. A finite set $\mathcal{C} \subset \mathcal{P}(D) \times \mathcal{P}(D)$ of related clusters is called a clustering of $D \times D$, iff*

$$\mathcal{X}_\lambda(\mathcal{Y}_{\lambda'}(k)) = \sum_{(\sigma,\tau)\in\mathcal{C}} \mathcal{X}_\lambda(\mathcal{Y}_{\lambda'}(k1_{\sigma\times\tau})).$$

For each $c \in \mathcal{C}$ let the relative size η_c be defined by

$$\eta_{(\sigma,\tau)} := \begin{cases} \dfrac{\check{r}_\sigma + \check{r}_\tau}{\|\check{c}_\sigma - \check{c}_\tau\|}, & \text{if } \check{c}_\sigma \neq \check{c}_\tau, \\ \infty & \text{otherwise.} \end{cases}$$

A far field $\mathcal{F} := \mathcal{F}_\mathcal{C}$ of $\mathcal{C}$ is a subset of $\{c \in \mathcal{C} : \eta_c < 1\}$ and the corresponding near field $\mathcal{N} := \mathcal{N}_\mathcal{C}$ the complement $\mathcal{C}\backslash\mathcal{F}_\mathcal{C}$. Far and near field imply a bipartition of $D \times D$:

$$D_\mathcal{F} := \bigcup_{(\sigma,\tau)\in\mathcal{F}} \sigma \times \tau, \quad D_\mathcal{N} := (D \times D)\backslash D_\mathcal{F}. \tag{18.20}$$

The grain of a far field is defined by $\eta_\mathcal{F} := \max_{c\in\mathcal{F}} \eta_c$.

In order to define precisely a cluster approximation of $\boldsymbol{K}$, we introduce the following notation for the restriction of subsets $\mathcal{T}$ of product sets $\mathcal{S} \times \mathcal{S}'$ to one of its components:

$$(\mathcal{T})_1 := \{s \in \mathcal{S} : \exists s' \in \mathcal{S}'(s,s') \in \mathcal{T}\}, \quad (\mathcal{T})_2 := \{s \in \mathcal{S}' : \exists s \in \mathcal{S}(s,s') \in \mathcal{T}\}. \tag{18.21}$$

Definition 18.2 *Let $m \in \mathbb{N}_0$ and $\mathcal{F}$ be a far field of $D \times D$. Let, in addition, $\tilde{k}$ be an approximation according to Assumption 18.1. Then, a cluster approximation $\widetilde{\boldsymbol{K}} := \widetilde{\boldsymbol{K}}(\mathcal{F}, m, \tilde{k})$ of $\boldsymbol{K}$ is defined by*

$$\widetilde{\boldsymbol{K}} := \boldsymbol{N} + \sum_{(\sigma,\tau)\in\mathcal{F}} \boldsymbol{X}_\sigma^T \boldsymbol{F}_{\sigma\tau} \boldsymbol{Y}_\tau \tag{18.22}$$

with $\boldsymbol{N} \in \mathbb{R}^{N \times N}$, $\boldsymbol{X}_\sigma \in \mathbb{R}^{(\mathcal{I}_m)_1 \times N}$, $\boldsymbol{Y}_\tau \in \mathbb{R}^{(\mathcal{I}_m)_2 \times N}$ and $\boldsymbol{F}_{\sigma\tau} \in \mathbb{R}^{(\mathcal{I}_m)_1 \times (\mathcal{I}_m)_2}$ given by [3]

$$(\boldsymbol{N})_{\lambda\lambda'} := \mathcal{X}_\lambda \left(\mathcal{Y}_{\lambda'}(k 1_{\mathcal{D}_N}) \right), \quad (\boldsymbol{X}_\sigma)_{\mu,\lambda} := \mathcal{X}_\lambda (X_\mu(\cdot, \check{c}_\sigma) 1_\sigma), \quad (\boldsymbol{Y}_\tau)_{\nu,\lambda'} := \mathcal{Y}_{\lambda'}(Y_\nu(\cdot, \check{c}_\tau) 1_\tau),$$

$$(\boldsymbol{F}_{\sigma\tau})_{\mu,\nu} := \begin{cases} \kappa^m_{(\mu,\nu)}(\check{c}_\sigma, \check{c}_\tau) & \text{if } (\mu,\nu) \in \mathcal{I}_m, \\ 0 & \text{otherwise}. \end{cases}$$

Then we have the following error bounds [26, 19, 18].

Theorem 18.1 *Suppose the linear functionals $\mathcal{X}_\lambda$ and $\mathcal{Y}_\lambda$ satisfy $\mathcal{X}_\lambda \phi \leq \mathcal{X}_\lambda \phi'$ and $\mathcal{Y}_\lambda \phi \leq \mathcal{Y}_\lambda \phi'$, respectively, for all $\lambda \in \Lambda$ and $0 \leq \phi \leq \phi'$ and consider a cluster approximation $\boldsymbol{K}(\mathcal{F}, m, \tilde{k})$ of $\boldsymbol{K}$ according to Definition 18.2. Let*

$$\rho := C(\eta_\mathcal{F}) \, m > 0 \tag{18.23}$$

with $\eta_\mathcal{F}$ the grain of $\mathcal{F}$ and $C(\cdot)$ as in (18.12). Then, the following error bounds are satisfied:

$$\|\widetilde{\boldsymbol{K}} - \boldsymbol{K}\|_\infty \leq C \, e^{-\rho} \, M_\mathcal{X}, \quad \|\widetilde{\boldsymbol{K}} - \boldsymbol{K}\|_1 \leq C \, e^{-\rho} \, M_\mathcal{Y},$$
$$\|\widetilde{\boldsymbol{K}} - \boldsymbol{K}\|_2 \leq C \, e^{-\rho} \, \sqrt{M_\mathcal{X} M_\mathcal{Y}} \tag{18.24}$$

where C is independent of m and Γ, and

$$M_\mathcal{X} := \max_{\lambda \in \Lambda} \sum_{\lambda' \in \Lambda} \mathcal{X}_\lambda \left(\mathcal{Y}_{\lambda'} \left(\frac{1_{\mathcal{D}_\mathcal{F}}}{|y - x|^s} \right) \right), \quad M_\mathcal{Y} := \max_{\lambda' \in \Lambda} \sum_{\lambda \in \Lambda} \mathcal{X}_\lambda \left(\mathcal{Y}_{\lambda'} \left(\frac{1_{\mathcal{D}_\mathcal{F}}}{|y - x|^s} \right) \right). \tag{18.25}$$

One verifies that for (18.24) to be bounded by ε, m in (18.25) must be of the order $|\log \varepsilon|$.

Remark 18.1 If $k(x,y) = e(x,y)$, the error in the kernel approximation (18.10) is completely independent of the surface Γ and its parametrization as well as of the distribution of the integration points on Γ. If $k(x,y)$ is a derivative of $e(x,y)$, as e.g. $k(x,y) = \partial_{n(y)} e(x,y)$ in the double layer potential, (18.10) can be derived in two ways: i) by applying $\partial_{n(y)}$ to an approximation $\tilde{e}(x,y)$ of e [4], and ii) by expanding $\partial_{n(y)} e(x,y)$ directly. *We emphasize that with option i) the constant C in the clustering errors (18.24) is independent of the complexity of Γ, resp. its parametric representation. This is not so for option ii) and for the wavelet methods below. Moreover, having a cluster approximation of the single layer potential only the $\boldsymbol{Y}_\tau$ matrices have to be recalculated to obtain a cluster approximation of the double layer potential when using option i).*

18.2.4 Clustering Algorithms

Essential to the efficiency of the algorithm is (i) the construction of a partition $\mathcal{C}$ such that the near field matrix $\boldsymbol{N}$ is a sparse matrix, i.e., contains only $O(N)$ entries, and (ii) the fast evaluation of the approximate far field contribution, in particular the fast evaluation of the matrix vector product

$$v = \sum_{(\sigma,\tau) \in \mathcal{F}} \boldsymbol{X}_\sigma^T \boldsymbol{F}_{\sigma\tau} \boldsymbol{Y}_\tau \, u. \tag{18.26}$$

---

[3]Note that identical formulations arise in the presence of numerical integration. In this case, the functionals $\mathcal{X}_\lambda, \mathcal{Y}_{\lambda'}$ are suitable quadrature formulas.

[4]The application of $\partial_{n(y)}$ can then be considered part $\mathcal{Y}_{\lambda'}$

The key is a hierarchical organization of clusters. Let $\mathcal{P}$ denote the given panelization of Γ. We subdivide $\mathcal{P}$ into two about equally large sets recursively until the subsets contain $O(1)$ panels. This defines a binary tree with root $\mathcal{P}$. Each node of the tree represents a subset of $\mathcal{P}$ which in turn implies a subset of Γ, i.e. the binary tree defines a hierarchical decomposition of Γ into clusters.

Let $0 < \eta < 1$. By traversing the tree a clustering $\mathcal{C} = \mathcal{F} \cup \mathcal{N}$ can be constructed:

```
partition (σ, τ, F, N) {
    if (η(σ,τ) < η) then
        F ← {(σ, τ)} ∪ F
    else if (σ is a leaf) or (τ is a leaf) then
        N ← {(σ, τ)} ∪ N
    else if (řσ < řτ) then
        for all children τ' of τ   partition(σ, τ', F, N)
    else
        for all children σ' of σ   partition(σ', τ, F, N)
}
```

The grain of the far field $\mathcal{F}$ will be bounded by η.

The matrix vector product (18.26) is evaluated in three steps:

1. evaluate for all τ: $\boldsymbol{u}_\tau := \boldsymbol{Y}_\tau \boldsymbol{u}$,

2. evaluate for all σ: $\boldsymbol{v}_\sigma := \begin{cases} \boldsymbol{F}_{\sigma\tau}\boldsymbol{u}_\tau & \text{for } (\sigma, \tau) \in \mathcal{F}, \\ 0 & \text{otherwise}, \end{cases}$

3. evaluate $\boldsymbol{v} = \sum_\sigma \boldsymbol{X}_\sigma^T \boldsymbol{v}_\sigma$.

The steps 1 and 3 could be accelerated by using so-called shift operations:

$$\boldsymbol{Y}_\tau = \sum_{\tau'\text{child of }\tau} \boldsymbol{D}_{\tau\tau'} \boldsymbol{Y}_{\tau'}, \tag{18.27}$$

with matrices $\boldsymbol{D}_{\tau\tau'}$, i.e.,

$$\boldsymbol{u}_\tau = \begin{cases} \boldsymbol{Y}_\tau \boldsymbol{u} & \text{for } \tau \text{ a leaf}, \\ \sum_{\tau'\text{ child of }\tau} \boldsymbol{D}_{\tau\tau'} \boldsymbol{u}_{\tau'} & \text{otherwise}. \end{cases} \tag{18.28}$$

Hence, to evaluate $\boldsymbol{u}_\tau$ for all τ we only have to assemble matrices $\boldsymbol{Y}_\tau$ if τ is a leaf. These matrices are sparse. In the case of the multipole ansatz, for example, they contain only $O(|\mathcal{J}_m|) = O(m^2)$ entries. The products $\boldsymbol{D}_{\tau\tau'}\boldsymbol{u}_{\tau'}$ are handled by efficient algorithms without assembling $\boldsymbol{D}_{\tau\tau'}$ explicitly [16, 34]. The same holds for step 3. With matrices $\boldsymbol{C}_{\sigma\sigma^*}$ defined by

$$\boldsymbol{X}_{\sigma^*}^T = \sum_{\sigma\text{ child of }\sigma^*} \boldsymbol{X}_\sigma^T \boldsymbol{C}_{\sigma\sigma^*}, \tag{18.29}$$

and vectors $\bar{\boldsymbol{v}}_\sigma := \boldsymbol{v}_\sigma + \boldsymbol{C}_{\sigma\sigma^*}\cdot\bar{\boldsymbol{v}}_{\sigma^*}$, σ child of σ^*, it follows that

$$\boldsymbol{v} = \sum_\sigma \boldsymbol{X}_\sigma^T \boldsymbol{v}_\sigma = \sum_{\sigma\text{ a leaf}} \boldsymbol{X}_\sigma^T \bar{\boldsymbol{v}}_\sigma. \tag{18.30}$$

Again, only matrices $\boldsymbol{X}_\sigma$ for leaves σ must be assembled.

An analysis of the complexity (cf. [26], [18]) shows that using the multipole evaluation the number of operations necessary to perform the matrix vector product (18.26) is of order $O(m^4 N)$, with N the number of unknowns.[5] The memory requirements are of order $O(m^2 N)$. To ensure that the error of the far field approximation is asymptotically equal to the order of the discretization error, we have to choose $m = O(\log N)$.

18.2.5   Numerical experiment

Here we present some numerical experiments in order to show the performance of the method. The goals of the experiments are to investigate the error dependence on the order m of the cluster expansion, and validation of the $O(N(\log N)^4)$ complexity of the algorithm.

We therefore consider the problem (18.1) in the unit shere Ω where the "true" potential $U(x)$ is given by

$$U(x) = |x|^{-1} + x_1 x_2 |x|^{-5} \tag{18.31}$$

and the boundary condition is $\partial_n U = f$ with $n(x)$ the exterior unit normal vector to Γ.

We approximated the unit sphere by planar triangles. Continuous, piecewise linear polynomials have been used as trial and test functions. The numerical quadrature for the near field integrals has been done using special quadrature techniques [22, 27].

The results were obtained on a *SUN* Ultra-Enterprise 4000/5000 on a single processor (*UltraSPARC*, 248MHz), 2 GB RAM using the *SUN* C++ 4.2 Compiler and the class library *Concepts-1.3* for boundary elements.

The linear system of equations was solved using a GMRES solver without any pre-conditioning. About 30 iterations were necessary to keep the error lower than the dis-cretization error, independent of the number of unknowns. For our cluster algorithm the matrix-vector operations for the calculation of the far field contribution have been done in every iteration step. The necessary information about the $\boldsymbol{X}_\sigma$, $\boldsymbol{Y}_\tau$ and $\boldsymbol{F}_{\sigma\tau}$ matrices have been stored in core on the workstation. The quality of the solution has been checked at a grid of points with distance 0.5 to the surface of the unit sphere.

Figure 18.1 shows the CPU-time for the matrix assembly for the standard BEM (dashed line) and our fast algorithm (solid lines). The latter depends on the order m of the multipole expansion. The computations have been done for $m = 3\ldots7$. The results are shown as function of the number of unknowns, i.e., of the resolution. The finest resolution contains 65538 unknowns, i.e., 131072 panels. The dependence of the CPU-time on the expansion order m is minor, because N dominates. Compared with the standard method a speed-up of up to 3 orders of magnitude is realized for the finest resolution.

---

[5]With a more sophisticated approach to evaluate the products $\boldsymbol{F}_{\sigma\tau}\boldsymbol{u}_\tau$ using exponential expansions this could be reduced to $O(m^3 N)$ [16].

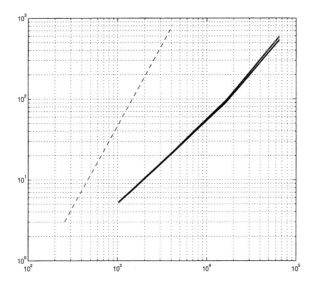

Figure 18.1: CPU-time for matrix assembly $(m = 3, 4, 5, 6, 7)$(in seconds) versus number N of panels: standard BEM (dashed line) versus fast algorithm (solid lines)

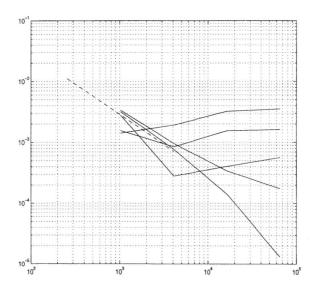

Figure 18.2: Relative mean absolute error in a set of points with distance 0.5 from the surface of the unit sphere: standard BEM (dashed line) versus fast algorithm for $m = 3, 4, 5, 6, 7$ (solid lines)

Figure 18.2 shows the relative mean absolute error in the potential at exterior points located at a distance of 0.5 from the surface $\partial\Omega$ for various m. The solid lines represent the cluster-BEM solution for $m = 3\ldots7$, the dashed line represents the standard-BEM solution. Only for $m = 6,7$ do we observe an almost monotone decreasing error with increasing number of unknowns. This indicates that small values of m corresponding to low expansion orders produce approximation errors that dominate the total error budget if the discretization becomes finer. At a certain discretization level $m = 5$ gives a better accuracy than $m = 7$. This can be explained by the influence of the discretization error which dominates at this discretization level the total error budget.

Figure 18.3 shows the compression rate as a function of the number of unknowns. A compression factor of 0.01 means that the total of entries to store the necessary information of the $\boldsymbol{X}_\sigma$, $\boldsymbol{F}_{\sigma\tau}$, $\boldsymbol{Y}_\tau$ matrices is equal to 1% of the entries of the dense stiffness matrix $\boldsymbol{A}$.

In Figure 18.4 we show the number of necessary matrix entries for the cluster-BEM and the standard-BEM as a function of the potential error in exterior points. It clearly shows that the higher the accuracy requirements are the more storage could be saved with the cluster-BEM.

18.3   WAVELET BEM

Wavelet-BEM are collocation or Galerkin BEM (18.7b), (18.7c) where the FE basis $\{\varphi_\lambda : \lambda \in \Lambda\}$ of V^N has been changed into a wavelet basis $\{\psi_\lambda : \lambda \in \Lambda\}$[6]. The basic idea is simple: the wavelets have *vanishing moments* in local surface coordinates and this causes most entries in the stiffness matrix $\boldsymbol{A}^\psi$ with respect to the wavelet basis ψ to be negligible. Let us illustrate this by the simplest surface wavelet.

18.3.1   Haar Multiwavelets on Γ

We assume each $\Gamma_k \subset \Gamma$ to be triangular and generate a dyadic sequence of meshes $\{\mathcal{M}_k^\ell\}_{0\leq\ell\leq L}$ on Γ_k by halving the sides of each triangle $T \in \mathcal{M}_k^\ell$. The corresponding mesh on Γ is then $\mathcal{M}^\ell = \cup_{k=1}^{N_0}\mathcal{M}_k^\ell$ with $N_\ell = N_0\,4^\ell$ triangles T_j^ℓ, $j = 0,\ldots,N_\ell-1$. Let φ_j^ℓ denote the characteristic function of T_j^ℓ, then we may write any piecewise constant function u^L on M_L as

$$u^L = \sum_{j=0}^{N_L-1} \{\boldsymbol{u}_\varphi^L\}_j\, \varphi_j^L \ . \tag{18.32}$$

A wavelet basis for $V^L = \mathrm{span}\{\varphi_j^L\}$ is obtained by adding only "new", incremental unknowns when refining from $\mathcal{M}^\ell \to \mathcal{M}^{\ell+1}$. This concept can also be described in terms of function spaces on Γ. Let V^ℓ be the space of piecewise constants on $\mathcal{M}^\ell$. Clearly,

$$V^0 \subset V^1 \subset \cdots \subset V^\ell \subset \ldots \tag{18.33}$$

---

[6]Since V^N was assumed to be a finite element space of piecewise, continuous or discontinuous, polynomials on Γ, $V^N = \mathrm{span}\{\psi_\lambda : \lambda \in \Lambda\}$ means that we consider here only *spline-wavelets*; the fractal, fully orthogonal Daubecies wavelets (e.g. [12]), could in principle also be used, but do not seem suitable to us for BEM since they cause difficulties in numerical integration.

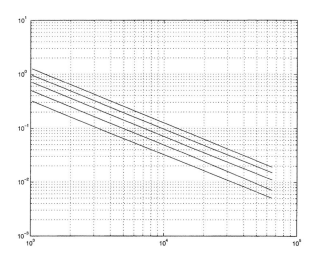

Figure 18.3: Compression of the stiffness matrix for $m = 3, 4, 5, 6, 7$.

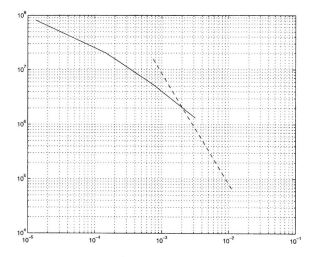

Figure 18.4: Number of necessary matrix entries as function of the potential error in exterior points: standard BEM (dashed line) versus fast algorithm ($m = 7$) (solid line).

and the sequence $\{V^\ell\}_\ell$ is dense in $V = L^2(\Gamma)$. Define $W^0 = V^0$ and

$$W^\ell = \{\psi \in V^\ell : \langle \psi, \varphi \rangle = 0 \quad \forall \varphi \in V^{\ell-1}\}, \quad \ell = 1, 2, 3, \ldots \tag{18.34}$$

where $\langle \cdot, \cdot \rangle$ denotes the L^2-inner product on Γ. Then

$$\dim W^\ell = \dim V^\ell - \dim V^{\ell-1} = N_\ell - N_{\ell-1} = N_0(4^\ell - 4^{\ell-1}) = 3N_0 \, 4^{\ell-1}\ .$$

For $\ell = 1$ there are 3 shape functions ψ_j^1 on each Γ_k, the so-called "motherwavelets", which are piecewise constant on $\mathcal{M}^1$ and have vanishing mean. A basis for W^ℓ is obtained by replication, i.e. translation and scaling, from the $\{\psi_j^1\}$. Clearly,

$$V^L = W^L \oplus W^{L-1} \otimes \cdots \otimes W^0 \tag{18.35}$$

and $u^L \in V^L$ can equivalently be expressed in the basis $\{\psi_j^\ell\}$:

$$u^L = \sum_{j=0}^{N_L-1} \{u_\varphi^L\}_j \, \varphi_j^L = \sum_{j=0}^{N_0-1} \{u_\psi^0\}_j \, \psi_j^0 + \sum_{\ell=0}^{L-1} \sum_{j=0}^{3N_\ell-1} \{u_\psi^{\ell+1}\}_j \, \psi_j^{\ell+1}\ . \tag{18.36}$$

Translation between the coefficient vectors $\{u_\varphi^L\}$ and $\{u_\psi^\ell\}_{\ell \le L}$ is achieved in $O(N_L)$ operations using the pyramid scheme:

$$\{u_\varphi^{\ell+1}\} = H_\ell\{u_\varphi^\ell\} + G_\ell\{u_\psi^{\ell+1}\} \tag{18.37}$$

where

$$H_\ell = \text{blockdiag} \left\{ \begin{pmatrix} 1 \\ 1 \\ 1 \\ 1 \end{pmatrix} \right\} \in \mathbb{R}^{N_{\ell+1} \times N_\ell}, \ G_\ell = \text{blockdiag} \left\{ \begin{pmatrix} 1 & -1 & -1 \\ -1 & 1 & -1 \\ -1 & -1 & 1 \\ 1 & 1 & 1 \end{pmatrix} \right\} \in \mathbb{R}^{N_\ell \times 3N_\ell}\ .$$

It is clear that higher order, discontinuous wavelets can be generated completely analogously (see [31] for degree 1 multiwavelets on triangles and [7, Section 9.3], for the general construction valid also on quads). Such wavelets satisfy an analog of Parseval's equality in $L^2(\Gamma)$: let $u \in L^2(\Gamma)$ be expanded in wavelets:

$$u = \sum_\lambda \langle u, \psi_\lambda \rangle \, \psi_\lambda\ , \tag{18.38}$$

then

$$\|u\|_{L^2(\Gamma)}^2 = \sum_\lambda |\langle u, \psi_\lambda \rangle|^2\ . \tag{18.39}$$

In particular, the Haar multiwavelets are fully orthogonal, i.e.

$$\langle \psi_\lambda, \psi_{\lambda'} \rangle = \delta_{\lambda\lambda'} \quad \forall \lambda\lambda'\ , \tag{18.40}$$

and the wavelets of (piecewise) degree p have vanishing moments of order $d = p + 1$, i.e.[7]

$$\langle p, \psi_\lambda \rangle = 0 \quad \forall p \in P_d(\Gamma), \ |\lambda| > 0\ , \tag{18.41}$$

where P_d is the set of piecewise polynomials on Γ of total degree $< d$.

---

[7]for curved surfaces Γ, (18.36) must hold for the pullback of ψ_λ in local coordinates.

18.3.2 Other wavelet families

The Haar-type multiwavelets are only suitable for second kind BIEs. For hypersingular BIEs (18.2) posed in $V = H^{1/2}(\Gamma)$, evidently discontinuous ψ_λ are not admissible. Several continuous, biorthogonal wavelets have been proposed. As a rule, to achieve the analog of (18.39) in $H^{1/2}(\Gamma)$ is difficult, since the $H^{1/2}(\Gamma)$ norm is, unlike the $L^2(\Gamma)$ norm, not additive on the Γ_k. The same applies to $H^{-1/2}(\Gamma)$ - here discontinuous wavelets are admissible, but one has to have a higher number of vanishing moments than for the Haar wavelets. Let in (18.4) $r = s - 2$ the order of A, and assume the vanishing moment condition

$$\langle p, \psi_\lambda \rangle = 0 \quad \forall p \in P_{\tilde{d}}(\Gamma), \ |\lambda| > 0 \tag{18.42}$$

where ψ_λ is a spline wavelet of degree $< d$. In order to achieve $O(N)$-algorithms $\tilde{d}$ must satisfy ([39])

$$d < \tilde{d} + r \Longleftrightarrow \tilde{d} > d - r \tag{18.43}$$

If equality holds in (18.43), as in the case of Haar-wavelets for second kind BIEs ($d = \tilde{d} = p + 1, r = 0$), only $O(N(\log N)^a)$, $0 < a \leq 2$ complexity is achievable.

Biorthogonal spline wavelets satisfying (18.43) for $r = -1, 0, 1$ are available on $\mathbb{R}^2$ (see [11], [7], [6]). On polyhedral surfaces, some delicate compatibility conditions at the edges must be satisfied; this can be done (see [9], [10] for these constructions) at the expense of a larger support. We also mention [32], [33] for C^0-wavelets and collocation of second kind BIEs. There, in particular, the first fully discrete schemes with nonanalytic surfaces Γ were obtained. The computational performance of these wavelets on surfaces is as yet open.

18.3.3 Wavelet BEM

For ease of notation, we combine the indices (ℓ, j) of ψ_j^ℓ into a multiindex $\lambda = (\ell, j)$, $|\lambda| = \ell$; then $V^L = \mathrm{span}\{\psi_\lambda : \lambda \in \Lambda\}$. Assume that A in (18.2) is a Fredholm operator of the second kind, i.e. $s \leq 2$ in (18.4). A wavelet version of the classical panel method can be obtained by letting x_λ be the barycenter of the $T^\lambda = T_j^\ell \in \mathcal{M}_\ell$, and $\mathcal{X}_\lambda \varphi = \varphi(x_\lambda)$, $\mathcal{Y}_{\lambda'} \varphi = \langle \varphi, \psi_{\lambda'} \rangle$. In the Galerkin variant also $\mathcal{X}_\lambda \varphi = (\varphi, \psi_\lambda)$, and we denote the resulting stiffness matrix by $\mathbf{A}^L$ to emphasize dependence on the level L. The problem (18.5) reads:

$$\mathbf{A}^L \mathbf{u}^L = \mathbf{g}^L . \tag{18.44}$$

18.3.4 Compression

The main effect of using a wavelet basis is the smallness of farfield entries of $\mathbf{A}^L$. Let

$$U_\lambda = \mathrm{supp}\ \mathcal{X}_\lambda, \quad U_{\lambda'} = \mathrm{supp}\ \mathcal{Y}_{\lambda'}, \quad \lambda, \lambda' \in \Lambda ,$$

and assume that the $\mathcal{X}_\lambda, \mathcal{Y}_{\lambda'}$ have *vanishing moments*:

$$\mathcal{X}_\lambda p = 0 \quad \forall p \in \mathcal{P}_{\tilde{d}}, \quad \mathcal{Y}_\lambda p = 0 \quad \forall p \in \mathcal{P}_{\tilde{d}'} , \tag{18.45}$$

where $\mathcal{P}_d$ denotes the polynomials of total degree $< d$ in local coordinates. Then

$$|A^L_{\mathcal{X}y}| \le C\,(d_{\lambda\lambda'})^{-s-\tilde{d}-\tilde{d}'} \tag{18.46}$$

whenever $d_{\lambda\lambda'} := \operatorname{dist}(U_\lambda, U_{\lambda'}) \ge c\,2^{-\min(|\lambda|,|\lambda'|)}$, i.e. *each vanishing moment of either* $\mathcal{X}_\lambda$ *or* $\mathcal{Y}_{\lambda'}$ *gives one extra order of decay in the farfield.* The constant C in (18.46) depends, in general, on the normalization of the ψ_λ. The decay estimate (18.46) is then a direct consequence of (18.4) and (18.45) ([7, 30, 39, 25]). The *matrix-compression* of $\boldsymbol{A}^L$ is:

$$\tilde{\boldsymbol{A}}^L_{\lambda,\lambda'} := \begin{cases} \boldsymbol{A}^L_{\lambda,\lambda'} & \text{if } d_{\lambda\lambda'} < \tau_{\lambda\lambda'} \\ 0 & \text{else} \end{cases} \tag{18.47}$$

where $\tau_{\lambda\lambda'}$ are judiciously chosen truncation parameters [8]). For second kind BIEs we have $s = 2$ in (18.4), and, for Galerkin BEM with Haar-multiwavelets of degree p, (18.45) holds with $\tilde{d} = \tilde{d}' = p+1$, and we find [25, 30] the rule

$$\tau_{\lambda\lambda'} = \max\{a2^{-|\lambda|}, a2^{-|\lambda'|}, a2^{L-|\lambda|-|\lambda'|}\} \tag{18.48}$$

which was used in our numerical simulations below. Note that for collocation, or operators of nonzero order, other truncation parameters have to be selected (see [7, 32, 33, 39] for more).

The estimate (18.46) indicates that vanishing moments of both, $\mathcal{X}_\lambda$ and $\mathcal{Y}_{\lambda'}$, contribute to the compression. Galerkin BEM (18.7c) are preferable from this point of view, since for the collocation case $\mathcal{X}_\lambda(\varphi) = \varphi(x_\lambda)$ has no vanishing moments at all. This can, however, be compensated by judicious linear combinations of Dirac's (see [32], Sect. 3.3 for an example).

18.3.5 Compression error estimate

We restrict ourselves to operators A of order zero and to Galerkin-BEM using the discontinuous Haar multiwavelets ψ_λ of degree $p \ge 0$ on triangulations. We have the

Theorem 18.2 *[30, 25]: Under the above assumptions, if a in (18.48) is sufficiently large, the compressed stiffness matrix $\tilde{\boldsymbol{A}}^L$ in (18.47) is stable, i.e. $\operatorname{cond}(\tilde{\boldsymbol{A}}^L) \le C < \infty$ for all L. Moreover, the error in the BEM solution $\tilde{u}^L$ obtained from the compressed matrix $\tilde{\boldsymbol{A}}^L$ can be estimated by*

$$\|u - \tilde{u}^L\|_{L^2(\Gamma)} \le C\,N_L^{-s/2}\,L^\nu\|u\|_{H^s(\Gamma)} = C\,h^s\,|\log h|^\nu\,\|u\|_{H^s(\Gamma)} \tag{18.49}$$

for $0 \le s \le p+1$ with $\nu = 0$ if $s < p+1$, $\nu = 1$ if $s = p+1$. Moreover, for the approximate solution $\tilde{U}^L$ of (18.1) holds: $\forall x \in \Omega$ ex. $C(x) > 0$ such that

$$|U(x) - \tilde{U}^L(x)| \le C(x)\,h^{-(s+\tilde{s})}|\log h|^{\nu(s)+\nu(\tilde{s})}\,\|u\|_{H^s(\Gamma)}\|g\|_{H^{\tilde{s}}(\Gamma)} \tag{18.50}$$

where $0 \le s, \tilde{s} \le p+1$ and $g \in H^{\tilde{s}}(\Gamma)$, $A^*\varphi = g \implies \varphi \in H^{\tilde{s}}(\Gamma)$ and $\nu(t) = 0$ for $0 \le t < p+1$, $\nu(p+1) = 3/2$. The compressed stiffness matrix $\tilde{\boldsymbol{A}}^L$ can be obtained using $O(N_L(\log N_L)^4)$ work and $O(N_L(\log N_L)^2)$ memory.

---

[8]In addition, also certain elements in the near field can be neglected, if the distance between U_λ and $U^{\text{sing}}_{\lambda'} = \operatorname{sing supp}(\mathcal{Y}_{\lambda'})$ gets large, see [7], (9.28) and [39].

There are analogs of the present theorem also for first kind, weakly and hypersingular BIEs as well as for collocation schemes; see [7, 39]. The estmate (18.50) shows that the superconvergence of potentials at interior points in the Galerkin-BEM is preserved under compression, *provided $Au = f$ and $A^*\varphi = g$ have sufficiently regular solutions for smooth f and g.* This is unrealistic on polyhedra due to edge- and vertex singularities. In many cases, one has (18.49), (18.50) only with $0 \leq s, \widetilde{s} < 1$, which is why the Haar-Galerkin wavelet with $p = 0$ is close to optimal *if uniform mesh refinement is used.* To exploit higher order wavelets on polyhedra, *adaptivity* must come into play. Recent, very technical results indicate that adaptive wavelet algorithms can be designed (see [5]) which realize optimal convergence rates at $O(N)$ complexity, but numerical experience has yet to be gained with these algorithms.

18.3.6   Numerical Experiments

In this section, we present the results of three numerical experiments obtained with the described implementation of the multiscale scheme. In a polyhedron $\Omega \subset \mathbb{R}^3$ we considered (18.1). The double layer ansatz $U(x) = \langle k(x, \cdot), u \rangle$ where the double layer kernel is given by

$$k(x, y) = -\frac{1}{4\pi} \frac{\langle n(y), y - x \rangle}{|y - x|^3} \tag{18.51}$$

leads with the jump relations to the second kind boundary integral equation

$$u \in L^2(\Gamma): \quad \langle v, Au \rangle = \langle v, f \rangle \quad \forall v \in L^2(\Gamma) \tag{18.52}$$

with

$$(Au)(x) = -\frac{1}{2} u(x) + \int_\Gamma k(x, y) u(y) \, ds_y \tag{18.53}$$

defined almost everywhere on Γ. We solved (18.52) on several polyhedral domains with quite similar performance. Here, we only report the results obtained with a tetrahedron defined by four equilateral triangles with vertices on the unit sphere $(\gamma \sim 1)$ and the right hand side

$$f(x) = |x - x_0|^{-1}, \qquad x_0 := (1, 1, 1)^T \text{ in the exterior of } \Omega. \tag{18.54}$$

For the discretization constant test and trial functions $(d = 0)$ were used. We did not make use of the fact that entries in the stiffness matrix corresponding to panels located in the same face of the polyhedron Ω vanish. All results were obtained on a *SUN Ultra-Enterprise 4000/5000* on a single processor (*UltraSPARC*, 248 MHz) and 2 GB RAM using the *SUN C++ 4.2* Compiler.

In the first experiment we kept the parameters a and $\alpha = \alpha'$ of the thresholds (18.47) controlling the compression fixed and solved the problem on various levels up to about 260000 unknowns (Table 18.1). On the finest mesh the compressed matrix consists of only 0.1% of the entries of the dense stiffness matrix. In addition, it can be observed that

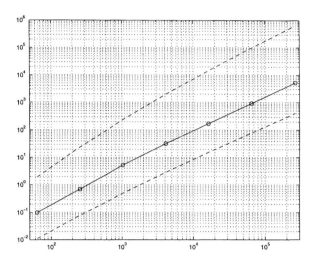

Figure 18.5: Time for assembly and compression of the matrix.

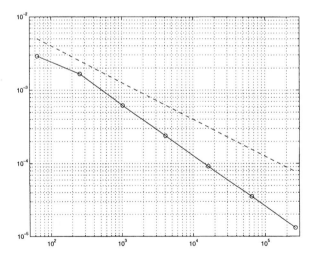

Figure 18.6: $|\,\|u^l\|_0 - \|u\|_0\,|$ versus N_L.

Table 18.1: First experiment: a, α = threshold parameters, *time* = time for assembly and solution, *mem* = memory required to store the compressed matrix inclusive management overhead, *it* = number of iterations, *cpr* = memory consumption with respect to a dense matrix.

level	N_L	a, α	time[s]	mem[MB]	it	cpr
2	64	0.3, 1.0	0.1	0.03	16	0.941
3	256	0.3, 1.0	0.8	0.21	18	0.424
4	1024	0.3, 1.0	5.9	1.17	19	0.146
5	4096	0.3, 1.0	36.7	5.80	19	0.045
6	16384	0.3, 1.0	193.2	27.44	18	0.013
7	65536	0.3, 1.0	1045.1	126.29	18	0.004
8	262144	0.3, 1.0	5744.9	570.43	19	0.001

the number of iterations used by the solver (GMRes without restart) is almost constant validating the bounded condition numbers of the compressed matrices.

In Figure 18.5 the time of assembly and compression is depicted. Here, the upper dashed line corresponds to the bound $O(N_L(\log N_L)^4)$ in Theorem 18.2 [30, 25]. The plot indicates that the influence of the higher order logarithmic terms on the computing time seems to be negligible compared to the $O(N_L(\log N_L)^2)$ term illustrated by the lower dashed line. Roughly speaking, on average a nearly constant number of operations is used to evaluate an entry of the stiffness matrix.

In all numerical experiments the time for GMRes accounts only for less than 10% of the total time shown in the tables. Therefore, with the present method the BEM-paradigm that most of the work is spent for quadrature is still valid and a speed up similar to the one for dense matrices can be achieved with the parallelization of the matrix assembly.

Figure 18.6 and Figure 18.7 show the behaviour of the L^2-error of the density u on the boundary and the average error in several interior points of the solution U, respectively. The L^2-error is approximated by the difference of the norm of the discrete density and the norm of the exact density. Since an exact solution is not available we computed an approximate value by higher order quadrature on a higher level with a parallelized implementation. According to Theorem 18.2, the expected rate of convergence is determined by regularity properties of A and its adjoint A^*. From the known edge and vertex singularities of the Laplacean in polyhedra it can be verified that in the example under consideration here both operators admit solutions belonging to $H^1(\Gamma)$ for smooth right hand sides. This means that we have Theorem 18.2 with $s = \tilde{s} = 1$ and expect essentially $O(N_L^{-1/2})$ convergence in the $L^2(\Gamma)$-norm and $O(N_L^{-1})$ convergence at an interior point (note that collocation or Nyström schemes do not display this kind of superconvergence at an interior point and would require H^2-regularity on Γ and $d = 1$ to achieve $O(N_L^{-1})$ convergence at an interior point).

Again, the dashed line in Figure 18.6 illustrates the expected behaviour of essentially $O(N_L^{-1/2})$. For the error in interior points, Figure 18.7, twice the convergence rate should be observed, hence essentially $O(N_L^{-1})$ (lower dashed line) or, according to Theorem 18.2,

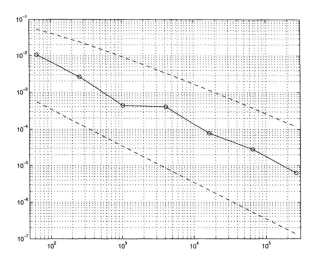

Figure 18.7: Error at interior points versus N_L.

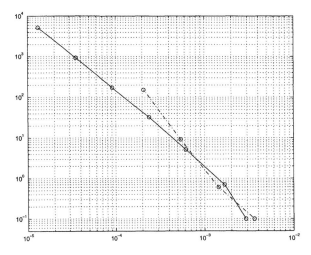

Figure 18.8: CPU-Time in sec versus $|\, \|u^l\|_0 - \|u\|_0 \,|$.

Table 18.2: Second experiment

level	N_L	a, α	time[s]	mem[MB]	it	cpr
8	262144	0.1, 1.0	5606.5	424.57	19	0.0008
8	262144	0.2, 1.0	5669.8	490.47	19	0.0009
8	262144	0.4, 1.0	5842.8	672.14	18	0.0013
8	262144	0.7, 1.0	6224.1	1009.25	18	0.0019
8	262144	1.0, 1.0	6778.2	1450.70	18	0.0028

Table 18.3: Third experiment.

level	N_L	a, α	time[s]	mem[MB]	it	cpr
8	262144	1.0, 0.2	5702.2	663.75	19	0.0013
8	262144	1.0, 0.4	5774.8	753.93	18	0.0014
8	262144	1.0, 0.6	5944.6	920.13	18	0.0018
8	262144	1.0, 0.8	6256.1	1137.70	18	0.0022
8	262144	1.0, 1.0	6778.2	1450.70	18	0.0028

$O(N_L^{-1}(\log N_L)^2)$ (upper dashed line).

Finally, we compared our method with a standard boundary element implementation generating the fully populated stiffness matrix with an optimized quadrature rule. For both methods the time and memory used to generate a solution satisfying a given L^2-error are depicted in Figure 18.8 and Figure 18.9, respectively, where the dashed line corresponds to the standard approach. It turns out that already for moderate accuracy, "moderate" with respect to our model problem, the wavelet method beats the standard approach: assuming an error of about 10^{-4} the wavelet method is 10 times faster. Moreover, in this case it saves about 98% of the memory. In addition, it can be observed that the memory consumption is always less than in the standard case although additional information to recover the structure of the compressed matrix (tags) must be stored.

The second experiment investigates the behaviour of the method when the amount of compression driven by the parameter a changes (Table 18.2). The constant number of iterations shows that even for a high compression the algorithm remains stable. The convergence rates, in addition, are in all cases preserved as indicated by the error in interior points shown in Figure 18.10: when the influence of the coarser meshes, where practically no compression is possible, vanishes, the lines corresponding to different values of a fan out. Nevertheless, they finally take the same slope. However, if the amount of compression is reduced by means of parameter α instead of a, this is not the case as predicted by Theorem 18.2 and observed in the last experiment (Table 18.3).

We point out that the influence of the amount of compression on the computing time, in particular the time of assembly, is small compared to the influence on the memory consumption (Tables 18.2,18.3). The reason for this is that the time to evaluate an entry of the stiffness matrix depends, via the quadrature order, on the distance of the supports of

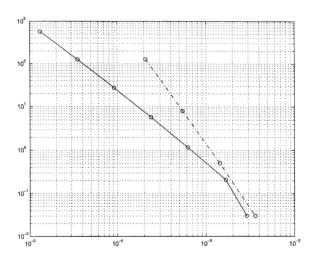

Figure 18.9: Memory versus $|\,\|u^l\|_0 - \|u\|_0\,|$.

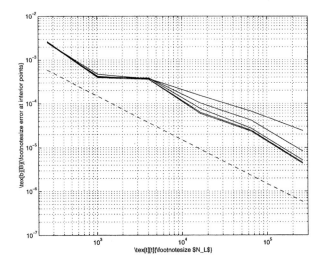

Figure 18.10: Error at interior points versus N_L, $a = 0.1, 0.2, 0.4, 0.7, 1.0$, $\alpha = 1.0$.

Table 18.4: Assembly of the system matrix using threads.

level	N_L	a, α	sequential time(s)	4 processors time(s) (speed up)
4	1024	0.3, 1.0	5.2	1.3 (4.0)
5	4096	0.3, 1.0	32.3	8.1 (4.0)
6	16384	0.3, 1.0	171.5	42.9 (4.0)
7	65536	0.3, 1.0	933.3	232.6 (4.0)
8	262144	0.3, 1.0	5200.3	1296.1 (4.0)
9	1048576	0.3, 1.0		4102.9

the related wavelets whereas the amount of memory to store the value is always the same. Increasing the thresholds means adding entries to the matrix with more or less distant support, which can be computed very fast compared to the entries near the diagonal. The time of solution, however, increases as fast as the memory.

To conclude, we report on an implementation of the matrix assembly using threads which were assigned to different processors on the *SUN Ultra-Enterprise 4000/5000*. Each thread processes rows and columns of the compressed matrix with a local cache. Rows and columns are assigned cyclically to ensure a static load balance. With this implementation the considered problem was solved up to level 9, i.e. with more than a million unknowns, on four processors. The CPU-times compared to the sequential version are shown in Table 18.4. A nearly perfect speed up is observed.

REFERENCES

[1] K. Atkinson, (1997). The numerical solution of integral equations of the second kind, Cambridge Univ. Press.

[2] M. Bonnet, (1999). Boundary integral equation methods for solid and fluids, Wiley, Chichester.

[3] G. Beylkin, R. Coifman and V. Rokhlin, (1991). Fast wavelet transforms and numerical algorithms, *Comm. Pure Appl. Math.* **44**, 141–183.

[4] W. C. Chew, J.-M. Jin, C.-C. Lu, E. Michielssen and J. M. Song (1997). Fast solution methods in eletromagnetics, *IEEE Trans. Ant. Propagation* **45**, 533–543.

[5] A. Cohen, W. Dahmen and R. DeVore (1998). Adaptive Wavelet Methods for elliptic Operator Equations, Convergence Rates, Preprint, IGPM, RWTH Aachen.

[6] A. Cohen, I. Daubecies and , J.-C. Feauveau (1992). Biorthogonal bases of compactly supported wavelets, *Comm. Pure. Appl. Math.* **45**, 485–560.

[7] W. Dahmen (1997). Wavelet and multiscale methods for operator equations, *Acta Numerica* **6**, 55–228.

[8] W. Dahmen, S. Prössdorf and R. Schneider (1994). Wavelet approximation methods for pseudodifferential equations on smooth manifolds, Proc. of the Int. Conf. on wavelets (C.K. Chui, L. Montefusco and L. Puccio, eds.), Academic Press, 385–424.

[9] W. Dahmen and R. Schneider (1996). Composite wavelet bases for operator equations, Preprint SFB393/96-19, TK-Chemnitz.

[10] W. Dahmen and R. Schneider (1999/2000). Wavelets on manifolds I: Construction and Domain Decomposition, Preprint.

[11] W. Dahmen and R. Stevenson (1999/2000). Element-by-element construction of wavelets satisfying stability and moment conditions, to appear in *SIAM J. Numer. Anal.*

[12] I. Daubecies (1992): 10 Lectures on wavelets, SIAM Publ. Philadelphia.

[13] Y. Fu, K. J. Klimkowski, G. J. Rodin, E. Berger, J. C. Browne, J. K. Singer, R. A. van de Geijn, and K. S. Vemaganti (1998). "A fast solution method for three-dimensional many-particle problems of linear elasticity" *Int. J. Numer. Meth. Engng.* 42, 1215–1229.

[14] K. Giebermann (1997). Schnelle Summationsverfahren zur numerischen Lösung von Integralgleichungen für Streuprobleme in $\mathbb{R}^3$, (in german), Ph.D. Dissertation, Dept. Mathematics, Univ. of Karlsruhe.

[15] I. G. Graham, Hackbusch, W. and S.A. Sauter (1998). Fast integration techniques in 3-d BEM, Preprint 98-29, School of Math. Sciences, Univ. Bath, U.K.

[16] L. Greengard and V. Rokhlin, V (1997). A new version of the fast multipole method for the Laplace equation in three dimensions, *Acta Numerica* **6**, 229–269, Cambridge University Press.

[17] M. Guiggiani, G. Krishnasamy, T.J. Rudolphi, F. Rizzo (1992). A general algorithm for the numerical solution of hypersingular BIEs, *Trans. ASME J. Appl. Mech.* **59**, 604–614.

[18] W. Hackbusch and Z. P. Nowak (1989). On the fast matrix multiplication in the BEM by panel clustering, *Numer. Math.* **54**, 463–491.

[19] W. Hackbusch (1995). Integral equations, Birkhäuser Publ. Basel.

[20] K. Hayami. and S.A. Sauter (1999). Application of panel clustering to three-d elastostatics, pp. 625–634 in Proc. of the 19th Intl. Conf. on BEM, Rome, Sept. 1999, M. Marchetti, C.A. Brebbia and M.H. Aliabadi (Eds.), Comp. Mech. Publ., Southampton, U.K.

[21] G. Hsiao and W. L. Wendland (1999/2000). Variational methods for boundary integral equations, Springer Verlag (to appear).

[22] Ch. Lage (1995). Softwareentwicklung zur Randelementmethode: Analyse und Entwurf effizienter Techniken (in german), Dissertation Univ. Kiel, Germany.

[23] Ch. Lage and C. Schwab (1996). A wavelet-Galerkin boundary element method on polyhedral surfaces in $\mathbb{R}^3$, in: Lecture Notes in Numerical Fluid Mechanics, Vol. 55 (W. Hackbusch and G. Wittum, eds.), Vieweg Publ. Braunschweig, 194–206.

[24] Ch. Lage and C. Schwab (1998). On the implementation of a fully discrete multiscale Galerkin BEM, Proc. of the BEM XIX Conference, Rome, Sp. 9-12, 1997. Comp. Mechanics Publications, Southampton, UK, 635–644.

[25] Ch. Lage and C. Schwab (1999). Wavelet Galerkin algorithms for boundary integral equations, (in press in SIAM J. Scient. Stat. Computing).

[26] Ch. Lage (1999/2000). Fast evaluation of singular kernel functions by clustermethods, (in preparation).

[27] Ch. Lage and S.A. Sauter (1999). General black box quadrature in BEM, Math. Comp., to appear.

[28] J. C. Nedelec (1990). Integral equations associated with elliptic boundary value problems in $\mathbb{R}^3$, pp. 114 ff, in J.L. Lions & R. Dautray, eds., Mathematical Analysis and Numerical Methods in Science and Technology, Vol. 4, Springer Verlag.

[29] T. von Petersdorff. and C. Schwab (1996). Wavelet discretization of first kind boundary integral equations on polygons, *Numer. Math.* **74**, 479–519.

[30] T. von Petersdorff and C. Schwab (1997). Fully discrete multiscale Galerkin BEM, in: Multiresolution Analysis and Partial Differential Equations, W. Dahmen, P. Kurdila and P. Oswald (eds.), Wavelet Analysis and its applications, Vol. 6, Academic Press, New York, 287–346.

[31] T. von Petersdorff, R. Schneider and Schwab, C. (1997). Multiwavelets for second kind integral equations, *SIAM J. Num. Anal.* **34** (6), 2212–2227.

[32] A. Rathsfeld (1998). A wavelet algorithm for the solution of a singular integral equation over a smooth 2-d manifold, *J. Int. Eq. Appl.* **10**.

[33] A. Rathsfeld and S. Ehrich (1998). Piecewise linear wavelet collocation on triangular grids, Preprint 434, WIAS, Berlin.

[34] V. Rokhlin (1993). Diagonal forms of translation operators for the Helmholtz equation in three dimensions. *Appl. Comp. Harmonic Anal.* **1**, 82–93.

[35] Boundary Elements (1996). Implementation and Analysis of advanced algorithms, Notes on Num. Fluid Mech. 54 (W. Hackbusch and G. Wittum, eds.), Vieweg Publ. Braunschweig, Germany.

[36] S.A. Sauter and A. Krapp (1996). On the effect of numerical integration in the Galerkin BEM, *Numer. Math.* **74**, 337–359.

[37] S.A. Sauter and C. Schwab (1997). Quadrature for hp-Galerkin BEM in $\mathbb{R}^3$, *Numer. Math.* **78**, 211–258.

[38] S.A. Sauter and C. Schwab (1999/2000). Semianalytic integration in BEM I: collocation methods, (in preparation).

[39] R. Schneider (1998). Multiskalen und Wavelet-Matrixkompression (in german), Teubner Publ. Stuttgart.

[40] C. Schwab (1994). Variable Order Composite Quadrature for singular and nearly singular integrals, Computing **53**, 173–194.

[41] C. Schwab and W.L. Wendland (1992). Kernel properties and representations of boundary integral operators, *Math. Nachr.* **156**, 187–218.

[42] Y. Yamada and K. Hayami (1995). A multipole BEM for 2-d elastostatics, Report METR 95-07, Fac. of Engg., University of Tokyo, Bunkyo-ku, Japan.

[43] W.L. Wendland (1999). On boundary integral equations and applications, Acc. Naz. dei Lincei, Atti dei Convegni Lincei **147**.

19 $\mathcal{H}$-MATRIX APPROXIMATION ON GRADED MESHES

Wolfgang Hackbusch and Boris N. Khoromskij

Max–Planck–Institut für *Mathematik in den Naturwissenschaften*
Inselstr. 22–26, D-04103 Leipzig, Germany

ABSTRACT

In a preceding paper [6], a class of matrices ($\mathcal{H}$-matrices) has been introduced which are data-sparse and allow an approximate matrix arithmetic of almost linear complexity. Several types of $\mathcal{H}$-matrices were analysed in [6, 7, 8, 9] which are able to approximate integral (nonlocal) operators in FEM and BEM applications in the case of quasi-uniform unstructured meshes.

In the present paper, the special class of $\mathcal{H}$-matrices on graded meshes is analysed. We investigate two types of separation criteria for the construction of the cluster tree which allow one to optimise either the approximation power (cardinality–balancing strategy) or the complexity (distance–balancing strategy) of the $\mathcal{H}$-matrices under consideration. For both approaches, we prove the optimal complexity and approximation results in the case of composite meshes and tensor–product meshes with polynomial/exponential grading in $\mathbb{R}^d$, $d = 1, 2, 3$.

Key words. Fast algorithms, hierarchical matrices, data–sparse matrices, mesh refinement, BEM, FEM.

19.1 INTRODUCTION

Consider the h-version of the Galerkin FE method for approximation of the integral operator $A \in \mathcal{L}(W, W')$ defined in the Sobolev space $W = H^r(\Sigma)$. In typical BEM applications, we deal with integral operators of the form

$$(Au)(x) = \int_\Sigma s(x, y)u(y)\, dy, \qquad x \in \Sigma,$$

with s being the fundamental solution (singularity function) associated with the *pde* under consideration or with s replaced by a suitable directional derivative Ds of s. Here Σ is either a bounded $(d - 1)$-dimensional manifold (surface) or a bounded domain in $\mathbb{R}^d$, $d = 2, 3$. In this paper, we confine ourselves to the case of an ansatz space

307

$W_h := \text{span}\{\varphi_i\}_{i \in I} \subset W$ of piecewise constant/linear basis functions with respect to the graded tensor–product meshes (and the associated triangulation if necessary) on the computational domain $\Sigma = (0,1)^{d_\Sigma}$. Therefore, we specify $H^r(\Sigma) = H^r_{00}(\Sigma)$ and $d_\Sigma = d-1$ in the BEM applications, while $W = H^{2r}(\Sigma)$, $d_\Sigma = d$ for the volume integral calculations. The extension of our approach to the case of closed surfaces is rather straightforward.

We assume that the singularity function s is asymptotically smooth[1], i.e.,

$$|\partial_x^\alpha \partial_y^\beta s(x,y)| \le c(|\alpha|,|\beta|)|x-y|^{-|\alpha|-|\beta|}g(x,y) \qquad \text{for all } |\alpha|,|\beta| \le m \qquad (19.1)$$

and for all $x,y \in \mathbb{R}^d$, $x \ne y$, where α, β are multi-indices with $|\alpha| = \alpha_1 + \ldots + \alpha_d$. We consider two particular choices of the function $g \ge 0$ defined on $\Sigma \times \Sigma$. The first case $g(x,y) = s(x,y)$, $s(x,y) \ge 0$ is discussed in [7]. The second variant to be considered is $g(x,y) = |x-y|^{1-d-2r}$. Here $2r \in \mathbb{R}$ is the order of the integral operator $A : H^r(\Sigma) \to H^{-r}(\Sigma)$ in the BEM applications. Similar smoothness prerequisites are usually required in the wavelet or multi-resolution techniques, see also the related mosaic–skeleton approach in [12] as well as [3].

We analyse a data–sparse $\mathcal{H}$-matrix approximation for the integral operators A with asymptotically smooth kernels, see (19.1). The construction consists of three essential stages:

(a) the *admissible* block-partitioning P_2 of the tensor product index set $I \times I$, where I is isomorphic to the set of supports of the FE basis functions from W_h;

(b) the construction of an approximate integral operator $A_\mathcal{H} \in \mathcal{L}(W,W')$ with the kernel $s_\mathcal{H}(\cdot,\cdot)$ defined on each block $X(\sigma) \times X(\tau) \in \Sigma \times \Sigma$, $\sigma \times \tau \in P_2$ by a *separable expansion* $s_{\tau,\sigma} = \sum_{i \le k} a_i(x)c_i(y)$ of the order $k << n = \dim W_h$;

(c) the setup of the approximating Galerkin $\mathcal{H}$-matrix $\mathcal{A}_\mathcal{H} = \langle A_\mathcal{H}\varphi_i, \varphi_j \rangle_{i,j \in I}$ for the operator $A_\mathcal{H}$, where the near-field (resp. far-field) components are evaluated with the exact (resp. approximate) kernel.

Consider the important step (a) in more details. Let I be the index set of unknowns (e.g., the FE-nodal points). For each $i \in I$, the support of the corresponding basis function φ_i is denoted by $X(i) := \text{supp}(\varphi_i)$. The *cluster tree* $T(I)$ is characterised by the following properties: (i) all vertices of $T(I)$ are subsets of I, (ii) $I \in T(I)$ is the root; (iii) if $\tau \in T(I)$ contains more than one element, the set $S(\tau)$ of sons of τ consists of at least 2 disjoint subsets satisfying $\tau = \bigcup_{\sigma \in S(\tau)} \sigma$; (iv) the leaves of the tree are $\{i\}$ for all $i \in I$. For $\tau \in T(I)$ we extend the definition of $X(\cdot)$ by $X(\tau) = \bigcup_{i \in \tau} X(i)$.

In the standard quasiuniform FE application, the cluster tree $T(I)$ is obtained by a recursive division of I into subsets of almost equal size having a diameter as small as possible. In the quasiuniform case, the term "almost equal size" can be understood in a geometrical sense (i.e., $\text{diam}(X(\tau')) \approx \text{diam}(X(\tau''))$ as well as with respect to the cardinality $\#\tau' \approx \#\tau''$. An appropriate construction of $T(I)$ will fulfil both criteria. However, in the *non-quasiuniform* case, these two properties cannot be satisfied in parallel. The remedy is that the first property can be substituted by $\text{diam}(X(\tau')) \approx \text{dist}(X(\tau'),X(\tau''))$ due to admissibility condition (19.3) below.

The matrix entries belong to the index set $I \times I$. In a canonical way (cf. [7]), a block-cluster tree $T(I \times I)$ can be constructed from $T(I)$, where all vertices $b \in T(I \times I)$ are of the

---

[1]The estimate (19.1) with $g = s(x,y)$ is valid in many situations, e.g., in the case of the singularity function $\frac{1}{4\pi}|x-y|^{-1}$ for $d = 3$.

form $b = \tau \times \sigma$ for $\tau, \sigma \in T(I)$. Given a matrix $M \in \mathbb{R}^{I \times I}$, the block-matrix corresponding to $b \in T(I \times I)$ is denoted by $M^b = (m_{ij})_{(i,j) \in b}$. A *block partitioning* $P_2 \subset T(I \times I)$ is a set of disjoint blocks $b \in T(I \times I)$, whose union equals $I \times I$. A block partitioning P_2 determines the H-matrix format. We use the following explicit definition of H-matrices.

Definition 19.1 *Let a block partitioning P_2 of $I \times I$ and $k << n$ be given. The set of real H-matrices induced by P_2 and k is*

$$\mathcal{M}_{\mathcal{H},k}(I \times I, P_2) := \{M \in \mathbb{R}^{I \times I} : \forall\, b \in P_2, \text{ there holds } \operatorname{rank}(M^b) \leq k\}. \tag{19.2}$$

The admissibility conditions are used to incorporate the singularity location of the kernel function $s(x,y)$, $(x,y) \in \Sigma \times \Sigma$, in order to balance the size of matrix–blocks b and their distance from the singularity points, see [8] for more details. For the BEM applications, we assume that the following admissibility condition

$$\min\{\operatorname{diam}(\sigma), \operatorname{diam}(\tau)\} \leq 2\eta \operatorname{dist}(\sigma, \tau) \tag{19.3}$$

holds for all $\sigma \times \tau \in P_2$, where $\eta \leq 1$ is a fixed parameter. We estimate the approximation

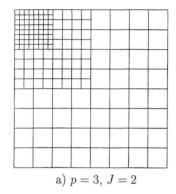 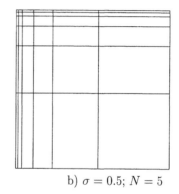

a) $p = 3$, $J = 2$ b) $\sigma = 0.5$; $N = 5$

Figure 19.1: Composite and tensor–product graded meshes

error $\|A - A_{\mathcal{H}}\|_{W \to W'}$ and the global perturbation of the solution arising at the stage (**b**) above as well as the computational complexity of certain H-formats in the following cases:
(i) J–level composite meshes characterised by $h_0 = 2^{-p}$ and $h_{\min} = 2^{-J} h_0$ (see Fig. 19.1a);
(ii) polynomially graded tensor-product meshes $\{\omega_i\}^d$, where $\omega_i \sim \left(\frac{i}{N}\right)^{\beta}$, $\beta \geq 1$ (see Fig. 19.1b), with $N^d = \dim W_h$;
(iii) exponentially graded tensor-product geometric meshes $\{\omega_i\}^d$, with $\omega_i \sim \sigma^{N-i}$, $\sigma < 1$.
    Note that in the cases (i) and (ii) the proper h-version of the Galerkin BEM is directly applied in the setup phase (**c**). A possible extension of this scheme to the case of hp–version of FEs is based on the corresponding construction for the exponentially graded meshes, see item (iii) above. However, in this paper we shall not discuss approximations with higher order finite elements.
    We develop two different strategies for constructing the H–matrices on graded meshes. The first approach is as close as possible to the structure of the *block partitioning* P_2 of

$I \times I$ on the uniform mesh with $h_i = h$. It is based on the cardinality–balanced cluster tree and yields almost linear complexity. The analysis of the geometrical partition of $\Sigma \times \Sigma$ which arises leads to the optimal approximation result. The second concept is based on a separation criterion providing a cluster tree with well balanced geometrical decompositions of Σ on each level. The corresponding admissible block partitioning P_2 is built using a ternary tree $T(I)$ with $\#I = O(3^p)$. However, in the case $h_{i-1} \ll h_i$, the ternary tree approaches a binary one, see figure in Section 19.3.

19.2   CARDINALITY–BALANCED PARTITIONS

In this section, we introduce the cardinality–balanced separation strategy for constructing the cluster tree. We show that the corresponding hierarchical matrices are dense enough in the case of graded meshes, i.e., they lead to the same asymptotically optimal approximations as the exact FE/BE Galerkin schemes. Consider the tensor–product grid $\omega = \{\omega_i\}^d$, where the grid points ω_i, $i = 0, 1, ..., N$, are defined by a sequence of mesh parameters $\{h_i\}_{i \in I_1}$, $h_i > 0$, $I_1 := \{1, ..., N\}$,

$$\omega_0 = 0, \quad \omega_N = 1, \quad \omega_i = \omega_{i-1} + h_i, \qquad i \in I_1.$$

The associated tensor–product index set I is given by

$$I := \{\mathbf{i} = (i_1, ..., i_d) : 1 \le i_k \le N, \ k = 1, ..., d\}, \qquad N = 2^p,$$

where each multi–index $\mathbf{i} \in I$ corresponds to the box $\delta_{\mathbf{i}} := [\omega_{i_k}, \omega_{i_k-1}]_{k=1}^{d_\Sigma} \subset [0, 1]^{d_\Sigma}$. In the case of Sobolev spaces $W = H_{00}^r(\Sigma)$ with negative index $r < 0$, we use the ansatz–space W_h of piecewise constant FEs on a rectangular mesh, while for $r \ge 0$ the linear elements on the associated triangulation are involved. For each $t \in \mathbb{R}$ define the weight–function $\mu(x) \in L^2(\Sigma)$ by

$$\mu(x) := h_{\mathbf{i}}^t, \quad \text{at } x \in \delta_{\mathbf{i}}, \text{ with } h_{\mathbf{i}} = \min_{1 \le k \le d}\{h_{i_k}\}.$$

Assumption 19.1 *(Inverse inequality) For $t \in [0, 1]$, the following holds:*

$$||\mu(x) \cdot v||_{0,\Sigma} \le c||v||_{-t,\Sigma}, \qquad v \in W_h. \tag{19.4}$$

The estimate (19.4) was discussed in [2] in an equivalent form for rather general nonuniform grids, see conditions (A1)–(A3) there. These conditions remain valid for the grids under consideration, see (i)-(iii) in Introduction.

In the case of quasi–uniform meshes (i.e., there are constants c_1, $c_2 > 0$ such that $c_1 h \le h_i \le c_2 h$, $i \in I_1$ with $h = N^{-1}$), a class of hierarchical matrix formats $\mathcal{M}_{\mathcal{H},k}(I \times I, P_2)$ was shown to have almost linear complexity and optimal approximation in the BEM applications. Our goal here is an extension of this result to the situation with strongly refined grids. Set $d = 1$ for the moment. Let $T(I)$ be the binary cluster tree of the uniform depth p, where $\#I = 2^p$, which has the root $I_1^0 = I$ on level $\ell = 0$. It is built by successive splitting of each vertex into two parts of equal cardinality. $T(I)$ contains the subsets I_j^ℓ, $0 \le \ell \le p$, $1 \le j \le 2^\ell$, on each level ℓ such that at level p, we reach the one–element sets (leaves) $I_1^p = \{1\}, ..., I_n^p = \{n\}$. The sons $I_{j_1}^{\ell+1}$, $I_{j_2}^{\ell+1}$ of I_j^ℓ are defined by

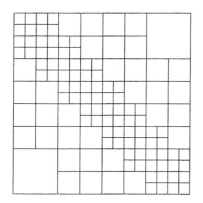 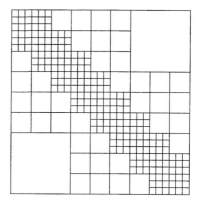

Figure 19.2: Block–partitionings using binary (a) and ternary (b) trees

$$\underset{I_1^1 \qquad\qquad I_2^1}{\rule{4cm}{0pt}}$$

$$\underset{I_1^1 \quad I_2^1 \qquad\qquad I_3^1}{\rule{4cm}{0pt}}$$

(a) $\#I_1^1 = \#I_2^1$ (b) I_k^1, $k = 1, 2, 3$ satisfy (19.13)

Figure 19.3: Cardinality- (a) and distance-balanced (b) separation criteria.

the separation criteria $\#I_{j_1}^{\ell+1} = \#I_{j_2}^{\ell+1}$, see Fig. 19.3a. The cluster tree $T_2 := T(I \times I)$ has the following set of vertices, $\mathbf{I}_{ij}^\ell := I_i^\ell \times I_j^\ell$ for $0 \le \ell \le p$, $1 \le i, j \le 2^\ell$. The set $S_2(t)$ of sons for $t = \mathbf{I}_{ij}^\ell \in T_2$ is given by $S_2(t) := \{\tau \times \sigma : \tau \in S_1(I_i^\ell),\ \sigma \in S_1(I_j^\ell)\}$, where $S_1(f)$ denotes the set of sons belonging the parent cluster $f \in T(I)$. Finally, we obtain the explicit block partitioning $P_2 := \cup_{\ell=2}^p P_2^\ell$, where $P_2^2 = \{\mathbf{I}_{14}^2\} \cup \{\mathbf{I}_{41}^2\}$ and

$$P_2^\ell = \{\mathbf{I}_{ij}^\ell \in T_2 : \quad |i - j| \ge 1 \text{ and } \mathbf{I}_{ij}^\ell \cap P_2^{\ell'} = \emptyset,\ \ell' < \ell\} \quad \text{for} \quad \ell = 3, ..., p.$$

The construction for $d = 2, 3$ is completely similar. We make a technical assumption which is connected with a certain kind of monotonicity of the refinement.

Assumption 19.2 *Let μ and t be as in (19.4). For each $\tau \times \sigma \in P_2^\ell$, $\ell = 1, ..., p$, there holds*

$$\iint\limits_{X(\tau) \times X(\sigma)} \mu^{-2}(x)\mu^{-2}(y)\, dxdy \le c\, dist(\tau, \sigma)^{2(d_\Sigma + 2t)} \qquad \text{for} \quad t = -\frac{1}{2} \text{ and } t = 0. \qquad (19.5)$$

Assumption 19.2 may be varified for the mesh–refinements considered. There are many opportunities to build separable expansions of the form

$$s_{\tau,\sigma}(x, y) = \sum_{j=1}^k a_j(x) c_j(y), \qquad (x, y) \in X(\tau) \times X(\sigma) \qquad (19.6)$$

for each cluster $\tau \times \sigma \in P_2$, where $k = O(\log^{d-1} N)$ is the order of expansion. Let x, y vary in the respective sets $X(\tau)$ and $X(\sigma)$ corresponding to the admissible clusters $\tau, \sigma \in T(I)$ and assume without loss of generality that $\mathrm{diam}(X(\sigma)) \leq \mathrm{diam}(X(\tau))$. The optimal centre of expansion is the Chebyshev centre[2] y_* of $X(\sigma)$, since then $\|y - y_*\| \leq \frac{1}{2} \mathrm{diam}(X(\sigma))$ for all $y \in X(\sigma)$. We recall the familiar approximation result (see, e.g., [7] for the proof) based on the Taylor expansion with respect to y.

Lemma 19.1 *Assume that (19.1) is valid and that the admissibility condition (19.3) holds with η satisfying $c(0,1)\eta < 1$. Then for $m \geq 1$, the remainder of the Taylor expansion satisfies the estimate*

$$|s(x,y) - \sum_{|\nu|=0}^{m-1} \frac{1}{\nu!}(y_* - y)^\nu \frac{\partial^\nu s(x,y_*)}{\partial y^\nu}| \leq \frac{c(0,m)}{m!}\eta^m \max_{y \in X(\sigma)} |g(x,y)|, \qquad (19.7)$$

for $x \in X(\tau), y \in X(\sigma)$.

Let $A_{\mathcal{H}}$ be the integral operator with s replaced by the Taylor expansion $s_{\sigma,\tau}$ for $(x,y) \in X(\tau) \times X(\sigma)$ provided that $\tau \times \sigma \in P_2$ is an admissible block and no leaf. Construct the Galerkin system matrix from $A_{\mathcal{H}}$ instead of A. The perturbation of the matrix induced by $A_{\mathcal{H}} - A$ yields a perturbed discrete solution of the original variational equation

$$\langle (\lambda I + A)u, v \rangle = \langle f, v \rangle \qquad \forall v \in W := H^r(\Sigma), \ r \leq 1,$$

where $\lambda \in R$ is a given parameter. The effect of this perturbation in the panel clustering methods is studied in several papers (cf. [10]). Here, we give the consistency error estimate for the $\mathcal{H}$-matrix approximation. Define the integral operator $\widehat{A}$ with the kernel

$$\widehat{s}(x,y) := \begin{cases} \rho(\sigma,\tau) \max_{y \in \sigma} |g(x,y)| & \text{for } (x,y) \in X(\tau) \times X(\sigma), \ |\tau|, |\sigma| \geq m^{d_\Sigma}, \\ 0 & \text{otherwise,} \end{cases} \qquad (19.8)$$

where $\rho(\sigma,\tau) = (\frac{\mathrm{diam}(\sigma)}{\mathrm{dist}(\sigma,\tau)})^m$, $|\sigma| := \#\sigma$. For the given ansatz space $W_h \subset W$ of piecewise constant/linear FEs, consider the perturbed Galerkin equation for $u_{\mathcal{H}} \in W_h$,

$$\langle (\lambda I + A_{\mathcal{H}})u_{\mathcal{H}}, v \rangle = \langle f, v \rangle \qquad \forall v \in W_h.$$

Theorem 19.1 *Assume that (19.1) is valid and set $\eta = \frac{\sqrt{d}}{2}$. Suppose that the operator $\lambda I + A \in \mathcal{L}(W, W')$ is W-elliptic. Then there holds[3]*

$$\|u - u_{\mathcal{H}}\|_W \lesssim \inf_{v_h \in V_h} \|u - v_h\|_W + \frac{c(0,m)}{m!}\eta^m \|\widehat{A}\|_{W_h \to W_h'} \|u\|_W. \qquad (19.9)$$

Under Assumptions 19.1 and 19.2 the norm of $\widehat{A}$ is estimated by

$$\|\widehat{A}\|_{W_h \to W_h'} \lesssim \begin{cases} \|A\| & \text{if } g = s(x,y) \wedge s(x,y) \geq 0 \\ c\, N^{d_\Sigma/2} & \text{if } g = |x - y|^{1-d-2r}, \end{cases} \qquad (19.10)$$

for the range of Sobolev index $-1 \leq 2r \leq 1 + d_\Sigma - d$.

---

[2]Given a set X, the centre of the minimal sphere containing X is called the *Chebyshev centre*.

[3]$A \lesssim B$ means that $A \leq CB$, where C is a constant which does not depend on any of the parameters of the discretisation.

Proof. The continuity and strong ellipticity of A imply

$$||u - u_{\mathcal{H}}||_W \lesssim \inf_{v \in W_h} ||u - v||_W + \sup_{u,v \in W_h} \frac{|\langle (A - A_{\mathcal{H}})u, v \rangle|}{||u||_W ||v||_W} ||u_{\mathcal{H}}||_W$$

(cf. first Strang lemma). On the other hand, under the assumption (19.1), Lemma 19.1 yields

$$|\langle (A - A_{\mathcal{H}})u, v \rangle| \lesssim \frac{c(0, m)}{m!} \eta^m \sum_{\tau \times \sigma \in P_2} \iint_{X(\tau) \times X(\sigma)} |\widehat{s}(x, y) u(y)\, v(x)|\, dx dy \qquad (19.11)$$

$$\lesssim \frac{c(0, m)}{m!} \eta^m ||\widehat{A}||_{W_h \to W_h'} ||u||_W ||v||_W \qquad \forall u, v \in W_h.$$

Now, assuming that $\dfrac{c(0, m)}{m!} \eta^m ||\widehat{A}||_{W_h \to W_h'}$ is sufficiently small, the estimate (19.10) and $\eta < 1$ imply the strong ellipticity of the discrete Galerkin operator yielding the stability $||u_{\mathcal{H}}||_W \leq c ||u||_W$. This implies (19.9). In the case $g = s(x, y)$, the first assertion in (19.10) follows from $\rho(\sigma, \tau) \leq 1$ and from the bound $|| |u| ||_W \leq ||u||_W$ for all $u \in W_h$. Consider the case $g = |x - y|^{1-d-2r}$. If $r \geq 0$, the standard L^2–norm estimate combined with the imbedding $H^r(\Sigma) \subset L^2(\Sigma)$ implies

$$|\langle \widehat{A}u, v \rangle| \leq \left(\iint_\Sigma \widehat{s}(x, y)^2\, dx dy \right)^{1/2} ||u||_{r,\Sigma} ||v||_{r,\Sigma}.$$

Setting $\varepsilon = 1 - d - 2r$, we then have that

$$||\widehat{A}||_{W_h \to W_h'}^2 \lesssim \iint_\Sigma \widehat{s}(x, y)^2\, dx dy \leq \sum_{\sigma \times \tau \in P_2} \iint_{X(\sigma) \times X(\tau)} \rho^2(\sigma, \tau)(\text{dist}(\sigma, \tau))^{2\varepsilon}\, dx dy$$

$$\leq \sum_{\ell=2}^p \sum_{\sigma \times \tau \in P_2^\ell} (\text{dist}(\sigma, \tau))^{2\varepsilon} |X(\sigma)||X(\tau)| \leq \sum_{\ell=2}^p \sum_{\sigma \times \tau \in P_2^\ell} (\text{dist}(\sigma, \tau))^{2(1+d_\Sigma - d - 2r)}$$

$$\leq \sum_{\ell=2}^p \sum_{\sigma \times \tau \in P_2^\ell} 1 \leq \sum_{\ell=2}^p 2^{\ell d_\Sigma} \leq c N^{d_\Sigma},$$

where the first estimate in the last line is based on the property of the admissible partitioning: $\#P_2^\ell = O(2^{d_\Sigma \ell})$.

In the case $r < 0$, we first obtain the bound in the weighted L^2–norm and then apply the inverse inequality on graded meshes, see (19.5). It is enough to consider only the value $r = -\frac{1}{2}$. For such a choice there holds

$$|\langle \widehat{A}u, v \rangle| \leq \left(\iint_\Sigma \mu^{-2}(x) \mu^{-2}(y) \widehat{s}(x, y)^2\, dx dy \right)^{1/2} ||u||_{r,\Sigma} ||v||_{r,\Sigma};$$

$$||\widehat{A}||_{W_h \to W_h'}^2 \lesssim \iint_\Sigma \mu^{-2}(x) \mu^{-2}(y) \widehat{s}(x, y)^2\, dx dy$$

$$\leq \sum_{\ell=2}^{p} \sum_{\sigma \times \tau \in P_2^\ell} (\text{dist}(\tau, \sigma))^{2(1-d-2r)} \iint_{X(\sigma) \times X(\tau)} \mu^{-2}(x)\mu^{-2}(y)\,dx dy$$

$$\leq \sum_{\ell=2}^{p} \sum_{\sigma \times \tau \in P_2^\ell} (\text{dist}(\tau, \sigma))^{2(1+d_\Sigma-d)} \leq c\, N^{d_\Sigma}.$$

This completes our proof. $\square$

By the construction, our P_2 partitioning generates the same block–structure of $\mathcal{H}$–matrices as the corresponding one for the case of uniform meshes. Then, following to [8], we obtain the almost linear complexity bound.

Proposition 19.1 *Let $d \in \{1, 2, 3\}$, $A \in \mathcal{M}_{\mathcal{H},k}(I \times I, P_2)$ and $\eta = \frac{\sqrt{d}}{2}$. Then the storage and matrix-vector multiplication expenses are bounded by*

$$\mathcal{N}_{st} \leq (2^d - 1)(\sqrt{d}\eta^{-1} + 1)^d pkN, \qquad \mathcal{N}_{MV} \leq \mathcal{N}_{st}, \tag{19.12}$$

where the cost unit of $\mathcal{N}_{MV}$ is one addition and one multiplication. Both estimates are asymptotically sharp.

The local Rk-approximations in the Galerkin method may be computed as follows. The block entry $\mathcal{A}_{\mathcal{H}}^{\tau \times \sigma}$ of the Galerkin matrix $\mathcal{A}_{\mathcal{H}} := \{\langle A_{\mathcal{H}}\varphi_i, \varphi_j \rangle\}_{i,j=1}^{N}$ associated with each cluster $\tau \times \sigma$ on level ℓ may be presented as a rank–k matrix $\mathcal{A}_{\mathcal{H}}^{\tau \times \sigma} = \sum_{|\nu|=0}^{m-1} a_\nu * b_\nu^T$, where

$$k := \binom{d_\Sigma + m - 1}{m - 1} = O((m-1)^{d_\Sigma}) \text{ is the number of terms. In turn,}$$

$$a_\nu = \left\{ \int_{X(\tau)} (y - y_*)^\nu \varphi_i(y)\,dy \right\}_{i=1}^{N_\tau}, \qquad b_\nu = \left\{ \int_{X(\sigma)} \frac{\partial^\nu s(x, y_*)}{\partial y^\nu} \varphi_j(x)\,dx \right\}_{j=1}^{N_\sigma},$$

where $N_\tau = \#\tau = O(2^{d_\Sigma(p-\ell)})$ (resp. $N_\sigma = \#\sigma = O(2^{d_\Sigma(p-\ell)})$) is the cardinality of τ (resp. σ).

19.3   ON DISTANCE–BALANCED PARTITIONS

In this section, we study the algorithm based on the concept of balanced geometrical partitionings. The main topic to be discussed here is an argument for the linear complexity of the resulting $\mathcal{H}$-matrices. For ease of presentation consider the one dimensional case. In contrast to the *cardinality–balanced partitioning*, here we use a ternary cluster tree $T_1 = T(I)$ with $N = 3^p$. Starting with the root $\{I\}$ on level $\ell = 0$, we introduce the triple of sons I_1^1, I_2^1 and I_3^1 on level $\ell = 1$, see Fig. 1.3b), satisfying the *separation criteria*

$$(a) \quad \#I_1^1 = \#I_3^1; \qquad (b) \quad \text{diam}I_1^1 = \text{diam}I_2^1, \tag{19.13}$$

where $i < i < k$ for all $i \in I_1^1$, $j \in I_2^1$, $k \in I_3^1$. To analyse the complexity of the admissible P_2 partitioning, we first construct the nonary tree $T_2 = T(I \times I) := \{\mathbf{I}_{ij}^\ell\}_{\ell=0}^{p}$, where $\{\mathbf{I}_{ij}^\ell\} = I_i^\ell \times I_j^\ell$, $i, j \leq 3^\ell$, and then build P_2 with respect to the admissibility condition (19.3). Note that the intermediate cluster I_2^1 plays here an artificial role because in the case of strong refinement we have $\#I_2^1 \ll \#I_1^1$. Therefore, the corresponding branch of the block–cluster tree will be the shorter the stronger the refinement is. It it easy to see that for monotonous refinement the conditions (19.13a,b) imply that the clusters $\sigma = I_1^1$ and $\tau = I_3^1$ satisfy the standard admissibility requirement (19.3) (similar on each level ℓ). The latter immediately yields the local approximation property, see Lemma 19.1, for each admissible block $\sigma \times \tau \in P_2$. Then the global error estimate similar to Theorem 19.1 holds true. At this point, the number of blocks from P_2^ℓ is bounded from above and below by the corresponding one for the nonary and quad trees, respectively, constructed on the uniform grids. On the other hand, the ratio of the latter two values is uniformly bounded with respect to the problem size. In the case of quasi–uniform meshes,

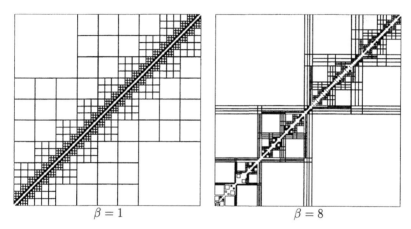

$$\beta = 1 \qquad\qquad \beta = 8$$

the complexity result is completely similar to those from Theorem 19.1 above, see [8] for the case of binary tree.

Lemma 19.2 *Let* $d \in \{1, 2, 3\}$, $A \in \mathcal{M}_{\mathcal{H},k}(I \times I, P_2)$, $N = 3^{d_{\Sigma^p}}$ *and* $\eta = \frac{\sqrt{d}}{2}$. *Then we have that*

$$\mathcal{N}_{st} \leq (3^d - 1)(\sqrt{d}\eta^{-1} + 1)^d \, kN \log_3 N; \qquad \mathcal{N}_{MV} \leq \mathcal{N}_{st} \qquad (19.14)$$

for the storage and matrix-vector multiplication (the cost unit of $\mathcal{N}_{MV}$ is one addition and one multiplication). Both estimates are asymptotically sharp.

In the general case, the complexity analysis is based on the observation that due to assumption $h_i \leq h_{i+1}$, we obtain larger admissible blocks in P_2 on each level compared with those arising on the basis of the balanced nonary tree T_2 for uniform grids. Thus, we have $\mathcal{N}_{st,g} \leq c\mathcal{N}_{st}$ and $\mathcal{N}_{MV,g} \leq c\mathcal{N}_{MV}$, where the abbreviation "*g*" denotes graded meshes. Moreover, if $h_i \ll h_{i+1}$ the corresponding nonary tree approaches the *quad–tree* and the induced $P_{2,g}$ partitioning becomes very close to the simplest one based on the binary tree for $n = 2^p$ (see the figure in this section drawn for the case of $1D$ polynomially graded mesh with $n = 512$, $\beta = 1$ and $\beta = 8$).

Remark 19.1 The analysis of $\mathcal{H}$–matrix approximations on the composite grids of Fig. 1.1a is a particular case of the arguments above. In the case of piecewise linear elements, a standard modification of the FE space by using slave nodes is required.

Acknowledgements The authors want to thank Klaus Giebermann for his assistance in producing the figure in Section 3 by MATLAB 4.0/4.1.

REFERENCES

[1] K. Giebermann (1999): A New Version of Panel Clustering for the Boundary Element Method. Preprint University of Bonn.

[2] I. G. Graham, W. Hackbusch and S. A. Sauter (1998): *Hybrid Galerkin Boundary Elements on Degenerate Meshes.* Bath Mathematics Preprint 98/22, University of Bath.

[3] L. Greengard and V. Rokhlin (1997): A new version of the fast multipole method for the Laplace equation in three dimensions. *Acta Numerica*, 229-269.

[4] W. Hackbusch (1990): *The panel clustering algorithm.* In The Mathematics of Finite Elements and Applications, MAFELAP, J. R. Whiteman, ed.. Academic Press, London, pp. 339–348.

[5] W. Hackbusch (1995): *Integral Equations. Theory and Numerical Treatment.* ISNM 128. Birkhäuser, Basel.

[6] W. Hackbusch (1999): *A Sparse Matrix Arithmetic based on $\mathcal{H}$-Matrices. Part I: Introduction to $\mathcal{H}$-Matrices. Computing* **62**, 89-108.

[7] W. Hackbusch and B. N. Khoromskij (1999): A Sparse $\mathcal{H}$-Matrix Arithmetic. Part II. Application to Multi-Dimensional Problems. Preprint MPI, Leipzig, No.22.

[8] W. Hackbusch and B. N. Khoromskij (1999): A sparse $\mathcal{H}$-matrix arithmetic: General complexity estimates. Submitted to NA2000.

[9] W. Hackbusch, B. N. Khoromskij and S. Sauter (1999): On $\mathcal{H}^2$-matrices. Preprint MPI, Leipzig, No. 50.

[10] W. Hackbusch and Z. P. Nowak (1989): *On the fast matrix multiplication in the boundary element method by panel clustering. Numer. Math.* **54** 463–491.

[11] W. Hackbusch and S. A. Sauter (1993): *On the efficient use of the Galerkin method to solve Fredholm integral equations. Applications of Mathematics* **38** 301–322.

[12] E. E. Tyrtyshnikov (1996): *Mosaic-skeleton approximations.* Calcolo **33** 47–57.

20 BOUNDARY INTEGRAL FORMULATIONS FOR STOKES FLOWS IN DEFORMING REGIONS

Luiz C. Wrobel[a], Ana R.M. Primo[b] and Henry Power[c]

[a]Department of Mechanical Engineering, Brunel University,
Uxbridge UB8 3PH, UK

[b]Department of Mechanical Engineering, Federal University of Pernambuco,
Recife, PE, Brazil

[c]Wessex Institute of Technology, Ashurst Lodge, Ashurst,
Southampton SO40 7AA, UK

ABSTRACT

This paper describes four novel boundary integral formulations for dealing with the deformation of a slow viscous flow driven by surface tension, and related to viscous sintering problems. The formulations are valid, respectively, for simply connected viscous regions, viscous masses with air bubbles, drops of fluids of different viscosities inside a viscous fluid, and a single solid particle inside a viscous fluid. The first two formulations are appropriate for general viscous sintering problems, while the last two are appropriate for studying the effects of impurities in the sintering process.

The aim of the present research is to increase the theoretical understanding of the viscous sintering process through elegant mathematical formulations requiring a simple computational implementation.

Key words. Boundary element method, Stokes flow, viscous sintering, completed integral formulations

20.1 INTRODUCTION

Sintering is a process in which a granular compact of metals, ionic crystals or glasses consisting of many particles, is heated to such a temperature that the mobility of the material is sufficient to make contiguous particles coalesce. In the production of aerogels and glassy materials, the material transport can be modelled as a viscous incompressible Newtonian flow, driven solely by surface tension, and the process is known as viscous sintering.

A good theoretical understanding of the densification kinetics is needed to produce dense and homogeneous compacts. Because of the complexity of the phenomenum, scientists studying sintering have long been interested in the behaviour of simple systems which are representative of contacting particles. More recently, numerical techniques have been developed for viscous sintering problems.

The boundary element method (BEM) was initially applied by Kuiken [1] and van de Vorst *et al.* [2, 3] to obtain a description of the sintering of simple two-dimensional and axisymmetric systems. Kuiken [1] defined the viscous flow problem in terms of the stream function-vorticity formulation and found the solution using a system of two coupled integral equations, for harmonic and biharmonic functions. However, numerical problems occurred due to inaccuracies in computing the derivative of the curvature, which was required in that particular integral formulation.

Van de Vorst *et al.* [2, 3] defined the problem in terms of the primitive variables and found the solution using Lorentz's direct integral representation formulae for the Stokes flow, given by the sum of a single- and a double-layer potential whose densities are the surface traction and velocity, respectively. By directly prescribing the surface traction, a Fredholm integral equation of the second kind for the unknown surface velocity was obtained which is free from the problem of computing the tangential derivative of the curvature. This formulation allowed the extension of Kuiken's range of solutions.

The present paper introduces four novel indirect boundary integral formulations for the deformation of bounded viscous regions, driven by surface tension. These formulations represent an extension of the completed double-layer boundary integral equation method of Power and Miranda [4], and basically introduce extra terms in the proposed solution that do not alter the nature of the problem and complete the deficient range of the integral operator. The formulations deal with a simply-connected viscous fluid region, and multiply-connected viscous masses containing air bubbles, drops of fluids of different viscosities and a solid inclusion, respectively. All formulations have a direct application to viscous sintering, with the last two corresponding to the problem of viscous sintering with impurities.

The first formulation will be presented in more detail, including its numerical implementation, in order to show that the completed boundary integral equation method is an elegant tool for studying deforming viscous flows and leads to a simple computational implementation.

20.2   INDIRECT BOUNDARY INTEGRAL FORMULATIONS

20.2.1   Deformation of a Simply-Connected Viscous Fluid Region

Consider a Stokes flow in a fluid region Ω bounded by a surface S (Figure 20.1). The problem will be defined here in terms of primitive variables, similar to van de Vorst *et al.* [2, 3]. The flow is represented by the Stokes equations:

$$\mu \frac{\partial^2 u_i}{\partial x_j \partial x_j} - \frac{\partial p}{\partial x_i} = 0 \qquad (20.1)$$

$$\frac{\partial u_i}{\partial x_i} = 0 \qquad (20.2)$$

for any point $\mathbf{x} \in \Omega(t)$, where $\mathbf{u}$ is the velocity vector, p is the pressure and μ is the dynamic viscosity.

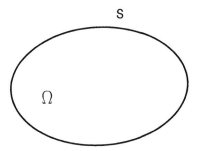

Figure 20.1: A viscous fluid region Ω bounded by a surface S.

The interior flow field has to satisfy the following boundary condition:

$$\sigma_{ij} n_j = \gamma \kappa n_i \qquad \text{on } S(t) \qquad (20.3)$$

where

$$\sigma_{ij} = -p\delta_{ij} + \mu \left(\frac{\partial u_i}{\partial x_j} + \frac{\partial u_j}{\partial x_i} \right) \qquad (20.4)$$

is the stress tensor, $\sigma_{ij} n_j$ is the surface traction, γ is the surface tension coefficient, κ is the surface curvature and $\mathbf{n}$ is the unit normal vector outward to the viscous drop.

The rate of deformation is determined by the kinematic boundary condition at S, which states that the normal component of the fluid velocity at a point $\boldsymbol{\xi}$ of the drop's surface is equal to the normal component of the surface velocity at that point:

$$\frac{dx_i}{dt} n_i = u_i n_i \qquad \text{at } \boldsymbol{\xi} \in S(t) \qquad (20.5)$$

Taking into account the quasistatic character of the present description, the boundary-value problem (20.1)-(20.3) can be solved for a given drop shape $S(t)$. With the computed surface velocity $u_i(\boldsymbol{\xi}, t)$ and a time step Δt, the shape of the deformed drop $S(t + \Delta t)$ can be determined using the kinematic condition (20.5). The scheme starts with a given initial drop shape $S(t = 0)$; then, at each instant in time, equations (20.1)-(20.3) define a second-kind boundary-value problem for the Stokes equations.

The motion of the drop's surface cannot be arbitrary since at each instant in time the force and torque yielded by the surface traction, given by the boundary condition (20.3), have to be zero upon the surface $S(t)$, *i.e.*

$$\int_{S(t)} \sigma_{ij}(\mathbf{u}, p) n_j \varphi_i^l dS = \int_{S(t)} \gamma \kappa n_i \varphi_i^l dS = 0 \tag{20.6}$$

where $\varphi^l, l = 1, 2, 3$, are the three linearly-independent rigid body vectors.

The solution of the present second-kind boundary-value problem is sought in terms of a single-layer potential alone, with unknown vector density $\mathbf{\Phi}$:

$$u_i(\mathbf{x}) = -\int_S u_i^j(\mathbf{x}, \mathbf{y}) \Phi_j(\mathbf{y}) dS_y \tag{20.7}$$

for every $\mathbf{x} \in \Omega$, where

$$u_i^j(\mathbf{x}, \mathbf{y}) = -\frac{1}{4\pi\mu} \left(\delta_{ij} \ln(\frac{1}{r}) + \frac{(x_i - y_i)(x_j - y_j)}{r^2} \right); \qquad r = |\mathbf{x} - \mathbf{y}| \tag{20.8}$$

is the two-dimensional fundamental solution of the Stokes system of equations, known as a Stokeslet. The pressure field is given by the pressure corresponding to the single-layer potential :

$$p(\mathbf{x}) = -\frac{1}{2\pi} \int_S \frac{\partial}{\partial x_j} \left(\ln \frac{1}{r} \right) \Phi_j(\mathbf{y}) dS_y \tag{20.9}$$

Applying the boundary condition (20.3) to the flow field described by (20.7) and (20.9), and using the discontinuity property of the surface forces of a single-layer potential across the density-carrying surface S, the following linear Fredholm integral equation of the second kind is obtained for the unknown density $\mathbf{\Phi}$:

$$\frac{1}{2}\Phi_i(\boldsymbol{\xi}) - \int_S K_{ji}(\mathbf{y}, \boldsymbol{\xi}) \Phi_j(\mathbf{y}) dS_y = \gamma \kappa(\boldsymbol{\xi}) n_i(\boldsymbol{\xi}) \tag{20.10}$$

for every $\boldsymbol{\xi} \in S$, where the kernel is given by:

$$K_{ji}(\mathbf{y}, \boldsymbol{\xi}) = -\frac{1}{\pi} \frac{(y_i - \xi_i)(y_j - \xi_j)(y_k - \xi_k)}{r^4} n_k(\boldsymbol{\xi}) \tag{20.11}$$

However, the solution of equation (20.10) is non-unique, since we can add to it any linear combination of the eigensolutions $\mathbf{\Psi}^l, l = 1, 2, 3$, which are the three linearly independent solutions of the homogeneous form of equation (20.10). To remove these eigenfunctions, it is necessary to add to the original equation (20.10) a term which is linearly proportional to the rigid body vector $\varphi^l, l = 1, 2, 3$, i.e.

$$\frac{1}{2}\Phi_i(\boldsymbol{\xi}) - \int_S K_{ji}(\mathbf{y}, \boldsymbol{\xi}) \Phi_j(\mathbf{y}) dS_y + \varphi_i^l(\boldsymbol{\xi})\beta_l = \gamma \kappa(\boldsymbol{\xi}) n_i(\boldsymbol{\xi}) \tag{20.12}$$

with $\boldsymbol{\beta}$ chosen as:

$$\beta_l = \int_S \Phi_i \varphi_i^l dS; \qquad l = 1, 2, 3 \tag{20.13}$$

The addition of this new term in equation (20.12) does not alter the nature of the problem since for any admittable non-homogeneous term, i.e. satisfying the compatibility condition (20.6), the vector $\boldsymbol{\beta}$ will end up being zero. A similar approach has been proposed by Botte and Power [5] in order to complete the deficient range of the double-layer integral operator when the solution of the interior first-kind boundary-value problem for the Stokes system of equations is represented in terms of a double-layer potential alone. In such a case the original second-kind Fredholm integral equation, coming from the jump property of the double-layer potential, is completed by adding a term linearly proportional to the surface normal vector. The coefficient of proportionality is identically equal to zero when the non-homogeneous term, *i.e.* the given surface velocity, satisfies the corresponding compatibility condition, *i.e.* the no-flux condition. Although for any given admittable surface velocity the original integral equation and the completed one are identical, the former does not have a unique solution since its corresponding homogeneous equation has a non-trivial solution, whilst the completed one does since its corresponding homogeneous equation only admits a trivial solution.

A theoretical proof of uniqueness of the solution of the Fredholm integral equation (20.12) can be found in Primo, Wrobel and Power [6].

20.2.2 *Deformation of a Viscous Fluid Region Containing Air Bubbles*

Consider a bounded Stokes flow containing several air bubbles, as shown in Figure 20.2. Let Ω be the finite fluid domain exterior to the bubbles and interior to the drop surface S_0. Therefore, the fluid region consists of a multiply-connected domain bounded externally by S_0 and internally by the surfaces $S_1...S_M$. The symbol S denotes the entire surface, *i.e.* $S = S_0 \cup S_1 \cup ...S_M$.

The flow field $(\mathbf{u}, p)$ is governed by the Stokes creeping flow equations (20.1) and (20.2), and the following boundary condition:

$$\sigma_{ij}(\mathbf{u}(\boldsymbol{\xi}), p(\boldsymbol{\xi}))n_j(\boldsymbol{\xi}) = \pm\gamma\kappa(\boldsymbol{\xi})n_i(\boldsymbol{\xi}) \qquad \text{for every } \boldsymbol{\xi} \in S_l \qquad (20.14)$$

with $l = 0, 1, 2, ...M$, where the $+$ sign corresponds to the exterior surface S_0 and the $-$ sign to each of the internal surfaces $S_1, S_2, ...S_M$. The normal vector $\mathbf{n}$ is exterior to the flow domain at the surface S_0, but interior to Ω at the surfaces $S_1, S_2, ...S_M$. The flow field is also subject to the solvability condition:

$$\int_{S_0(t)} \sigma_{ij}(\mathbf{u}, p)n_j\varphi_i^l dS = \int_{S(t)-S_0(t)} \sigma_{ij}(\mathbf{u}, p)n_j\varphi_i^l dS; \qquad l = 1, 2, 3 \qquad (20.15)$$

If the flow field is solely represented in terms of a single-layer potential with a density-carrying surface S, the resulting system of linear Fredholm integral equations of the second kind obtained by imposing the surface traction boundary conditions (20.14) does not have a unique solution. It can also be shown that the problem has a further M eigenvalues, each associated to one bubble surface [6].

The solution of the present second-kind boundary-value problem is given in the form of a single-layer potential over the entire surface S plus M source harmonic potentials, each of them located at the interior of each of the bubbles enclosed by the viscous drop, *i.e.*

$$u_i(\mathbf{x}) = -\int_S u_i^j(\mathbf{x}, \mathbf{y})\Phi_j(\mathbf{y})dS_y + \sum_{n=1}^{M} \frac{\alpha^n(x_i - x_i^n)}{2\pi R_n^2} \qquad \text{for every } \mathbf{x} \in \Omega \qquad (20.16)$$

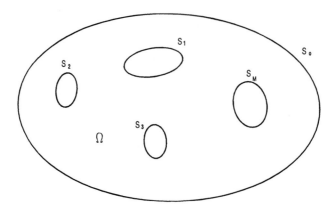

Figure 20.2: A viscous fluid region of surface S_0 containing air bubbles.

with $R_n = |x_i - x_i^n|$ and x_i^n the coordinates of a point inside each of the bubbles. The last M terms correspond to the velocity field due to the harmonic potentials $\ln(R_n)/2\pi$, i.e. $u_i^S = (1/2\pi)\partial \ln R_n/\partial x_i$.

The source intensities are chosen to be linearly dependent upon the single-layer density at the corresponding bubble surface in the following manner:

$$\alpha^n = \int_{S_n} \Phi_i(\mathbf{y})n_i(\mathbf{y})dS_y \qquad n = 1, 2, ...M \tag{20.17}$$

Applying the boundary condition (20.14) to the above flow field, using the known jump property of the stresses of the single layer, and adding to the surface traction equation at the exterior surface S_0 a term proportional to the rigid body vectors φ^l, $l = 1, 2, 3$, which, as before, does not to alter the nature of the problem, leads to the following system of linear Fredholm integral equations of the second kind:

$$\frac{1}{2}\Phi_i(\boldsymbol{\xi}) - \int_{S_0} K_{ji}(\mathbf{y}, \boldsymbol{\xi})\Phi_j(\mathbf{y})dS_y - \int_{S-S_0} K_{ji}(\mathbf{y}, \boldsymbol{\xi})\Phi_j(\mathbf{y})dS_y +$$

$$\sum_{n=1}^{M} \alpha^n \sigma_{ij}(\mathbf{u}^{S,n}(\boldsymbol{\xi}))n_j(\boldsymbol{\xi}) + \varphi_i^l(\boldsymbol{\xi})\beta_l = \gamma\kappa(\boldsymbol{\xi})n_i(\boldsymbol{\xi}) \qquad \text{for } \boldsymbol{\xi} \in S_0 \tag{20.18}$$

and

$$-\frac{1}{2}\Phi_i(\boldsymbol{\xi}) - \int_{S_m} K_{ji}(\mathbf{y}, \boldsymbol{\xi})\Phi_j(\mathbf{y})dS_y - \int_{S-S_m} K_{ji}(\mathbf{y}, \boldsymbol{\xi})\Phi_j(\mathbf{y})dS_y +$$

$$\sum_{n=1}^{M} \alpha^n \sigma_{ij}(\mathbf{u}^{S,n}(\boldsymbol{\xi}))n_j(\boldsymbol{\xi}) = -\gamma\kappa(\boldsymbol{\xi})n_i(\boldsymbol{\xi}) \qquad \text{for } \boldsymbol{\xi} \in S_m \tag{20.19}$$

where

$$\sigma_{ij}(\mathbf{u}^{S,n}(\boldsymbol{\xi})) = \sigma_{ij}\left(\frac{\boldsymbol{\xi} - \boldsymbol{\xi}^n}{2\pi R_n^2}\right) = \frac{1}{\pi}\left(\frac{\delta_{ij}}{R_n^2} - \frac{2(\xi_i - \xi_i^n)(\xi_j - \xi_j^n)}{R_n^4}\right) \tag{20.20}$$

and β has been chosen as before (equation (20.13)).

A proof that the above system of Fredholm integral equations possesses a unique solution is presented in Primo, Wrobel and Power [6].

20.2.3 Deformation of a Viscous Fluid Region Containing Viscous Drops

A compound drop is now considered, consisting of a fluid drop with viscosity μ_2 completely coated by another drop of viscosity μ_1 (Figure 20.3). The present analysis can be directly extended to a finite number of internal drops of different viscosities. Consider that the external drop Ω_1 is bounded by the surface S_1. The internal region Ω_2 is bounded by the surface S_2, which represents the interface between the two fluids. Let γ_2 be the interfacial surface tension between fluids 2 and 1, and γ_1 between fluid 1 and the exterior environment. The governing equations for the fluid velocity $\mathbf{u}$ and pressure p are the Stokes equations (20.1) and (20.2).

The flow fields have to satisfy the following boundary conditions at every point $\boldsymbol{\xi} \in S_1$:

$$\sigma_{ij}^1(\boldsymbol{\xi})n_j(\boldsymbol{\xi}) = \gamma_1\kappa(\boldsymbol{\xi})n_i(\boldsymbol{\xi}) \tag{20.21}$$

and the following matching conditions at every point $\boldsymbol{\xi} \in S_2$:

$$u_i^1(\boldsymbol{\xi}) = u_i^2(\boldsymbol{\xi}) \tag{20.22}$$

and

$$\sigma_{ij}^1(\boldsymbol{\xi})n_j(\boldsymbol{\xi}) - \sigma_{ij}^2(\boldsymbol{\xi})n_j(\boldsymbol{\xi}) = -\gamma_2\kappa(\boldsymbol{\xi})n_i(\boldsymbol{\xi}) \tag{20.23}$$

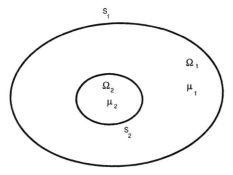

Figure 20.3: A compound drop.

The motion of the drop's surface cannot be arbitrary since at each instant in time the force and torque yielded by the surface tractions have to satisfy the following compatibility conditions:

$$\int_{S_1} \sigma_{ij}^1 n_j \varphi_i^l dS - \int_{S_2} \sigma_{ij}^1 n_j \varphi_i^l dS = 0 \qquad (20.24)$$

and

$$\int_{S_2} \sigma_{ij}^2 n_j \varphi_i^l dS = 0 \qquad (20.25)$$

where φ^l, $l = 1, 2, 3$, are the three linearly-independent rigid body vectors.

Adding equations (20.24) and (20.25), using the boundary condition (20.21) and the matching condition (20.23), leads finally to the solvability condition

$$\int_{S_1} \gamma_1 \kappa(\boldsymbol{\xi}) n_i(\boldsymbol{\xi}) \varphi_i^l(\boldsymbol{\xi}) dS = - \int_{S_2} \gamma_2 \kappa(\boldsymbol{\xi}) n_i(\boldsymbol{\xi}) \varphi_i^l(\boldsymbol{\xi}) dS \qquad (20.26)$$

where the normal vector $\mathbf{n}$ has been defined outwardly to the domain bounded by the corresponding surface boundary, *i.e.* S_1 and S_2.

In order to guarantee continuity of the velocity field across the surface S_2, equation (20.22), the velocity field inside the domain $\Omega_1 \cup \Omega_2$ is defined by the following single-layer potential:

$$u_i(\mathbf{x}) = - \int_{S_1 \cup S_2} u_i^j(\mathbf{x}, \mathbf{y}) \Phi_j(\mathbf{y}) dS_y \qquad (20.27)$$

for every $\mathbf{x} \in \Omega_1$ or Ω_2.

For convenience, the dynamic viscosity of each fluid is defined in the corresponding pressure field. Therefore, at each fluid domain:

$$p^l(\mathbf{x}) = \mu_l \int_{S_1 \cup S_2} q^j(\mathbf{x}, \mathbf{y}) \Phi_j(\mathbf{y}) dS_y \qquad (20.28)$$

Using the above representation for the flow field in each fluid region and letting a point $\mathbf{x} \in \Omega_1$ approach a point $\boldsymbol{\xi} \in S_1$, taking the boundary condition (20.21) into consideration, a system of second kind Fredholm integral equations for the unknown surface density $\boldsymbol{\Phi}$ over $S_1 \cup S_2$ is obtained, which does not have a unique solution due to the eigenvalues related to the exterior surface [7]. To obtain a unique solution, a complementary term

$$\varphi_i^l(\boldsymbol{\xi}) \int_{S_1} \Phi_j(\mathbf{y}) \varphi_j^l(\mathbf{y}) dS_y$$

is added to the solution (see [7]) in order to remove the three eigenvalues on S_1, leading to the equations:

$$T_1 \kappa(\boldsymbol{\xi}) n_i(\boldsymbol{\xi}) = \frac{1}{2} \Phi_i(\boldsymbol{\xi}) -$$

$$\left[\int_{S_1} K_{ji}(\mathbf{y}, \boldsymbol{\xi}) \Phi_j(\mathbf{y}) dS_y + \int_{S_2} K_{ji}(\mathbf{y}, \boldsymbol{\xi}) \Phi_j(\mathbf{y}) dS_y \right] + \varphi_i^l(\boldsymbol{\xi}) \int_{S_1} \Phi_j(\mathbf{y}) \varphi_j^l(\mathbf{y}) dS_y \qquad (20.29)$$

when $\boldsymbol{\xi} \in S_1$, and

$$-T_2 \kappa(\boldsymbol{\xi}) n_i(\boldsymbol{\xi}) = -\frac{(1 + \lambda)}{2} \Phi_i(\boldsymbol{\xi}) -$$

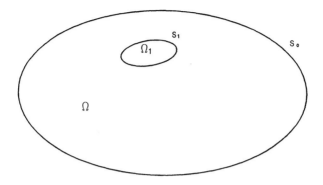

Figure 20.4: A rigid inclusion inside a finite viscous drop.

$$(1 - \lambda) \left[\int_{S_2} K_{ji}(\mathbf{y}, \boldsymbol{\xi}) \Phi_j(\mathbf{y}) dS_y + \int_{S_1} K_{ji}(\mathbf{y}, \boldsymbol{\xi}) \Phi_j(\mathbf{y}) dS_y \right] \qquad (20.30)$$

when $\boldsymbol{\xi} \in S_2$, where $T_l = \gamma_l / \mu_1$, $l = 1, 2$ and $\lambda = \mu_2 / \mu_1$.

Again, it can be shown that the addition of the extra terms in equations (20.29) and (20.30) does not alter the nature of the problem since for any admittable non-homogeneous term, *i.e.* satisfying the compatibility conditions (20.24) and (20.25), the extra integral in equation (20.29) will end up being zero. The coefficient of proportionality is identically equal to zero when the non-homogeneous term, *i.e.* the given surface velocity, satisfies the corresponding compatibility condition, *i.e.* the no-flux condition. Although for any given admittable surface velocity the original integral equation and the completed one are identical, the former does not have a unique solution whilst the latter does. A mathematical proof of uniqueness of the solution of the above system can be found in Primo, Wrobel and Power [7].

20.2.4   *Deformation of a Viscous Fluid Region Containing a Solid Particle*

Consider now a rigid inclusion inside a finite viscous drop. Let Ω be the finite fluid domain exterior to the inclusion and interior to the drop surface S_0, and let Ω_1 be the rigid domain bounded by S_1, as shown in Figure 20.4. The viscosity of the fluid drop is μ while γ is the interfacial surface tension between the fluid and the exterior environment. The governing equations for the fluid velocity $\mathbf{u}$ and pressure p are the Stokes equations (20.1) and (20.2). The flow field has to satisfy the following boundary condition for any point $\mathbf{x} \in S_0(t)$ at the exterior boundary

$$\sigma_{ij} n_j = \gamma \kappa n_i \qquad (20.31)$$

and, for any point $\mathbf{x} \in S_1(t)$ on the solid particle

$$u_i = C_k \varphi_i^k \qquad (20.32)$$

where C_k are unknown constants and φ_i^k correspond to the rigid body vectors.

The rate of deformation of the viscous drop is determined by the kinematic boundary condition at S_0, given by equation (20.5). The solid region does not deform but moves with the velocity given by equation (20.32).

The problem defined by equations (20.1), (20.2), (20.31) and (20.32) has a quasistatic character. Starting with an initial drop shape $S_0(t)$ and a given initial inclusion location $S_1(t)$, the surface velocity of the drop contour $u_i(\boldsymbol{\xi}, t)$ is obtained, as well as the translational and rotational velocities of the solid particle. The shape of the deformed drop $S_0(t + \Delta t)$, as well as the new inclusion position $S_1(t + \Delta t)$, can be determined using boundary conditions (20.31) and (20.32) and the kinematic condition (20.5).

The solution of the present problem is sought as a combination of a single- and a double-layer potential, with unknown vector densities $\boldsymbol{\Phi}$ and $\boldsymbol{\Psi}$, plus a Stokeslet and a Rotlet located at the centre of the particle $\mathbf{x}^c$, with intensities α_j and β, respectively:

$$u_i(\mathbf{x}) = V_i(\mathbf{x}, \boldsymbol{\Phi}) + W_i(\mathbf{x}, \boldsymbol{\Psi}) + \alpha_j v_i^j(\mathbf{x}, \mathbf{x}^c) + \beta r_i(\mathbf{x}, \mathbf{x}^c) \tag{20.33}$$

where the single-layer potential is given by:

$$V_i(\mathbf{x}, \boldsymbol{\Phi}) = -\int_{S_0} v_i^j(\mathbf{x}, \mathbf{y}) \Phi_j(\mathbf{y}) dS_y \tag{20.34}$$

for every $\mathbf{x} \in \Omega$, and $v_i^j(\mathbf{x}, \mathbf{y})$ is the Stokeslet given by equation (20.8). The pressure field of the single-layer potential is given by equation (20.9).

The double-layer potential is given by:

$$W_i(\mathbf{x}, \boldsymbol{\Psi}) = \int_{S_1} K_{ij}(\mathbf{x}, \mathbf{y}) \Psi_j(\mathbf{y}) dS_y \tag{20.35}$$

with the kernel $K_{ij}(\mathbf{x}, \mathbf{y})$ given by equation (20.11). The pressure field of the double-layer potential is given by:

$$p(\mathbf{x}, \boldsymbol{\Psi}) = -\mu \int_S \left[\frac{1}{\pi r^2} \left(-\delta_{jk} + \frac{2(x_j - y_j)(x_k - y_k)}{r^2} \right) \right] \Psi_k(\mathbf{y}) dS_y \tag{20.36}$$

The Stokeslet $v_i^j(\mathbf{x}, \mathbf{x}^c)$ in equation (20.33) is given by:

$$v_i^j(\mathbf{x}, \mathbf{x}^c) = -\frac{1}{4\pi\mu} \left(\delta_{ij} \ln(\frac{1}{R}) + \frac{(x_i - x_i^c)(x_j - x_j^c)}{R^2} \right) \tag{20.37}$$

and the Rotlet is given by

$$r_i(\mathbf{x}, \mathbf{x}^c) = \frac{\varepsilon_{i3k} (x_k - x_k^c)}{R^2} \tag{20.38}$$

where ε_{i3k} is the permutation symbol, $R = |\mathbf{x} - \mathbf{x}^c|$ is the distance between the geometric centre $\mathbf{x}^c$ of the particle where the Stokeslet and Rotlet are being applied and a point $\mathbf{x}$ in the flow field.

The vector α_k and the constant β represent the magnitude of the Stokeslet $v_i^j(\mathbf{x}, \mathbf{x}^c)$ and the magnitude of the Rotlet $r_i(\mathbf{x}, \mathbf{x}^c)$, respectively. For convenience, α_k and β are chosen as:

$$\alpha_k = \int_{S_1} \Psi_i(\mathbf{y})\varphi_i^k(\mathbf{y})dS_y \qquad k = 1, 2 \tag{20.39}$$

$$\beta = \int_{S_1} \Psi_i(\mathbf{y})\varphi_i^3(\mathbf{y})dS_y \tag{20.40}$$

Applying the boundary condition (20.31) to the flow field given by equation (20.33) and using the discontinuity property of the surface forces of a single-layer potential across the density-carrying surface S_0, the following linear Fredholm integral equation of the second kind is obtained for the unknown density $\boldsymbol{\Phi}$, when a point $\mathbf{x}$ tends to a point $\boldsymbol{\xi} \in S_0$:

$$\frac{1}{2}\Phi_i(\boldsymbol{\xi}) - \int_{S_0} K_{ji}(\mathbf{y}, \boldsymbol{\xi})\Phi_j(\mathbf{y})dS_y + \int_{S_1} T_{ji}(\mathbf{y}, \boldsymbol{\xi})\Psi_j(\mathbf{y})dS_y +$$

$$+\alpha_j K_{ij}(\boldsymbol{\xi}, \mathbf{x}^c) + \beta t_i(\boldsymbol{\xi}, \mathbf{x}^c) = \gamma\kappa(\boldsymbol{\xi})n_i(\boldsymbol{\xi}) \tag{20.41}$$

where the kernels of the integrals over S_0 and S_1 are given by:

$$K_{ji}(\mathbf{y}, \boldsymbol{\xi}) = \frac{1}{\pi} \frac{(\xi_i - y_i)(\xi_j - y_j)(\xi_k - y_k)}{r^4} n_k(\boldsymbol{\xi}) \tag{20.42}$$

and

$$T_{ji}(\mathbf{y}, \boldsymbol{\xi}) = -\frac{1}{\pi r^4}\left[-\frac{8}{r^2}(\xi_i - y_i)(\xi_j - y_j)(\xi_k - y_k)(\xi_l - y_l)n_k(\mathbf{y})n_l(\boldsymbol{\xi}) +\right.$$

$$+(\xi_i - y_i)(\xi_j - y_j)n_k(\mathbf{y})n_k(\boldsymbol{\xi}) + (\xi_i - y_i)(\xi_k - y_k)n_k(\boldsymbol{\xi})n_j(\mathbf{y}) +$$

$$\left.+(\xi_j - y_j)(\xi_l - y_l)n_l(\mathbf{y})n_i(\boldsymbol{\xi}) + (\xi_l - y_l)(\xi_k - y_k)n_l(\boldsymbol{\xi})n_k(\mathbf{y})\delta_{ij} + r^2 n_i(\mathbf{y})n_j(\boldsymbol{\xi})\right] \tag{20.43}$$

The surface forces of the two-dimensional Stokeslet $v_i^j(\mathbf{x}, \mathbf{x}^c)$ are given by:

$$K_{ij}(\boldsymbol{\xi}, \mathbf{x}^c) = -\frac{1}{\pi} \frac{(\xi_i - x_i^c)(\xi_j - x_j^c)(\xi_k - x_k^c)}{R^4} n_k(\boldsymbol{\xi}) \tag{20.44}$$

and the surface forces of the two-dimensional Rotlet $r_i^j(\boldsymbol{\xi}, \mathbf{x}^c)$ are given by:

$$t_i(\boldsymbol{\xi}, \mathbf{x}^c) = -\frac{2}{R^4}\left[\epsilon_{i3k}\left(\xi_j - x_j^c\right) + \epsilon_{j3k}\left(\xi_i - x_i^c\right)\right]\left(\xi_k - x_k^c\right)n_j(\boldsymbol{\xi}) \tag{20.45}$$

Now the boundary condition (20.32) is applied to the velocity field given by equation (20.33). Using the discontinuity property of the double-layer potential across the density-carrying surface S_1, the following linear Fredholm integral equation of the second kind is obtained for the unknown density $\boldsymbol{\Psi}$:

$$C_k\varphi_i^k(\boldsymbol{\xi}) = -\frac{1}{2}\Psi_i(\boldsymbol{\xi}) + \int_{S_1} K_{ij}(\boldsymbol{\xi}, \mathbf{y})\Psi_j(\mathbf{y})dS_y - \int_{S_0} v_i^j(\boldsymbol{\xi}, \mathbf{y})\Phi_j(\mathbf{y})dS_y +$$

$$+\alpha_j v_i^j(\boldsymbol{\xi}, \mathbf{x}^c) + \beta r_i(\boldsymbol{\xi}, \mathbf{x}^c) \tag{20.46}$$

The above system of equations, subject to the following orthogonality condition:

$$\int_{S_0} \Phi_i(\mathbf{y})\varphi_i^k(\mathbf{y})dS_y = 0 \qquad k = 1, 2, 3 \tag{20.47}$$

has a unique solution, as demonstrated in Primo, Wrobel and Power [8].

20.3   NUMERICAL PROCEDURES

20.3.1   Indirect Boundary Element Method

The numerical procedures will only be described for the first boundary integral formulation, since all others have similar implementations.

Therefore, discretizing a generalized form of equation (20.12) for any boundary point ξ gives:

$$c_{ij}(\xi)\Phi_j(\xi) - \sum_{d=1}^{NE} \Phi_j^k \int_{S_d} K_{ji}(\mathbf{y},\xi)N_k(\mathbf{y})dS_y + \varphi_i^l(\xi)\sum_{d=1}^{NE} \Phi_j^k \int_{S_d} \varphi_j^l(\mathbf{y})N_k(\mathbf{y})dS_y = \gamma\kappa(\xi)n_i(\xi)$$

$$(20.48)$$

where NE is the number of boundary elements, c_{ij} is the characteristic term [9] and Φ_j, the unknown density function, was approximated by using quadratic interpolation functions.

Carrying out the integrations over each element and adding up those terms multiplying the same nodal value Φ_j^k produces the coefficients:

$$h_{ij,\xi d}^k = -\int_{S_d} K_{ji}(\mathbf{y},\xi)N_k(\mathbf{y})dS_y + \varphi_i^l(\xi)\int_{S_d} \varphi_j^l(\mathbf{y})N_k(\mathbf{y})dS_y \qquad (20.49)$$

It should be noted that the coefficients related to nodes at the extremity of each element will involve the summation of two contributions coming from the integrals over the adjoining elements. This superposition of influences is taken into consideration when assembling the system matrix.

Equation (20.48) for a collocation point ξ can then be expressed in matrix form as:

$$\mathbf{c}_\xi\Phi_\xi + \left[\hat{\mathbf{H}}_{\xi 1}\hat{\mathbf{H}}_{\xi 2}...\hat{\mathbf{H}}_{\xi P}\right]\begin{bmatrix}\Phi_1 \\ \Phi_2 \\ \vdots \\ \Phi_P\end{bmatrix} = \mathbf{T}_\xi \qquad (20.50)$$

where the characteristic term $\mathbf{c}_\xi$ is a 2×2 matrix, each $\hat{\mathbf{H}}_{\xi n}$ is a 2×2 matrix obtained by assembling the coefficients $h_{ij,\xi d}^k$ and each Φ_n is a vector with components Φ_1, Φ_2; $\mathbf{T}$ is a vector representing the surface tension, and P is the total number of boundary nodes.

The above equation can also be written in the form:

$$\sum_{\chi=1}^{P} \mathbf{H}_{\xi\chi}\Phi_\chi = \mathbf{T}_\xi \qquad (20.51)$$

where

$$\mathbf{H}_{\xi\xi} = \hat{\mathbf{H}}_{\xi\xi} + \mathbf{c}_\xi \qquad \text{and} \qquad \mathbf{H}_{\xi\chi} = \hat{\mathbf{H}}_{\xi\chi} \qquad \text{for } \chi \neq \xi \qquad (20.52)$$

Equation (20.50) is collocated at each node generating a $2P \times 2P$ system of equations which can be expressed in matrix form as:

$$\mathbf{H\Phi} = \mathbf{T} \tag{20.53}$$

The above system of equations for the vector density $\mathbf{\Phi}$ was solved by Gauss elimination.

The next step is to find the boundary velocities using equation (20.7) which is also valid at boundary points since the single-layer potential $u_i^j(\mathbf{\xi}, \mathbf{y})$ is continuous across the density carrying surface S. Discretizing equation (20.7) gives:

$$u_i(\mathbf{\xi}) = -\sum_{d=1}^{NE} \int_{S_d} u_i^j(\mathbf{\xi}, \mathbf{y})\Phi_j(\mathbf{y})dS_y \tag{20.54}$$

Collocating the above equation at all boundary nodes, using the same quadratic interpolation for the density function, the following equation is obtained:

$$\mathbf{U} = \mathbf{L\Phi} \tag{20.55}$$

with $\mathbf{L}$ a square matrix obtained from the boundary integrals. Matrix $\mathbf{L}$ is then multiplied by the now known vector density $\mathbf{\Phi}$, providing the velocity at the boundary nodes.

20.3.2 Calculation of Surface Tension

The surface tension driving the sintering process is expressed as a product of the curvature by the normal, as given in equation (20.3). The curvature and the normal vector were calculated at each nodal point by fitting a fourth-order Lagrangian polynomial.

20.3.3 Boundary Deformation

After calculating the velocity at the nodal points, the deformation was obtained by taking into account the Lagrangian representation of the fluid, following equation (20.5), giving the path followed by the particle:

$$\frac{\Delta \mathbf{x}}{\Delta t} = \mathbf{u}(\mathbf{\xi}) \qquad \mathbf{\xi} \in S \tag{20.56}$$

where $\mathbf{u}(\mathbf{\xi})$ is the computed velocity and Δt is a fixed interval of time. A simple Euler scheme was adopted in which the variation of the position of the fluid particle was found by multiplying the fixed time interval by the computed velocity. Van de Vorst *et al.* [2] reported possible stiffness problems in the system of ordinary differential equations for more complex geometrical shapes, and used backward difference formulae to solve the ODEs. The simple Euler scheme provided sufficient accuracy for all cases studied here.

20.4 RESULTS AND DISCUSSION

Quadratic boundary elements were used in all simulations. The physical parameters were made dimensionless, following Kuiken [1], by defining a characteristic velocity u_c, a characteristic pressure p_c and a characteristic time t_c:

$$u_c = \frac{\gamma}{\mu}; \quad p_c = \frac{\gamma}{l}; \quad t_c = \frac{l\mu}{\gamma} \tag{20.57}$$

where l is a characteristic length.

Kuiken [1] studied the deformation, due to surface tension, of an elliptical fluid region $(x^2 + 10y^2 = 1)$. The steady-state circular shape was achieved at a dimensionless time of $t = 10$. A similar deformation process can be seen in Figure 20.5, obtained using the present formulation with 20 quadratic elements evenly distributed on the elliptical boundary, with deformations calculated at time steps of $\Delta t = 0.1$.

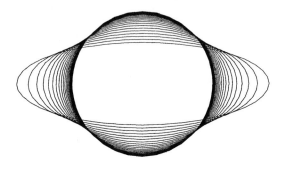

t = 0.0(0.1)10.0

Figure 20.5: Deformation of an ellipse into a circle.

As the fluid is assumed to be incompressible, the area of the fluid region should be conserved at all times. The present formulation produced acceptable values for the area conservation. The initial ellipse has an area $\pi ab = 0.9934$ while the final circular area is $\pi r^2 = 0.9747$, which is 1.88% smaller. Using the same discretization of 20 quadratic elements but reducing the time step size improves area conservation. For $\Delta t = 0.01$, the radius of the final circular form is 0.5617, its area 0.9912 and the error 0.22%; for $\Delta t = 0.001$, the radius is 0.5622, the area 0.9929 and the error 0.05%. Keeping the time step constant and doubling the number of elements produced virtually no improvement in the results.

The second boundary integral formulation given in this work, dealing with the deformation of a fluid region containing air bubbles, was applied to study the deformation of a fluid cylinder with nine cylindrical pores (Figure 20.6). The calculations were carried out with the same nondimensional values used by van de Vorst [3]: radius of the cylinder equal to 1.3 and radius of each pore equal to 0.2. The shrinkage was calculated at time steps of $\Delta t = 0.01$ and shown at steps of $\Delta t = 0.05$. As can be seen in Figure 20.6, the holes have almost vanished at $t = 0.3$.

Figure 20.7 shows an application of the third formulation to study the deformation of nine drops of different geometrical forms (circular and elliptical), and different densities. The central drop has $\lambda = 10.0$ while the others have $\lambda = 5.0$, $\lambda = 7.5$ or $\lambda = 10.0$. As expected, elliptical drops show motion and deformation while circular drops only move. Although there are some elliptical drops, the morphological configuration is symmetrical, making the central drop stay at rest. As no comparison with other authors was possible, checks of mass conservation were carried out showing very good results.

The formulation given by equations (20.41), (20.46) and (20.47) allows the presence of a single solid particle to be considered inside a viscous mass. Figure 20.8 shows an elliptical inclusion ($a = 0.20, b = 0.10$) generated by the points ($a\cos(\theta + \pi/4), b\sin(\theta)$), with its centre at the initial position $x = -0.175, y = -0.175$. The particle displacement and rotation are clearly visible, and mass was properly conserved as the external boundary deformed into a circle.

20.5 CONCLUSIONS

Four indirect boundary integral formulations were presented for analysing the deformation of slow viscous fluids driven by surface tension. The common feature is the use of completed integral equations which guarantee uniqueness of solutions. The formulations have a direct application to viscous sintering problems.

We advocate the use of indirect boundary integral formulations for the problem since they provide simpler and more rigorous mathematical analyses of the existence and uniqueness of solutions of the resulting Fredholm integral equations. The system of equations generated by indirect formulations is very stable and more amenable to fast iterative solvers using multipole expansions.

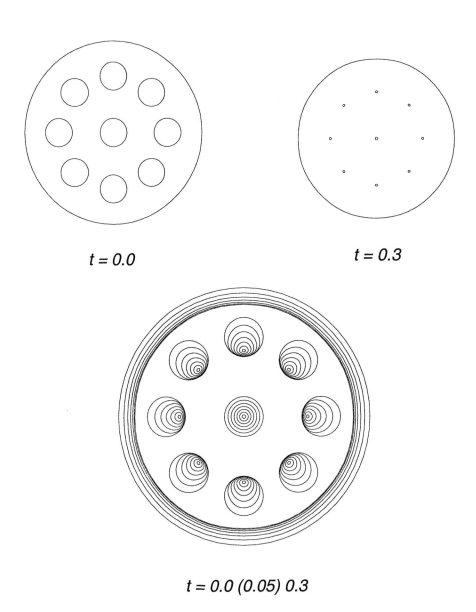

$t = 0.0$

$t = 0.3$

$t = 0.0\ (0.05)\ 0.3$

Figure 20.6: Shrinkage of a fluid cylinder with nine equally-sized cylindrical pores.

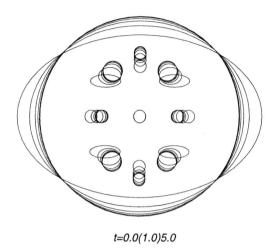

t=0.0(1.0)5.0

Figure 20.7: Nine circular and elliptical drops of different viscosity ratios inside an elliptical region. The central drop has $\lambda = 10.0$ while the other drops have $\lambda = 5.0$, $\lambda = 7.5$ or $\lambda = 10.0$.

t=0.0(0.5)6.0

Figure 20.8: An elliptical solid inclusion with centre at initial position $x = -0.175, y = -0.175$ inside elliptical viscous region. Inclusion axes are rotated by -20° with respect to those of the exterior ellipse.

REFERENCES

[1] H.K. Kuiken (1990). Viscous sintering: the surface-tension-driven flow of a liquid form under the influence of curvature gradients at its surface. *J. Fluid Mech.*, 214:503–515.

[2] G.A.L. van de Vorst, R.M.M. Mattheij and H.K. Kuiken (1992). A boundary element solution for two-dimensional viscous sintering. *J. Comp. Phys.*, 100:50–63.

[3] G.A.L. van de Vorst (1993). Integral method for a two-dimensional Stokes flow with shrinking holes applied to viscous sintering. *J. Fluid Mech.*, 257:667–689.

[4] H. Power and G. Miranda (1987). Second kind integral equation formulation of Stokes flows past a particle of arbitrary shape. *SIAM J. Appl. Math.*, 47:689–698.

[5] V. Botte and H. Power (1995). A second kind integral equation formulation for three-dimensional interior flows at low Reynolds number. *Bound. Elem. Comm.*, 6:163–166.

[6] A.R.M. Primo, L.C. Wrobel and H. Power (1999a). An indirect boundary element method for slow viscous flow in a bounded region containing air bubbles. *J. Eng. Math.*, in press.

[7] A.R.M. Primo, L.C. Wrobel and H. Power (1999b). Low Reynolds number deformation of viscous drops in a bounded flow region under surface tension. *Math. Comp. Mod.*, in press.

[8] A.R.M. Primo, L.C. Wrobel and H. Power (1999c). Boundary integral formulation for slow viscous flow in a deforming region containing a solid inclusion. *Eng. Anal. with Boundary Elements*, in press.

[9] F. Hartmann (1989). *Introduction to Boundary Elements*. Berlin: Springer-Verlag.

21 SEMI-LAGRANGIAN FINITE VOLUME METHODS FOR VISCOELASTIC FLOW PROBLEMS

T. N. Phillips[a] and A. J. Williams[b]

[a]Department of Mathematics, University of Wales, Aberystwyth SY23 3BZ.
[b]Mathematics and Statistics Group, Middlesex University, London N11 2NQ.

ABSTRACT

A semi-Lagrangian finite volume scheme for solving viscoelastic flow problems is presented. A staggered grid arrangement is used in which the dependent variables are located at different mesh points in the computational domain. The convection terms in the momentum and constitutive equations are treated using a semi-Lagrangian approach in which particles on a regular grid are traced backwards over a single time-step. The method is applied to the 4:1 planar contraction problem for an Oldroyd B fluid under inertial flow conditions. The development of vortex behaviour with increasing values of the Weissenberg number We is analyzed.

Key words. finite volume method, viscoelasticity, semi-Lagrangian method, SIMPLER scheme

21.1   INTRODUCTION

Over the last decade there has been tremendous progress in the development of efficient and stable numerical methods for solving viscoelastic flow problems. Although the range of Weissenberg numbers over which converged solutions are attained has been extended, a limit nevertheless still exists. Beyond this limit many schemes suffer from numerical instabilities. This may be due to a number of factors including the presence of geometric singularities and boundary layers in the flow, the inability of the method to resolve them, and the dominance of the nonlinear terms in the equations for high values of the Weissenberg number.

In this paper we examine the planar flow of an Oldroyd B fluid in a 4:1 contraction geometry which is one of the standard benchmark problems in viscoelastic flow. Although the shortcomings of the Oldroyd B model are well understood it is suitable for presenting the novel features of the scheme introduced in this paper and for making comparisons with many results in the literature. The behaviour of the vortex in the salient corner is investigated and comparisons made with other results in the literature. The size and intensity of the salient corner vortex decreases with increasing We. A weak lip vortex is detected when $We = 2.5$.

The hyperbolic nature of the constitutive equation is responsible for many of the difficulties associated with the numerical simulation of viscoelastic flows. Many techniques introduce some form of upwinding in order to obtain numerical solutions at high values of the Deborah number when the flow becomes convection-dominated (see Xue et al. [10], for example). In fact the treatment of convection together with the positioning of the mesh points are the features which distinguish one finite volume method from another.

The Semi-Lagrangian approach described here, which is a generalization of the work of Scroggs and Semazzi [9], is a natural way of treating the convective terms in the constitutive equation without resorting to upwinding. This scheme circumvents the problems associated with upwind biased interpolation schemes, possesses less restrictive stability requirements, and combines the advantages of fixed grids inherent in Eulerian methods with modifications to the location of grid points at previous time-steps based on the Lagrangian approach. The remaining terms in the governing equations are treated implicitly and discretized by integrating over the appropriate control volume. The discrete equations are solved using the SIMPLER algorithm [6]. Therefore, this approach may be viewed as a time-splitting scheme in which the different operators in the governing equations are discretized by appropriate techniques.

21.2 GOVERNING EQUATIONS

We consider the flow of an Oldroyd B fluid through a planar 4:1 contraction. A schematic diagram of the lower half of this geometry is shown in Fig. 21.1. The downstream half channel width is denoted by L. Upstream of the contraction we impose parabolic Poiseuille flow and we assume that the downstream exit length is chosen long enough so that the flow is parabolic once again. Since the flow is symmetric about $y = 0$, it is only necessary to seek a solution for $y \leq 0$.

The governing equations, in dimensionless form, are

$$Re \left(\frac{\partial \mathbf{v}}{\partial t} + \mathbf{v}.\nabla \mathbf{v} \right) = -\nabla p + \nabla.\boldsymbol{\tau}_1 + \beta \nabla^2 \mathbf{v}, \qquad (21.1)$$

$$\nabla.\mathbf{v} = 0, \qquad (21.2)$$

$$\boldsymbol{\tau}_1 + We \overset{\triangledown}{\boldsymbol{\tau}}_1 = 2(1-\beta)\mathbf{d}, \qquad (21.3)$$

where ρ is the density, p is an arbitrary isotropic pressure, $\boldsymbol{\tau}_1$ is the viscoelastic contribution to the extra-stress tensor, and $\mathbf{v}$ is the velocity vector, Re is the Reynolds number and We is the Weissenberg number. The parameter β is the ratio of the retardation and relaxation times of the fluid and is taken to be $1/9$ throughout this paper. The extra-stress tensor $\boldsymbol{\tau}$ is given by

$$\boldsymbol{\tau} = \boldsymbol{\tau}_1 + 2(1-\beta)\mathbf{D},$$

where the rate-of-deformation tensor $\mathbf{D}$ is given by

$$\mathbf{D} = \frac{1}{2}(\nabla\mathbf{v} + \nabla\mathbf{v}^T).$$

To solve the system of partial differential equations (21.1) - (21.3) we need to impose

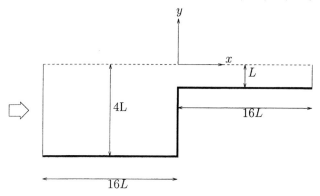

Figure 21.1: The contraction flow geometry.

suitable boundary and initial conditions. At inflow we impose fully developed flow. No-slip conditions are imposed on solid boundaries and symmetry conditions are specified on the axis of symmetry. At outflow a fully developed velocity profile is imposed as well as homogeneous Neumann boundary conditions for the extra-stress. The initial conditions are chosen to be either the solution to the corresponding Stokes problem or the solution obtained for a lower value of the Weissenberg number.

21.3 FINITE VOLUME DISCRETIZATION

The finite volume method is generally applied to a system of differential equations written in conservation form, e.g.,

$$\frac{\partial\mathbf{w}}{\partial t} + \frac{\partial\boldsymbol{f}}{\partial x} + \frac{\partial\boldsymbol{g}}{\partial y} = \mathbf{S}, \qquad (21.4)$$

where $\mathbf{w}$ is the vector of unknowns, $\boldsymbol{f}$ and $\boldsymbol{g}$ are vector functions of $\mathbf{x} = (x, y)$, $\mathbf{w}$ and $\nabla\mathbf{w}$, and $\mathbf{S}$ is the source term. In this paper we consider cell centre finite volume methods. These methods are closely related to finite difference methods.

Each of the governing equations (21.1)-(21.3) can be cast into the general conservative form

$$\delta\frac{\partial\phi}{\partial t} + \frac{\partial}{\partial x}\left(u\phi\theta - \Gamma\frac{\partial\phi}{\partial x}\right) + \frac{\partial}{\partial y}\left(v\phi\theta - \Gamma\frac{\partial\phi}{\partial y}\right) = S_\phi, \qquad (21.5)$$

where δ and Γ are constants, ϕ and S_ϕ are functions that are defined depending on the particular equation under consideration (see Table 21.1).

Table 21.1: Definition of the constants and functions in the general equation (21.5)

Equation	ϕ	δ	θ	Γ	S_ϕ
u-momentum	u	Re	Re	β	$-\frac{\partial p}{\partial x}+\frac{\partial \tau_{xx}}{\partial x}+\frac{\partial \tau_{xy}}{\partial y}$
v-momentum	v	Re	Re	β	$-\frac{\partial p}{\partial y}+\frac{\partial \tau_{xy}}{\partial x}+\frac{\partial \tau_{yy}}{\partial y}$
τ_{xx} normal stress	τ_{xx}	We	We	0	$2(1-\beta)\frac{\partial u}{\partial x}+\left(2We\frac{\partial u}{\partial x}-1\right)\tau_{xx}+2We\tau_{xy}\frac{\partial u}{\partial y}$
τ_{yy} normal stress	τ_{yy}	We	We	0	$2(1-\beta)\frac{\partial v}{\partial y}+\left(2We\frac{\partial v}{\partial y}-1\right)\tau_{yy}+2We\tau_{xy}\frac{\partial v}{\partial x}$
shear stress	τ_{xy}	We	We	0	$(1-\beta)\left(\frac{\partial v}{\partial x}+\frac{\partial u}{\partial y}\right)-\tau_{xy}+We\frac{\partial v}{\partial x}\tau_{xx}+We\frac{\partial u}{\partial y}\tau_{yy}$
continuity	1	0	1	0	0

21.3.1 Computational Grid

A grid is placed on the computational domain and a control or finite volume is associated with each unknown on the grid. This grid, called the reference grid, remains fixed in space for all time. Each component equation of (21.4) is integrated over an appropriate control volume. Finite difference type approximations are then used to approximate line integrals over each side of the control volume. The property of conservation of physical quantities, which is preserved by the discrete system, is one of the attractions of the finite volume method.

A staggered grid is used in which the different dependent variables are approximated at different mesh points (see Fig. 21.2). This mesh ensures that the solution is not polluted by spurious pressure modes. The normal stresses are located in the same position as

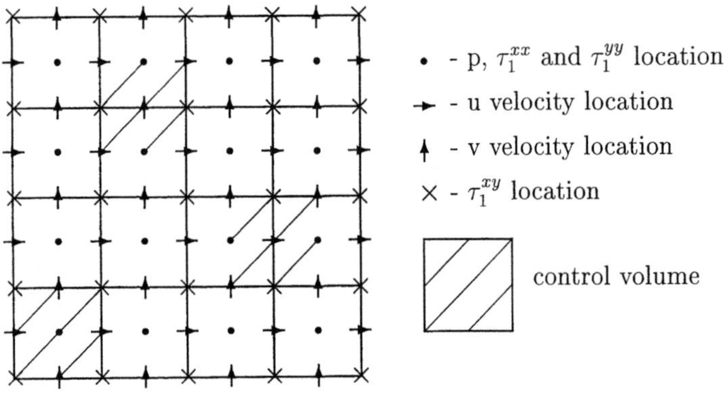

• - p, τ_1^{xx} and τ_1^{yy} location
- u velocity location
- v velocity location
× - τ_1^{xy} location
control volume

Figure 21.2: A grid with stress located at different positions.

the pressure variables, with the shear stresses located at the corners of the mesh cells. This arrangement has been used by Darwish, Whiteman and Bevis [1] and, for example,

Mompean and Deville [4]. It avoids the need for interpolation of the shear stresses in the momentum equations but introduces a shear stress value at the corner singularity.

21.3.2 Discretization

The approach adopted here is to decouple the treatments of convection and diffusion and to use appropriate methods of discretization for each subproblem [7]. At each time step the convection problems

$$\frac{\partial \mathbf{v}}{\partial t} + \mathbf{v}.\nabla \mathbf{v} = 0, \quad \frac{\partial \boldsymbol{\tau}}{\partial t} + \mathbf{v}.\nabla \boldsymbol{\tau} = 0, \tag{21.6}$$

are solved using a particle tracking method. The idea is to determine the position at the previous time step of the particle paths which pass through the vertices of a control volume at the current time. The values of the velocity ($\mathbf{v}^*$) and extra-stress ($\boldsymbol{\tau}^*$) in these deformed control volumes are determined using interpolation from known values on the reference grid at the previous time step. This is followed by the solution of an unsteady generalized Stokes problem to determine the pressure and the new velocity, and the solution of an algebraic problem to update the stresses. Thus, we may write the time discretized form of the equations (21.1)-(21.3):

$$Re\left(\frac{\mathbf{v}^{n+1} - \mathbf{v}^{*n}}{\Delta t}\right) - \beta \nabla^2 \mathbf{v}^{n+1} + \nabla p^{n+1} = \nabla.\boldsymbol{\tau}_1^n, \tag{21.7}$$

$$\nabla.\mathbf{v}^{n+1} = 0, \tag{21.8}$$

$$We\left(\frac{\boldsymbol{\tau}_1^{n+1} - \boldsymbol{\tau}_1^{*n}}{\Delta t}\right) - \mathbf{L}^n\boldsymbol{\tau}_1^{n+1} - \boldsymbol{\tau}_1^{n+1}(\mathbf{L}^n)^T + \boldsymbol{\tau}_1^{n+1} = 2(1-\beta)\mathbf{D}^n, \tag{21.9}$$

where $\mathbf{L}$ is the velocity deformation gradient.

21.4   TREATMENT OF CONVECTION

Consider the mesh associated with one of the dependent variables, ϕ say, where $\phi = u, v, \tau^{xx}, \tau^{xy}$ or τ^{yy}. Let $C_{i,j}$ be one such control volume. Let the positions of the corners of cell $C_{i,j}$ be located at the points $\mathbf{X}_{i\pm1/2,j\pm1/2} = (x_{i\pm1/2}, y_{j\pm1/2})$. Particles which arrive at these four corner points at time $t = t^{n+1}$ were located at the vertices of some cell, which is to be determined as part of the solution process at time $t = t^n$. In general, this will be a deformed control volume which may lie anywhere on the underlying grid or indeed outside the domain if the time step is sufficiently large. We approximate this cell by a quadrilateral $C_{i,j}^{*n}$, formed by joining the departure points by straight line segments.

Associated with each cell $C_{i,j}$ at each time $t_n = n\Delta t$ we introduce an approximation, denoted by $\phi_{i,j}^n$, to the cell average of $\phi(x, y, t_n)$, i.e.,

$$\phi_{i,j}^n \approx \frac{1}{\Delta x_i \Delta y_j} \iint_{C_{i,j}} \phi(x, y, t_n) \, dx dy, \tag{21.10}$$

where $\Delta x_i = x_{i+1/2} - x_{i-1/2}$, $\Delta y_j = y_{j+1/2} - y_{j-1/2}$. Note that $\phi_{i,j}^n$ will, in general, be distinct from the pointwise approximation to $\phi(x_i, y_j, t_n)$.

Thus, there are two stages to the numerical calculation at each time step. In the first stage the departure points are determined. These are the vertices of the cells $C_{i,j}^{*n}$. In the second stage the cell average values of ϕ^{*n} are determined from a knowledge of the cell average values of ϕ at time $t = t_n$ on the reference grid. These values are then inserted in equations (21.7) and (21.9) in order to determine the values of velocity, pressure and stress at the new time step.

21.4.1 Calculation of Departure Points

The departure points at time t_n of each point on the reference grid are determined by solving the following particle following transformation

$$\frac{d\xi(t)}{dt} = u(\xi(t), \eta(t), \tau), \quad \frac{d\eta(t)}{dt} = v(\xi(t), \eta(t), \tau), \quad \frac{d\tau(t)}{dt} = 1, \qquad (21.11)$$

for $t \in [t_n, t_{n+1}]$, subject to

$$\xi(t^{n+1}) = x_{i+\frac{1}{2}, j+\frac{1}{2}}, \quad \eta(t^{n+1}) = y_{i+\frac{1}{2}, j+\frac{1}{2}}, \quad \tau(t^{n+1}) = t^{n+1}. \qquad (21.12)$$

If $C_{i,j}$ represents a control volume for the u component of velocity then the following one-step method may be used to determine the positions of the corners of $C_{i,j}^{*n}$:

$$
\begin{aligned}
x_{i+\frac{1}{2}, j+\frac{1}{2}}^{*n} &= x_{i+\frac{1}{2}, j+\frac{1}{2}}^{n+1} - \frac{\Delta t}{4}(u_{i,j}^n + u_{i+1,j}^n + u_{i,j+1}^n + u_{i+1,j+1}^n), \\
y_{i+\frac{1}{2}, j+\frac{1}{2}}^{*n} &= y_{i+\frac{1}{2}, j+\frac{1}{2}}^{n+1} - \Delta t v_{i+\frac{1}{2}, j+\frac{1}{2}}^n,
\end{aligned}
\qquad (21.13)
$$

where subscripts indicate grid locations and superscripts indicate the time step. The above scheme is first order in time.

21.4.2 Calculation of Area-Weighting Coefficients

At the beginning of each time step the values of $\phi_{i,j}^n$ are known in all control volumes. Given the location of the departure points of the reference grid at time t^n the values $\phi_{i,j}^{*n}$ must be determined. This approximation is generated by means of an area-weighting technique which uses a weighted sum of the values of ϕ^n over the control volumes on the reference grid which overlap with cell $C_{i,j}^{*n}$. Area-weighting techniques are not new and they have been demonstrated to possess attractive stability properties. They were originally developed by users of particle-in-cell methods [2]. In the Lagrange-Galerkin finite element method the evaluation of inner products using non-exact integration must be performed with great care [5]. In the present application of the technique the control volumes are allowed to move with distortion and rotation. The first-order area-weighting scheme of Scroggs and Semazzi [9] for determining the value of $\phi_{i,j}^{*n}$ is

$$\phi_{i,j}^{*n} = \frac{1}{\Delta x_i \Delta y_j} \sum_{I, J \in Z} \omega_{i,j}^{I,J} \bar{\phi}_{I,J}^n, \qquad (21.14)$$

where $\omega_{i,j}^{I,J}$ is the common area between $C_{i,j}^{*n}$ and $C_{I,J}$, i.e., the area of $C_{I,J} \cap C_{i,j}^{*n}$, and Z is the set of indices of all the points in the computational domain.

21.5 NUMERICAL RESULTS

The algorithm for solving (21.7)-(21.9) is based on SIMPLER [6]. When a converged velocity field has been obtained within the time-step the constitutive (21.9) equations are solved for $\boldsymbol{\tau}$. We then check for a converged velocity field at the end of the time-step and if convergence has not been reached the process is repeated.

Numerical computations are performed on a series of meshes in order to ensure that the solutions obtained are independent of the mesh parameters. Nonuniform meshes are used in which the mesh spacing varies geometrically from the reentrant corner. In this way we can ensure a greater density of mesh points in the region where the solution changes most rapidly. The mesh characteristics of three meshes which have been used to solve the 4:1 contraction problem are shown in Table 21.2. Note that although Mesh 2 and Mesh 3 contain the same number of control volumes they differ in the way the nonuniform mesh spacing varies. Mesh 3 is more refined around the reentrant corner than Mesh 2. Calculations have been performed for a range of values of We for $Re = 1$. The length

Table 21.2: Mesh characteristics.

Mesh	Control Volumes	Total Degrees of Freedom	Δx_{min}	Δy_{min}
Mesh 1	2,800	16,800	0.027	0.067
Mesh 2	4,000	24,000	0.017	0.045
Mesh 3	4,000	24,000	0.009	0.038

of the salient corner vortex is defined to be the distance between the point where the separation line meets the bottom of the channel and the salient corner. The results on all meshes have been obtained with the time step $\Delta t = 8 \times 10^{-5}$. This was the largest value of Δt which ensured convergence for the most refined meshes. We observe that the value of L_1 does not vary significantly for the different meshes for a given value of We although the smallest values are obtained for Mesh 3 which is more refined near the reentrant corner.

Close examination of the results obtained in the outflow region showed that fully developed flow had been reached and that the downstream channel length of 16 units was long enough for the range of values of We used in this work. All the following results have been calculated using Mesh 3 for $0 \leq We \leq 2.5$.

Numerical convergence was lost for values of We much above 2.5. Further investigation showed that loss of convergence was associated with the failure of the numerical solution to preserve a property that the true solution is endowed with. For an Oldroyd B fluid the tensor

$$\boldsymbol{T}_A = \boldsymbol{\tau}_1 + \frac{1}{We}\boldsymbol{I},$$

must remain positive definite. This follows from the integral equivalent of the model and the fact that the inverse of the relative Cauchy strain tensor is positive definite. The positive definiteness of $\boldsymbol{T}_A$ was tested at the centre of each grid cell in the domain. We found that for each converged solution obtained when $0 \leq We \leq 2.5$, for $Re = 1$, $\boldsymbol{T}_A$ was positive definite. When $We = 2.7$, $\boldsymbol{T}_A$ eventually lost positive definiteness when $Re = 1$, with the loss first occurring in the neighbourhood of the corner singularity.

The streamline plot for $We = 2.5$ is given in Fig. 21.3 It is shown that the size of the salient corner vortex decreases as We increases, and that for $We = 2.5$ a weak lip vortex appears, which is separated from the salient corner vortex. The strength of the lip vortex decreases as the mesh is refined, which again suggests that the lip vortex is sensitive to mesh refinement. In the results of Matallah et al [3] and Sato and Richardson [8] a weak lip vortex was only detected for $We \geq 2.0$. They also suggested that the the lip vortex was sensitive to mesh size.

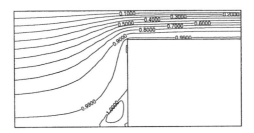

Figure 21.3: The streamlines for $We = 2.5$ when $Re = 1$.

In Table 21.3 the values of L_1 and ψ_{max}, the maximum absolute value of the stream function in the salient corner, are presented for $0 \leq We \leq 2.5$. We see that as We increases from 0 to 2.0 the size and intensity of the corner vortex decrease, with the size of the corner vortex continuing to decrease up to $We = 2.5$ while the intensity increases slightly between $We = 2.0$ and $We = 2.5$. The values of L_1 which have been obtained for

Table 21.3: Dependence of L_1 and ψ_{max} on We when $Re = 1$.

We	L_1	ψ_{max}
0	1.164	1.00050
0.1	1.142	1.00048
0.5	1.093	1.00041
1.0	1.048	1.00035
1.5	1.013	1.00031
2.0	0.986	1.00029
2.5	0.974	1.00030

$Re = 1$ agree qualitatively and quantitatively with those obtained by Sato and Richardson [8], particularly for $0 \leq We \leq 1.0$ although they do not show as great a decrease as those of the latter for $We > 1.0$.

In Fig. 21.4 the components of the normal extra stress τ_{xx} are plotted along the line $y = -1$ for $We = 0.5$, $We = 1.0$, $We = 1.5$, $We = 2.0$ and $We = 2.5$. The contours are smooth throughout the domain. The peak values of the stresses are given in Table 21.4.

The peak values of the normal stresses increase as We increases. However, the peak value of the shear stress decreases as We increases.

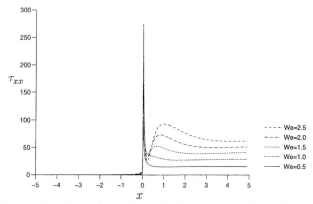

Figure 21.4: The values of τ_{xx} along the line $y = -1$ when $Re = 1$.

Table 21.4: The peak values of the shear and normal stresses for various We numbers when $Re = 1$

We	Max τ_{xy}	Max τ_{xx}	Max τ_{yy}
0.5	8.40	106.44	5.81
1.0	8.10	194.75	10.53
1.5	7.37	240.08	15.33
2.0	6.70	263.56	20.43
2.5	6.14	275.05	24.41

21.6 CONCLUSIONS

A semi-Lagrangian finite volume method for solving time-dependent viscoelastic flow problems is presented. The convection terms in the momentum and constitutive equations are treated using a semi-Lagrangian approach in which the vertices of a control volume at the new time level are traced back in time over a single time step using a particle following transformation. The values of the dependent variables in the transformed control volumes at the previous time level are determined using an area weighting technique. The remaining terms in the equations are solved using the standard SIMPLER method. This approach results in a stable scheme and circumvents problems associated with upwind biased schemes.

The flow of an Oldroyd-B fluid through a 4:1 planar contraction has been simulated numerically for $Re = 1$ (where $\frac{\lambda_2}{\lambda_1} = \frac{1}{9}$ throughout). Results have been presented for $0 \le We \le 2.5$,

ACKNOWLEDGEMENT

The authors would like to thank the University of Wales Aberystwyth for their financial support of this work through the award of a research grant.

REFERENCES

[1] M. J. Bevis, M. S. Darwish and J. R. Whiteman (1992). Numerical modelling of viscoelastic liquids using a finite volume method. *J. Non-Newtonian Fluid Mech.*, 45:311–337.

[2] F. H. Harlow (1964). The particle in cell computing method for fluid dynamics. In B. Adler, S. Fernbach and M. Rotenberg, editors, *Methods in Computational Physics*, volume 3. Academic Press, New York.

[3] H. Matallah, P. Townsend and M. F. Webster (1998). Recovery and stress-splitting schemes for viscoelastic flows. *J. Non-Newtonian Fluid Mech.*, 75:139–166.

[4] G. Mompean and M. Deville (1997). Unsteady finite volume simulation of Oldroyd-B fluid through a three-dimensional planar contraction. *J. Non-Newtonian Fluid Mech.*, 72:253–279.

[5] K. W. Morton, A. Priestley and E. Süli (1988). Stability analysis of the Lagrange-Galerkin method with non-exact integration. M^2AN, 22:625–653.

[6] S. V. Patankar (1981). *Numerical Heat Transfer and Fluid Flow*. Hemisphere Publishing, New York.

[7] T. N. Phillips and A. J. Williams (submitted 1998). A semi-Lagrangian finite volume method for the Navier-Stokes equations. *SIAM J. Sci. Comput.*

[8] T. Sato and S. M. Richardson (1994). Explicit numerical simulation of time-dependent viscoelastic flow problems by a finite element/finite volume method. *J. Non-Newtonian Fluid Mech.*, 51:249–275.

[9] J. S. Scroggs and F. H. M. Semazzi (1995). A conservative semi-Lagrangian method for multidimensional fluid dynamics applications. *Numer. Meth. Partial Differential Equations*, 11:445–452.

[10] S. C. Xue, N. Phan-Thien and R. I. Tanner (1995). Numerical study of secondary flows of viscoelastic fluid in straight pipes by an implicit finite volume method. *J. Non-Newtonian Fluid Mech.*, 59:191–213.

22 A FINITE VOLUME METHOD FOR VISCOUS COMPRESSIBLE FLOWS IN LOW AND HIGH SPEED APPLICATIONS

Jan Vierendeels, Kris Riemslagh, Bart Merci and Erik Dick

Department of Flow, Heat and Combustion Mechanics,
University of Gent, St.-Pietersnieuwstraat 41, B-9000 Gent, Belgium

ABSTRACT

An AUSM (Advection Upstream Splitting Method) based discretization method, using an explicit third-order discretization for the convective part, a line-implicit central discretization for the acoustic part and for the diffusive part, has been developed for incompressible and low speed compressible flow. The lines are chosen in the direction of the gridpoints with shortest connection. The semi-implicit line method is used in multistage form because of the explicit third-order discretization of the convective part. Multigrid is used as an acceleration technique. Due to the implicit treatment of the acoustic and the diffusive terms, the stiffness otherwise caused by high aspect ratio cells is removed. Low Mach number stiffness is treated by a preconditioning technique. To ensure physically correct behaviour of the discretization for vanishing Mach number, extreme care has been taken. For vanishing Mach number, stabilization terms are added to the mass flux. Pressure and temperature stabilization terms are necessary. The coefficients of these terms are chosen so that correct scaling with Mach number is obtained. A blend of the low speed algorithm with the original AUSM algorithm developed for high speed applications has been constructed so that the resulting algorithm can be used at all speeds.

Key words. AUSM, low Mach, preconditioning, all speed flows

22.1 INTRODUCTION

Preconditioning of the incompressible [7, 9] and compressible Navier-Stokes equations [1, 4, 7, 9] is used by many authors in order to accelerate convergence, especially for low

345

Mach number flow. However, this technique does not always provide good results on high aspect ratio grids, because of the stiffness introduced by the numerically-anisotropic behaviour of the diffusive and acoustic terms.

In our work, the stiffness due to the grid aspect ratio is removed by the use of a line method. The low Mach number stiffness is avoided by the use of an appropriate discretization and a local preconditioning technique. Multigrid is used as a convergence accelerator. The disctretization is based on AUSM (Advection Upstream Splitting Method) as developed by Liou and Steffen [6] and further extended to low Mach number applications by Edwards and Liou [3]. The preconditioning technique of Weiss and Smith [9] is employed.

The purpose of the work is to construct a scheme which reaches high quality and high efficiency independent of Mach number. Several examples of flows are discussed. It is demonstrated that the convergence is very good, independent of grid aspect ratio and Mach number.

22.2   INCOMPRESSIBLE FLOW

22.2.1   *Governing equations*

The two-dimensional steady Navier-Stokes equations in conservative form for an incompressible fluid are

$$\frac{\partial F_c}{\partial x} + \frac{\partial F_a}{\partial x} + \frac{\partial G_c}{\partial y} + \frac{\partial G_a}{\partial y} = \frac{\partial F_v}{\partial x} + \frac{\partial G_v}{\partial y},$$

where F_c and G_c are the convective fluxes, F_a and G_a are the acoustic fluxes and F_v and G_v are the viscous fluxes :

$$F_c = \begin{bmatrix} 0 \\ u^2 \\ uv \end{bmatrix}, \quad F_a = \begin{bmatrix} u \\ p' \\ 0 \end{bmatrix}, \quad F_v = \begin{bmatrix} 0 \\ \nu\frac{\partial u}{\partial x} \\ \nu\frac{\partial v}{\partial x} \end{bmatrix}, \quad G_c = \begin{bmatrix} 0 \\ uv \\ v^2 \end{bmatrix}, \quad G_a = \begin{bmatrix} v \\ 0 \\ p' \end{bmatrix}, \quad G_v = \begin{bmatrix} 0 \\ \nu\frac{\partial u}{\partial y} \\ \nu\frac{\partial v}{\partial y} \end{bmatrix},$$

where u and v are the Cartesian components of velocity, p' is the kinematic pressure ($p' = p/\rho$), p is the pressure, ρ is the density, and ν is the kinematic viscosity.

A distinction is made between the convective and acoustic parts of the inviscid flux vector because a different spatial and temporal discretization will be used for these parts.

22.2.2   *Discretization*

Figure 22.1 shows part of a rectangular grid with constant mesh size Δx and Δy (Cartesian grid). The control volumes are centred around the vertices of the grid.

The discretization of the convective flux is based on velocity upwinding :

$$F_{c_{i+1/2}} = u_{i+1/2} \; [0 \;\; u \;\; v]^T_{L/R}, \quad G_{c_{j+1/2}} = v_{j+1/2} \; [0 \;\; u \;\; v]^T_{L/R},$$

where

$$(\cdot)_{L/R} = \begin{cases} (\cdot)_L & \text{if } u_{1/2} > 0 \\ (\cdot)_R & \text{otherwise} \end{cases} \quad \text{and} \quad u_{i+1/2} = \frac{u_i + u_{i+1}}{2}, \quad v_{j+1/2} = \frac{v_j + v_{j+1}}{2}.$$

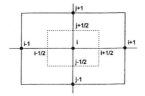

Figure 22.1: Vertex-centred control volume

We use an abbreviated notation for the subscripts. The subscript which is not shifted with respect to i or j is omitted (figure 22.1). The left (L) and right (R) values are computed with the Van Leer-κ approach :

$$q_L = q_i + \frac{1}{4}\left[(1+\kappa)\left(q_{i+1} - q_i\right) + (1-\kappa)\left(q_i - q_{i-1}\right)\right],$$

$$q_R = q_{i+1} - \frac{1}{4}\left[(1+\kappa)\left(q_{i+1} - q_i\right) + (1-\kappa)\left(q_{i+2} - q_{i+1}\right)\right],$$

with $\kappa=1/3$ for third order accuracy, where q stands for any of the state variables p, u or v.

The acoustic flux is discretized in the central way :

$$\boldsymbol{F}_{a_{i+1/2}} = [u \quad p \quad 0]^T_{i+1/2}, \quad \boldsymbol{G}_{a_{j+1/2}} = [v \quad 0 \quad p]^T_{j+1/2}.$$

The discretization of the convective and acoustic terms corresponds with the original AUSM scheme [6] if the energy equation is omitted, a constant density is assumed and for Mach number tending to zero. The viscous flux is also discretized in the central way.

Since the pressure term is discretized in the central way, pressure stabilization is needed. In the original AUSM scheme there is a lack of pressure stabilization for Mach number tending to zero. Therefore a pressure-velocity coupling was added in a newer version of the AUSM scheme suitable for all speeds [3]. This pressure-velocity coupling consists of an artificial diffusion added to the mass flux. As discussed by Edwards et al. [3] the scaling of this term should be inverse to the velocity magnitude. The same scaling of pressure-velocity coupling is present in the preconditioned Roe scheme [9] or in the flux difference splitting for incompressible Navier-Stokes equations [2]. This pressure-velocity coupling introduces a pressure diffusion in the continuity equation and an upwinding of the pressure derivative in the momentum equations. However, the upwinding is only needed for compressible flow, since the whole flux vector must be upwinded for supersonic flow. For incompressible flow, only an artificial dissipation term for pressure is needed. The terms added in the mass flux in x- and y-directions respectively, are :

$$\delta\frac{p'_{i+1} - p'_i}{\beta_x} \quad \text{and} \quad \delta\frac{p'_{j+1} - p'_j}{\beta_y},$$

where β_x and β_y have the dimension of velocity. We have taken

$$\beta_x = w_r + 2\nu/\Delta x, \quad \beta_y = w_r + 2\nu/\Delta y,$$

where w_r is in our application the local velocity prevented from being zero and $\delta = 1/2$. For inviscid flow (i.e. $\nu = 0$), this term corresponds with the dissipation term introduced in the continuity equation by the flux-difference splitting method for incompressible flow [2]. According to Weiss and Smith [9], β_x and β_y are treated in such a way that they scale with the local diffusion velocities $\nu/\Delta x$ and $\nu/\Delta y$ when these terms become important.

22.2.3   Time marching method

We use a time marching method in order to reach the steady state solution of the incompressible Navier-Stokes equations. Since the discretization of the fluxes will be partly explicit and partly implicit, we do not consider a transformation into the primitive form of the equations. Applying the pseudo-compressibility method to the conservative form of the inviscid part of the Navier-Stokes equations gives :

$$\Gamma\frac{\partial Q}{\partial \tau} + \frac{\partial F_c}{\partial x} + \frac{\partial F_a}{\partial x} + \frac{\partial G_c}{\partial y} + \frac{\partial G_a}{\partial y} = 0.$$

Q is the vector of variables $[p, u, v]^T$. The preconditioning matrix Γ is given by

$$\Gamma = \begin{bmatrix} \dfrac{1}{\beta^2} & 0 & 0 \\ 0 & 1 & 0 \\ 0 & 0 & 1 \end{bmatrix},$$

where β has the dimension of velocity. The eigenvalues of the inviscid part of the preconditioned system are given by

$$\lambda\left(\Gamma^{-1}\frac{\partial\left(n_x F + n_y G\right)}{\partial Q}\right) = w, w \pm c, \tag{22.1}$$

where $w = n_x u + n_y v$, $c = \sqrt{w^2 + \beta^2}$ and n_x and n_y denote an arbitrary direction with $n_x^2 + n_y^2 = 1$. If β is of the same order of magnitude as the convective speed, all eigenvalues will be properly scaled at least in one direction.

The multistage stepping has four stages with standard coefficients $\{1/4, 1/3, 1/2, 1\}$ and Courant-Friedrichs-Luwy number (*cfl*) equal to 1.8.

This multistage semi-implicit method is accelerated with the multigrid method. A full approximation scheme is used in a W-cycle with four levels of grids. The computation is started on the finest grid in order to show the full performance of the multigrid method. For the restriction operator, full weighting is used. The prolongation is done with bilinear interpolation. Two pre- and post-relaxations are done. This results in a cost of 30.75 work units for each multigrid cycle, when one work unit consists of a residual evaluation and an update, or a residual evaluation together with a restriction and source calculation on the coarser grid and a prolongation.

22.2.4 Determination of the pseudo-time step

Consider a uniform Cartesian mesh with constant Δx, Δy. The time step $\Delta \tau$ for a cell on this mesh is computed for an explicit method as

$$\Delta \tau = \frac{1}{\dfrac{u + c_x}{\Delta x} + \dfrac{v + c_y}{\Delta y}}, \quad \text{with} \ \ c_x = \sqrt{(u^2 + \beta^2)}, \ \ c_y = \sqrt{(v^2 + \beta^2)}.$$

Assume that the flow is inviscid and aligned to the x-direction, i.e. $v = 0$. If β is chosen with the order of u, then all three eigenvalues (22.1) have the same order of magnitude in the x-direction and all waves are convected into this direction with a *cfl* number in the order of unity.

When the allowable time step becomes smaller, the convergence will break down. This happens in the case of large grid aspect ratios. We define the grid aspect ratio g_{ar} for the Cartesian grid as $g_{ar} = \Delta x / \Delta y$. If g_{ar} is very large, then the allowable time step $\Delta \tau$ is equal to $\Delta y / c_y$ and the maximum allowable *cfl* number in the x-direction is

$$cfl_x = \frac{(u + c_x)\Delta \tau}{\Delta x} = \frac{u + c_x}{c_y} \frac{1}{g_{ar}} \approx \frac{1}{g_{ar}}.$$

This will lead to a breakdown of the convergence. If the acoustic fluxes in y-direction are discretized implicitly, the time step definition is changed into

$$\Delta \tau = \frac{1}{\dfrac{u + c_x}{\Delta x} + \dfrac{\omega_1 v}{\Delta y}},$$

where ω_1 is a scaling factor. If the flow is aligned to the x-direction, then cfl_x will be equal to unity. If the flow is aligned to the y-direction, a good value for ω_1 is about 2. If the viscous terms become important, the maximum allowable time step is determinated by the von Neumann number $\Delta \tau = \Delta y^2 / (2\nu)$ and cfl_x becomes small.

If the viscous terms are also treated with a line implicit method in y-direction, then this von Neumann restriction on the time step disappears, and the cfl_x number is again of the order of unity. Therefore, our strategy is a combination of an explicit local preconditioning method and an implicit line method for the acoustic and viscous terms in the direction of the smallest grid distances.

This semi-implicit line method for a grid with small cell dimensions in the y-direction is given by

$$\Gamma \frac{\partial \boldsymbol{Q}}{\partial \tau} + \frac{\partial \boldsymbol{F}_c}{\partial x}^n + \frac{\partial \boldsymbol{F}_a}{\partial x}^n + \frac{\partial \boldsymbol{G}_c}{\partial y}^n + \frac{\partial \boldsymbol{G}_a}{\partial y}^{n+1} - (A_v + A_d) \left(\boldsymbol{Q}^n_{i-1,j} - 2\boldsymbol{Q}^{n+1}_{i,j} + \boldsymbol{Q}^n_{i+1,j} \right)$$
$$- (B_v + B_d) \left(\boldsymbol{Q}^{n+1}_{i,j-1} - 2\boldsymbol{Q}^{n+1}_{i,j} + \boldsymbol{Q}^{n+1}_{i,j+1} \right) = 0,$$

where A_v, B_v, A_d and B_d are

$$A_v = \operatorname{diag} \left\{ 0, \frac{\nu}{\Delta x^2}, \frac{\nu}{\Delta x^2} \right\}, \quad B_v = \operatorname{diag} \left\{ 0, \frac{\nu}{\Delta y^2}, \frac{\nu}{\Delta y^2} \right\},$$

$$A_d = \frac{\delta}{\beta_x \Delta x} \operatorname{diag} \left\{ 1, u, v \right\}, \quad B_d = \frac{\delta}{\beta_y \Delta y} \operatorname{diag} \left\{ 1, u, v \right\},$$

with $\beta = \sqrt{u^2 + v^2} + \frac{2\nu}{\Delta x}$.

For use on grids without high aspect ratio cells, we also consider the semi-implicit point method. For this method the equations are given by

$$
\Gamma \frac{\partial \mathbf{Q}}{\partial \tau} + \frac{\partial \mathbf{F}_c}{\partial x}^n + \frac{\partial \mathbf{F}_a}{\partial x}^n + \frac{\partial \mathbf{G}_c}{\partial y}^n + \frac{\partial \mathbf{G}_a}{\partial y}^n - (A_v + A_d)\left(\mathbf{Q}_{i-1,j}^n - 2\mathbf{Q}_{i,j}^{n+1} + \mathbf{Q}_{i+1,j}^n\right)
$$
$$
- (B_v + B_d)\left(\mathbf{Q}_{i,j-1}^n - 2\mathbf{Q}_{i,j}^{n+1} + \mathbf{Q}_{i,j+1}^n\right) = 0,
$$

with $\beta = \sqrt{u^2 + v^2} + \frac{2\nu}{\Delta x} + \frac{2\nu}{\Delta y}$.

Since the acoustic terms are centrally discretized, these terms are treated explicitly in the semi-implicit point method.

22.3   LOW SPEED COMPRESSIBLE FLOW

The method described above can easily be extended for viscous low Mach number flow. The pseudo-compressibility method causes the convective and pseudo-acoustic wave speeds to be of the same order of magnitude. Therefore, for compressible flow, any preconditioner can be used which also scales the convective and the acoustic speeds. We used the preconditioner of Weiss and Smith [9] :

$$
\Gamma = \begin{bmatrix} \Theta & 0 & 0 & \rho_T \\ \Theta u & \rho & 0 & \rho_T u \\ \Theta v & 0 & \rho & \rho_T v \\ \Theta H - 1 & \rho u & \rho v & \rho_T H + \rho C_p \end{bmatrix}, \tag{22.2}
$$

where $\Theta = \frac{1}{\beta^2} - \frac{\rho_T}{\rho C_p}$. This preconditioner is used to update the so called viscous variables : $\mathbf{Q}_v = [p \ u \ v \ T]^T$, where T denotes the temperature and ^{T} the transposed vector. ρ_T is the derivative of ρ with respect to T.

Again a semi-implicit line method is used in the direction of the smallest grid distances.

22.3.1   Discretization

As for incompressible flows, the convective part of the momentum equation is discretized with velocity upwinding :

$$
\mathbf{F}_{c_{i+1/2}} = u_{i+1/2}[0 \ \ \rho u \ \ \rho v \ \ 0]_{L/R}^T, \quad \mathbf{G}_{c_{j+1/2}} = v_{j+1/2}[0 \ \ \rho u \ \ \rho v \ \ 0]_{L/R}^T.
$$

The pressure term in the momentum equations and the velocity terms in the continuity and energy equations are treated in the same way as the pseudo-acoustic part for incompressible flow and are discretized centrally.

The preconditioner (22.2) shows that the continuity equation causes an update for pressure and temperature. We use a dissipation term for both pressure and temperature with the ratio of both terms inspired by the preconditioner. For the x-direction, this is

$$
\delta \frac{p_{i+1} - p_i}{\beta_x} + |u| \rho_T (T_{i+1} - T_i).
$$

A similar term is used in the y-direction.

The discretization in pseudo time is done with the preconditioner (22.2). The acoustic flux is treated implicitly in the direction of the shortest grid distances. Since this flux is nonlinear in the compressible case, a linearization is needed. The acoustic flux on time level $n + 1$ is written as

$$G_{a_{j+1/2}}^{n+1} = \begin{bmatrix} \rho^n v^{n+1} \\ 0 \\ p^{n+1} \\ \rho^n H^n v^{n+1} \end{bmatrix}_{j+1/2}.$$

Finally, the normal viscous fluxes are treated implicitly. The tangential fluxes are treated explicitly.

22.4   VISCOUS LOW SPEED FLOW

The method is tested on a backward facing step problem The height of the step is chosen as one third of the channel height. We consider two grids. The first grid has 81×49 nodes and the second grid has 81×193 nodes. Both grids have the same distribution of points in the x-direction. In the y-direction the second grid has four times more cells than the first one. The highest aspect ratio on the first grid is about 35 and on the second grid 140.

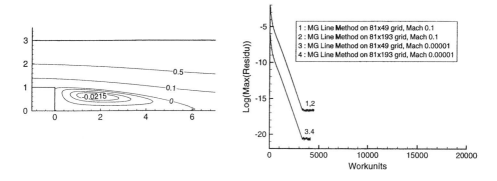

Figure 22.2: Left : Streamline pattern for the backward facing step problem, obtained at the finest grid. Right : Convergence results for the compressible backward facing step problem, comparison of the MG line method on two different grids for different Mach numbers

The left panel of Figure 22.2 shows the streamline pattern obtained on the first grid, for a Reynolds number $Re_h = (U_{max} h) / \nu = 150$, where h is the height of the step and U_{max} is the maximum value of the velocity at the inlet section. The right panel shows the convergence history for the MG line method on the two grids with different grid aspect ratios and for different inflow Mach numbers. It is clear that neither the grid aspect ratio nor the Mach number has any influence on the performance of the method.

The method was also tested on a thermally driven flow problem. The results are shown in [8]. Physical solutions for different Rayleigh numbers were obtained. Also for this case, the convergence behaviour was almost independent of the Rayleigh number.

22.5   BLENDING TO THE AUSM+ SCHEME

The AUSM+ scheme [3] has been used as a discretization scheme for high speed flows. In this method the inviscid interface flux $\boldsymbol{F}_{i+1/2}$ in the x-direction is split into a convective contribution $\boldsymbol{F}_{c_{i+1/2}}$ plus a pressure contribution $\boldsymbol{F}_{p_{i+1/2}}$:

$$
\boldsymbol{F}_c = \dot{m}_{i+1/2}
\begin{bmatrix} 1 \\ u \\ v \\ H \end{bmatrix}_{L/R}
\quad , \quad
\boldsymbol{F}_p =
\begin{bmatrix} 0 \\ p_{i+1/2} \\ 0 \\ 0 \end{bmatrix}
$$

where the state L is chosen if $\dot{m}_{i+1/2}$ is non-negative and state R is chosen if $\dot{m}_{i+1/2}$ is negative. The interface quantities $\dot{m}_{i+1/2}$ and $p_{i+1/2}$ are expressed in terms of sets of polynomials in the Mach number, defined as

$$
M_i = \frac{u_i}{a_{i+1/2}},
$$

where $a_{i+1/2}$ is an interface speed of sound [5]. Three sets of polynomials are required :

$$
\mathcal{M}_{(1)}^{\pm} = \frac{1}{2}(M \pm |M|),
$$

$$
\mathcal{M}_{(4)}^{\pm} =
\begin{cases}
\pm\frac{1}{4}(M \pm 1)^2 \pm \frac{1}{8}(M^2 - 1)^2, & |M| < 1 \\
\mathcal{M}_{(1)}^{\pm} & \text{otherwise}
\end{cases}
$$

and

$$
\mathcal{P}_{(5)}^{\pm} =
\begin{cases}
\frac{1}{4}(M \pm 1)^2(2 \mp M) \pm \frac{3}{16}M(M^2 - 1)^2, & |M| < 1 \\
\frac{1}{M}\mathcal{M}_{(1)}^{\pm} & \text{otherwise}
\end{cases}
$$

The numerals in the subscripts of $\mathcal{M}$ and $\mathcal{P}$ indicate the degree of the polynomials. With these, the interface quantities are defined as

$$
\dot{m}_{i+1/2} = a_{i+1/2}(\rho_L m_{i+1/2}^{+} + \rho_R m_{i+1/2}^{-}),
$$
$$
p_{i+1/2} = \mathcal{P}_{(5)}^{+}(M_L)p_L + \mathcal{P}_{(5)}^{-}(M_R)p_R
$$

with

$$
m_{i+1/2} = \mathcal{M}_{(4)}^{+}(M_L) + \mathcal{M}_{(4)}^{-}(M_R),
$$
$$
m_{i+1/2}^{\pm} = \frac{1}{2}(m_{i+1/2} \pm |m_{i+1/2}|).
$$

In order to operate the scheme at all Mach numbers, a blending of the low speed flux F^{LS} and high speed flux $F^{\text{AUSM+}}$ definitions is needed :

$$
F = (1 - \alpha)F^{\text{AUSM+}} + \alpha F^{\text{LS}},
$$

with

$$\alpha = \begin{cases} 1 - \left|m_{i+1/2}\right|/M_{\text{blend}} + \frac{1}{2\pi}\sin(2\pi\left|m_{i+1/2}\right|/M_{\text{blend}}), & \left|m_{i+1/2}\right| < M_{\text{blend}} \\ 0 & \text{otherwise} \end{cases}$$

M_{blend} is chosen as 0.5. With this blending function the first and second derivatives of F are continuous in the whole Mach number range.

22.6   INVISCID FLOW PAST A BUMP IN A CHANNEL

Figure 22.3 provides an indication of the behaviour of second-order implementations of the scheme for a range of Mach numbers. For all computations, the semi-implicit point method, one pre- and post-relaxation and *cfl*=1 are used. The scheme responds similarly for subsonic flows (figures 22.3abc) but differs for transonic flow (figures 22.3d). Convergence histories are shown in Figure 22.4. The low subsonic flow regimes show a flow-independent convergence rate. The M_∞=0.001 calculation levels out due to roundoff errors after a residual reduction of 7 orders of magnitude. The convergence rate of the higher subsonic regime is somewhat different. For subsonic flow the $\kappa = 1/3$ approach is used to achieve higher order accuracy. The transonic flow calculation has the slowest convergence rate. For this case a minmod limiter function was used to achieve second-order accuracy.

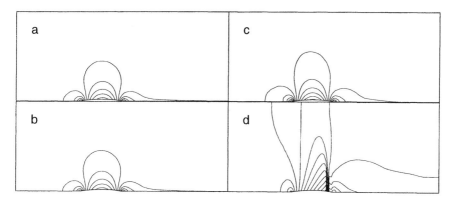

Figure 22.3: Mach number contours for the 2nd order scheme : a. M_∞=0.001, b. M_∞=0.1, c. M_∞=0.4, d. M_∞=0.85.

22.7   CONCLUSION

A method for calculating flow at all speed regimes and on high aspect ratio grids has been presented in this work. Results for low speed flow indicate excellent accuracy and case independent convergence behaviour. For transonic flow, a lower convergence rate is observed.

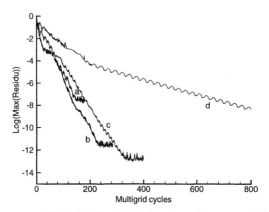

Figure 22.4: Convergence histories for the 2nd order scheme : a. $M_\infty=0.001$, b. $M_\infty=0.1$, c. $M_\infty=0.4$, d. $M_\infty=0.85$.

REFERENCES

[1] Y. Choi and C. Merkle (1993). The application of preconditioning in viscous flows. *J. Comp. Phys.*, 105:207–223.

[2] E. Dick and J. Linden (1992). A multigrid method for steady incompressible Navier-Stokes equations based on flux difference splitting. *Int. J. Numer. Methods in Fluids*, 14:1311–1323.

[3] J. Edwards and M.-S. Liou (1998). Low-diffusion flux splitting methods for flows at all speeds. *AIAA Journal*, 36(9):1610–1617.

[4] D. Lee, B. van Leer and J. Lynn (June 1997). A local Navier-Stokes preconditioner for all Mach and cell Reynolds numbers. In *Proc. 13th AIAA CFD conference, Snowmass, CO*, pages 842–855. AIAA press, Washington. AIAA-97-2024. ISBN 1-56347-233-3.

[5] M.-S. Liou (June 1997). Probing numerical fluxes: mass flux, positivity, and entropy-satisfying property. In *Proc. 13th AIAA CFD conference, Snowmass, CO*, pages 943–954. AIAA press, Washington. AIAA-97-2035. ISBN 1-56347-233-3.

[6] M.-S. Liou and C.J. Jr. Steffen (1993). A new flux splitting scheme. *J. Comp. Phys.*, 107:23–39.

[7] E. Turkel (1987). Preconditioned methods for solving the incompressible and low speed compressible equations. *J. Comp. Phys.*, 72(2):277–298.

[8] J. Vierendeels, K. Riemslagh and E. Dick. A multigrid semi-implicit line-method for viscous incompressible and low Mach number flows on high aspect ratio grids. *J. Comp. Phys.* Accepted for publication.

[9] J. Weiss and W. Smith (1995). Preconditioning applied to variable and constant density flows. *AIAA*, 33(11):2050–2057.

23 ON FINITE ELEMENT METHODS FOR COUPLING EIGENVALUE PROBLEMS

Hennie De Schepper and Roger Van Keer

Department of Mathematical Analysis, Faculty of Engineering,
University of Ghent, Galglaan 2, B-9000 Ghent, Belgium

ABSTRACT

We consider second-order elliptic eigenvalue problems on a composite structure, consisting of polygonal domains in the plane, where the interaction between the domains is expressed through nonlocal coupling conditions of Dirichlet type. We study the finite element approximation without and with numerical quadrature, by adapting the operator method, outlined in [9]. In view of the error analysis, a crucial point is the definition and error estimation of a suitably modified vector Lagrange interpolant on the mesh. Compared to the results in [9], the same order of convergence in terms of the mesh parameter is achieved, however under a higher regularity assumption for the exact eigenfunctions.
Key words. eigenvalue problems, nonlocal transition conditions, imperfect interpolation

23.1 INTRODUCTION

Standard Dirichlet or Robin/Neumann type boundary conditions (BCs) turn out to be inadequate in the mathematical modelling of various modern engineering applications. For instance, depending on what can be measured at the boundary of the physical domain, in some situations, neither the solution, nor the flux can be prescribed pointwise. Instead, an average value of either of these unknown quantities may be given, leading to nonstandard BCs of nonlocal type, see, e.g., [7], [8]. A similar argument holds – a fortiori – for so-called coupling problems in composite structures, where the interaction between the subdomains may lead to complicated nonlocal transition conditions (TCs) at the interfaces, see, e.g., [1], [4].

In this chapter, we consider *nonlocal coupling conditions* in the context of eigenvalue problems (EVPs) in multi-component domains. An outline of the chapter is now in order.

An abstract setting for the variational EVPs considered is stated in Section 23.2, together with some concrete examples in differential form. In Section 23.3 we first introduce suitable approximation spaces and next we define and analyse the *imperfect* Lagrange interpolant. Moreover, some preliminary results for the error analysis are proved. The consistent mass and the numerical quadrature FEMs are considered in Section 23.4.

23.2   PROBLEM SETTING

23.2.1   General formulation

Let $\Omega_1, \ldots, \Omega_M$ be bounded, convex polygonal domains in the plane – possibly *overlapping*, with respective boundaries $\partial \Omega_i$, and suppose that Ω, defined by $\overline{\Omega} = \bigcup_{i=1}^{M} \overline{\Omega_i}$, is again a convex polygonal, with boundary $\partial \Omega$ (see Fig. 23.1 for examples). By Γ_i and Γ_i' we denote disjoint parts of $\partial \Omega_i$, each consisting of a number of sides $(i = 1, \ldots, M)$.

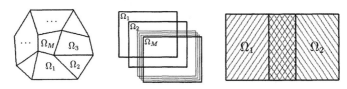

Figure 23.1: Examples of the multi-component domain Ω:
(a) adjacent subdomains; (b) identical subdomains; (c) overlapping subdomains

First, let $H^1(\Omega_i)$ be the usual first order Sobolev space on Ω_i with norm $\|.\|_{1,\Omega_i}$ and let

$$V_i = \left\{ v \in H^1(\Omega_i) \mid \gamma v = 0 \text{ on } \Gamma_i \right\}, \quad i = 1, \ldots, M.$$

Here, γ denotes the usual trace operator. [For notational simplicity we rewrite γ_i as γ.] Next, starting from these spaces, we introduce the product Hilbert space

$$\tilde{V} \equiv \prod_{i=1}^{M} V_i = \left\{ v = (v_1, \ldots, v_M) \mid v_i \in V_i, \ i = 1, \ldots, M \right\},$$

endowed with its natural product norm, denoted by $\|.\|_{\widehat{1,\Omega}}$. Likewise, we consider the product Lebesgue space $\tilde{H} = L_2(\Omega_1) \times \cdots \times L_2(\Omega_M)$, endowed with the natural product norm $|.|$, derived from the inner product $(.,.)$.

Finally, let V be a closed subspace of $\tilde{V}$ which incorporates a set of essential, *nonlocal* coupling or transition conditions, explicitly

$$V = \left\{ v \in \tilde{V} \mid I_j(v) \equiv \int_{\text{Int}_j} \left(\sum_{i \in \mathcal{N}^{(j)}} c_i^{(j)} v_i \right) d\sigma = 0, \ j = 1, \ldots, J \right\}. \tag{23.1}$$

By Int_j, $j = 1, \ldots, J$, we denote the respective 1D or 2D "interfaces", shared by two or more subdomains. Further, $\mathcal{N}^{(j)} = \{i, \ i = 1, \ldots, M \mid \text{Int}_j \subset \overline{\Omega_i}\}$. The scalars $c_i^{(j)}$, $i \in \mathcal{N}^{(j)}$, $j = 1, \ldots, J$, appearing in (23.1), are given.

V is seen to be a Hilbert space for the $(.,.)_{\widehat{1,\Omega}}$-inner product, and, moreover, is compactly embedded in H, H being the closure of V in $\tilde{H}$.

We consider the variational EVP: find $\lambda \in \mathbb{R}$ and $u \in V$ such that

$$a(u, v) = \lambda(u, v) \qquad \forall v \in V, \tag{23.2}$$

with

$$a(u, v) = \sum_{i=1}^{M} \left[\int_{\Omega_i} \left(\sum_{l,m=1}^{2} a_{lm}^{(i)} \frac{\partial u_i}{\partial x_l} \frac{\partial v_i}{\partial x_m} + a_0^{(i)} u_i v_i \right) dx + \int_{\Gamma_i'} \sigma^{(i)} u_i v_i \, dx \right]. \tag{23.3}$$

We assume that the coefficient functions, appearing in (23.3), are such that the bilinear form $a(.,.)$ is bounded, symmetric and strongly coercive on $V \times V$, cf. [9]. Hence the problem (23.2) fits into the general abstract framework of [6, §6.2]. Thus, we have

Theorem 23.1 *(1) The EVP (23.2) has an infinite sequence of eigenvalues, all being strictly positive and having finite multiplicity, showing no finite accumulation point. We arrange them as*

$$0 < \lambda_1 \leq \lambda_2 \leq \cdots \to +\infty,$$

where each eigenvalue appears as many times as given by its multiplicity.
(2) The corresponding eigenfunctions $\left(u_l\right)_{l=1}^{\infty}$ can be chosen to be orthonormal in H, the sequence $\left(\lambda_l^{-\frac{1}{2}} u_l\right)_{l=1}^{\infty}$ then being orthonormal in V w.r.t. $a(.,.)$. They constitute a Hilbert basis for V as well as for H.

23.2.2 Examples

In this subsection, we describe some of the concrete differential EVPs, the variational formulation of which fits into the above setting.

(a) Consider the domain, shown in Fig. 23.1a, where Γ_i and Γ_i' are chosen to be complementary parts of $\partial\Omega_i \cap \partial\Omega$. Let, for $i = 1, \ldots, M$,

$$\mathcal{N}_i = \{\sigma \mid \text{meas}_1(\partial\Omega_i \cap \partial\Omega_\sigma) > 0\}; \quad \Gamma_{i\sigma} = \begin{cases} \partial\Omega_i \cap \partial\Omega_\sigma, & \sigma \in \mathcal{N}_i \\ \emptyset, & \sigma \notin \mathcal{N}_i \end{cases}$$

The second-order EVP for $[\lambda; u_1, \ldots, u_M] \in \mathbb{R} \times \left(\prod_{i=1}^{M} H^2(\Omega_i)\right)$, consisting of the differential equations

$$-\sum_{l,m=1}^{2} \frac{\partial}{\partial x_l} \left(a_{lm}^{(i)} \frac{\partial u_i}{\partial x_m} \right) + a_0^{(i)} u_i = \lambda u_i, \qquad \text{in } \Omega_i, \qquad i = 1, \ldots, M, \tag{23.4}$$

along with the classical Robin (possibly, Neumann) and Dirichlet BCs

$$\frac{\partial u_i}{\partial \nu_{a^i}} + \sigma^{(i)} u_i = 0 \text{ on } \Gamma_i', \quad \text{and} \quad u_i = 0 \text{ on } \Gamma_i, \qquad i = 1, \ldots, M, \tag{23.5}$$

and along with the nonlocal Dirichlet TCs

$$\int_{\Gamma_{i\sigma}} u_i \, ds = \int_{\Gamma_{i\sigma}} u_\sigma \, ds, \qquad \forall \sigma \in \mathcal{N}_i, \ i = 1, \ldots, M,$$

accompanied of

$$\frac{\partial u_i}{\partial \nu_{a^i}} = -\frac{\partial u_\sigma}{\partial \nu_{a^\sigma}} = c_{i\sigma} \ \text{(unknown constant) on } \Gamma_{i\sigma}, \qquad \forall \sigma \in \mathcal{N}_i, \ i = 1, \dots, M,$$

has a variational formulation of the form (23.2), with

$$V = \Big\{ v \in \tilde{V} \ \Big| \int_{\Gamma_{i\sigma}} \gamma v_i \, ds = \int_{\Gamma_{i\sigma}} \gamma v_\sigma \, ds, \ \forall \sigma \in \mathcal{N}_i, \ i = 1, \dots, M \Big\}.$$

(b) Now, take $\Omega_1 = \cdots = \Omega_M = \Omega$, as in Fig. 23.1b, and let Γ be a side of Ω, while Γ_i and Γ_i' are complementary parts of $\partial \Omega_i \setminus \Gamma$, $i = 1, \dots, M$. Then the EVP, consisting of the DEs (23.4), the classical Robin and Dirichlet BCs (23.5) and the coupling BCs

$$\sum_{i=1}^{M} \int_{\Gamma} u_i \, ds = 0,$$

$$\frac{\partial u_1}{\partial \nu_{a^i}} = \dots = \frac{\partial u_M}{\partial \nu_{a^i}} = \ \text{constant (unknown) on } \Gamma,$$

results in a variational EVP of the form (23.2), with

$$V = \Big\{ v \in \tilde{V} \ \Big| \sum_{i=1}^{M} \int_{\Gamma} \gamma v_i \, ds = 0 \Big\}.$$

(c) Finally, in the situation of two *partially overlapping* domains, as in Fig. 23.1c, we may consider the differential equations

$$-\sum_{l,m=1}^{2} \frac{\partial}{\partial x_l} \Big(a_{lm}^{(i)} \frac{\partial u_i}{\partial x_m} \Big) + a_0^{(i)} u_i + (-1)^i \chi_{\Omega_1 \cap \Omega_2} C = \lambda u_i, \qquad \text{in } \Omega_i, \qquad i = 1, 2,$$

($\chi_{\Omega_1 \cap \Omega_2}$ denoting the characteristic function of $\Omega_1 \cap \Omega_2$), along with the classical Robin and Dirichlet conditions (23.5), now on $\partial \Omega_i$, $(i = 1, 2)$, and the nonlocal coupling condition

$$\int_{\Omega_1 \cap \Omega_2} \big(u_1 - u_2 \big) \, dx = 0.$$

By the latter condition, the unknown constant C becomes a linear and homogeneous *integral* expression over $\Omega_1 \cap \Omega_2$ and over $\partial(\Omega_1 \cap \Omega_2)$, in terms of u_1, u_2 and of their partial derivatives, respectively. Here, the proper space of trial and test functions is

$$V = \Big\{ v \in \tilde{V} \ \Big| \int_{\Omega_1 \cap \Omega_2} \big(u_1 - u_2 \big) \, dx = 0 \Big\}.$$

In all three examples, the closedness of V in $\tilde{V}$ can readily be verified. In fact, this property may directly be shown for the general space V, (23.1), as well.

23.3 PRELIMINARY RESULTS

23.3.1 The approximation space V_h

We consider a regular family of triangulations $\left(\mathcal{T}_{h_i}^i\right)_{h_i}$, see, e.g., [2, p.132], of each component $\overline{\Omega_i}$, $i = 1, \ldots, M$, consisting of either triangular or rectangular elements. We assume that the families $\left(\mathcal{T}_{h_i}^i\right)_{h_i}$ are quasi-uniform in the sense of [2, (3.2.28)], as well as mutually quasi-uniform in the sense of [9, (3.4)].

With a triangulation $\mathcal{T}_{h_i}^i$ and a fixed number $k \in \mathbb{N}_0$, we associate the finite dimensional subspace $X_{h_i}^i$ of $H^1(\Omega_i)$, $X_{h_i}^i = \{v \in C^0(\overline{\Omega_i}) \mid v|_K \in P(K), \forall K \in \mathcal{T}_{h_i}^i\}$, where h_i is the mesh parameter. $P(K)$ stands for the polynomial spaces $P_k(K)$ or $Q_k(K)$, in the case of a triangular or of a rectangular mesh, respectively.

We introduce the product spaces

$$X_h = \{v = (v_1, \ldots, v_M) \mid v_i \in X_{h_i}^i, \ i = 1, \ldots, M\},$$
$$Y_h = \{v \in X_h \mid v_i = 0 \text{ on } \Gamma_i, \ i = 1, \ldots, M\} \subset \tilde{V},$$
$$V_h = \{v \in Y_h \mid I_j(v) = 0, \ j = 1, \ldots, J\} \subset V, \tag{23.6}$$

where $h = \max_{1 \le i \le M} h_i$ is the overall mesh parameter, taken to be sufficiently small.

23.3.2 An imperfect Lagrange interpolant

In what follows, we will denote $\prod_{i=1}^M H^s(\Omega_i)$ by $\widehat{H^s(\Omega)}$, $s \in \mathbb{N}_0$, endowed with the product norm $\|.\|_{\widehat{s,\Omega}}$. The product seminorm is denoted as $|.|_{\widehat{s,\Omega}}$.

Let $v \in V \cap \widehat{H^2(\Omega)}$, notice that $H^2(\Omega_i) \hookrightarrow C^0(\overline{\Omega_i})$, $(i = 1, \ldots, M)$, and consider the vector piecewise Lagrange interpolant $\Pi_h v$ of v on the global mesh, i.e.,

$$\Pi_h v \equiv (\Pi_{h_1} v_1, \ldots, \Pi_{h_M} v_M) \in X_h ; \ \Pi_{h_i} v_i(a_l^{(i)}) = v_i(a_l^{(i)}), \ l = 1, \ldots, N_i, \ i = 1, \ldots, M,$$

where $\bigcup_{l=1}^{N_i} \{a_l^{(i)}\}$ is the usual set of nodes, associated with $X_{h_i}^i$, $i = 1, \ldots, M$, and being chosen as in [2, pp.43–60].

Clearly, by the definition of V_h, in general $\Pi_h v \notin V_h$. Hence, for the error analysis of the FEMs to be considered, we can no longer rely upon standard results from interpolation theory. To overcome this difficulty, we define an *imperfect* interpolant $\tilde{\Pi}_h v$, by suitably modifying the nodal value at one single node per interface. To this end, we need some additional notations. First, for $j = 1, \ldots, J$ and $i \in \mathcal{N}^{(j)}$, we denote

$$\alpha^{(j)}(a_l^{(i)}) = \int_{\text{Int}_j} \varphi_l^{(i)} \, d\sigma, \quad \text{when } a_l^{(i)} \in \overline{\text{Int}_j} \, .$$

Next, for $j = 1, \ldots, J$, we put $i_j = \max_{i \in \mathcal{N}^{(j)}} i$, and we choose one single node $A_j \in \overline{\Omega_{i_j}}$, belonging to $\overline{\text{Int}_j} \setminus \left(\bigcup_{\substack{\bar{j}=1 \\ \bar{j} \ne j}}^J \overline{\text{Int}_{\bar{j}}} \bigcup \overline{\Gamma_{i_j}}\right)$. The set of all these special nodes is denoted by $\text{Imp}(\Omega)$.

Definition 23.1 Let $v \in \widehat{C^0(\overline{\Omega})} \equiv \prod_{i=1}^{M} C^0(\overline{\Omega}_i)$. Then, we define its imperfect interpolant $\tilde{\Pi}_h v \equiv (\tilde{\Pi}_{h_1} v_1, \ldots, \tilde{\Pi}_{h_M} v_M) \in X_h$ by

$$\tilde{\Pi}_{h_i} v_i(a_l^{(i)}) = v_i(a_l^{(i)}), \quad \text{when } a_l^{(i)} \notin \mathrm{Imp}(\Omega), \; l = 1, \ldots, N_i, \; i = 1, \ldots, M\,;$$

$$\tilde{\Pi}_{h_{i_j}} v_{i_j}(A_j) = v_{i_j}(A_j) - \frac{I_j(\Pi_h v)}{c_{i_j}^{(j)} \alpha^{(j)}(A_j)}, \quad j = 1, \ldots, J.$$

We have, by construction, that $\tilde{\Pi}_h v \in V_h$, $\forall v \in V \cap \widehat{H^2(\Omega)}$. For the error in the imperfect interpolation, we obtain

Proposition 23.1 Let $r = 1, \ldots, k$, then there exists a constant $C = C(k) > 0$, independent of h, such that, for $m = 0, \ldots, r + 1$

$$\left(\sum_{i=1}^{M} \sum_{K \in \mathcal{T}_{h_i}^i} |v_i - \tilde{\Pi}_K v_i|_{m,K}^2 \right)^{\frac{1}{2}} \leq C\, h^{r+1-m} \|v\|_{\widehat{r+2,\Omega}} \quad \forall v \in V \cap \widehat{H^{r+2}(\Omega)}. \tag{23.7}$$

$(|.|_{m,K}$ denotes the m-th order Sobolev semi-norm on K.)

Proof. Notice that, for $i \in \{1, \ldots, M\}$, $\tilde{\Pi}_K v_i (\equiv \tilde{\Pi}_{h_i} v_i|_K)$ and $\Pi_K v_i (\equiv \Pi_{h_i} v_i|_K)$ only differ when, for some $j \in \{1, \ldots, J\}$, $i = i_j$ and $A_j \in K$. Consider such an i and such an element K. When h is sufficiently small, K will contain only one node from $\mathrm{Imp}(\Omega)$, say $A_{\hat{j}}$. Then, from Definition 23.1,

$$\Pi_K v_i - \tilde{\Pi}_K v_i = \left[v_i(A_{\hat{j}}) - \tilde{\Pi}_K v_i(A_{\hat{j}}) \right] \psi_{A_{\hat{j}}}^K = \frac{I_{\hat{j}}(\Pi_h v)}{c_i^{(\hat{j})} \alpha^{(\hat{j})}(A_{\hat{j}})} \psi_{A_{\hat{j}}}^K,$$

where $\psi_{A_{\hat{j}}}^K$ denotes the shape function on K, associated to $A_{\hat{j}}$. Hence

$$|\Pi_K v_i - \tilde{\Pi}_K v_i|_{m,K} \leq \frac{C}{|\alpha^{(\hat{j})}(A_{\hat{j}})|} |\psi_{A_{\hat{j}}}^K|_{m,K} \int_{\mathrm{Int}_{\hat{j}}} \sum_{s \in \mathcal{N}(\hat{j})} |\Pi_{h_s} v_s - v_s|\, d\sigma\,, \quad \forall v \in V.$$

We provide estimates for the three factors entering the right hand side. First, applying [2, §2.3] we get $|\psi_{A_{\hat{j}}}^K|_{m,K} \leq Ch^{1-m}$, $m = 0, \ldots, k+1$. Next, denoting the dimension of $\mathrm{Int}_{\hat{j}}$ by $\hat{n} \equiv \hat{n}(\hat{j})$, and invoking the mutual quasi-uniformity of the families $(\mathcal{T}_{h_i}^i)_{h_i}$, we have $|\alpha^{(\hat{j})}(A_{\hat{j}})| \geq Ch^{\hat{n}}$. Finally, applying results from interpolation theory, see [2, §3.2], and, when $\hat{n} = 1$, also invoking the continuous imbedding $H^{r+2}(\Omega_s) \hookrightarrow H^{r+1}(\partial \Omega_s)$, cf. [5, Theorem 6.7.8], we obtain

$$\int_{\mathrm{Int}_{\hat{j}}} \sum_{s \in \mathcal{N}(\hat{j})} |\Pi_{h_s} v_s - v_s|\, d\sigma \leq Ch^{r+\hat{n}} \|v\|_{\widehat{r+2,\Omega}} \quad \forall v \in V \cap \widehat{H^{r+2}(\Omega)}, \; r = 1, \ldots, k.$$

Hence, $|\Pi_K v_i - \tilde{\Pi}_K v_i|_{m,K} \leq Ch^{r+1-m} \|v\|_{\widehat{r+2,\Omega}}$, $\forall v \in V \cap \widehat{H^{r+2}(\Omega)}$, $r = 1, \ldots, k$. Combining this estimate with a standard result for $|v_i - \Pi_K v_i|_{m,K}$, see [2, §3.2], we arrive at (23.7). $\square$

Remark 23.1 The higher regularity of the original function v, required in (23.7), viz $\widehat{H^{r+2}(\Omega)}$ instead of the usual $\widehat{H^{r+1}(\Omega)}$, cf. [2, Theorem 3.2.1], will be reflected in the estimates below.

23.3.3 The elliptic projector

Leaning upon (23.7), we can prove

Proposition 23.2 *The finite element space $V_h \subset V$ satisfies the approximation property*

$$\inf_{v_h \in V_h} \left\{ |v - v_h| + h\, |v - v_h|_{\widehat{1,\Omega}} \right\} \leq C\, h^{r+1}\, \|v\|_{\widehat{r+2,\Omega}} \quad \forall v \in V \cap \widehat{H^{r+2}(\Omega)},\ r = 1, \ldots, k,$$

(23.8)

where the constant $C = C(k)$ is independent of h.

Next, we define the elliptic projector $P : V \to V_h$ by

$$a(v - Pv, w) = 0, \qquad \forall v \in V,\ \forall w \in V_h.$$

(23.9)

Adapting the arguments of [9, §3.3], *now* invoking (23.8), we can prove

Proposition 23.3 *For the elliptic projector P, (23.9), the following estimates hold:*

$$\|v - Pv\|_{\widehat{1,\Omega}} \leq C h^r \|v\|_{\widehat{r+2,\Omega}}, \qquad \forall v \in V \cap \widehat{H^{r+2}(\Omega)},\ r = 1, \ldots, k,$$

$$\left(\sum_{i=1}^{M} \sum_{K \in \mathcal{T}_{h_i}^i} |(Pv)_i|_{r,K}^2 \right)^{\frac{1}{2}} \leq C \|v\|_{\widehat{r+1,\Omega}} \qquad \forall v \in V \cap \widehat{H^{r+1}(\Omega)},\ r = 1, \ldots, k+1,$$

where the constant $C = C(k)$ is independent of h.

23.3.4 A density property

Proposition 23.4 $\widehat{H^3(\Omega)} \cap V$ *is dense in V*

Proof. Consider $v \in V$ and $\varepsilon > 0$ arbitrary. We will construct a function $\varphi^* \in \widehat{C^\infty(\overline{\Omega})} \cap V$, fulfilling $\|v - \varphi^*\|_{\widehat{1,\Omega}} < \varepsilon$. [Here, $\widehat{C^\infty(\overline{\Omega})}$ stands for $\prod_{i=1}^{M} C^\infty(\overline{\Omega_i})$.]

To this end, keep $i \in \{1, \ldots, M\}$ fixed and let $\mathcal{N}_{(i)} = \{j,\ j = 1, \ldots, J \mid \mathrm{Int}_j \subset \overline{\Omega_i}\}$, with $n_i = \#\mathcal{N}_{(i)}$. Further, denote $a_j = \int_{\mathrm{Int}_j} v_i\, d\sigma$ (omitting the dependence on the chosen i for notational simplicity). Invoking a density argument of [10, p.92], we infer, for $t = 1, 2, \ldots$, the existence of a function $\varphi_i^{(t)} \in C^\infty(\overline{\Omega_i}) \cap V_i$, such that $\|v_i - \varphi_i^{(t)}\|_{1,\Omega_i} < \varepsilon^{(t)}$, with

$$\varepsilon^{(1)} = \frac{\varepsilon}{2(n_i + 1)\sqrt{M}}; \quad \varepsilon^{(t)} < \min_{j \in \mathcal{N}_{(i)}} \left\{ \frac{(\mathrm{meas}\ \mathrm{Int}_j)^{-\frac{1}{2}}}{C} \left\{ \frac{|a_j - a_j^{(t-1)}|}{2}, \frac{d^{(t-1)}}{\sqrt{n_i}} \right\} \right\}, \quad t = 2, 3, \ldots,$$

where $a_j^{(t)} = \int_{\mathrm{Int}_j} \varphi_i^{(t)}\, d\sigma$, C denotes the maximum of the constants, appearing in the trace inequalities on Ω_i, $i = 1, \ldots, M$, and $d^{(t)}$ stands for the distance in $\mathbb{R}^{n_i}$ between $a = \left(a_j \right)_{j \in \mathcal{N}_{(i)}}$ and the $(t-1)$-dimensional subspace $\mathcal{A}^{(t)}$, spanned by $a^{(s)} \equiv \left(a_j^{(s)} \right)_{j \in \mathcal{N}_{(i)}}$, $s = 1, \ldots, t$. This process is repeated until a is contained in $\mathcal{A}^{(t)}$, which occurs at last when $t = n_i + 1$. Thus, we have constructed a set of functions $\left(\varphi_i^{(t)}\right)_{t=1}^{t_0+1}$, with $t_0 \leq n_i$, such that

$$\|v_i - \varphi_i^{(t)}\|_{1,\Omega_i} \leq \frac{2^{-t}\varepsilon}{(n_i+1)\sqrt{M}}; \ |a_j - a_j^{(t)}| < \min\left\{\frac{|a_j - a_j^{(t-1)}|}{2}, \frac{d^{(t-1)}}{\sqrt{n_i}}\right\}, \ j \in \mathcal{N}_{(i)}. \tag{23.10}$$

Here, we have assumed that $a_j^{(t)} \neq a_j$, $\forall j \in \mathcal{N}_{(i)}$, $t = 1, \ldots, t_0$. In case one of these assumptions is not fulfilled, a modified, but analogous construction is required, similar to the one in [3, §3.2].

From (23.10) it can be argued that a is contained in the intersection of $\mathcal{A}^{(t_0+1)}$ and the n_i-dimensional product interval

$$\prod_{j=1}^{n_i} \left] a_j^{(t_0+1)} - \min_{t\in\{1,\ldots,t_0\}} |a_j^{(t_0+1)} - a_j^{(t)}|, \ a_j^{(t_0+1)} + \min_{t\in\{1,\ldots,t_0\}} |a_j^{(t_0+1)} - a_j^{(t)}| \right[,$$

and hence also in the union of the convex hulls of the $(t_0 + 1)$-tuples of points $\left(A_t\right)_{t=1}^{t_0+1}$, given by $A_t = a^{(t_0+1)} \pm (a^{(t_0+1)} - a^{(t)})$, $t = 1, \ldots, t_0$ and $A_{t_0+1} = a^{(t_0+1)}$. Hence, there exist constants $\left(C_t\right)_{t=1}^{t_0+1}$, with $|C_t| \leq t_0 + 1$ and $\sum_{t=1}^{t_0+1} C_t = 1$, such that $a = \sum_{t=1}^{t_0+1} C_t a^{(t)}$.

The function

$$\varphi_i^* = \sum_{t=1}^{t_0+1} C_t \varphi_i^{(t)}$$

belongs to $C^\infty(\overline{\Omega_i}) \cap V_i$, while by construction $\int_{\mathrm{Int}_j} \varphi_i^* \, d\sigma = a_j$, $\forall j \in \mathcal{N}_{(i)}$, and finally, $\|v_i - \varphi_i^*\|_{1,\Omega_i} < \varepsilon/\sqrt{M}$.

Repeating the procedure for the $M - 1$ other components of v, we arrive at a vector function $\varphi^* = (\varphi_1^*, \ldots, \varphi_M^*) \in \widehat{C^\infty(\overline{\Omega})} \cap V$, for which $\|v - \varphi^*\|_{\widehat{1,\Omega}} < \varepsilon$. $\square$

23.4   FINITE ELEMENT APPROXIMATIONS

23.4.1   The consistent mass EVP

The variational consistent mass EVP, approximating (23.2) is: find $[\lambda_h, u_h] \in \mathbb{R} \times V_h$ such that

$$a(u_h, v_h) = \lambda_h(u_h, v_h) \qquad \forall v_h \in V_h. \tag{23.11}$$

From the results of the previous section, the convergence of the FEM is assured, when rephrasing the arguments of [6, §6.5]. As compared to the error estimates in [9, §4] for the approximation of eigenpairs of EVPs in a composite structure with *local* TCs, the assumed order of regularity of the exact eigenfunctions must be increased by one unit on account of (23.7), to retain the optimal order of convergence. A typical result will be formulated in the context of the numerical quadrature FEM.

23.4.2   The numerical quadrature EVP

In practice, instead of (23.11), we will consider the EVP: find $[\tilde{\lambda}_h, \tilde{u}_h] \in \mathbb{R} \times V_h$ such that

$$a_h(\tilde{u}_h, v_h) = \tilde{\lambda}_h (\tilde{u}_h, v_h)_h \qquad \forall v_h \in V_h, \tag{23.12}$$

where $a_h(.,.)$ results from $a(.,.)$ by the Gauss–Legendre quadrature per element (and per edge of an element) and $(.,.)_h$ corresponds to $(.,.)$ when using the Lobatto quadrature rule, see, e.g., [9, §5.1].

On account of the preliminary results, proved in Section 23.3, the operator method of [9] can be rephrased for the present type of EVPs and of approximation spaces. Again, optimal estimates hold, similar to the ones in [9, §5.6-5.8], when increasing the assumed order of regularity of the exact eigenfunctions by one unit. In particular, for the case of a multiple exact eigenvalue, we obtain, for instance:

Theorem 23.2 *Assume that $a_{lm}^{(i)} \in W^{k+1,\infty}(\Omega_i)$, $l, m = 1, 2$, that $a_0^{(i)} \in W^{k,\infty}(\Omega_i)$, and that $\sigma^{(i)} \in W^{k+1,\infty}(\Gamma_i')$, $i = 1, \ldots, M$. Let λ_l be an $(L+1)$-fold exact eigenvalue, i.e. $\lambda_{l-1} < \lambda_l = \lambda_{l+1} = \cdots = \lambda_{l+L} < \lambda_{l+L+1}$, $(L \neq 0)$. Moreover let $\tilde{\lambda}_{l+t,h}$, $t = 0, \ldots, L$, be the corresponding approximate eigenvalues of (23.12), with associated eigenfunctions $\tilde{u}_{l+t,h}$, orthonormalised in H. Assume that the eigenspace associated with the exact eigenvalue λ_l belongs to $\widehat{H^{k+2}(\Omega)}$. Then, there exists a set of exact eigenfunctions $(U_{l+t}(h))_{t=0}^L$, associated to λ_l and orthonormal in H, such that*

$$\|U_{l+t}(h) - u_{l+t,h}\|_{\overline{1,\Omega}} \le Ch^k, \quad t = 0, \ldots, L, \tag{23.13}$$

where C is a constant, independent of h.

Moreover, estimates similar to (23.13) can also be formulated for a *fixed* set of exact eigenfunctions as well, however for a particular sequence of values of h, $(h \to 0)$, cf. [9, Theorem 5.6].

23.4.3   Some computational aspects

It remains to identify a suitable basis of V_h, (23.6), in order to derive an algebraic EVP, equivalent to (23.11) or (23.12), the stiffness and mass matrices of which have a relatively transparent structure.

To this end, we start from a cardinal basis of each space $X_{h_i}^i$, denoted by $\left(\varphi_l^{(i)}\right)_{l=1}^{N_i}$, $(i = 1, \ldots, M)$. Keep $i \in \{1, \ldots, M\}$ fixed. To each node $a_l^{(i)} \in \overline{\Omega}_i \setminus \left(\overline{\Gamma}_i \cup \{A_j \mid j \in \mathcal{N}_{(i)} \text{ and } i = i_j\}\right)$, we will associate a vector basis function, proceeding as follows. First, we denote

$$\mathcal{N}_{(i)}(l) = \left\{ j \in \mathcal{N}_{(i)} \mid a_l^{(i)} \in \overline{\text{Int}_j} \right\}.$$

Now, when $\mathcal{N}_{(i)}(l) = \emptyset$, we may simply take the corresponding vector basis function to be

$$\big[\underbrace{0, \ldots, 0}_{i-1}, \varphi_l^{(i)}, \underbrace{0, \ldots, 0}_{M-i}\big]. \tag{23.14}$$

However, when $\mathcal{N}_{(i)}(l) \neq \emptyset$, we need to construct a function from V_h, which fulfills in a *nontrivial* way the nonlocal coupling conditions over Int_j, $j \in \mathcal{N}_{(i)}(l)$. Denoting this function by $[\Phi_1, \ldots, \Phi_M]$ – omitting the dependence on i and l – we take

$$
\Phi_i = \begin{cases} \varphi_l^{(i)}, & \text{when } i \neq i_j, \ \forall j \in \mathcal{N}_{(i)}(l), \\ \varphi_l^{(i)} - \displaystyle\sum_{\substack{j \in \mathcal{N}_{(i)}(l) \\ i_j = i}} \frac{\alpha^{(j)}(a_l^{(i)})}{\alpha^{(j)}(A_j)} \varphi_{A_j}^{(i)}, & \text{else}, \end{cases} \tag{23.15a}
$$

and for $m \in \{1, \ldots, M\}$, $m \neq i$,

$$
\Phi_m = \begin{cases} 0, & \text{when } m \neq i_j, \ \forall j \in \mathcal{N}_{(i)}(l), \\ -\displaystyle\sum_{\substack{j \in \mathcal{N}_{(i)}(l) \\ i_j = m}} \frac{c_i^{(j)}\alpha^{(j)}(a_l^{(i)})}{c_m^{(j)}\alpha^{(j)}(A_j)} \varphi_{A_j}^{(m)}, & \text{else}. \end{cases} \tag{23.15b}
$$

Proposition 23.5 *The functions (23.14)–(23.15), for $i = 1, \ldots, M$, form a basis for V_h.*

The mass and stiffness matrices in the resulting algebraic EVP take a *modified* diagonal block structure, resulting from appropriate linear combinations of the rows and colums of a block diagonal matrix, which itself corresponds to individual EVPs on the domains Ω_i, with Dirichlet/Robin BCs on $\Gamma_i \cup \Gamma_i'$ and Neumann BCs on $\partial\Omega_i \setminus (\Gamma_i \cup \Gamma_i')$, $i = 1, \ldots, M$.

The perturbation of this block structure reflects the *coupling* character of the EVP, through the interactions of the functions (23.15), both mutually and with the functions (23.14).

For an illustrative example of the type (a) from Section 23.2.2, we refer to [3].

REFERENCES

[1] F. Ali Mehmeti and S. Nicaise (1993). Nonlinear interaction problems. *Nonlinear Anal.*, 20(1):27–61.

[2] P.G. Ciarlet (1978). *The finite element method for elliptic problems.* North Holland Publishing Company, Amsterdam.

[3] H. De Schepper and R. Van Keer (1999). A finite element method for elliptic eigen-value problems in a multi–component domain in 2D with nonlocal Dirichlet transition conditions. *J. Comp. Appl. Math.*, 112 (to appear).

[4] O. Hansen (1997). A global existence theorem for two coupled semilinear diffusion equations from climate modeling. *Discrete Contin. Dynam. Systems*, 3(4):541–564.

[5] A. Kufner *et al.* (1977). *Function spaces.* Noordhoff International Publishing, Leiden.

[6] P.A. Raviart and J.M. Thomas (1993). *Introduction à l'analyse numérique des équations aux dérivées partielles (2ième tirage).* Masson, Paris.

[7] H. Gerke *et al.* (1999). *Optimal control of soil venting: Mathematical modeling and applications*. Birkhäuser, Basel.

[8] R. Van Keer *et al.* (1999). Computational methods for the evaluation of electromagnetic losses in electrical machinery. *Archives Comp. Meth. Engrg.*, 5(4):385–443.

[9] M. Vanmaele and R. Van Keer (1995). An operator method for a numerical quadrature finite element approximation for a class of second-order elliptic eigenvalue problems in composite structures. *RAIRO - Math. Mod. Num. Anal.*, 29(3):339-365.

[10] A. Zenišek (1990). *Nonlinear elliptic and evolution problems and their finite element approximations*. Academic Press, New York.

24 MESH SHAPE AND ANISOTROPIC ELEMENTS: THEORY AND PRACTICE

Th. Apel[a], M. Berzins[b], P.K. Jimack[b], G. Kunert[a], A. Plaks[c], I. Tsukerman[c] and M. Walkley[b]

[a]Fak. Mathematik, TU Chemnitz, D–09107 Chemnitz, Germany.

[b]CPDE Unit, Computer Studies, Univ. of Leeds, LS2 9JT, UK.

[c]Electrical Eng Dept, The University of Akron, OH 44325-3904, USA.

ABSTRACT

The relationship between the shape of finite elements in unstructured meshes and the error that results in the numerical solution is of increasing importance as finite elements are used to solve problems with highly anisotropic and, often, very complex solutions. This issue is explored in terms of *a priori* and *a posteriori* error estimates, and through consideration of the practical issues associated with assessing element shape quality and implementing an adaptive finite element solver.

Key words. anisotropic finite elements, *a priori* estimates, *a posteriori* estimates, element shape, maximum angle condition

24.1 INTRODUCTION

The solutions of many important partial differential equations (PDEs) possess features whose accurate resolution using finite element (FE) trial functions requires local refinement of the underlying computational mesh. Frequently however these solution features are strongly directional, with the principal length scale in one direction being significantly smaller than in others. Examples of such features include boundary layers, shocks and edge singularities. The most efficient FE trial spaces for representing these solutions are defined by the use of anisotropic meshes whose elements have an orientation and geometry which reflect the nature of the solution itself. In this paper we present a brief overview of some of our work towards better understanding the practical issues associated with using such anisotropic meshes. This begins by considering how one might define anisotropic elements in 2D and 3D and the observation that, in order to realize improved *a priori* estimates using such elements, interpolation error estimates which make explicit use of anisotropy need to be derived. This is discussed in Section 24.2, which also shows how improved *a priori* estimates may be made for certain model problems when an appropriate sequence of anisotropic meshes is considered. Section 24.3 of the paper then considers *a posteriori* error estimates on anisotropic meshes. It is observed that conventional *a posteriori* estimates are not generally appropriate for anisotropic elements and a number of modified estimators are considered for two different types of PDE.

Having introduced a fundamental theory which is applicable to appropriately aligned anisotropic meshes in Sections 24.2 and 24.3, the final parts of the paper discuss some of the algorithmic issues associated with mesh shape and anisotropic mesh adaptivity respectively. In Section 24.4 a practical means of quantifying the quality of the shape of a tetrahedron is introduced and its use in estimating interpolation error is illustrated. This is followed by a brief comparison with some other known element quality indicators. Finally, in Section 24.5 the adaptive solution of a 3D hyperbolic test problem is considered. It is demonstrated that the use of anisotropic elements can improve the solution process through the addition of a very simple "*r*-refinement" strategy to a conventional isotropic "*h*-refinement" algorithm.

24.2   A PRIORI ERROR ESTIMATION

Classical finite element theory assumes that the aspect ratio of the finite elements is bounded. In contrast to this, the aspect ratio of anisotropic finite elements is large, and may even be unbounded in the limit. If we wish to use such elements we must be aware that the whole finite element theory has then to be reassessed. In order to convince ourselves of the necessity of anisotropic mesh refinement consider the following two examples.

Example 1 Consider the Poisson equation, $-\Delta u = f$ in Ω, with Dirichlet boundary conditions, $u = g$ on $\partial\Omega$, in a polyhedral domain $\Omega \subset \mathbb{R}^3$. It is well known that the solution u has singularities of r^λ type near edges with interior angle ω, with $\lambda = \pi/\omega < 1$ for $\omega > \pi$. Since r is here the distance to the edge and the solution is more regular in the direction of the edge, the solution behaves anisotropically.

As a result of this singularity, the finite element method with piecewise linear shape functions on a quasi-uniform family of tetrahedral meshes converges with order h^λ in the energy norm. To recover the optimal convergence order h, an anisotropically refined finite element mesh is suggested, see Figure 24.1 for an example mesh in the neighbourhood of an edge. Such meshes have been described, and the optimal convergence order has been proved by Apel *et al.* [1, 2, 5] under different assumptions on the data and also for higher order shape functions. Several refinement strategies are compared in [4].

Example 2 The solution of the singularly perturbed reaction-diffusion problem $-\varepsilon^2 \Delta u + cu = f$ in $\Omega = (0,1)^2$, $u = 0$ on $\partial\Omega$, behaves like $1 - e^{-r/\varepsilon}$, r being here the distance to $\partial\Omega$. Therefore the ℓth derivative in the normal direction grows with $\varepsilon^{-\ell}$ in a boundary layer of width $\mathcal{O}(\varepsilon|\ln \varepsilon|)$, while derivatives in the tangential direction are not affected by $\varepsilon \ll 1$.

Consequently, the finite element method with trial functions of degree k converges in the energy norm like $\||\, u - u_h \,\|| \leq Ch^k \varepsilon^{1/2-k}$ on a family of quasi-uniform meshes.

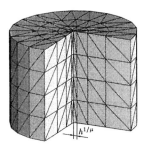

Figure 24.1: Anisotropic mesh near edge.

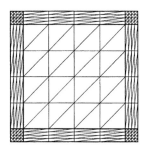

Figure 24.2: Anisotropic mesh in a boundary layer.

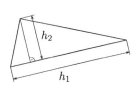

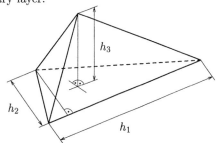

Figure 24.3: Illustration of the definition of the element sizes.

Anisotropic refinement in the boundary layer, as shown in Figure 24.2, leads to the the optimal error estimate $\| u - u_h \| \leq Ch^k(\varepsilon^{1/2-\delta} + h)$ [1, 3], $\delta > 0$ arbitrarily small. We note that the number of unknowns is of the order h^{-d}, d being the space dimension, in both examples. Moreover, an optimal error estimate that is uniform in the perturbation parameter ε, cannot be obtained with isotropic mesh refinement.

One of the basic tasks in finite element theory is to estimate the local interpolation error. Such error estimates are needed for *a priori* estimates of the finite element error, for the theory of error estimators, and in the multilevel theory for the solution of the resulting systems of equations. For simplicity, let us focus here on Lagrange interpolation with linear trial functions on simplicial elements as one of the simplest interpolation operators, though this operator is not suited for interpolating functions with low regularity.

Classical finite element theory leads to the estimate

$$|u - I_h u|_{1,p,K} \leq Cah|u|_{2,p,K},$$

where we denote by a the aspect ratio of the element K and by $|\cdot|_{m,p,K}$ the usual seminorm in $W^{m,p}(K)$. The dependence on the aspect ratio was removed in the fifties ([17]) and seventies ([6, 10]), but the resulting error estimate was still not satisfactory because only the diameter h of the element appears in the estimate.

In order to compensate large norms of certain directional derivatives of the solution u by a small element size in this direction, we need sharper estimates, like

$$|u - I_h u|_{1,p,K} \leq C \sum_{i=1}^{d} h_i \left| \frac{\partial u}{\partial x_i} \right|_{1,p,K}, \tag{24.1}$$

where h_i are suitably defined element sizes (see Figure 24.3 for a possible definition). We emphasize that a refined estimate on the reference element is necessary for the proof of such estimates, see [1, Example 2.1]. Moreover, the geometry of the elements is restricted by a maximal angle condition (see also Section 24.4 below). Anisotropic local interpolation error estimates may now be derived. Indeed, such estimates have been proved by Apel *et al.* for several element types (triangles, quadrilaterals, tetrahedra, triangular prisms, hexahedra, including some subparametric and non-conforming elements), for trial functions of arbitrary order, and under various smoothness assumptions on the function to be interpolated (including functions from weighted Sobolev spaces) [1, 2].

We stress that anisotropic elements must be treated carefully. For example, estimate (24.1) is valid for simplicial elements and linear trial functions under the following assumptions on p. In the 2D case (24.1) is valid for the whole range of p, $p \in [1, \infty]$. However, in the 3D case the estimate is valid for $p \in (3/2, \infty]$ for isotropic elements but only for $p \in (2, \infty]$ for anisotropic elements. There is a counter example for $p \leq 2$, see [2]. For the interpolation of less regular functions one can use modified Scott–Zhang interpolants [1].

24.3   A POSTERIORI ERROR ESTIMATION

The fundamental requirements of an adaptive algorithm that is able to exploit the use of anisotropic elements are that information on the *stretching direction* of the anisotropic elements, the *stretching ratio* (or *aspect ratio*) of the elements and the *size* of the elements should be utilized. None of these issues have yet been fully understood however. For example, the stretching direction and ratio are often determined (heuristically) by investigation of the Hessian matrix, but other approaches may equally well be employed (see § 24.5 for example). The question of the appropriate element size is closely related to *a posteriori error estimation*.

Research into error estimators for anisotropic meshes has been intensified in recent years. So far, many applications of anisotropic elements utilize heuristic arguments and lack rigorously analyzed error estimators. Some strictly mathematically-based estimators have appeared recently however, due to Siebert [16], Kunert [13, 14], Kunert and Verfürth [15] and Dobrowolski *et al.* [9]. Before discussing these estimators in more detail, let us comment on an important feature that seems to be inherent in anisotropic error estimators. The theory of these estimators is not as complete as for isotropic elements since, at this time, no error estimator is known that bounds the error reliably from *above and below*, independently of the mesh $\mathcal{T}_h$ and the solutions u and u_h. In other words, the effectivity index cannot be guaranteed to be $\mathcal{O}(1)$. This rather unsatisfactory situation can be interpreted in two ways.

- If the error is to be bounded from above and below *without further assumptions* on $u, u_h, \mathcal{T}_h$, then the two error bounds have to contain different terms. Hence *two* error estimators would be required, one for each bound.

- If a *single error estimator* is to give upper and lower error bounds, then the mesh $\mathcal{T}_h$ and the solutions u, u_h have to correspond in some way. For example, the anisotropy of $\mathcal{T}_h$ may have to be aligned with the anisotropy of u to result in an effectivity index

$\mathcal{O}(1)$. Such assumptions on $u, u_h, \mathcal{T}_h$ can be seen in the aforementioned work, and are briefly discussed below.

In the following paragraphs we briefly describe existing approaches to anisotropic *a posteriori* error estimation. Rather than giving detailed formulae we attempt to present the main ideas and results for each of the estimators. We begin by investigating the Poisson problem (see e.g. Example 1 of the previous section).

Siebert [16] considers anisotropic *rectangular, cuboidal or prismatic* meshes and derives a *residual error estimator* by measuring and weighting the residuals (i.e. gradient jumps and element residual). Two specific assumptions on $u, u_h, \mathcal{T}_h$ guarantee a global upper and a local lower error bound in the energy norm.

In Kunert [14], a *residual error estimator* for *tetrahedral or triangular* meshes is presented, thus giving a greater geometrical flexibility. Moreover, it improves Siebert's estimator by means of better weights of the gradient jumps. Hence one of Siebert's assumptions is superfluous. The remaining assumption on $u, u_h, \mathcal{T}_h$ is expressed by a so-called *matching function* $m_1(v, \mathcal{T}_h)$ which measures how well an anisotropic mesh $\mathcal{T}_h$ is aligned with an anisotropic function v (see [13] for details). The matching function $m_1(u - u_h, \mathcal{T}_h)$ enters the upper bound of the error: the better the mesh $\mathcal{T}_h$ is suited to the problem, the sharper the error bound will be. Note that although $m_1(u - u_h, \mathcal{T}_h)$ cannot be calculated exactly, it can be approximated numerically ([13]). In Kunert and Verfürth [15] a modification of the previous *residual error estimator* is analyzed. The difference being that the gradient jumps alone suffice to define the estimator, i.e. the element residuals are omitted. All conclusions and results remain valid. Kunert has also derived a recent *error estimator* based on the solution of a *local problem* which is given for tetrahedral meshes. The appropriate choice of local problem appears to be more critical than for the isotropic case, leading to a result very similar to that obtained for the residual error estimation.

Zienkiewicz–Zhu (ZZ) type error estimators utilize a postprocessing procedure (like an averaged gradient) to estimate the error. On anisotropic meshes, such estimators are much less developed. Furthermore, on these meshes the analysis is hindered by an apparent lack of superconvergence. Nevertheless some initial promising attempts of ZZ estimators are presented in Kunert [13]. There, the equivalence to the residual error estimator of [15] is proven for specific meshes. Also in [13], Kunert derives and analyzes an L_2 *residual error estimator* for tetrahedral meshes that bounds the error (in the L_2 norm) from above and below. Additionally, the face residuals (i.e. the gradient jumps) alone suffice to define this estimator, see Kunert and Verfürth [15] for further details.

Dobrowolski *et al.* [9] also investigate the Poisson problem on *triangular* meshes. Applying the methodology of Bank and Weiser, they derive a *global error estimator* by solving a *global* problem. The (global) error bound relies on a saturation assumption that again requires a suitable anisotropic mesh. Note that no *local* error bound is obtained.

We summarize by noting that recent and ongoing research is ensuring that anisotropic *a posteriori* error estimators are becoming increasingly well understood. As well as for Poisson's equation discussed here it should also be noted that other problems are also currently under investigation. These include singularly perturbed reaction–(convection)–diffusion problems which lend themselves naturally to anisotropic boundary or interior layers. A model equation $-\varepsilon \Delta u + u = f$ is investigated by Kunert [13, 15] on tetrahedral meshes (see also Example 2 of the previous section). A *residual error estimator* is derived that bounds the error (in the ε dependent energy norm) locally from below and globally from above. Note that the upper error bound contains exactly the same matching function $m_1(u - u_h, \mathcal{T}_h)$ as in the Poisson case. Despite this recent progress however it is clear that the incorporation of these, and possible future estimators, into practical adaptive anisotropic strategies is still a significant challenge. Some of the issues associated with this challenge are discussed in the remainder of this paper.

24.4 INTERPOLATION ERRORS AND ELEMENT SHAPE

Nedelec-Raviart-Thomas edge and face elements play an increasingly important role in computational electromagnetism, incompressible flow problems, and other areas. This section presents, as a somewhat unexpected application of edge elements, a new H^1-interpolation error bound for first order tetrahedra. This estimate is in terms of the minimum singular value of a particular rectangular matrix. The derivation (see [19]) is based on the well known fact ([8]) that the standard Lagrange P^1-interpolation (with operator Π_{P^1}) of a scalar field u on a tetrahedral mesh M is equivalent to the edge element interpolation Π_{edge} of the respective conservative field $v = \nabla u$. (Π_{edge} preserves edge circulations, so that $\int_{edge}(\Pi_{edge}v) \cdot dl = \int_{edge} v \cdot dl$ over each tetrahedral edge.)

24.4.1 The Minimum Singular Value Condition for Tetrahedra

The shape of a tetrahedron can be characterized by six unit vectors $e_1, \ldots, e_6$ directed along the edges (in either of two possible directions). The element '**edge shape matrix**' E [19] is a 3×6 matrix whose columns are the (Cartesian) vectors e_i. The E^T-matrix governs the transformation between the Cartesian and edge components of an arbitrary vector ξ in $\mathbb{R}^3$: $\xi_{edge} = E^T \xi_{cart}$, where $\xi_{edge} \in \mathbb{R}^6$ and $\xi_{cart} \in \mathbb{R}^3$.

A key role in our analysis is played by the singular value decomposition $E = P\Sigma Q^T$ of the edge shape matrix. In particular, the *minimum singular value*[1] $\sigma_{\min}(E) = \lambda_{\min}^{1/2}(EE^T)$ characterizes, algebraically, the level of linear independence of the unit edge vectors e_i, or, geometrically, their proximity to one plane. More precisely ([19]),

$$\sigma_{\min}(E) = \min_{\|\xi_{cart}\|=1} \|\xi_{edge}\|_{\mathbb{R}^6} = \min_{\|\xi_{cart}\|=1} \|E^T\xi_{cart}\|_{\mathbb{R}^6}, \qquad (24.2)$$

with minimization achieved when ξ_{cart} is the eigenvector $\xi_{\min}$ corresponding to $\lambda_{\min}(EE^T)$. Note that $\sigma_{\min}(E) = 0$ if and only if all six edge vectors lie in one plane (perpendicular to $\xi_{\min}$), i.e. the tetrahedron is degenerate. Moreover, the following error bound ([19]) shows that $\sigma_{\min}(E)$ may be considered as a governing factor for interpolation errors:

$$\|\Pi_{P^1}u - u\|_{H^1(\Omega)}^2 = \|\Pi_{edge}v - v\|_{(L_2(\Omega))^3}^2 \leq C(v) \cdot \sum_{K_i \in M} h_i^2(\sigma_{\min}^{-2}(E(K_i))V_i . \quad (24.3)$$

Here $v = \nabla u$ is a (sufficiently smooth) conservative field, K_i is a tetrahedral element, $h_i = \text{diam}(K_i)$, $V_i = \text{meas}(K_i)$. This is a global estimate, but each term in the sum represents the square of the element-wise interpolation error.

---

[1]Which is obviously independent of the choice of the Cartesian system in $\mathbb{R}^3$.

24.4.2   Links with Maximum Angle, Křížek and Jamet Conditions

In the illustrative case of a triangular element with angles ϕ_1, ϕ_2, ϕ_3 the minimum singular value of the 2×3 edge shape matrix

$$E \; = \; \left[\begin{array}{ccc} 1 & \cos \phi_1 & -\cos \phi_2 \\ 0 & \sin \phi_1 & \sin \phi_2 \end{array} \right] \tag{24.4}$$

can be explicitly evaluated via the trace of $(EE^T)^{-1}$, which ultimately yields

$$\frac{1}{3}(\sin^2 \phi_1 + \sin^2 \phi_2 + \sin^2 \phi_3) \; \leq \; \sigma_{\min}^2(E) \; \leq \; \frac{2}{3}(\sin^2 \phi_1 + \sin^2 \phi_2 + \sin^2 \phi_3) \,. \tag{24.5}$$

This is equivalent to the maximum angle condition [17, 6] as, in (24.5), $\sigma_{\min}$ may only tend to zero on a sequence of elements if one angle tends to π and the other two to zero.

Possibly the most common geometric characteristic of a *tetrahedral* element is the ratio of radius r of the inscribed sphere to the maximum element edge length h. Tsukerman [19] shows that the singular value criterion is less stringent than the r/h ratio.

Křížek [12] introduced a sufficient convergence condition requiring that all dihedral angles, as well as all face angles, be bounded away from π. A rather simple lower bound of $\sigma_{\min}$ in terms of the dihedral and face angles can be given and will be presented elsewhere. The minimum singular value and Křížek conditions are equivalent as asymptotic criteria of convergence of piecewise-linear interpolation on a family of tetrahedral meshes.

Jamet [10] has obtained interpolation error bounds under quite general assumptions. For tetrahedral elements, the governing factor $\cos \theta$ in Jamet's estimate can be obtained from the rightmost expression in (24.2) for $\sigma_{\min}$ by simply replacing the 2-norm in $\mathbb{R}^6$ with the ∞-norm.

24.5   A CASE STUDY OF 3D ANISOTROPIC REFINEMENT

We conclude this paper with a simple example which demonstrates the practical use of anisotropic adaptation using an hr-refinement scheme in which the nodes are moved according to an edge-based error indicator and comparing against standard h-refinement results. The test problem considered is the steady 3D hyperbolic equation $\mathbf{a} \cdot \nabla u = 0$, and a standard SUPG finite element method [11] is used, based on an unstructured tetrahedral mesh with linear basis functions $\phi_i(\mathbf{x})$ and test functions $\psi_i(\mathbf{x})\phi_i + \tau \mathbf{a} \cdot \nabla \phi_i$ defined at each of the N_P nodes $\mathbf{x}_i$. The parameter τ is defined as an element quantity, $\tau^K = \alpha h^K/|\mathbf{a}|$ for some measure of the element length h^K, taken here to be the minimum element height, and the resulting linear system is solved using the ILU preconditioned GMRES method.

In order to compare the two adaptive algorithms being considered, standard finite element **h-refinement** is driven by solving a problem with a known solution and using the L^1 norm of the *exact* error $e(\mathbf{x})$ on each element. The total L^1 error, e, may then be split into its contributions e^K from each of the N_E elements. To achieve a final L^1 error of e^* the error in each cell is reduced to below e^*/N_E using nested isotropic h-refinement of an initial unstructured tetrahedral base mesh.

As well as using local mesh refinement, the **hr-refinement** algorithm makes use of a simple node movement scheme designed to steer nodes towards regions of sharp variation in the solution. This is motivated by recent work of Berzins, [7], in which the interpolation error is estimated by assuming that the exact solution can be approximated in a locally quadratic form on each tetrahedral element and then considering the difference between this quadratic function and the linear finite element interpolant:

$$\int_K e_{lin}^2 \, d\Omega \;=\; \frac{6}{4} V^K \frac{2}{7!} \left(\left(\sum_{s=1}^6 d_s \right)^2 - d_1 d_4 - d_2 d_5 - d_3 d_6 + \sum_{s=1}^6 d_s^2 \right) \qquad (24.6)$$

where V^K is the volume of element K. Here, d_s denotes the directed edge second derivative for an edge $s = s(\mathbf{x}_i, \mathbf{x}_j)$ that connects the nodes $\mathbf{x}_i$ and $\mathbf{x}_j$. It is given by

$$d_s := \left(\nabla u(\mathbf{x}_i) - \nabla u(\mathbf{x}_j) \right) \cdot (\mathbf{x}_i - \mathbf{x}_j) \cdot L_{ij}^{-2} \qquad \text{with} \qquad L_{ij} := \|\mathbf{x}_i - \mathbf{x}_j\| \, . \qquad (24.7)$$

A node movement scheme is then developed by using the edge second derivatives (24.7) as weights in the following weighted average expression for the node position,

$$\mathbf{x}_i^{av} \;=\; \frac{\sum_j L_{ij} |d_s| \mathbf{x}_j}{\sum_j L_{ij} |d_s|} \qquad \text{with} \qquad s = s(\mathbf{x}_i, \mathbf{x}_j) \, . \qquad (24.8)$$

The nodal position is updated by $\mathbf{x}_i \rightarrow (1 - \gamma_i)\mathbf{x}_i + \gamma_i \mathbf{x}_i^{av}$ where γ_i is a safety factor at each node $\mathbf{x}_i$ that prevents the mesh from becoming tangled. Several such iterations are performed at each r-refinement stage (which is undertaken prior to h-refinement).

Further details of the test problem are shown in Figure 24.4 (a 2D form of the problem is also included to aid comprehension). At the inflow boundary, where $\mathbf{a} \cdot \mathbf{n} < 0$, the imposed solution is defined to have a thin vertical layer across which u varies linearly from 1 to 0. The solution within the domain is therefore defined by the convection of this layer in the direction $\mathbf{a}$, as illustrated. Here, $\mathbf{a} = (2, 1, 1)$ and the layer has thickness 0.025 on a unit cube domain.

The initial mesh is a uniform discretisation of the unit cube into 1331 nodes (11^3) and 6000 tetrahedral elements. Figure 24.5 compares the total error for the h- and hr-refinement schemes. The hr-refinement scheme significantly reduces the error on a given grid, leading to a similar level of error with approximately 20000 nodes to that requiring over 100000 nodes using isotropic h-refinement. Figure 24.6 shows two tetrahedra from the final hr-refined mesh in the region $(x, y, z) \in [0.3, 0.4] \times [0.7, 0.8] \times [0.1, 0.2]$. It is seen that the upper tetrahedron is aligned with the layer, although large internal angles have been produced. In contrast, the lower tetrahedron is less well-aligned.

24.6   DISCUSSION

This paper reflects the main topics presented at the Minisymposium on Anisotropic Finite Elements at the 1999 MAFELAP conference. Whilst it is impossible to cover all aspects relating to element shape and anisotropy in such a brief exposition, we have attempted to present an overview of what we believe to be some of the most important issues: *a priori* and *a posteriori* error estimation, mesh quality, and mesh adaptivity.

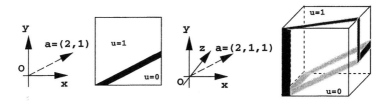

Figure 24.4: Model steady solution of the 3D convection equation

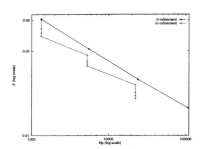

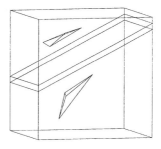

Figure 24.5: Errors on the adapted grid sequences for h- and hr-refinement

Figure 24.6: Zoom of the hr-refined tetrahedra near the layer

ACKNOWLEDGEMENTS

Collaboration between the groups at Chemnitz and Leeds has been supported by the DAAD and British Council under the ARC programme. The work of TA and GK (Sections 24.2 and 24.3) was partially supported by the DFG (SFB 393). The work of IT and AP (Section 24.4) was supported in part by the NSF. The work of MW (Section 24.5) was supported by the EPSRC.

REFERENCES

[1] Th. Apel (1999). *Anisotropic finite elements: Local estimates and applications*. Advances in Numerical Mathematics. Teubner, Stuttgart.

[2] Th. Apel and M. Dobrowolski (1992). Anisotropic interpolation with applications to the finite element method. *Computing*, 47:277–193.

[3] Th. Apel and G. Lube (1998). Anisotropic mesh refinement for a singularly perturbed reaction diffusion model problem. *Appl. Numer. Math.*, 26:415–433.

[4] Th. Apel and F. Milde (1996). Comparison of several mesh refinement strategies near edges. *Comm. Numer. Methods Engrg.*, 12:373–381.

[5] Th. Apel and S. Nicaise (1998). The finite element method with anisotropic mesh

grading for elliptic problems in domains with corners and edges. *Math. Methods Appl. Sci.*, 21:519–549.

[6] I. Babuška and A.K. Aziz (1976). On the angle condition in the finite element method. *SIAM J. Numer. Anal.*, 13:214–126.

[7] M. Berzins (1998). A solution-based triangular and tetrahedral mesh quality indicator. *SIAM J. Sci. Comp.*, 19:2051–1060.

[8] A. Bossavit (1988). A rationale for 'edge elements' in 3-D fields computations. *IEEE Trans. Magn.*, 24:74–79.

[9] M. Dobrowolski, S. Gräf and C. Pflaum (1999). On a posteriori error estimators in the finite element method on anisotropic meshes. *Elec. Trans. Numer. Anal.*, 8:36–45.

[10] P. Jamet (1976). Estimations de l'erreur pour des éléments finis droits preque dégénérés. *Rairo Anal. Numer.*, 10:43–60.

[11] C. Johnson (1987). *Numerical solution of partial differential equations by the finite element method.* Cambridge University Press.

[12] M. Křížek (1992). On the maximum angle condition for linear tetrahedral elements. *SIAM J. Numer. Anal.*, 29:513–520.

[13] G. Kunert (1999). *A posteriori error estimation for anisotropic tetrahedral and triangular finite element meshes.* Ph.D. Thesis, TU Chemnitz (http://www.tu-chemnitz.de/pub/1999/0012/index.html).

[14] G. Kunert (1999). An a posteriori residual error estimator for the finite element method on anisotropic tetrahedral meshes. To appear in *Numer. Math.*

[15] G. Kunert and R. Verfürth (1999). Edge residuals dominate a posteriori error estimates for linear finite element methods on anisotropic triangular and tetrahedral meshes. To appear in *Numer. Math.*

[16] K.G. Siebert (1996). An a posteriori error estimator for anisotropic refinement. *Numer. Math.*, 73:373–398.

[17] J.L. Synge (1957). *The hypercircle in mathematical physics; a method for the approximate solution of boundary value problems.* Cambridge University Press.

[18] I.A. Tsukerman (1998). A general accuracy criterion for finite element approximation. *IEEE Trans. Magn.*, 34:2425–1428.

[19] I.A. Tsukerman (1998). Approximation of conservative fields and the element 'edge shape matrix'. *IEEE Trans. Magn.*, 34:3248–3251.

25 ON THE TREATMENT OF PROPAGATING MODE-1 CRACKS BY VARIATIONAL INEQUALITIES

Michael Bach

Institute of Mathematics A, University of Stuttgart, Pfaffenwaldring 57,
D-70569 Stuttgart, Germany

ABSTRACT

The propagation of a plane mode-1 crack under both the Irwin and the Griffith criterion, is considered. Using the method of matched asymptotic expansions, the first terms of the asymptotics for the stress intensity factor and for the total energy of the crack at the quasistatic time t are constructed. Both criteria yield a variational inequality the solution of which describes the propagation of the crack. For numerical purposes, the inequalities are penalized and discretized by a Galerkin-Bubnov method. A numercial example for a penny-shaped crack is presented.

Key words. mode-1 crack, Griffith criterion, Irwin criterion, asymptotic expansion, variational inequality, penalization, Galerkin method, penny crack

25.1 INTRODUCTION

We consider the propagation of a plane crack G in the elastic isotropic space $\mathbb{R}^3$ and G is bounded by the simple closed smooth contour $\Gamma = \partial G$, the crack front. In the following we avoid distinguishing between 2–D sets on $\partial \mathbb{R}^3_+$ and their immersions into $\mathbb{R}^3$. We assume that a symmetric normal loading P acts to the crack surfaces $G^\pm$ and opens the crack. Therefore the stress intensity factor (SIF) K_1 is a positive function on the crack front Γ and we have the situation of a normal rupture with mode–1 stress–strain state. Note that $K_1 \in C^\infty(\Gamma)$ if the function P is smooth in the vicinity of Γ. With the help of the Papkovich–Neuber representation (see, e.g. [11]), the 3–D elasticity problem in $\mathbb{R}^3 \setminus \overline{G}$ can be reduced to the following scalar mixed boundary value problem in the halfspace $\mathbb{R}^3_+ = \{x = (y,z) : z > 0\}$:

Find the harmonic function v in $\mathbb{R}^3_+$ which vanishes at infinity and satisfies homogeneous Dirichlet conditions on $\mathbb{R}^2_+ \setminus \overline{G}$ and inhomogeneous Neumann conditions on G:

$$
\begin{aligned}
\Delta v(y, z) &= 0 & &\text{for } (y, z) \in \mathbb{R}^3_+ , \\
v(y, 0) &= 0, & &\text{for } y \in \mathbb{R}^2_+ \setminus \overline{G} , \\
-\partial_z v(y, 0) &= p(y) = \alpha P(y), & &\text{for } y \in G,
\end{aligned}
\tag{25.1}
$$

with $\alpha = 2\mu(\lambda + \mu)(\lambda + 2\mu)^{-1}$ the Lamé constants μ and λ.
We are interested in the propagation of the crack G and seek the shape of the growing crack $G(t)$ with crack front

$$
\Gamma(t) = \{ y \in \partial\mathbb{R}^3_+ \cap U : r = t\, h(s) \},
$$

where $t\, h(s)$ describes the depth of the propagation along the normal to Γ at the point with arc length s. Here U is some neighbourhood of Γ and $t \geq 0$ is a fixed time–like parameter describing the increment of the loading

$$
P(t; y) = P(y) + t\, P'(y) ,
\tag{25.2}
$$

where $P(y)$ is the critical loading, so that the positive increment $t\, P'(y)$ leads to the propagation of the crack. Both t and $h(s)$ are assumed to be small. Regarding the crack $G(t)$ as a perturbation of the initial crack $G := G(0)$, the perturbed problem is given by

$$
\begin{aligned}
\Delta v(t; y, z) &= 0 & &\text{for } (y, z) \in \mathbb{R}^3_+ , \\
v(t; y, 0) &= 0, & &\text{for } y \in \mathbb{R}^2_+ \setminus \overline{G(t)} , \\
-\partial_z v(t; y, 0) &= p(t; y) = \alpha P(t; y), & &\text{for } y \in G(t) .
\end{aligned}
\tag{25.3}
$$

We introduce curvilinear coordinates (r, φ, s) in the neighbourhood U of the contour Γ, where (r, φ) are polar coordinates in the planes perpendicular to Γ ($r = \text{dist}\{x, \Gamma\}$ and $\varphi \in (-\pi, \pi)$).
Since the stress σ_{33} coincides on $\partial\mathbb{R}^3_+$ with $\alpha^{-1}\partial_z v$, one has the local representation

$$
\partial_z v(y, 0) = (2\pi r)^{\frac{1}{2}} \big(K_1(s) + r k_1(s) \big) + \alpha p_\Gamma(s) + \mathcal{O}(r^2), \quad y \in \mathbb{R}^2_+ \setminus \overline{G},
$$

where K_1 denotes the first stress intensity factor and $k_1(s)$ the 'junior' stress intensity factor, i.e., the factor corresponding to the low-order singularity; p_Γ implies the trace of p on Γ.

Written in the curvilinear coordinates defined above, the energy space solution $v \in H^1(\mathbb{R}^3_+)$ of problem (25.1) has the following asymptotic form near to the edge Γ ($r \to 0$), see e.g. [14, 15]:

$$
\begin{aligned}
v(x) &= \alpha \left(2\pi^{-1}\right)^{\frac{1}{2}} \left(r^{\frac{1}{2}} K_1(s) \sin\frac{\varphi}{2} + \frac{1}{3} r^{\frac{3}{2}} k_1(s) \sin\frac{3\varphi}{2} \right) \\
&\quad + \alpha \left(2\pi^{-1}\right)^{\frac{1}{2}} \kappa(s) K_1(s) r^{\frac{3}{2}} \left(\frac{1}{4}\sin\frac{\varphi}{2} - \frac{1}{12}\sin\frac{3\varphi}{2} \right) \\
&\quad + r\, p_\Gamma(s) \sin\varphi + \mathcal{O}(r^2) ,
\end{aligned}
\tag{25.4}
$$

where $\kappa(s)$ denotes the curvature of Γ at the point s. In addition, we need the asymptotic expansion

$$V(H, x) \;=\; (2\pi r)^{-\frac{1}{2}} H(s) \sin\left(\frac{\varphi}{2}\right) + \alpha \left(\frac{2r}{\pi}\right)^{\frac{1}{2}} \mathcal{M}[H](s) \sin\left(\frac{\varphi}{2}\right) \qquad (25.5)$$

$$-\kappa(s)\left(\frac{r}{2\pi}\right)^{\frac{1}{2}} H(s) \left(\frac{1}{4}\sin\left(\frac{3\varphi}{2}\right) - \frac{3}{4}\sin\left(\frac{\varphi}{2}\right)\right) + \mathcal{O}(r^2) \qquad (25.6)$$

of the weight function V, which is an element of the weighted Sobolev space $V_{2,\beta,\gamma}^{\ell+1}(\mathbb{R}_+^3)$, $|\gamma - \ell| < \frac{1}{2}$, $\frac{1}{2} < \beta - \ell < \frac{3}{2}$ (see [6, 14, 15]). This is a non-energy space solution of problem (25.1) with a homogeneous Neumann condition, and has the square root singularity distributed with the density H over Γ. The 'stress intensity factor' of V, of course, depends on H, and according to [14] this dependence is described by the integral operator

$$\mathcal{M}[H](z) := \int_\Gamma (H(s) - H(z))\, K(s, z)\, ds + \mathcal{K}(z) H(z) - \frac{3}{8\alpha} \kappa(z) H(z) \qquad (25.7)$$

with the symmetric positive kernel $K(s, z)$, given by

$$K(z, s) := \alpha^{-1} (2\pi)^{-1} |z - s|^{-2} + \mathcal{O}(|\ln|s - z||) . \qquad (25.8)$$

Both the kernel K and the factor $\mathcal{K}$ depend on the shape of the crack G.

25.2   RUPTURE CRITERIA

We first describe the Irwin criterion [9], which is a stress criterion and based on the comparison of the SIF K_1 with the critical SIF K_{1c} as material parameter. Due to the irreversibility of the fracture process we observe that

$$h(s) \geq 0 \qquad \forall s \in \Gamma.$$

Assuming the quasistatic propagation of the crack, i.e. neglecting dynamic effects, the application of the Irwin criterion (see [14, 17]) leads to the relations

$$\begin{aligned} h(s) = 0 &\implies K_1(t; s) \leq K_{1c} , \\ h(s) > 0 &\implies K_1(t; s) = K_{1c} , \end{aligned} \qquad (25.9)$$

where $K_1(t; s)$ denotes the SIF at $\Gamma(t)$ for the loading $P(t; y)$ on the crack surfaces $G^\pm(t)$. These relations mean that the crack $G(t)$ appears to be in equilibrium, i.e. $K_1(t; s) \leq K_{1c}$ at all points on its front $\Gamma(t)$, and that the SIF $K_1(t; s)$ takes the critical value on the new parts of the crack front. Reformulating (25.9) as a variational inequality we get the following variational problem:

Find the non–negative function $H(s) = K_1(s)\, h(s)$ such that for every $X \in C^\infty(\Gamma)$ with $X \geq 0$ the following inequality is valid:

$$\langle K_1(t; s) - K_{1c}, H(s) - X(s) \rangle_\Gamma \geq 0 . \qquad (25.10)$$

The second criterion which we will use is the Griffith criterion [8]. It is based on the calculation of the total energy and implies that the variation of the crack-front will be such that the total energy of the cracked body is minimized. Let us denote by Π_t the total energy, by U_t the potential energy and by S_t the surface energy at the time t. Then we have to solve the minimization problem

$$\Pi_t(h) := U_t(h) + S_t(h) \longrightarrow \min , \tag{25.11}$$

and get for the first variation the condition

$$\frac{d}{d\delta}\Pi_t(H + \delta(H - X))\big|_{\delta=0} \geq 0 \tag{25.12}$$

with $H = K_1 h$ and $X \in C^\infty(\Gamma)$. (25.12) defines a variational inequality whose solution H describes the propagation of the crack.

25.3   VARIATIONAL INEQUALITIES

In order to consider the variational inequalities further we need to calculate the stress intensity factor $K_1(t; s)$ and the total energy $\Pi_t(h)$, or at least their asymptotic expansions. To do this we use the method of matched asymptotic expansions to get the outer and inner representations of the solution $v(t; x)$ of problem (25.1), corresponding to the crack G_t and the loading $P(t; y)$ (using the asymptotics of v and V). The inner representation gives the first two-terms of the asymptotic expansion of the SIF $K_1(t; s)$ and the outer representation gives the first three terms of the asymptotic expansion of the energy Π_t (see [3, 10, 14]). Inserting these asymptotics in the variational inequalities (25.10) and (25.12) and denoting by the index 'I'('G') the solution/data in case of the Irwin (Griffith) criterion, we have to solve the variational problem(s):

Find $H_{I(G)} \in W_{2,+}^{\frac{1}{2}}(\Gamma) := \{X \in W_2^{\frac{1}{2}}(\Gamma) \mid X \geq 0\}$ such that for all $X \in W_{2,+}^{\frac{1}{2}}(\Gamma)$

$$-\langle b_{I(G)}H_{I(G)}, X - H_{I(G)}\rangle - \langle \mathcal{B}[H_{I(G)}], X - H_{I(G)}\rangle \geq \langle f_{I(G)}, X - H_{I(G)}\rangle . \tag{25.13}$$

The integral operator $\mathcal{B}$ is given by

$$\mathcal{B}[H_{I(G)}](s) := \alpha \left(\mathcal{M}[H_{I(G)}](s) - \mathcal{K}(s)H_{I(G)}(s) + \frac{3}{8\alpha}\kappa(s)H_{I(G)}(s) \right) , \tag{25.14}$$

and the functions $b_{I(G)}$, $f_{I(G)}$, both depending on the data of the problem for the crack in the initial position, are defined by

$$
\begin{aligned}
b_I(s) &:= 2^{-1} K_1^{-1}(s) k_1(s) + \alpha \mathcal{K}(s) - \frac{3}{8}\kappa(s) , \\
b_G(s) &:= 2^{-1} K_1^{-1}(s) k_1(s) + 2^{-1}\kappa(s) \left(1 - \frac{K_{1c}^2}{K_1^2(s)} \right) + \alpha \mathcal{K}(s) - \frac{3}{8}\kappa(s) , \\
f_I(s) &:= t^{-1} (K(s) - 1) + K_{1c}^{-1}(s) K_1^*(s) , \\
f_G(s) &:= t^{-1} (K(s) - 1)(K_1(s) + K_{1c})(2K_1(s))^{-1} + K_{1c}^{-1}(s) K_1^*(s) .
\end{aligned}
$$

(K_1^* is the first SIF of the solution of the problem (25.1) for the crack in the initial position with the loading $\partial_t P(0; s)$ as Neumann condition.)

Theorem 25.1 *([2]) Let $b \in \{b_I, b_G\}$, $b > 0$ on Γ, $f \in \{f_I, f_G\}$ and $H \in \{H_I, H_G\}$. Then for every $f \in L_2(\Gamma)$ there exists a unique solution $H \in W_{2,+}^{\frac{1}{2}}(\Gamma)$ of the problem (25.1) and the estimate*

$$\|H; W_{2,+}^{\frac{1}{2}}(\Gamma)\| \le c_1 \|f; L_2(\Gamma)\| \tag{25.15}$$

is valid, where c_1 is a constant independent of f. If in addition, $f \in L_p(\Gamma), 2 \le p < +\infty$, then the solution $H \in W_2^{\frac{1}{2}}(\Gamma)$ of the problem (25.1) belongs to the Sobolev space $W_p^1(\Gamma)$ and satisfies the estimate

$$\|H; W_p^1(\Gamma)\| \le c_2 \|f; L_p(\Gamma)\| \tag{25.16}$$

with a constant c_2 independent of f.

25.4 VARIATIONAL EQUATION

There exist different ways of determining the solution of the variational inequality (25.13) under the assumptions of Theorem 25.1. Here we use the penalization method to transform the inequality in an easier to handle equation. We should mention that we do not distinguish in this section between them the variational inequality we get from the Irwin criterion and the variational inequality we get from the Griffith criterion, because both have the same outer form and the only difference between lies in the data b and f ([16]).

Let $X \in W_2^{\frac{1}{2}}(\Gamma)$. Then we define the penalty function β by

$$\beta(s, X(s)) := X_-(s) = \begin{cases} 0 & \text{for } X(s) \ge 0, \\ X(s) & \text{for } X(s) < 0, \end{cases} \tag{25.17}$$

and the corresponding Nemyzki operator by

$$\mathcal{B}_\beta[X](s) := \beta(s, X(s)) . \tag{25.18}$$

Instead of the inequality (25.13) we have to solve the following variational equation:
Find $H_\lambda \in W_2^{\frac{1}{2}}(\Gamma)$ such that for all $X \in W_2^{\frac{1}{2}}(\Gamma)$

$$\langle \mathcal{A}H_\lambda, X \rangle_\Gamma + \frac{1}{\lambda} \langle \mathcal{B}_\beta H_\lambda, X \rangle_\Gamma = \langle f, X \rangle_\Gamma , \tag{25.19}$$

where $\mathcal{A} := -\mathcal{B} - b$. $\lambda > 0$ denotes the penalty parameter.

Lemma 25.1 *([1])*
  (i) *The Nemyzki operator $\mathcal{B}_\beta$ is Lipschitz continuous (with a Lipschitz constant $L > 0$), i.e.*

$$\|\mathcal{B}_\beta X_1 - \mathcal{B}_\beta X_2\|_0 \le L \|X_1 - X_2\|_0 \qquad \forall X_1, X_2 \in L_2(\Gamma) . \tag{25.20}$$

  (ii) *The Nemyzki operator $\mathcal{B}_\beta$ is monotone, i.e.*

$$\langle \mathcal{B}_\beta X_1 - \mathcal{B}_\beta X_2, X_1 - X_2 \rangle_\Gamma \ge 0 \qquad \forall X_1, X_2 \in L_2(\Gamma) . \tag{25.21}$$

(iii) The operator $\mathcal{A}_\lambda := \mathcal{A} + \frac{1}{\lambda}\mathcal{B}_\beta : W_2^{\frac{1}{2}}(\Gamma) \to W_2^{\frac{1}{2}}(\Gamma)$ *is strongly monotone (with a constant $C > 0$), i.e.*

$$\langle \mathcal{A}_\lambda X_1 - \mathcal{A}_\lambda X_2, X_1 - X_2 \rangle_\Gamma \geq C \, \|X_1 - X_2\|_{\frac{1}{2}}^2 \qquad \forall X_1, X_2 \in W_2^{\frac{1}{2}}(\Gamma) \,. \qquad (25.22)$$

Theorem 25.2 *([1]) Let $b < 0, \lambda > 0$ and $f \in L_2(\Gamma)$. Then the penalized variational equation (25.19) has a unique solution $H_\lambda \in W_2^{\frac{1}{2}}(\Gamma)$.*

Sketch of Proof. The proof is based on a relaxation method and the application of the fix-point criterion. First we use the Bessel potential operator $\mathcal{A}_p : W_2^{-\frac{1}{2}}(\Gamma) \to W_2^{\frac{1}{2}}(\Gamma)$ with norm 1 to build the strongly monotone and Lipschitz continuous operator

$$\mathcal{A}_{p\lambda} := \mathcal{A}_p \circ \mathcal{A}_\lambda : W_2^{\frac{1}{2}}(\Gamma) \to W_2^{\frac{1}{2}}(\Gamma) \,.$$

As the relaxation method we use

$$H_\lambda^{n+1} = H_\lambda^n - \alpha(\mathcal{A}_{p\lambda} H_\lambda^n - \mathcal{A}_p f)$$

with $\alpha > 0$ and an arbitrary start value $H_\lambda^0 \in W_2^{\frac{1}{2}}(\Gamma)$. Together with the properties of the operator $\mathcal{A}_{p\lambda}$ and some assumptions on the relaxation parameter α, the fix-point criterion gives us a Cauchy sequence H_λ^n converging to the solution $H_\lambda \in W_2^{\frac{1}{2}}(\Gamma)$ of (25.19). The uniqueness of the solution follows immediately from the strong monotonicity of $\mathcal{A}_\lambda$. $\square$

The variational equation (25.19) can be solved numerically with, for example, a Galerkin-Bubnov method. Therefore we use the orthogonal projection P_h of $W_2^0(\Gamma)$ onto the spline spaces S_N (for example the space of piecewise continuous linear splines, where N denotes the degree of freedom and h the mesh width), and determine the solution $H_{\lambda,h} \in S_N$ of the nonlinear discrete equations

$$\langle \mathcal{A} H_{\lambda,h} + \frac{1}{\lambda}\mathcal{B}_\beta H_{\lambda,h}, X_h \rangle_\Gamma = \langle f, X_h \rangle_\Gamma \qquad \forall X_h \in S_h \,. \qquad (25.23)$$

Theorem 25.3 *([1]) Let $b < 0, \lambda > 0$ and $f \in L_2(\Gamma)$. Then the discrete penalized variational equation (25.23) has a unique solution $H_{\lambda,h} \in S_N$.*

Sketch of Proof. The proof is very similiar to that of Theorem 25.2 and uses also a relaxation method in combination with the fix-point criterion. In addition we need the strong monotonicity of the operator $\mathcal{A}_{\lambda,h} := P_h \mathcal{A} + \frac{1}{\lambda}P_h \mathcal{B}_\beta : S_N \to S_N$, and also the inverse property [19] of the splines, which yields a quasi-uniform grid as a restriction for the descretization. $\square$

Theorem 25.4 *([1]) Let $b < 0, \lambda > 0$ with $\lambda = d\,h$, $d = const.$ and $f \in W_2^1(\Gamma)$. H should be the solution of (25.13) and $H_{\lambda,h}$ the solution of (25.23). Then the following asymptotic error estimate is valid:*

$$\|H - H_{\lambda,h}\|_{\frac{1}{2}} \leq c\,h \,. \qquad (25.24)$$

Sketch of Proof. It is necessary to get error estimates for

$$\|H - H_\lambda\|_{\frac{1}{2}} \qquad \text{and} \qquad \|H_\lambda - H_{\lambda,h}\|_{\frac{1}{2}} .$$

In treating the first error $\|H - H_\lambda\|_{\frac{1}{2}}$ we use the coercivity of the operator $\mathcal{A}$, together with the properties of the solutions H and H_λ (which are uniform bounded), and obtain the estimate

$$\|H_\lambda - H\|_{\frac{1}{2}} \leq c\lambda .$$

The second error $\|H_\lambda - H_{\lambda,h}\|_{\frac{1}{2}}$ can be estimated by

$$\|H_\lambda - H_{\lambda,h}\|_{\frac{1}{2}} \leq c(h^{\frac{3}{2}} + \frac{\tilde{c}}{\lambda}h^2) ,$$

where we have used the strong monotonicity of $\mathcal{A}_{\lambda,h}$, the continuity of $\mathcal{A}$ and the Lipschitz continuity of $\mathcal{B}_\beta$, in addition to the approximation property ([19]) of the spline spaces S_N. Together with $\lambda = d\,h$, $d =$const. the desired result (25.24) follows.

$\square$

25.5 NUMERICAL EXAMPLE

In the general case of an arbitrarily shaped, plane crack G it is not so easy to calculate the data which we need for the solution of the variational inequality (25.13) or variational equation (25.19). In particular the calculation of the operator $\mathcal{M}$ and the stress intensity factors K_1 and k_1 cause many troubles. In [1] an algorithm for the calculation of the integral operator $\mathcal{M}$ is derived, where the operator can be calculated numerically with help of an additional integral equation. There is a wide range of methods for the calculation of the stress intensity factors such as extrapolation techniques, augmented Galerkin methods, weighted integral methods and many others, see e.g. [4, 5, 13, 18].

Here we will restrict ourself on a simple example, the so called penny-shaped crack

$$G := \{y \,:\, |y| < 1/2\} ,$$

the surfaces of which are loaded by the stretching force P given as

$$P := \frac{K_{1c}(1 + t)}{\max \hat{K}_1}$$

with $\hat{K}_1$ a stress intensity factor corresponding to the loading $P = 1$ (this guarantees that the equilibrium is violated and that the crack will propagate). The material constants are $\mu = 77.52$, $\nu = 0.29$ and $K_{1c} = 0.06$.

In this special case it is possible to calculate the kernel function $K(s, z)$ using the Kelvin transformation. In this way we obtain

$$K(s, z) = \frac{1}{2\alpha\pi \sin^2(s - z)}$$

for the kernel function and

$$\mathcal{M}[H](z) = \frac{1}{2\alpha\pi} \int\limits_{0}^{\pi} \frac{H(s) - H(z)}{\sin^2(s - z)}\, ds - \frac{3}{4\alpha} H(z)$$

for the corresponding integral operator.

The calculations are done for the variational inequality based on the Irwin criterion. We should mention that the reference configuration is always the unperturbed penny-shaped crack, and that the propagated crack is presented with a magnification factor of 20 to make the new crack front visible.

Figure 25.1 shows the propagation of the penny-shaped crack for different times t and under an increasing load P concentrated at the origin of the crack.

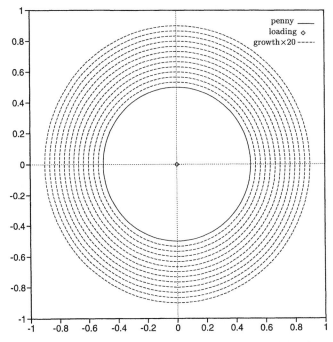

Figure 25.1: Symmetrical propagation under centered loading

Figure 25.2 shows the propagation of the penny-shaped crack for different times t and under the increasing load P concentrated in the point $Q = (0.007, 0)$. At the beginning the crack grows only on some parts of the front. In the limit case we have propagation everywhere except at one point. Finally the crack is circular once more, but now circular around the loading point Q.

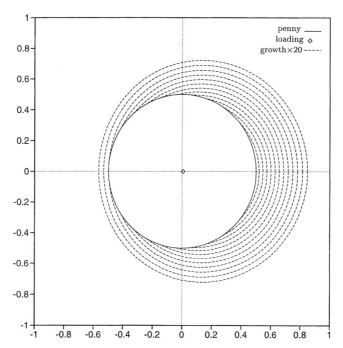

Figure 25.2: Asymmetrical propagation under non-centered loading

REFERENCES

[1] Bach, M. (1998). *Randvariationsungleichungen zur Beschreibung des quasistatischen Risswachstums eines ebenen 3D–Risses in isotropem Material.* Ph.D. thesis, Universität Stuttgart, Math. Inst. A.

[2] Bach, M., Nazarov, S.A. (1999). Smoothness properties of solutions to variational inequalities describing propagation of mode-1 cracks. In Bonnet, M., Sändig, A.-M., Wendland, W.L., editor, *Mathematical Aspects of Boundary Element Methods*, pages 23–32. Chapman & Hall/CRC Research Notes in Mathematics No. 414, Chapman & Hall / CRC London.

[3] Bach, M., Nazarov, S.A. (In preparation). Propagation of mode-1 cracks in an elastic space under the Griffith criterion.

[4] Blum, H. (1988). Numerical treatment of corner and crack singularities. In W. Wendland and E. Stein, editors, *Lecture Notes : Finite Element and Boundary Element Techniques from Mathematical and Engineering Point of View, CISM Courses and Lectures No. 301.* Springer-Verlag Wien-New York.

[5] Bueckner, H.F. (1970). A novel principle for the computation of stress intensity factors. *ZAMM*, 50:529–546.

[6] Bueckner, H.F. (1987). Weight functions and fundamental fields for the penny-shaped and the half-plane crack in three-space. *Int. J. Solids Structures*, 23(1):57–93.

[7] Dauge, M. (1988). *Elliptic Boundary Value Problems on Corner Domains*. Springer-Verlag, Berlin.

[8] Griffith, A.A. (1920). The phenomenon of rupture and flow in solids. *Philos. Trans. Roy. Soc. London Ser. A.*, 221:25–35.

[9] Irwin, G. (1951). Analysis of stresses and strains near the end of a crack traversing a plate. *J. Appl. Mech.*, 24:361–364.

[10] Kolton, L.H., Nazarov, S.A. (1992). Quasistatic propagation of a mode-1 crack in an elastic space. *C. R. Acad. Sci. Paris*, 315:1453–1457.

[11] Luré, A.I. (1970). *Theory of Elasticity*. Nauka Publ., Moscow.

[12] Mazya, V.G., Nazarov, S.A., Plamenevskii, B.A. (1991). *Asymptotische Theorie elliptischer Randwertaufgaben in singulär gestörten Gebieten, Teil I und II*. Akademie Verlag, Berlin.

[13] Mazya, V.G., Plamenevskii, B.A. (1976). On the coefficients in the asymptotics of solutions of elliptic boundary value problems near the edge. *Dokl. Akad. Nauk SSSR*, 229(4):970–974.

[14] Nazarov, S.A. (1989). Derivation of the variational inequality describing the mode-1 crack shape front increament (russ.). *Izvestia Academii Nauk SSSR, Mechanika Tverdogo Tela*, 2:152–160.

[15] Nazarov, S.A., Plamenevskii, B.A. (1991). *Elliptic Problems in Domains with Piecewise Smooth Boundary*. Nauka, Moscov. English transl., Walter de Gruyter, Berlin, 1994.

[16] Nazarov, S.A., Polyakova, O.R. (1992). On the equivalence of fracture criteria for a mode-1 crack in an elastic space. *Akademie der Wissenschaften, Mekh. Tverd. Tela*, 2:101–113.

[17] Nemat-Nasser, S., Keer, L.M., Sumi, Y. (1980). Unstable growth of tension cracks in brittle solids: Stable and unstable bifurkations, snap-through, and imperfection sensitivity. *Int. J. Solids Structures*, 16:1017–1035.

[18] Rice, J.R. (1968). A path independent integral and the approximate analysis of strain concentration by notches and cracks. *J. Appl. Mech.*, 35(2):379–386.

[19] Schatz, A.H., Thomée, V., Wendland, W.L. (1990). *Mathematical Theory of Finite and Boundary Element Methods*. Birkhäuser, Basel.

26 RECENT TRENDS IN THE COMPUTATIONAL MODELLING OF CONTINUA AND MULTI-FRACTURING SOLIDS

D.R.J. Owen[a], E.A. de Souza Neto[a], D. Perić[a] and M. Vaz Jr.[b]

[a]Department of Civil Engineering, University of Wales Swansea,
Singleton Park, Swansea SA2 8PP, United Kingdom

[b]Departmento de Engenharia Mecânica, Universidade do Estado de Santa Catarina,
89223-100 Joinville – SC, Brazil

ABSTRACT

This paper discusses some current trends in the mathematical modelling and computational simulation of large scale industrial problems involving finite straining of inelastic solids with possible occurence of material failure. On the theoretical side, the need for the rigorous development of thermodynamically sound continuum constitutive models of inelastic solids is emphasized. Attention is focussed on the formulation of rate-dependent and rate-independent finite elasto-plasticity models both in the isotropic and anisotropic (single crystal) cases. Attention is also given to the definition of fracturing criteria that model the continuum-discrete transition at the onset of macroscopic failure. On the computational side, the important issues of finite/discrete element modelling and mesh adaptivity, with particular attention to 3-D surface remeshing, are discussed. Current applications of such concepts within a finite/discrete element environment are discussed throughout the paper through the numerical solution of industrially relevant large scale problems.

Key words. Finite/discrete elements, multi-fracturing, adaptivity

26.1 INTRODUCTION

Over the last decade or so, computational methods in solid mechanics have matured considerably and, coupled with the remarkable advances in the power of computer hardware, permit the numerical simulation of large scale industrial and R&D problems. Progress has been made on several fronts and key developments in the treatment of continuum problems include a rational computational framework for finite strain plastic deformation, the modelling of frictional contact conditions, element technology for near incompressible flows and mesh adaptivity methods. These advances have been complemented by improved sparse matrix and iterative solution procedures for large scale systems, enabling simulation of industrially relevant multi-scale and/or multi-field problems.

Discrete element technology, which is based on the concept that individual material elements are considered to be separate and are (possibly) connected only along their boundaries by appropriate physically based interaction laws, has also progressed markedly within the last decade. Earlier work on these techniques assumed that each element was rigid, but the incorporation of deformation kinematics into the formulation has led naturally to combined finite/discrete element approaches. For situations involving multi-fracturing phenomena, such combined strategies offer a natural solution route to modelling the continuum/discontinuum transformation involved.

The paper discusses issues related to the computational treatment of continuum problems involving large plastic flows resulting in evolving geometries and continuum to discrete transformations brought about by multi-fracturing behaviour. Attention is also given to the particular case of single crystal plasticity. The relevance of the computational models presented is reinforced by application to industrial scale problems.

26.2   CONTINUUM MODELLING ISSUES

Crucial to the appropriate characterisation of material response in the finite strain range is the rigorous formulation of constitutive models within the framework of continuum thermodynamics. In industrial applications of practical interest, finite bulk deformations are almost invariably accompanied by large inelastic strains and the formulation of thermodynamically consistent finite elasto-plasticity models is necessary. Since the models are to be incorporated within a finite element environment suited to large scale simulations, the formulation of stable and accurate numerical integration algorithms for path-dependent constitutive equations, together with the associated consistent tangent operators have to be carefully addressed. In problems involving large deformations of metals, where inelastic deformations are volume preserving, appropriate finite element technology has to be adopted to produce accurate results.

26.2.1   Finite inelastic deformations. Constitutive characterisation

In line with current methodology for the formulation of finite inelastic constitutive models, the present approach to finite elasto-plasticity is based on the multiplicative split of the deformation gradient into elastic and plastic contributions, i.e.,

$$\boldsymbol{F} = \boldsymbol{F}^e \boldsymbol{F}^p . \qquad (26.1)$$

The multiplicative split of $\boldsymbol{F}$ admits the existence of a local unstressed *intermediate configuration* and finds rigorous physical justification in the slip theory of crystal plasticity.

In addition, the stresses are assumed to be obtained from a *hyperelastic potential*, $W(\boldsymbol{F}^e)$. The Kirchhoff stress tensor is then given as a function of $\boldsymbol{F}^e$ by:

$$\tau(\boldsymbol{F}^e) = \frac{\mathrm{d}W}{\mathrm{d}\boldsymbol{F}^e} \boldsymbol{F}^{eT}. \tag{26.2}$$

Isotropic finite plasticity

Under many situations of practical interest, particularly in problems involving the finite deformation of *polycrystals*, the assumption of plastic isotropy provides a reasonable approximation of the actual material behaviour. In such cases, the elasto-plastic model can be defined by adding to the above equations an evolution law for the plastic deformation gradient of the general form:

$$\bar{\boldsymbol{D}}^p = \dot{\gamma}\frac{\partial\Psi}{\partial\tau} \qquad \boldsymbol{W}^p = \boldsymbol{0}, \tag{26.3}$$

where Ψ is an isotropic *dissipation potential*, and $\bar{\boldsymbol{D}}^p$ is the *plastic contribution* to the rate of deformation tensor:

$$\bar{\boldsymbol{D}}^p = \boldsymbol{R}^e \operatorname{sym}[\dot{\boldsymbol{F}}^p \boldsymbol{F}^{p-1}] \boldsymbol{R}^{eT}, \tag{26.4}$$

and

$$\boldsymbol{W}^p = \operatorname{skew}[\dot{\boldsymbol{F}}^p \boldsymbol{F}^{p-1}] \tag{26.5}$$

is the *plastic spin*.

In the rate-independent case, the plastic multiplier $\dot{\gamma}$ satisfies the loading/unloading criterion:

$$\Phi \le 0 \qquad \dot{\gamma} \ge 0 \qquad \dot{\gamma}\Phi = 0, \tag{26.6}$$

where Φ is an isotropic *yield function* defining the onset of plastic flow at $\Phi(\tau, \boldsymbol{q}) = 0$. A typical example, appropriate for the description of metals, is the von Mises function whose finite strain generalisation in the present case reads:

$$\Phi(\tau, \tau_y) = \sqrt{\tfrac{1}{2}J_2(\tau')} - \tau_y, \tag{26.7}$$

where τ' is the Kirchhoff stress deviator and τ_y is the Kirchhoff uniaxial yield stress, whose value is usually a function of a suitable hardening internal variable.

In the rate-dependent case, $\dot{\gamma}$ is defined by a viscoplastic law such as

$$\dot{\gamma} = E_0 \left\langle \frac{\Phi(\tau, \tau_y)}{\tau_y} \right\rangle^{1/N}, \tag{26.8}$$

where $<>$ denote the *Macauley brackets* and the material constants E_0 and N are, respectively, the reference shear rate and the rate sensitivity parameter. The use of rate-dependent laws may become necessary at higher temperatures where phenomena such as creep and relaxation may not be disregarded without significant loss of accuracy.

Anisotropic plasticity. Single crystals

Macroscopic plastic deformation of metallic crystals is the result of sliding between crystal blocks along well defined directions in preferential crystallographic planes. The continuum plastic flow equation in this case is defined by:

$$\dot{\boldsymbol{F}}^p \boldsymbol{F}^{p-1} = \sum_{\alpha=1}^{n_{\text{act}}} \dot{\gamma}^\alpha \, \boldsymbol{s}_0^\alpha \otimes \boldsymbol{m}_0^\alpha \,, \tag{26.9}$$

where $\{\boldsymbol{s}^\alpha, \boldsymbol{m}^\alpha\}$ are unit vectors, respectively in the slip direction and normal to the corresponding slip plane, that define slip system α. The multiplier $\dot{\gamma}^\alpha$ is the plastic slip rate in system α and n_{act} denotes the number of *active* slip systems, i.e., the number of systems undergoing plastic slip.

In the rate-independent case, the multipliers satisfy the loading/unloading condition:

$$\phi^\alpha \leq 0 \qquad \dot{\gamma}^\alpha \geq 0 \qquad \phi^\alpha \dot{\gamma}^\alpha = 0 \,, \tag{26.10}$$

for $\alpha = 1, ..., 2n_{\text{syst}}$, with n_{syst} denoting the total number of slip systems. In this case, each physical slip system has been split into two mirrored systems:

$$\{\boldsymbol{s}_0^\alpha, \boldsymbol{m}_0^\alpha\} \qquad \text{and} \qquad \{\boldsymbol{s}_0^\beta, \boldsymbol{m}_0^\beta\} \equiv \{-\boldsymbol{s}_0^\alpha, \boldsymbol{m}_0^\alpha\} \,,$$

to allow the description of the plastic flow rule within the framework of multi-surface plasticity. Accordingly, $2 \times n_{\text{syst}}$ yield functions have been defined having the form:

$$\phi^\alpha(\tau^\alpha) \equiv \tau^\alpha - \tau_y \qquad \alpha = 1, ..., 2n_{\text{syst}} \,, \tag{26.11}$$

where τ^α is the *Schmid resolved stress* in slip-system α:

$$\tau^\alpha \equiv \boldsymbol{R}^{eT} \boldsymbol{\tau} \, \boldsymbol{R}^e : (\boldsymbol{s}_0^\alpha \otimes \boldsymbol{m}_0^\alpha) \,, \tag{26.12}$$

and the resolved yield stress, τ_y, may be a function of a hardening internal variable.

Under rate-dependent conditions, the slip rate $\dot{\gamma}^\alpha$ is given by a viscoplastic law such as:

$$\dot{\gamma}^\alpha = E_0 \left\langle \frac{\Phi^\alpha}{\tau_y} \right\rangle^{1/N} \,, \tag{26.13}$$

or the generalisation of the classical Norton creep law:

$$\dot{\gamma}^\alpha = E_0 \left(\frac{|\tau^\alpha|}{\tau_y} \right)^{1/N} \,. \tag{26.14}$$

26.2.2   Time-discrete constitutive equations

In the context of finite element analysis of solids modelled by path-dependent constitutive laws, the load path is followed incrementally and a numerical approximation to the constitutive equations is needed to update stresses as well as the relevant internal variables of the problem within each increment. Then, given the values of the variables $\{\boldsymbol{\sigma}_n, \boldsymbol{q}_n\}$ at the beginning of a generic increment $[t_n, t_{n+1}]$ an algorithm for integration of the evolution equations is required to obtain the updated values $\{\boldsymbol{\sigma}_{n+1}, \boldsymbol{q}_{n+1}\}$ at the end of the time (or pseudo-time) increment.

The exponential map integrator

With the plastic flow rule defined by the above laws, the evolution of the plastic deformation gradient can be suitably discretised by an implicit *exponential map integrator* whereas the remaining equations are, almost invariably, discretised by a standard *backward Euler* scheme. The resulting system of non-linear algebraic equations is usually solved by a Newton-Raphson algorithm.

The exponential map-based update equation for the plastic deformation gradient reads:

$$\boldsymbol{F}^p_{n+1} = \boldsymbol{F}^p_\Delta \, \boldsymbol{F}^p_n \qquad (26.15)$$

where, in the isotropic case, the incremental plastic deformation gradient is given by

$$\boldsymbol{F}^p_\Delta = \boldsymbol{R}^e \exp\left[\Delta\gamma \left.\frac{\partial\Psi}{\partial\boldsymbol{\tau}}\right|_{n+1}\right] \boldsymbol{R}^{eT}. \qquad (26.16)$$

Some crucially important properties result from the use of the exponential map in the discretisation of the plastic flow rule. Firstly, the incompressibility of the plastic flow associated with pressure insensitive flow potentials is carried over *exactly* to the incremental rule. As pointed out by Steinmann and Stein [24], this crucial property is directly linked to the accuracy of the stress update procedure. In addition, when the Hencky (logarithmic) strain-energy function is adopted to describe the hyperelastic behaviour, the exponential map is eliminated from the discretised equations resulting in an overall procedure in which the finite strain-related operations remain confined to the kinematic level. The essential stress updating in this case retains the classical small strain format of elastic predictor-return mapping schemes [14, 19, 23]

For the crystal model, the incremental plastic deformation gradient is given by:

$$\boldsymbol{F}^p_\Delta = \exp\left[\sum_{\alpha=1}^{n_{\text{act}}} \Delta\gamma^\alpha \, \boldsymbol{s}^\alpha_0 \otimes \boldsymbol{m}^\alpha_0\right], \qquad (26.17)$$

where, in contrast to the isotropic case, the argument of the tensor exponential function is *unsymmetric*. In this case, the tensor exponential appears *explicitly* in the discretised system of equations and is computed, as suggested by Miehe [9], by truncating its series representation [5]:

$$\exp[\boldsymbol{X}] = \sum_{n=0}^{\infty} \frac{1}{n!} \, \boldsymbol{X}^n. \qquad (26.18)$$

Obviously, the consistent linearisation of the field equations here will involve the *derivative* of the unsymmetric tensor exponential. A series representation for the exponential map derivative, suitable for computational implementation, has been introduced in [20, 22].

Application of the above methodology to the simulation of strain localisation in single crystal specimens is illustrated in Figure 26.1. In this case, a planar double-slip crystal model has been employed to describe the constitutive behaviour of a single crystal bar whose crystal lattice is misaligned with respect to the axis of loading. The characteristic unsymmetric strain localisation pattern observed in experiments under such conditions is reproduced in the numerical simulation. It should be emphasized that the present

(a)   (b)

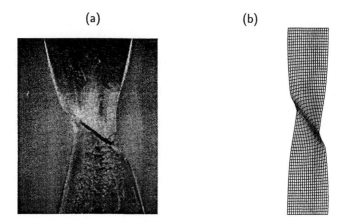

Figure 26.1: Strain localisation in a single crystal specimen. (a) Experiment, and (b) Numerical simulation.

approach can be used not only in the analysis of single crystal components but may also be employed in conjunction with averaging techniques to model the strain induced plastic anisotropy observed in (initially isotropic) polycrystals subjected to finite straining.

26.2.3 Finite element technology

A fundamental requirement in the simulation of the inelastic behaviour of metals is the use of appropriate elements which can model the isochoric nature of the plastic flow without exhibiting locking behaviour. The tendency of the solution to lock, i.e., to provide overstiff solutions with poor stress representation, becomes more predominant when large deformations are involved, as is the case in all forming operations. One effective solution to this problem is offered by the F-bar element [23], which is based on a multiplicative deviatoric/volumetric split in conjunction with the replacement of the compatible deformation gradient with an assumed modified counterpart. The element is free from hourglassing defects, exhibits a relatively large bowl of convergence and is especially suitable for adaptivity applications.

26.2.4 Frictional contact

The modelling of industrial forming problems invariably requires the treatment of frictional contact conditions between the workpiece and tools. The most recent numerical models for frictional sliding make use of the complete analogy that exists between contact/friction phenomena and the elasto-plastic deformation of materials. This approach has the advantage that the consistent linearisation procedures developed for the latter case can be employed in contact/friction models to provide robust and quadratically convergent numerical algorithms. A main development concerned with modelling frictional contact phenomena is the description of models in which the coefficient of friction varies during deformation and which are particularly suited to a range of forming processes. By introducing a hardening variable within the plasticity theory of friction, the dependence of the coefficient of friction on both the normal pressure and the degree of relative sliding between the tool and workpiece can be modelled [21].

26.3 MATERIAL FAILURE CRITERIA. CONTINUUM/DISCRETE TRANSITION

Internal damage is characterized by the presence and evolution of voids and microcracks which may, eventually, lead to material failure. In order to predict the localised continuum-discrete transition associated with the formation of a macroscopic crack, appropriate criteria have to be adopted. Such criteria should define transition conditions where the continuum constitutive laws addressed in Section 26.2 are no longer valid and new constitutive laws modelling the evolution of macroscopic cracks need be considered. Due to the complex interaction between the various phenomena that precede material failure the formulation of such criteria is not trivial. It should be emphasized that for many industrial applications, such as high speed machining and rock blasting, material failure is a crucial feature of the process and its appropriate representation is of paramount importance to produce meaningful numerical simulations.

Table 26.1 lists some of the ductile fracture criteria available from the literature which have been employed for conventional materials. Although one of the first indicators to be used in ductile fracture analysis, severe criticism has been raised on the reliability of plastic work based criteria to predict ductile fracture in bulk forming operations. More recent comparative analyses suggest that damage-based indicators are more reliable in predicting the correct site of fracture initiation [12]. By recognising the history dependency exhibited by the mechanisms of ductile fracture, the use of fracture criteria based on total damage work and uncoupled integration of Lemaitre's damage model for damaged and undamaged materials respectively offer considerable promise.

The simulation of progressive fracture necessitates inclusion of an element separation scheme to model crack propagation. The most general approaches permit fracture in an arbitrary direction within an element and rely on local adaptive remeshing employing the principles outlined in Section 26.5. Further classes of problems necessitate the introduction of element erosion concepts in which material is systematically removed due to abrasive wear or local failure. The failure indicators described in Table 26.1 can be again employed to monitor this erosion process.

Material characteristics can be significantly influenced when loading rates become high, resulting in high strain rates within the material. For material fracture, the key observations from high strain rate crack propagation experiments are [17, 18, 4]: (a) the crack initiation time is time dependent, (b) the crack velocity is dependent on the loading rate, (c) the stress intensity at the crack tip is independent of the loading rate at lower strain rates but becomes rate dependent at higher loading rates and (d) the time to fracture is dependent on the loading rate but converges asymptotically to a minimum time to fracture $t_{f\min}$.

Rate dependence effects can be incorporated within a strain softening based failure model by making both the failure stress and softening slope a function of the local strain

Table 26.1: Ductile fracture criteria

Criterion	Definition
Total plastic work	$\int \sigma_Y \mathrm{d}\varepsilon^p$
Max. pl. shear work	$\int \tau_{\max} \mathrm{d}\gamma^p_{\max}$
Lemaitre	$\int \left[\frac{2}{3}(1-\nu) + 3(1-2\nu)\left(\frac{\sigma_H}{\sigma_Y}\right)^2 \right] \mathrm{d}\varepsilon^p$
General Lemaitre	$\int \frac{\sigma_Y^{2s}}{2E} \left[\frac{2}{3}(1-\nu) + 3(1-2\nu)\left(\frac{\sigma_H}{\sigma_Y}\right)^2 \right]^s \mathrm{d}\varepsilon^p$
Cockcroft & Latham	$\int \sigma_1 \mathrm{d}\varepsilon^p$
Brozzo et al.	$\int \frac{2\sigma_1}{3(\sigma_1-\sigma_H)} \mathrm{d}\varepsilon^p$ r
McClintock	$\int \left[\frac{\sqrt{3}}{2(1-n)} \sinh\left(\frac{\sqrt{3}}{2}(1-n)\frac{\sigma_a+\sigma_b}{\sigma_Y} \right) \right.$ $\left. + \frac{3}{4}\left(\frac{\sqrt{3}}{2}(1-n)\frac{\sigma_a-\sigma_b}{\sigma_Y} \right) \right] \mathrm{d}\varepsilon^p$
Rice & Tracey	$\int 0.283 \exp\left[\frac{\sqrt{3}}{2}\frac{\sigma_H}{\sigma_Y} \right] \mathrm{d}\varepsilon^p$
Norris et al.	$\int \frac{1}{1-c\sigma_H} \mathrm{d}\varepsilon^p$
Atkins	$\int \frac{1+\mathrm{d}\varepsilon_2/\mathrm{d}\varepsilon_1}{2(1-c\sigma_H)} \mathrm{d}\varepsilon^p$
Oyane et al.	$\int \left(1 + \frac{\sigma_H}{A\sigma_Y} \right) \mathrm{d}\varepsilon^p$

rate as schematically illustrated in Fig 26.2. Under dynamic conditions the area under the softening slope is no longer equal to the fracture energy G_f due to the effects of inertia on the micromechanical response. As the strain rate increases the softening branch steepens, however at a saturation point, corresponding to the minimum fracture time, the softening branch again becomes more shallow.

The influence of strain rate effects becomes more complex when coupling with thermal phenomena is considered. A particular example of this type of interactive response is afforded by adiabatic shear localisation. The high loading rates involved in problems such as high speed machining (see example below) cause the energy produced by dissipation of inelastic work to remain localised, leading to a significant local rise in temperature, resulting in a reduction in the yield stress, creating conditions favourable not only to the development of shear bands, but also for material failure.

High-speed machining application

This example simulates the machining of Ti-6Al-4V titanium alloy. A rate of fracture indicator has been used as the error estimator for mesh adaptivity throughout this example with maximum and minimum element size restricted to 0.1 and 0.7 mm respectively. The tool is assumed to be rigid and element temperature has been computed over the workpiece only, using the thermo-mechanical model described by Owen and Vaz Jr. [12]. Thermoplastic shear localization was observed in all cases simulated for a cutting speed of 10 m/s, cutting depth of 0.5 mm and rake angles ranging from -9 to 9 deg. Figure 26.3

illustrates a typical mesh obtained using error estimate and adaptive remeshing procedures (see Section 26.5).

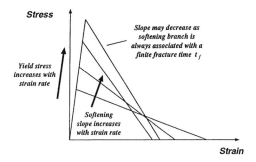

Figure 26.2: Strain rate dependent material model

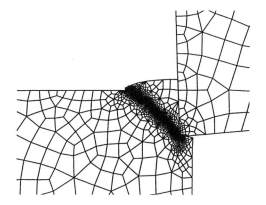

Figure 26.3: High-speed machining. Finite element mesh.

It is well-known that, with a local approach to material failure, the width of a shear band is strongly mesh dependent, but that regularisation of the solution can be accomplished, in several ways, through the introduction of an appropriate length scale. In the present example, localisation was found to be confined, in average, to a $30\mu m$ zone, corresponding to a region 5 elements wide, as depicted in Fig. 26.3. The distribution pattern of the equivalent plastic strain rate can change during the process due to the cyclic character of the mechanisms of chip formation. However, even at very early stages, $\dot{\varepsilon}^p$ was found to be significantly high, reaching values up to $1.2\times10^6\mathrm{s}^{-1}$ for this particular example as indicated in Fig. 26.4(a). A high strain rate causes localised plastic deformation which is reflected by the distribution of the equivalent plastic strain shown in Figure 26.4(b).

The very essence of thermoplastic shear localisation is presented in Fig. 26.4(c) and (d), which show the distribution of the temperature and yield stress. Highly localised plastic deformation causes the yield stress to decrease. Furthermore, the regions of low yield stress are more susceptible to larger plastic deformations. The cyclic response of *localised plastic deformation → temperature rise → yield stress reduction → large and localised deformation* instigates a significant upward movement of the portion of the chip above the shear band, which can eventually lead to its separation.

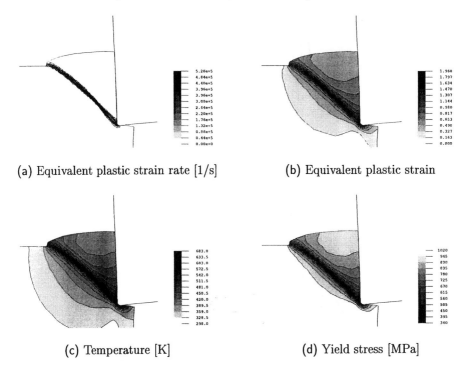

(a) Equivalent plastic strain rate [1/s]   (b) Equivalent plastic strain

(c) Temperature [K]   (d) Yield stress [MPa]

Figure 26.4: Thermoplastic strain localisation: $\alpha = -3°$, $v = 10$m/s and $t = 0.5$mm

Material failure in high-speed machining

Material failure in high speed machining is essentially due to phenomena associated with adiabatic strain localisation. Thus, failure occurs in the plastic strain localisation zone. The dominant question here is *when* does material failure initiate? According to the previous analysis of shear band progression, chip formation is of a cyclic character when machining Ti-4V-6A1. Therefore, in this example, fracture is assumed to initiate *before* the second cycle begins, corresponding to a tool advance of $U = 0.1037$mm.

Experimental observations suggest that a fracture strain, ε_f, may indicate failure onset due to void growth in problems involving plastic strain localisation [1, 2]. In the present study, fracture indicators based on the *equivalent plastic strain* and *uncoupled integration of Lemaitre's damage model* [7] are assessed in association with ε_f.

The chip breakage process for a fracture strain based on the equivalent plastic strain is illustrated in Fig. 26.5, which shows the elements undergoing a failure softening and fracture propagation. This criterion predicts that fracture initiates near the tool tip and progresses towards the free surface of the workpiece. The fracture strain based on Lemaitre's damage model, on the other hand, predicts that fracture initiates at the free surface of the workpiece and propagates towards the tool tip as illustrated in Fig. 26.6. Experimental observations indicate that this latter approach provides a better physical description of material failure for this class of problems.

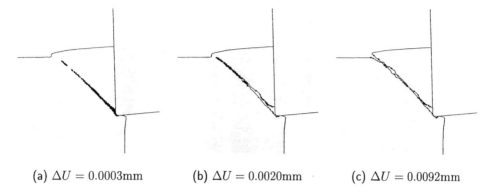

(a) $\Delta U = 0.0003$mm   (b) $\Delta U = 0.0020$mm   (c) $\Delta U = 0.0092$mm

Figure 26.5: Failure process based on *equivalent plastic strain*.

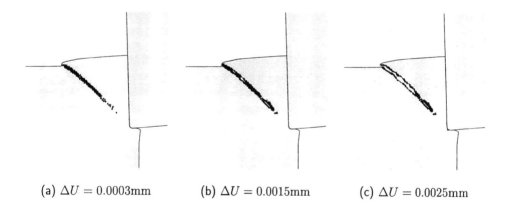

(a) $\Delta U = 0.0003$mm   (b) $\Delta U = 0.0015$mm   (c) $\Delta U = 0.0025$mm

Figure 26.6: Failure process based on *Lemaitre's damage model*.

26.4 FINITE/DISCRETE ELEMENT METHODS FOR MULTI-FRACTURED CONTINUA

Some classes of industrial problems are more naturally treated by the use of *discrete* element methods. The use of discrete elements originated in geotechnical and granular flow applications and are based on the concept that individual material elements are considered to be separate and are (possibly) connected only along their boundaries by appropriate physically based interaction laws.

In a finite/discrete element analysis the main issues which require consideration, for both dynamic and quasi-static behaviour [10, 11], are : (i) appropriate element modelling of the continuum and discrete regions with a view to incorporating the deformation mechanisms necessary to model stress, strain and interaction laws, (ii) the inclusion of finite strain elasto-plastic/viscoplastic behaviour, as discussed in Section 26.2 in both the finite and discrete elements, (iii) development of appropriate fracture criteria defining the continuum-discrete transition, (iv) detection procedures for monitoring contact between discrete elements and the continuum regions and (v) representation of interaction laws for contacting elements.

By far the most crucial operation which governs the efficiency of any discrete element analysis is the multi-body contact detection procedure. Algorithms which employ fast, bounding box search techniques, drastically reduce the number of local, geometrically based calculations required to define contact tractions between interacting bodies. In particular, the alternating digital tree (ADT) algorithm solves this problem with a computational expense proportional to $O = N \log(N)$.

26.4.1   Application to quasi-brittle materials. Rock blasting

Focusing discussion on quasi-brittle materials, macroscopic cracks evolve from the coalescence of microcracks within a distributed process zone. The process zone is thus a zone of micromechanical interaction and the stress field in this region dictates the crack response. Nonlocal constitutive models introduce a lengthscale that defines a finite region surrounding each continuum point in which micromechanical interaction is considered. The simple introduction of nonlocal field variables into the strain-softening formulation is not sufficient to eradicate the discretisation dependence. It is now accepted that the nonlocal concept can be successfully and conveniently implemented if positive monotonic state variables such as inelastic strain or damage are considered as nonlocal parameters. The modern implementation of the nonlocal concept is that of a nonlocal continuum with local strain, where the nonlocal averaging is only applied to those variables that cause strain-softening, while other field variables such as stress and strain remain local [3, 16].

When discrete elements are employed in the simulation of problems such as rock blasting operations it is necessary to couple the fragmentation process with the flow of detonation gases. For example, in bore-hole blasting processes, the detonation wave propagates from the bore-hole walls into the rock as a shock wave and, due to the decrease in tangential pressure, radial cracks propagate into the rock. The stress wave velocity is much higher than the fracture velocity, so that the detonation gases which enter the propagating cracks play a significant role in driving the fracture process, thus necessitating a coupled fluid-structure simulation capability. This coupled solution approach is illustrated schematically in Fig. 26.7. The interaction between the gas flow simulation of the detonation process, using a background Eulerian grid, and the finite/discrete element simulation of the consequent deformation and fracture of the rock, based on a Lagrangian approach, is achieved by use of a staggered fluid-structure coupled scheme.

Fig. 26.8 illustrates the fragmentation process resulting from the sequential detonation

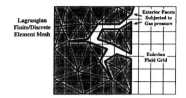

Figure 26.7: Staggered fluid-structure coupled algorithm.

of charges within the five bore holes shown. Commencing from a continuum description of the domain, the detonation of the explosive is modelled by the gas flow model. The pressure wave from the detonation gas causes damage evolution in the rock mass, resulting in discrete fractures, and gas flow into the fracture voids further drives the fracture process. The simulation provides information such as the distribution of particle sizes, throw distances, etc. allowing the improved design of blasting operations.

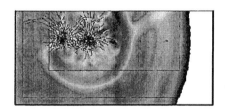
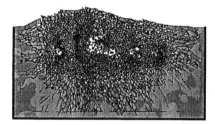

Figure 26.8: Simulation of a sequential detonation of charges within five bore holes: Two stages of the fragmentation process.

26.5 ADAPTIVITY

A general feature encountered in the finite element simulation of forming operations, such as forging, extrusion or deep drawing, is that the optimal mesh configuration changes continually throughout the deformation process. Therefore, the introduction of adaptive mesh refinement processes is crucial for the solution of large scale industrial problems, which necessitates: (i) a remeshing criterion, (ii) specification of an appropriate error estimation criterion, (iii) development of a strategy for adapting the mesh based on the error distribution and (iv) automatic mesh generation tools.

For forming problems, in which the geometric changes are usually large, the decision to update the finite element mesh is invariably based on element distortion parameters and the role of error estimation is then to decide the element size distribution in the new mesh.

The history dependent nature of the elasto-plastic deformation process necessitates transfer of all relevant problem variables from the old mesh to the new one, as successive remeshing is applied during the simulation. As the mesh is adapted, with respect to an appropriate error estimator, the solution procedure, in general, cannot be re-computed from the initial state, but has to be continued from the previously computed state. Hence, some suitable means for transferring the state variables between meshes, or *transfer operators*, needs to be defined. A class of transfer operators for large strain elasto-plastic problems is defined in this section.

Error indicators Extension of the error estimation based on the *plastic dissipation functional* and the rate of *plastic work* described by Perić *et al.* [15] for large strain elasto-plasticity has been found to be an appropriate solution to the adaptive solution of industrial problems. On the other hand, for situations involving strain localisation phenomena or material failure, error indicators based on damage considerations have shown particular advantages.

Mesh regeneration An unstructured meshing approach is used for the mesh generation and subsequent mesh adaptation. The algorithm employed is based on the *Delaunay triangulation* technique, which is particularly suited to local mesh regeneration, or a Advancing Front procedure. An extension of both schemes to quadrilateral elements in 2-D is also available and creates the possibility of employment of the low order F-bar elements mentioned earlier.

Transfer operations for evolving meshes After creating a new mesh, the transfer of displacement and history–dependent variables from the old mesh to a new one is required. Several important aspects of the transfer operation have to be addressed [6, 13]: (i) consistency with the constitutive equations, (ii) requirement of equilibrium and/or energy conservation, (iii) compatibility of the history–dependent internal variables transfer with the displacement field on the new mesh, (iv) compatibility with evolving boundary conditions, (v) minimisation of the numerical diffusion of transferred state fields.

To describe the transfer operation, let us define a state array $^h\Lambda_n = (^hu_n, ^he_n, ^h\sigma_n, ^hq_n)$ where $^hu_n, ^he_n, ^h\sigma_n, ^hq_n$ denote values of the displacement, strain tensor, stress tensor and a vector of internal variables at time t_n for the mesh h. Assume, furthermore, that the estimated error of the solution $^h\Lambda_n$ respects the prescribed criteria, while these are violated by the solution $^h\Lambda_{n+1}$. In this case a new mesh $h + 1$ is generated and a new solution $^{h+1}\Lambda_{n+1}$ needs to be computed. As the backward Euler scheme is adopted, the internal variables $^{h+1}q_n$ for a new mesh $h + 1$ at time t_n need to be evaluated. In this way the state $^{h+1}\widetilde{\Lambda}_n = (\bullet, \bullet, \bullet, ^{h+1}q_n)$ is constructed, where a symbol $^\sim$ is used to denote a reduced state array. It should be noted that this state characterises the history of the material and, in the case of a fully implicit scheme, provides sufficient information for computation of a new solution $^{h+1}\Lambda_{n+1}$. Conceptually, Fig 26.9 summarises a typical transfer operation that includes both, the mapping of the internal variables and mapping of the displacement field. The implementation of the given general transfer operation is performed in several steps and details are provided in references [6, 13].

While adaptive mesh refinement strategies for fully three dimensional cases follow those employed for their two dimensional counterparts, some serious implementation difficulties arise, particularly for problems whose geometry changes significantly during the loading

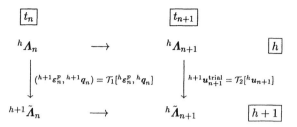

Figure 26.9: Transfer operator diagram

process. Firstly, for arbitrary three dimensional geometries automatic methods for mesh generation and subsequent mesh adaptation can only be currently accomplished by an unstructured meshing approach. Consequently, tetrahedral elements must be employed, as procedures for the generation of good quality hexahedral meshes for arbitrary geometries in 3D are as yet unavailable. For forming processes it is particularly important in 3D mesh generation to produce meshes which are free of sliver (near zero volume) elements, since when finite deformations are considered such elements can rapidly degenerate under further loading. An additional issue in mesh adaptivity for problems with evolving geometries is satisfaction of conformity between evolving surfaces defined by triangular facets and the appropriate geometric description.

26.5.1 Surface remeshing

Whenever a new finite element mesh is generated, attention is required to ensure that the discretised domain boundary preserves certain important properties of the original geometry of the body. If appropriate care is not taken at the boundaries, remeshing may result in unacceptable volume changes or even in geometric features such as corners being lost. To address the issue, a face-defined surface remeshing technique is adopted whereby a new (curved) surface mesh is generated by an advancing front method on a set of discretely defined faces and points (triangulation). The procedure involves four basic ingredients: (a) a geometric surface database, (b) the use of an advancing front technique to mesh the surface, (c) mesh quality improvement, (d) quadratic surface recovery. These are outlined in the following.

Geometric surface data base Given a discretely defined triangulation of the boundary surface, it is necessary at the outset to create a geometric surface data base. This data base contains the relevant information regarding the actual surface geometry. This information is used to ensure that geometric properties such as corners, edges, etc, are preserved whenever a new mesh is generated. The following steps are carried out to create the database: (a) create neighbour face table, (b) create boundary segment list, (c) create surface ridge list, (d) set up geometric point list, (e) set up geometric corner list, (f) set up geometric line data base, and (g) set up geometric surface data base.

Advancing front technique on the discretely defined triangulation With the geometric surface data base at hand, whenever a mesh is required, the advancing front technique is used to generate meshes using some prescribed mesh density for element size control. The important point to emphasize here is the fact that by making use of information on corners and surface ridges, the process will ensure that any newly generated mesh contains the important geometric features of the original surface. A schematic illustration of the process is given in Fig 26.10 where the typical positioning of a new node (x_p) is shown.

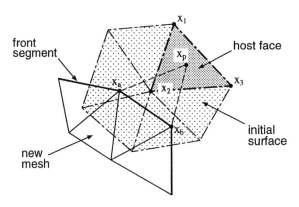

Figure 26.10: Positioning of a new node

Quality Improvement To ensure that elements are well-shaped, the so called '2.5-Dimension' Delaunay simplexification on curved surface meshes is performed, which implies that for each triangular element generated, no other mesh node in the resulting surface mesh lies inside its circumsphere. This may be obtained by locally swapping sides of neighbouring triangular elements. A surface mesh smoothing technique may also greatly improve the final mesh quality. The method consists of repeatedly averaging the position of all interior surface nodes by their surrounding surface nodes. Our experience shows that a total of four or five loops over the interior surface nodes is usually required to achieve a mesh of sufficient quality. In general, due to surface curvature, the averaged surface node does not lie on any of the initial faces and further node re-positioning is then needed. In the re-positioning process, each smoothed node is projected back onto its host face.

Quadratic surface recovery After surface mesh generation, all newly generated nodes lie on the initial discretely defined triangulation. Quadratic surface recovery is performed to reposition these new (interior) nodes onto a quadratic counterpart of the host face [8]. The procedure is schematically illustrated in Fig. 26.11.

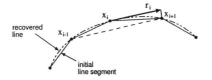

Figure 26.11: Quadratic surface recovery

26.5.2   Example

Application of the overall adaptive scheme together with the above free surface recovery technique is illustrated in Fig. 26.12 where the simulation of a three-dimensional stamping problem is presented. The initial tool/workpiece set up is shown in Fig. 26.12(a) together with the corresponding mesh. Figs. 26.12(b) and (c) show, respectively, the deformed geometries at an intermediate stage and at the end of the stamping operation. Needless to say, simulations of this nature, where extremely high strains and shape changes are present, are virtually impossible to carry out without an appropriate remeshing technique.

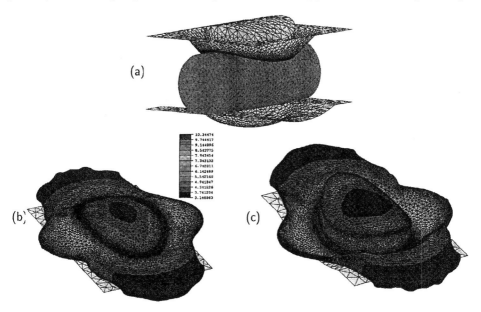

Figure 26.12: 3-D stamping. Adaptivity with surface remeshing. (a) Initial mesh, (b) Deformed mesh at an intermediate stage, and (c) Final deformed mesh.

26.6   CONCLUDING REMARKS

This paper has discussed some current issues related to the simulation of industrial problems where large inelastic flows with evolving geometries arise or where a continuum to discrete transformation due to multi-fracturing is of crucial importance. Consideration has been also given to single crystal plastic deformation, in which anisotropy plays an essential role. Other issues considered include mesh adaptivity procedures for 3D problems involving complex evolving geometries, where difficulties related to progressive fracture, surface remeshing and volume preservation are particularly acute. The paper only represents some progress in these challenging fields and considerable future work is required in many areas, such as the development of reliable error estimation and adaptation procedures for plastically deforming materials involving fracture and other local phenomena.

REFERENCES

[1] Y. Bai and B. Dodd (1992). *Adiabatic shear localization.* Pergamon Press, Oxford.

[2] Y. Bai, Q. Xue and L. Shen (1994). Characteristics and microstructure in the evolution of shear localisation in ti-6a1-4v alloy. *Mech. Mater.*, 17:155–164.

[3] Z.P. Bažant and J. Planas (1998). *Fracture and size effect in concrete and other quasibrittle materials.* CRC Press LLC.

[4] D.E. Grady and R.E. Hollenbach (1977). Rate-controlling mechanisms in the brittle fracture of rock. Technical Report SAND 76-0559, Sandia Laboratories.

[5] M.W. Hirsch and S. Smale (1974). *Differential Equations, Dynamical Systems, and Linear Algebra.* Academic Press.

[6] N.S. Lee and K.-J. Bathe (1994). Error indicators and adaptive remeshing in large deformation finite element analysis. *Fin. Elem. Anal. Design*, 16:99–139.

[7] J. Lemaitre (1985). A continuous damage mechanics model for ductile fracture. *J. Engng. Mat. Tech.*, 107:83–89.

[8] R. Lohner (1996). Regridding surface triangulations. *J. Computational Physics*, 126:1–10.

[9] C. Miehe (1996). Exponential map algorithm for stress updates in anisotropic multiplicative elastoplasticity for single crystals. *Int. J. Numer. Meth. Engng.*, 39:3367–3390.

[10] A. Munjiza, D.R.J. Owen and N. Bićanić (1995). A combined finite/discrete element method in transient dynamics of fracturing solids. *Engng. Comput.*, 12:145–174.

[11] D.R.J. Owen, D. Perić and M. Dutko (1997). Adaptive strategies for industrial forming problems with evolving geometries and multiple fracturing. In *Pro. of the Int. Conf. on Computational Engineering Science – ICES'97.* San Jose, Costa Rica.

[12] D.R.J. Owen and M. Vaz Jr. (1999). Computational techniques applied to high-speed machining under adiabatic strain localization conditions. *Comput. Meth. Appl. Mech. Engrg.*, 171:445–461.

[13] D. Perić, Ch. Hochard, M. Dutko and D.R.J. Owen (1996). Transfer operators for evolving meshes in small strain elasto-plasticity. *Comput. Meth. Appl. Mech. Engrg.*, 137:331–344.

[14] D. Perić, D.R.J. Owen and M.E. Honnor (1992). A model for finite strain elasto-plasticity based on logarithmic strains: Computational issues. *Comput. Meth. Appl. Mech Engrg.*, 94:35–61.

[15] D. Perić, J. Yu and D.R.J. Owen (1994). On error estimates and adaptivity in elastoplastic solids: Application to the numerical simulation of strain localization in classical and cosserat continua. *Int. J. Numer. Meth. Engng.*, 37:1351–1379.

[16] G. Pijaudier-Cabot and Z.P. Bažant (1997). Nonlocal damage theory. *J. Eng. Mech. ASCE*, 113:1512–1533.

[17] K. Ravi-Chandar and W.G. Knauss (1984). An experimental investigation into dynamic fracture. Part I: Crack initiation and arrest. *Int. J. Fract.*, 25:247–262.

[18] K. Ravi-Chandar and W.G. Knauss (1984). An experimental investigation into dynamic fracture. Part II: Microstructural aspects. *Int. J. Fract.*, 26:65–80.

[19] J.C. Simo (1992). Algorithms for static and dynamic multiplicative plasticity that preserve the classical return mapping schemes of the infinitesimal theory. *Comput. Meth. Appl. Mech. Engrg.*, 99:61–112.

[20] E.A. de Souza Neto (1999). The exact derivative of the exponential of an unsymmetric tensor. *Comput. Meth. Appl. Mech. Engrg.* (to appear).

[21] E.A. de Souza Neto, K. Hashimoto, D. Perić and D.R.J. Owen (1996). A phenomenological model for frictional contact accounting for wear effects. *Phil. Trans. R. Soc. Lond. A*, 354:1–25.

[22] E.A. de Souza Neto, D.R.J. Owen and D. Perić (1999). Computational modelling of problems in single crystal plasticity. In W. Wunderlich, editor, *Proceedings of the ECCM'99 – European Conference on Computational Mechanics, Munich.*

[23] E.A. de Souza Neto, D. Perić and D.R.J. Owen (1998). Continuum modelling and numerical simulation of material damage at finite strains. *Arch. Comput. Meth. Engng.*, 5:311–384.

[24] P. Steinmann and E. Stein (1996). On the numerical treatment and analysis of finite deformation ductile single crystal plasticity. *Comput. Meth. Appl. Mech. Engrg.*, 129:235–254.

INDEX

The index is compiled alphabetically word-by-word.

Italic numbers denote diagrams.

408

412

420

Printed and bound by CPI Group (UK) Ltd, Croydon, CR0 4YY

08/05/2025

01864849-0001